Flux-Corrected Transport

Scientific Computation

Editorial Board
J.-J. Chattot, Davis, CA, USA
P. Colella, Berkeley, CA, USA
R. Glowinski, Houston, TX, USA
Y. Hussaini, Tallahassee, FL, USA
P. Joly, Le Chesnay, France
J.E. Marsden, Pasadena, CA, USA
D.I. Meiron, Pasadena, CA, USA
O. Pironneau, Paris, France
A. Quarteroni, Lausanne, Switzerland
    and Politecnico of Milan, Milan, Italy
J. Rappaz, Lausanne, Switzerland
R. Rosner, Chicago, IL, USA
P. Sagaut, Paris, France
J.H. Seinfeld, Pasadena, CA, USA
A. Szepessy, Stockholm, Sweden
M.F. Wheeler, Austin, TX, USA

For further volumes:
www.springer.com/series/718

Dmitri Kuzmin · Rainald Löhner · Stefan Turek
Editors

Flux-Corrected Transport

Principles, Algorithms, and Applications

Second Edition

Editors
Prof. Dmitri Kuzmin
Institute of Applied Mathematics III
University of Erlangen-Nuremberg
Erlangen
Germany

Prof. Stefan Turek
Institute of Applied Mathematics, LS III
Dortmund University of Technology
Dortmund
Germany

Prof. Rainald Löhner
School of Computational Sciences
George Mason University
Fairfax, VA
USA

ISSN 1434-8322   Scientific Computation
ISBN 978-94-007-4037-2   ISBN 978-94-007-4038-9 (eBook)
DOI 10.1007/978-94-007-4038-9
Springer Dordrecht Heidelberg New York London

Library of Congress Control Number: 2012935964

© Springer Science+Business Media Dordrecht 2005, 2012
This work is subject to copyright. All rights are reserved by the Publisher, whether the whole or part of the material is concerned, specifically the rights of translation, reprinting, reuse of illustrations, recitation, broadcasting, reproduction on microfilms or in any other physical way, and transmission or information storage and retrieval, electronic adaptation, computer software, or by similar or dissimilar methodology now known or hereafter developed. Exempted from this legal reservation are brief excerpts in connection with reviews or scholarly analysis or material supplied specifically for the purpose of being entered and executed on a computer system, for exclusive use by the purchaser of the work. Duplication of this publication or parts thereof is permitted only under the provisions of the Copyright Law of the Publisher's location, in its current version, and permission for use must always be obtained from Springer. Permissions for use may be obtained through RightsLink at the Copyright Clearance Center. Violations are liable to prosecution under the respective Copyright Law.
The use of general descriptive names, registered names, trademarks, service marks, etc. in this publication does not imply, even in the absence of a specific statement, that such names are exempt from the relevant protective laws and regulations and therefore free for general use.
While the advice and information in this book are believed to be true and accurate at the date of publication, neither the authors nor the editors nor the publisher can accept any legal responsibility for any errors or omissions that may be made. The publisher makes no warranty, express or implied, with respect to the material contained herein.

Printed on acid-free paper

Springer is part of Springer Science+Business Media (www.springer.com)

Participants of the Workshop *30 years of FCT*

From left to right: G. Patnaik, R. Löhner, S.T. Zalesak,
D.L. Book, S. Turek, M. Möller, and D. Kuzmin

To SHASTA, 'a fluid transport algorithm that works'

Foreword

Jay P. Boris

Flux-Corrected Transport (FCT) was invented more than 32 years ago to improve the quality of numerical convection algorithms. The key word here is quality. Tasked to improve the quality of strong, multidimensional shock computations for an important new research program at NRL, we all knew that the traditional Computational Fluid Dynamics (CFD) approaches were not faring well, even qualitatively, because fluid quantities such as mass density and chemical species number densities would become negative—a physical impossibility. Given the importance of dynamic, high Mach-number shocks in many fields, a number of talented people were working on this problem throughout the world. Flux-Corrected Transport, NRL's qualitatively better solution, broke new ground in 1971. People often ask me how the idea for the nonlinear flux limiter arose and what were the considerations that led to the statement of the underlying principle—"The antidiffusion stage should generate no new maxima or minima in the solution, nor should it accentuate already existing extrema". Here is how it happened.

In the winter of 1968 I was finishing my PhD at the Plasma Physics Laboratory in Princeton when Dr. Keith Roberts, the Director of the Culham Laboratory in England, visited us. Before I knew what had happened, Keith, John Green (my advisor), and Klaus Hain, also visiting from NRL to discuss progress on CFD with Keith, had arranged a post-doctoral appointment at Culham Laboratory for me, to be executed in the coming year under Keith Robert's tutelage. During that year Keith became my mentor and friend and gave me a firm inventive foundation in CFD to go with my plasma physics and astrophysics training. Our work ranged over just about every computational topic imaginable during the year. However, toward year's end, I began preparing to move to Washington, DC to participate in a large theoretical/computational project that Professor John Dawson of Princeton had arranged with NRL. Thus Keith and I began looking at ways to improve multidimensional CFD—as we had promised Klaus. During some particularly discouraging tests, Keith introduced me to Godunov's 1959 "theorem" —that second- and higher-order algorithms could not preserve the physical positivity property. It was certainly clear that high-order schemes were not necessarily bringing greater accuracy so physics would have to step in to shore up the failing numerics.

At NRL in September of 1969 everyone was in a rush. Plasma physics, MHD, and CFD simulation techniques were all being pursued simultaneously for the big program. The strong, fluid-dynamic shock problem had become the number one computational roadblock by the fall of 1970 so I was urged to concentrate on the problem full time, finally developing the FCT convection algorithm in the winter. Called SHASTA (Sharp and Smooth Transport Algorithm), this continuity-equation solver was developed over weekends on computers at Princeton and Oak Ridge Laboratory. The results were sufficiently astonishing that both Keith and Klaus urged me to present the new capability at the conference/workshop "Computing as a Language of Physics," held in Trieste, Italy in the summer of 1971. This first paper "A fluid transport algorithm that works" was published by the organizers in book form through the IAEA.

In 1971 David Book joined the NRL melting pot and began working with me on FCT to extend the continuity equation solver to treat sets of coupled fluid dynamic equations and to generalize the formulation. Our 1973 article in the *Journal of Computational Physics* (submitted in November 1971) extended the Trieste paper to include strong shocks and other fluid flows. This first journal article in the FCT series of three was reprinted as the most cited article up to that time in the 25th Anniversary Issue of the JCP. This article was also acknowledged as the most cited NRL publication at the NRL 75th Anniversary Celebration a year later.

SHASTA was constructed as a layered set of corrections, each layer being added to mitigate the problems introduced by the previous layer. This general approach seldom works, as I have subsequently discovered (repeatedly), but this particular time it did. The starting point was general dissatisfaction with the negative densities and nonphysical wiggles near sharp gradients observed in leapfrog and Lax-Wendroff convection and a deep suspicion of the excess diffusion introduced in first order donor-cell algorithms. I also decided to work on a single, general continuity equation, trusting to earlier work with Keith that a numerical model for any set of fluid equations could be built up with such a building block.

Impressed by the potential of Arbitrary Lagrangian-Eulerian (ALE) algorithms, the first layer of SHASTA is a conservative Lagrangian displacement and compression of linear trapezoids of fluid. It is local, positive, and easy to program but leads to jagged sequences of mismatched trapezoids that quickly spawn most of the problems of a fully Lagrangian approach. The second layer of SHASTA, therefore, was a conservative Eulerian remap of the displaced, distorted trapezoids of fluid back onto the original fixed grid. This introduced a zero-order numerical diffusion whose coefficient, 1/8, was generally worse than donor cell diffusion! As a result, the third layer of SHASTA was an explicit, linear antidiffusion to subtract the excess diffusion introduced by the remap. Of course this resulted in solutions just as bad as the second-order Lax-Wendroff and leapfrog algorithms.

Rather than resign myself to adding a strong numerical viscosity based on derivatives that would be essentially meaningless near discontinuities and sharp gradients, I began looking in detail at the physics of the local profiles of the density profile being convected — before and after antidiffusion. In each specific case it was clear just how much of the antidiffusive correction flux could be used at each interface and how much had to be thrown away to prevent a particular cell value from being pushed past the monotonicity limits imposed by the neighboring values. However, the expression of this "flux correction," now usually called a flux limiter or slope limiter, was different in every case. Finally recognizing that the sign function of the density difference across a cell interface could be used to collapse the behavior near maxima and near minima into a single formula, the pivotal *max-min* expression emerged after a couple of days of fooling around during a marathon computing session at Oak Ridge in March 1971.

David recognized that the underlying linear convection algorithm could be just about anything; the SHASTA algorithm in the first working FCT code was just a vehicle and maybe not even a particularly good one. We tried variants of Lax-Wendroff and the other simple linear algorithms, resulting in a set of three papers in the *Journal of Computational Physics*, the second with Klaus Hain. Making the zeroth-order diffusion coefficient even larger, $1/6$, reduced the phase error in the convection from second order to fourth order with improved retention of profile structure. We found that the initial density profile could be made to emerge unscathed, when the interface velocities are zero, despite the diffusion and antidiffusion stages. David called this "Phoenical" FCT and we use this trick today. David also forced some mathematical rigor into the mix. Stung by a public criticism from Conrad Longmire and Greg Canavan at his first major FCT presentation that the results were faked, he did much experimentation and analysis to understand why FCT was working so well and insisted on a steady stream of peer-reviewed publications. Perhaps the best of these, if not the most cited, was the chapter "Solution of Continuity Equations by the Method of Flux-Corrected Transport" in volume 6 of *Methods in Computational Physics*. David recalls these early days in Chap. 1 below.

Many of NRL's fluid dynamicists made significant contributions in the 1970s. Steve Zalesak invented the multidimensional limiter, one of the main breakthroughs for FCT, and this was modified and extended by Rick DeVore to include MHD flows where $\nabla \cdot B = 0$ is an important consideration. Steve characteristically understates his own contributions in Chap. 2 below. Ed MacDonald showed us that FCT could be added to spectral algorithms, spawning an FFT FCT and a (linearly) reversible FCT algorithm. Elaine Oran and John Gardner contributed significantly to refinement of the time- and direction-split LCPFCT software package described in detail in the first and second editions of the book *Numerical Simulation of Reactive Flow*. This simple toolbox now has had tens of thousands of copies distributed worldwide. Kailas Kailasanath introduced boundary layer considerations for his ground breaking work on acoustic-vortex interactions and combustion instabilities. Ted Young, Niels Winsor, Sandy Landsberg, and Charles Lind played a major role in vectorizing and then parallelizing FCT, leading directly to NRL's current FAST3D generalgeometry CFD models. Rainald Löhner took the lead in carrying FCT into the Finite Element world with a practical general-geometry formulation that also provided a basis for adaptive meshing and moving geometry considerations. He and Joseph Baum consider the status and direction of FCT over the last thirty years in Chap. 5, focussing somewhat on blast and shock problems. Gopal Patnaik first extended FCT to an implicit formulation called BICFCT for slow but fully compressible reacting flows in performing the world's first dynamical, multidimensional flame simulations. A computer manufacturer, Texas Instruments, Inc. even got into the act. They added pairwise vector *min*, *max* and *sign* operations to the pipeline instruction space of their parallel processing Advanced Scientific Computer (ASC) at NRL's request so that the Flux-Corrected Transport in FAST3D would be even faster.

In the chapters following you will also see some of the other things done with these simple ideas in the ensuing three decades. My own focus at the Naval Research Laboratory has been on large-scale urban transport of contaminants. Chapter 4 by Gopal Patnaik, Jay Boris, Fernando Grinstein, and John Iselin considers this very important application, showing how to reduce the numerical dissipation still further for detailed urban and building airflows. 3D turbulence, the source of contaminant dispersion in these urban problems, is treated by Monotone Integrated Large Eddy Simulation (MILES), an LES formulation based implicitly on the monotonicity properties of FCT with stochastic backscatter in a 4th-order phase, time-accurate, finite volume model for detailed building and city aerodynamics. In retrospect, it should not be surprising that an algorithm designed for treating sharp gradients and discontinuities in compressible flow should work equally well for rotational flows. Nevertheless, it has taken fifteen years since its explanation in 1989 for the community to acknowledge the benefits of monotone (in our case FCT) algorithms for the treatment of turbulence. Chapter 3, by Fernando Grinstein and Christer Fureby, considers this originally unintended but greatly appreciated byproduct of FCT.

Chapters 6, 7 and 8, by Dmitri Kuzmin, Matthias Möller, and Stefan Turek complete the book, presenting a unified theoretical foundation for "Algebraic Flux-Correction" from the Finite Element perspective. This approach is intrinsically multidimensional and unifies FCT and TVD variants of monotonicity-preserving algorithms. The authors retrace the original progression of developments, beginning with scalar conservation laws, extending to the hyperbolic compressible Euler equations, and finishing with a methodology for treating incompressible flow problems.[1]

U.S. Naval Research Laboratory, Washington, DC, USA

---

[1] The unreferenced Chaps. 9, 10 and 11 were added in the second edition of this book.

Preface

By a remarkable coincidence, the second edition of this book will appear in the year that marks the 40th anniversary of Flux-Corrected Transport (FCT). The first decade of the 21st century has witnessed a renewed interest in applications of FCT to the equations of fluid dynamics. The first edition of the book has found many readers who used it to improve existing and design new FCT algorithms. Other readers recognized the advantages of FCT and applied it to challenging new problems. A recent comparative study of shock-capturing techniques for unsteady transport equations [16] has made FCT more popular in the finite element community. Being the oldest design tool for nonlinear high-resolution schemes, FCT is still superior to many modern methods when it comes to solving problems with steep gradients.

The revised and expanded second edition summarizes many recent advances in the field of FCT. Chapters 3–8 have been updated to reflect the current state of the art. Moreover, the second edition features three new chapters describing FCT-constrained data transfer in Arbitrary Lagrangian-Eulerian methods, an optimization-based approach to flux correction, and the implementation of an FCT algorithm for high-speed flows on structured overlapping grids. The research presented in the new chapters was done at three U.S. National Laboratories.

The Editors would like to thank all authors for their contributions that make this book the most complete source of information on modern FCT methods. The initiative of the Springer Verlag to publish the second edition is also gratefully acknowledged. Special thanks go to Tobias Schwaibold and Kirsten Theunissen who coordinated the review of the new book proposal and the publication process.

Erlangen, Germany Dmitri Kuzmin
Fairfax, VA, USA Rainald Löhner
Dortmund, Germany Stefan Turek
November 2011

Preface to the First Edition

In 1973, in the eighth year of its youth, the *Journal of Computational Physics* published the classic Boris and Book paper describing flux-corrected transport (FCT) [1]. Almost all of the monotonicity-preserving and non-oscillatory fluid transport algorithms of today trace their origins, ultimately, to ideas that first appeared in this paper.

Boris and Book's new and far-reaching idea was to locally replace formal truncation error considerations with conservative monotonicity enforcement in those places in the flow where the formal truncation error had lost its meaning, i.e., where the solution was not smooth and where formally high order methods would violate physically-motivated upper and lower bounds on the solution. This is today still the fundamental principle underlying the great bulk of the monotonicity-preserving and non-oscillatory algorithms that have appeared in more recent times. Occasionally this bit of history is lost in some of the more recent literature, in part due to the fact that the paper is now more than 30 years old (and the original publication [2] older still).

In [1], the authors applied this fundamental idea to a specific algorithm they termed SHASTA. They were able to show not only sharp monotone advection of linear discontinuities, but also sharp non-oscillatory gas dynamic shock waves. Included in [1] was a SHASTA calculation of a shock tube problem much more difficult than that used by Sod five years later [3], with nearly monotone results, and with no knowledge of the solution (e.g., Riemann solvers) built in to the algorithm. All of these calculations were the first of their kind with monotonicity-preserving algorithms of greater than first order accuracy. It was also in this paper that the term "flux-limiting" [1, p. 50] appeared in print for the first time.

In the years following 1973, Boris and Book and colleagues published two more FCT papers in the *Journal of Computational Physics* [4, 5], followed by a chapter in the *Methods in Computational Physics* book series [6] that summarized their work up through 1976. These works refined their ideas, generalizing the algorithms to a larger class of which SHASTA was just one member. Their emphasis was on the continuity equation as a scalar representative of systems of conservations laws, and upon advective phase error as a primary culprit in the elimination of the errors that remained after non-oscillatory behavior was eliminated via flux limiting. A more recent summary of FCT is given in the book by Oran and Boris [7].

Work on FCT algorithms has also thrived elsewhere in publications far too numerous to reference here. Two notable examples are the extension of FCT to fully multidimensional form by Zalesak in 1979 [8], and the generalization of FCT to finite element discretizations on unstructured grids (e.g., triangles and tetrahedra in two and three dimensions respectively) by Parrott and Christie in 1986 [9]. A general FEM-FCT methodology for the Euler and Navier-Stokes equations of fluid dynamics was introduced by Löhner et al. [10]. One of the consequences of this last development has been the ability to perform FCT calculations in extremely complex geometries. An example is the remarkable simulation of the 1993 World Trade Center blast which modeled in detail the garage of the building including all of the parked cars [11].

The response of the scientific computing community to FCT was and still is remarkably strong. Many new publications have emerged during the 1990s and early 2000s. The recent advances include but are not limited to: the use of FCT as an implicit subgrid scale model for *Monotonically Integrated Large Eddy Simulation* (MILES) [12], 'iterative and synchronous flux-correction' [13], monotonicity-preserving 'prelimiting' [14] as well as implicit FEM-FCT schemes based on a generalization of Zalesak's limiter [15]. Clearly the impact of the original paper [1] is still being felt long after its original publication. This impact is seen in an astounding number of citations and application of FCT to virtually every area of science, from aerodynamics and shock physics to atmospheric and ocean constituent transport, magnetohydrodynamics, kinetic and fluid plasma physics, astrophysics, and computational biology.

Most of the contributions compiled in the present volume are based on talks given at the Workshop "High-Resolution Schemes for Convection-Dominated Flows: 30 Years of FCT" which was held in September 2003 at the University of Dortmund, Germany. It was intended to provide a forum for discussion of the progress made in the development of numerical methods for fluid dynamics during the three decades that elapsed since the birth of FCT. The high caliber of the presented results and many fruitful discussions have made this informal meeting a remarkably successful one. This has led us to unite our efforts and describe the state of the art in this book. The Editors would like to express their sincere gratitude to Prof. Roland Glowinski (University of Houston) for the expert review of the manuscript and thank Prof. Wolf Beiglböck (Springer-Verlag Heidelberg) for the prompt publication.

Dortmund, August 2004 *The authors*

References

1. Boris, J.P., Book, D.L.: Flux-corrected transport. I. SHASTA, A fluid transport algorithm that works. J. Comput. Phys. **11**, 38 (1973).
2. Boris, J.P.: A fluid transport algorithm that works. In: Proceedings of the Seminar Course on Computing as a Language of Physics, International Centre for Theoretical Physics, Triest, Italy, 2–20 August 1971.
3. Sod, G.A.: A survey of several finite difference methods for systems of nonlinear hyperbolic conservation laws. J. Comput. Phys. **27**, 1 (1978).
4. Book, D.L., Boris, J.P.: Flux-corrected transport II: Generalizations of the method. J. Comput. Phys. **18**, 248 (1975).
5. Boris, J.P., Book, D.L.: Flux-corrected transport. III. Minimal error FCT algorithms. J. Comput. Phys. **20**, 397 (1976).
6. Boris, J.P., Book, D.L.: Solution of Continuity Equations by the Method of Flux-Corrected Transport. Methods in Computational Physics, vol. 16. Academic Press (1976).
7. Oran, E.S., Boris, J.P.: Numerical Simulation of Reactive Flow. Elsevier, New York (1987). 2nd edition: Cambridge University Press (2001).
8. Zalesak, S.T.: Fully multidimensional flux-corrected transport algorithms for fluids, J. Comput. Phys. **31**, 335 (1979).
9. Parrott, A.K., Christie, M.A.: FCT applied to the 2-D finite element solution of tracer transport by single phase flow in a porous medium. In: Proc ICFD Conf. on Numerical Methods in Fluid Dynamics, p. 609. Oxford University Press (1986).
10. Löhner, R., Morgan, K., Peraire, J., Vahdati, M.: Finite element flux-corrected transport (FEM-FCT) for the Euler and Navier-Stokes equations. Int. J. Numer. Methods Fluids **7**, 1093–1109 (1987).
11. Baum, J.D., Luo, H., Löhner, R.: Numerical simulation of the blast in the World Trade Center. AIAA-95-0085 (1995).
12. Boris, J.P., Grinstein, F.F., Oran, E.S., Kolbe, R.J.: New insights into Large Eddy Simulation. Fluid Dynamics Research **10**(4–6), 199–227 (1992).
13. Schär, C., Smolarkiewicz, P.K.: A synchronous and iterative flux-correction formalism for coupled transport equations. J. Comput. Phys. **128**, 101–120 (1996).
14. DeVore, C.R.: An improved limiter for multidimensional flux-corrected transport. NASA Technical Report AD-A360122 (1998).
15. Kuzmin, D., Turek, S.: Flux correction tools for finite elements. J. Comput. Phys. **175**, 525–558 (2002).
16. John, V., Schmeyer, E.: On finite element methods for 3D time-dependent convection-diffusion-reaction equations with small diffusion. Comput. Methods Appl. Mech. Engrg. **198**, 475–494 (2008).

Contents

Foreword ix
    Jay P. Boris

Preface xiii

Preface to the First Edition xv

The Conception, Gestation, Birth, and Infancy of FCT 1
    David L. Book

The Design of Flux-Corrected Transport (FCT) Algorithms for Structured Grids 23
    Steven T. Zalesak

On Monotonically Integrated Large Eddy Simulation of Turbulent Flows Based on FCT Algorithms 67
    Fernando F. Grinstein and Christer Fureby

Large Scale Urban Simulations with FCT 91
    Gopal Patnaik, Jay P. Boris, Fernando F. Grinstein, John P. Iselin, and Denise Hertwig

40 Years of FCT: Status and Directions 119
    Rainald Löhner and Joseph D. Baum

Algebraic Flux Correction I 145
    Dmitri Kuzmin

Algebraic Flux Correction II 193
    Dmitri Kuzmin, Matthias Möller, and Marcel Gurris

Algebraic Flux Correction III 239
    Stefan Turek and Dmitri Kuzmin

Algebraic Flux Correction and Geometric Conservation in ALE Computations 299
Guglielmo Scovazzi and Alejandro López Ortega

Constrained-Optimization Based Data Transfer 345
Pavel Bochev, Denis Ridzal, Guglielmo Scovazzi, and Mikhail Shashkov

An Evaluation of a Structured Overlapping Grid Implementation of FCT for High-Speed Flows 399
J.W. Banks and J.N. Shadid

Index 447

The Conception, Gestation, Birth, and Infancy of FCT

David L. Book

Abstract How Flux-Corrected Transport came to be, as recalled by one of the innovators: recollections of how FCT was developed and of the individuals responsible.

1 Conception

In 1970 there was essentially no reliable way to solve fluid equations numerically. By 1971 there was one. Flux-Corrected Transport (FCT) was the first method developed that yielded physically acceptable results for such equations. The present paper describes how Jay Boris and I developed FCT, with what I hope is an accurate account of our thinking at the time, the path we followed in the course of the development, including the missteps and blind alleys, and the roles of the other individuals who were involved.

Fluid or hydrodynamic equations are partial differential equations dominated by convective motion, that is, equations in which convective derivative terms of the form

$$\frac{\partial f}{\partial t} + \mathbf{v} \cdot \nabla f$$

play a decisive role, where f is one of the dependent fluid variables (density of mass, momentum, energy, or charge; pressure, entropy, species concentration, etc.), t is time, $\mathbf{v}$ is the flow velocity, $\nabla \equiv \frac{\partial}{\partial \mathbf{r}}$, and $\mathbf{r}$ is position. Examples are the continuity equation for a compressible medium with mass density ρ,

$$\frac{\partial \rho}{\partial t} + \nabla \cdot \rho \mathbf{v} = 0,$$

the Euler (momentum) equation in the presence of a scalar pressure p and a constant gravitational acceleration $\mathbf{g}$, which can be written

$$\frac{\partial (\rho \mathbf{v})}{\partial t} + \nabla \cdot \rho \mathbf{v} \mathbf{v} + \nabla p + \rho \mathbf{g} = 0,$$

D.L. Book (✉)
Enigmatics, Inc., P.O. Box 8610, Monterey, CA 93943, USA
e-mail: davidbook@enigmatics.com

and the Navier–Stokes equation (the Euler equation with the inclusion of viscosity),

$$\rho \frac{\partial \mathbf{v}}{\partial t} + \rho \mathbf{v} \cdot \nabla \mathbf{v} + \nabla p = \mu \nabla^2 \mathbf{v}.$$

Such equations are also called convective or hyperbolic, although strictly speaking the Navier–Stokes equation is parabolic in regions where the Laplacian term dominates. In general, the equation expressing the transport of any continuously distributed quantity, together with terms describing sources and losses, is of this form. This category includes all of the familiar conservation laws.

A third of a century ago computer resources were meager by comparison with today's technology, but most of the general computational approaches in use today were known, at least in broad outline, and the first steps had already been taken toward applying them. Finite differences and finite elements (the distinction between them then was somewhat blurred, though now people usually associate these terms with differencing schemes on structured and unstructured grids, respectively), characteristics, quasiparticles, and spectral methods had all been invented, and all of these tools were being applied to the problem of finding computational solutions to fluid equations. Dozens of Fortran codes based on each of these methods, or combinations of several of them, were in existence. They all fell short of what I feel is the goal of any numerical treatment of evolution equations: using limited, i.e., discretized, information about the dependent variables in order to predict the values of those variables at a later time *with the same accuracy or level of confidence*.

The presence of convective derivatives is what makes fluid equations difficult to solve numerically. Because of them the characteristics, the space–time trajectories along which the values of the fluid variables are constant, slope. In order to predict the values at a position $\mathbf{r}$ of the fluid variables at a time t' later than a time t for which their values are known, it is necessary to use information from the points through which the characteristics passed at time t, which are in general different from $\mathbf{r}$. Thus, to predict the values of the fluid variables at a particular point on, e.g., a finite-difference grid may require knowing their values at an earlier time from a location *that was not on the grid*.

In thinking about the implications of sloping characteristics I like to use an analogy with what I call the window problem. Suppose you are in a room with several windows looking out onto a nearby railroad track. When a train comes by each car in succession appears at a given window. Let's say that the train is moving from your left to your right. If you want to know what car is going to appear next at the window in front of you, you can look out the window to the left of it. You don't have to keep a continuous watch; it suffices to glance over at the second window at intervals, which correspond to the discrete timesteps in a finite-difference scheme. Of course, if the windows are too widely spaced, or if you are too close to them, there may be several cars hidden out of sight between the two windows. This is analogous to using too coarse a mesh in your difference scheme. Likewise, you may miss a car if the intervals between glances are too long, which corresponds to using too long a timestep Δt.

But the information you get is limited in another way, because the window is narrow and you can't see a whole railroad car at one time. In order to make good

Fig. 1 Pathological railcars

predictions you have to know something about the kinds of railroad cars that the train is allowed to have. A glimpse may be all you need to know that a car is a boxcar or a tank car or a flat car, but what if it's a car of a type you've never seen before, say, one carrying a crane to lift up wrecks? Or one with unusual dimensions or proportions (Fig. 1)? Your prediction is implicitly based on a bias or prejudice in favor of the kinds of cars you expect to see.

In the same way, a numerical technique must contain a built-in bias about the form solutions can take, because the available information is limited by discretization. No technique can handle all situations. Each one must be tailored to fit a particular class of problems.

The problems Jay and I were interested in were time-dependent fluid problems, especially those involving supersonic flow, and more generally, systems with discontinuities or steep gradients. These include not just shocks (which can occur only when there is supersonic flow), but also contact and tangential discontinuities and abrupt changes in temperature, species concentration, etc. We opted to employ a finite-difference approach because it simplified coding, particularly in multidimensions and at boundaries, and because that was what we were most comfortable with.

The naïve approach to finding a finite-difference approximation to differential equations on a mesh with a uniform spacing Δx is to expand the derivatives in Taylor series:

$$f(x \pm \Delta x) = f(x) \pm \frac{\partial f}{\partial x} \Delta x + \frac{1}{2} \frac{\partial^2 f}{\partial x^2} (\Delta x)^2 \pm \cdots.$$

Thus,

$$f(x + \Delta x) - f(x - \Delta x) = 2 \frac{\partial f}{\partial x} \Delta x + O(\Delta x)^3$$

or

$$\frac{\partial f(x,t)}{\partial x} = \frac{1}{2\Delta x} \left[f(x + \Delta x, t) - f(x - \Delta x, t) \right] + O(\Delta x)^2,$$

and similarly

$$\frac{\partial f(x,t)}{\partial t} = \frac{1}{2\Delta t} \left[f(x, t + \Delta t) - f(x, t - \Delta t) \right] + O(\Delta t)^2.$$

Hence a straightforward finite-difference approximation to the 1-D passive advection equation

$$\frac{\partial \rho}{\partial t} + u \frac{\partial \rho}{\partial x} = 0$$

is

$$\rho(x, t + \Delta t) = \rho(x, t - \Delta t) - \varepsilon [\rho(x + \Delta x, t) - \rho(x - \Delta x, t)],$$

where $\varepsilon = u \Delta t / \Delta x$ is the Courant number. The resulting finite-difference approximation is of course just second-order leapfrog, a useful scheme that has been widely adopted.

The Taylor-series approach has a number of strengths. It yields difference schemes (such as leapfrog) that are accurate for problems with slowly varying profiles, i.e., those in which discontinuities are absent. Also, it facilitates the analysis of amplitude and phase errors. Thus, assume a sinusoidal density profile

$$\rho_j^0 = \rho_0 \exp(2\pi i j \kappa / N),$$

where $j \Delta x$ is position, κ is the mode number and N is the number of mesh points. If

$$\rho_j^1 = \rho_1 \exp(2\pi i j \kappa / N)$$

is the corresponding profile found after one timestep, then the amplification factor is $A_\kappa = |\rho_1 / \rho_0|$ and the relative phase error (the error in the speed with which features are advected numerically, divided by the correct speed) is $R_\kappa = (\kappa \varepsilon)^{-1} \tan^{-1}(\operatorname{Im} \rho_1 / \operatorname{Re} \rho_0) - 1$.

But the Taylor-series expansion approach also has some notable deficiencies. It works only when "order" makes sense, i.e., when the scale of variation is large compared with the mesh spacing, so that the neglect of higher-order terms ("truncation errors") is justified. Consequently, it breaks down at discontinuities, where "dispersive" ripples make their appearance. (The finite-interval Gibbs effect, the analog for discrete Fourier transforms of the well-known Gibbs phenomenon, can also contribute to errors in the vicinity of a discontinuity. I will return to this topic below.)

The key insight (which to the best of my knowledge originated with Jay) is that the Taylor-series approach fails because it does not enforce *positivity*, a property sometimes called monotonicity. For physical reasons some variables can only take on positive values. Examples are mass and energy density (but not charge or momentum density), temperature, and pressure. Nothing in the Taylor-series approach ensures this. Positivity violations are found to be worst near discontinuities. At discontinuities formal high-order accuracy is less important than maintaining positivity.

This may be an appropriate time to try to spell out what is meant by the term "discontinuity." Shocks, contact discontinuities, and slip lines (tangential discontinuities) would be physically discontinuous in the absence of dissipation. Because dissipation can never be totally absent, however, no physical quantity is ever truly discontinuous, at least in classical physics. There are no discontinuities in nature.

In finite-difference approximations, on the other hand, all changes are discontinuous, but some are more discontinuous than others. A criterion is needed to determine when a discontinuity is "real." This criterion may involve calculating whether the relative or absolute change in a variable exceeds some threshold value, or it may be more complicated. The exact definition, as always, depends on the nature of the problem and one's expectations about the solution being sought.

In the absence of a specific criterion a finite-difference scheme will not be able to distinguish a weak physical discontinuity from a smooth feature or from noise. As I will show, there are several reasons why a profile that should be smoothly varying might develop ripples or "bumps" in a computational treatment. Hence in any scheme susceptible to such errors nonphysical features will be treated just like physical ones. The trick is to avoid, as much as possible, developing them in the first place.

The simplest system of equations that models shocks is that of ideal hydrodynamics. This consists of the continuity and Euler equations and an equation for the energy density

$$E = \frac{1}{2}\rho \mathbf{v}^2 + \frac{p}{\gamma - 1}.$$

The energy equation follows from the adiabatic law (pressure equation) in the form

$$\frac{\partial p}{\partial t} + \mathbf{v} \cdot \nabla p + \gamma p \nabla \cdot \mathbf{v} = 0,$$

where γ is the ratio of specific heats.

These three equations express the conservation of mass, momentum, and energy. That they are conservative is important; if the adiabatic law is used instead of the formally equivalent energy equation, then pressure remains positive but the Rankine–Hugoniot (jump) conditions are violated. If the energy equation is chosen as one of the fundamental evolution equations, then energy is conserved and shocks obey the Rankine–Hugoniot conditions, but pressure is a derived quantity which can become negative. The Rankine–Hugoniot conditions imply that the jump in entropy varies as $(M - 1)^3$, where M is the Mach number. Using the adiabatic law means that only weak shocks ($M \to 1$) are calculated correctly. This dilemma confronts all finite-difference schemes.

Physically, viscous dissipation is responsible for creating entropy at a shock front. This is what allows shocks to satisfy the Rankine–Hugoniot conditions. The equations of ideal hydrodynamics in conservative form contain no explicit viscosity terms, but nevertheless admit solutions that satisfy the Rankine–Hugoniot conditions. These so-called weak solutions represent a zero-viscosity limit. Physical shocks in systems with nonzero viscosity differ in having nonzero thickness; that is, the jump takes place over a region of finite extent.

Any numerical scheme must include *some* dissipation if it is going to allow the jump conditions to be satisfied. Before the invention of FCT the most successful finite-difference treatments of supersonic flow, developed at Los Alamos and widely employed elsewhere, introduced an artificial viscosity term, typically in the form of a velocity-dependent coefficient multiplying the second difference of the velocity. This forced the production of entropy at places where the velocity underwent abrupt change and allowed the Rankine–Hugoniot conditions to hold.

The trouble with this approach was that in order to generate enough entropy the coefficient had to be large. This in turn caused the shock to be spread out over several or many mesh spaces. That was fine if the physical viscosity in the problem was large, i.e., if the Reynolds number Re was of order unity. But for realistic problems of high-Reynolds-number flow—say, $Re > 100$—it was hopelessly inaccurate. To reduce the shock thickness to reasonable values would require being able to determine where shocks were located or were about to form or introducing thousands of grid points in each coordinate direction, which would have been prohibitively expensive. (Remember, this was in 1970.)

There is another downside to using artificial viscosity. The timestep limit in finite-difference problems arises because information can travel no faster than one mesh space per timestep (assuming three-point difference schemes), which means that the Courant number must not exceed unity. Violating this condition in explicit finite-difference treatments of the ideal hydrodynamic equations leads to catastrophic numerical instability. If the spatial mesh in a calculation is refined the timestep must shrink proportionately to stay within safe limits. But when diffusion terms of the form $\mu \nabla^2 \mathbf{v}$ dominate, then the timestep scales with the square of the mesh space. Consequently, in methods using artificial viscosity even refining the mesh locally near the shocks becomes much more expensive if the diffusion terms are differenced explicitly. (Implicit differencing has problems of its own.)

A shock wave heats the medium through which it is traveling, which causes the speed of sound to increase. Hence information propagates faster, so that signals in the region behind a shock tend to catch up with the shock. This means that shocks are self-steepening. Other discontinuities, such as contact surfaces, are not. As a result, shear surfaces and interfaces between two different media or between two regions in the same medium with different properties tend to be smeared out by numerical diffusion and are more difficult to model than shocks.

The passive advection equation models advection and propagates contact discontinuities, but in contrast to the system of hydrodynamic equations it has no self-steepening mechanism. Consequently, it is a more stringent test of numerical fluid-equation solvers. In addition it is simpler to work with than a set of nonlinear equations. Thus, it was natural for us to choose it as our test bench instead of the full set of hydrodynamic equations. We felt that if we could do a good job solving this equation numerically we could solve most fluid problems.

For our basic test problem we chose to propagate a square wave 20 mesh spaces in width across a grid 100 mesh spaces wide (Fig. 2) for 800 timesteps with a constant Courant number $\varepsilon = 0.2$, using periodic boundary conditions in order to al-

Fig. 2 Initial conditions for the square-wave test

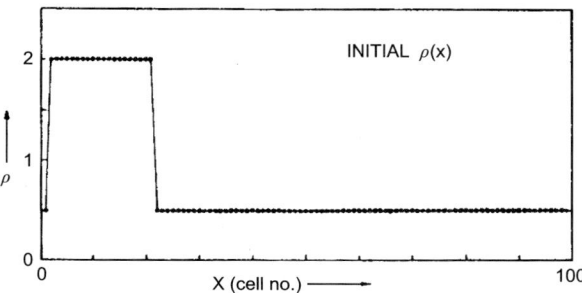

Fig. 3 Square-wave test of donor cell (first-order accuracy)

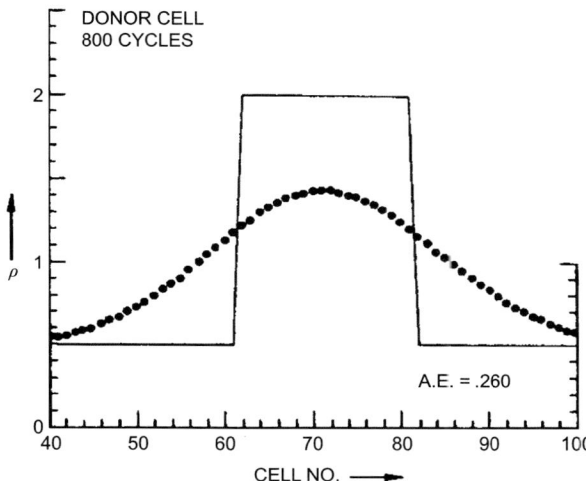

low the profile to reenter the system. As a quantitative figure of merit we used the average absolute value of the error (the L1 norm), abbreviated A.E. This test problem subsequently became a kind of universal standard in the computational physics community.

When various numerical methods are evaluated using this problem one can readily discern their strengths and shortcomings. Those that are inaccurate to zeroth order in terms of Taylor-series expansions in $k \Delta x$, the nondimensionalized wavenumber, fail to track the analytical solution correctly, yielding profiles that move either too fast or too slow. Those that are first-order accurate, such as donor-cell (upwind differencing), yield profiles that move at the right speed and maintain positivity, i.e., $\rho(x) > 0$ everywhere, but become smeared out over an ever-increasing portion of the grid (Fig. 3). In other words, they are highly *diffusive*.

Methods that are second-order accurate, such as leapfrog or Lax–Wendroff, yield profiles that develop multiple ripples (Fig. 4). These arise because the various Fourier harmonics that make up the square wave propagate at different speeds. The long-wavelength components propagate at nearly the right speed, while the short-wavelengths usually lag behind. In other words, the errors are *dispersive*. The rip-

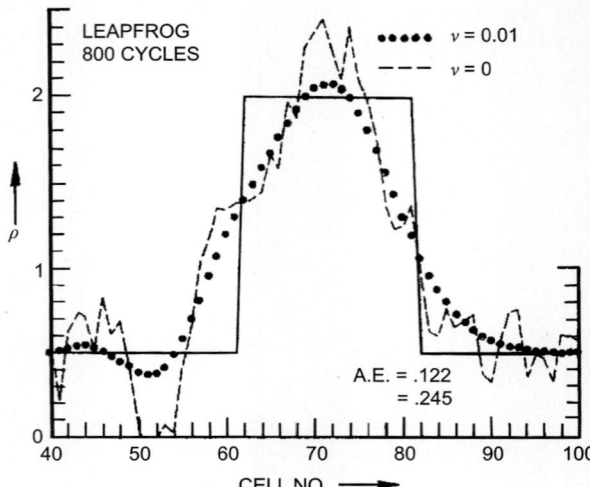

Fig. 4 Square-wave test of leapfrog (second-order accuracy)

ples grow in amplitude until the profile can become negative in some places. Thus, second-order algorithms do not maintain positivity. Notice that introducing a small amount of smoothing ($\nu = 0.01$) not only eliminates these negative values but also reduces the A.E. The optimum choice of the smoothing coefficient ν is, however, problem-dependent.

It is difficult to say which is worse, diffusive or dispersive errors, Scylla or Charybdis. Going to higher than second order doesn't solve the problem. Every technique that can be expressed in terms of linear finite-difference operations on the dependent variable—including every finite-difference treatment of the passive advection equation and its hydrodynamic kin in existence prior to the invention of FCT—suffers from one or the other failing.

2 Gestation

FCT was the first *nonlinear* finite-difference technique. In my view there were three main steps in our thinking that led to its development: expressing all operations in terms of fluxes, certainly not a new idea at the time; a transport algorithm called SHASTA, which is highly diffusive even in the limit of zero velocity, suggesting the use of "antidiffusion" to cancel out the diffusive errors; and the idea of correcting (limiting) the antidiffusive fluxes in order to maintain positivity (the nonlinear ingredient).

Fluxes are quantities of an extensive variable (e.g., mass, momentum, energy) that pass from one cell or grid point to another. If a finite-difference algorithm can be expressed entirely in terms of fluxes then it is guaranteed to be conservative, because

Fig. 5 Finite-difference approximation represented by rectangles and by trapezoids

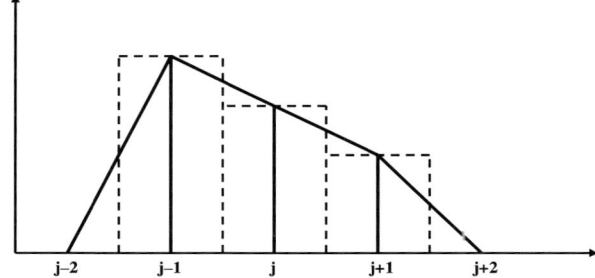

what is removed from one place reappears in another. Thus, advection (transport only) can be approximated using transportative fluxes:

$$\rho_j^T = \rho_j - \phi_{j+1/2}^T + \phi_{j-1/2}^T,$$

where

$$\phi_{j+1/2}^T = \varepsilon_{j+1/2} \rho_{j+1/2},$$

with $\varepsilon_{j+1/2} = u_{j+1/2} \Delta t / \Delta x_{j+1/2}$ and $\rho_{j+1/2}$ defined on the cell boundary, $x_{j+1/2} = \frac{1}{2}(x_j + x_{j+1})$, and $\Delta x_{j+1/2} = x_{j+1} - x_j$.

Similarly, diffusion can be expressed in terms of diffusive fluxes:

$$\rho_j^D = \rho_j + \phi_{j+1/2}^D - \phi_{j-1/2}^D,$$

where

$$\phi_{j+1/2}^D = \nu_{j+1/2}(\rho_{j+1} - \rho_j).$$

In both instances *what is subtracted from one cell is added to its neighbor*, so the total "mass" is conserved.

The second ingredient, SHASTA, was Jay's idea. He devised it by means of a geometric approach. Imagine a finite-difference approximation ρ_j to some continuous variable $\rho(x)$, represented by histograms or rectangles (the broken lines in Fig. 5). Connect the points denoting the values of ρ_j with straight lines to form trapezoids. The area contained in the resulting trapezoids is the same as that contained in the rectangles.

If this profile is then transported across the grid with some velocity $u(x)$, in general each trapezoidal packet of fluid undergoes advection together with compression or expansion. Each mesh point x_j moves to a new location x'_j. In one timestep the two vertical sides x_j and x_{j+1} of a trapezoid thus move in general by different distances, causing it to be deformed as well as translated. The condition $x'_j < x'_{j+1}$, which is necessary to ensure positivity, imposes the limitation $\varepsilon < 0.5$. At the end of each timestep the mass contained in the trapezoid is reassigned to the two rectangles it straddles: the portion to the left of the boundary between two cells is assigned to the left-hand cell and the portion to the right is assigned to the right-hand cell (Fig. 6).

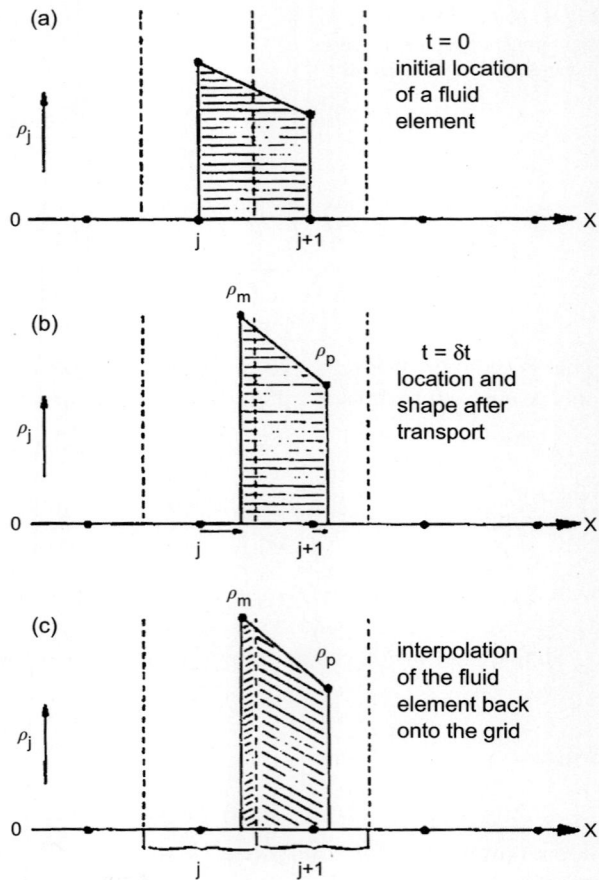

Fig. 6 SHASTA contains a zeroth-order diffusion

It is easy to see that the process of creating trapezoids and reassigning their mass is highly diffusive. In fact, in the limit when the velocity is identically zero the new value of ρ_j equals the old value plus a second difference with coefficient of 0.125:

$$\rho_j^{n+1} = \rho_j^n + \frac{1}{8}\left(\rho_{j+1}^n - 2\rho_j^n - \rho_{j-1}^n\right).$$

Since this algorithm is diffusive, the natural thing to do is to subtract the excess diffusion, or to put it another way, to apply antidiffusion. Antidiffusion is just diffusion with a negative coefficient:

$$\frac{\partial \rho}{\partial t} = -A\frac{\partial^2 \rho}{\partial x^2}, \quad A > 0.$$

Whereas diffusion erodes features, antidiffusion steepens them. Discontinuities become sharper and new extrema can occur. If this process is sufficiently drastic it can violate positivity (Fig. 7).

Fig. 7 Antidiffusion can violate positivity

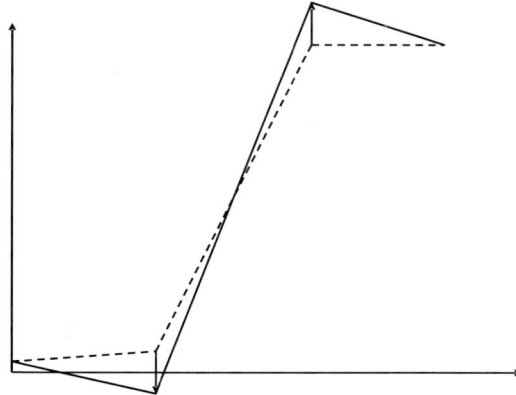

One way to look at this is to recognize that diffusion converts a second-order algorithm like Lax–Wendroff or leapfrog that has wiggles into one that is first-order like donor cell. Hence taking out the excess diffusion changes it back into a second-order algorithm and restores the wiggles.

Jay recognized that by correcting or *limiting* the antidiffusive fluxes before they are applied he could avoid restoring the wiggles. A flux large enough to push a point j down below its neighbors will create a new minimum that might go negative, so it is necessary to reduce the size of this flux. If the profile already has a minimum at j, then allowing it to be pushed further down is dangerous, so it is necessary to zero any flux that tends to do so. Likewise, it is necessary to avoid enhancing maxima on negative profiles. The two rules can be combined into one: replace the "raw" antidiffusive fluxes with fluxes "corrected" so that *no new extrema can form and existing extrema cannot grow*. Figure 8 illustrates the four different situations a limiter can face when dealing with the flux between the points j and $j+1$, assuming the gradient is positive there: (a) no pre-existing extrema; (b) a maximum present at $j+1$; (c) a minimum present at j; or (d) both.

This was the first flux limiter that gave satisfactory results. We eventually realized that there are other workable variants, but this one, in which each flux is corrected without reference to others, is arguably the simplest. Because its action sometimes amounts to overkill (in ways I will describe shortly) I called it "strong" flux limiting.

The idea of using a flux limiter was the crucial ingredient in FCT. Some of the credit for it should go to the late Klaus Hain, our colleague at the Naval Research Laboratory. At the time Klaus was also trying to develop an algorithm to solve convective equations. I believe he was the first to recognize that some sort of adjustment or correction in the fluxes was needed, but he had not yet found the right formulation. Jay picked up the idea from him and made it work.

In some ways we were Klaus's competitors more than his collaborators. Klaus was an extremely able numericist, and Jay clearly wanted to be the first to find a successful algorithm. Jay and I worked closely together, but Klaus worked almost entirely alone. I thought it was because Klaus, German-born and about two decades

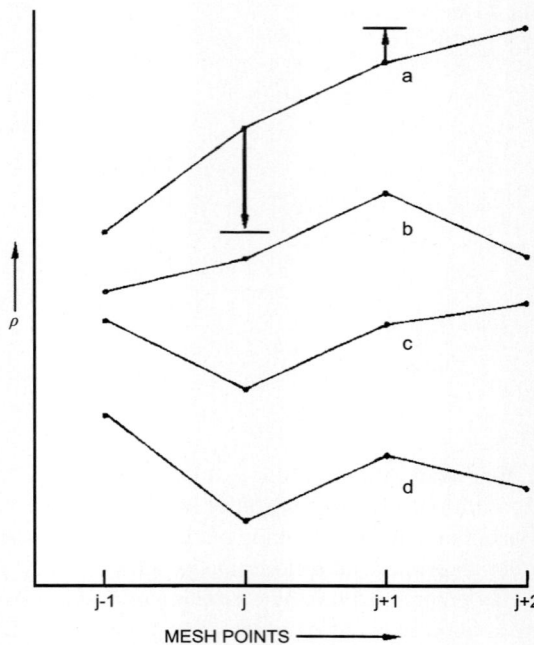

Fig. 8 Possible actions of a "strong" flux limiter on a positive flux

older than us, had trouble speaking and writing English. His wife Gertie, however, assured everyone that he was uncommunicative in German too.

3 Birth

By mid-1971 we had a working one-dimensional flux-corrected version of SHASTA. The name SHASTA supposedly came not from the famous mountain, but from the *nom de guerre* of a topless dancer. (I cannot confirm this from my own experience.) Possibly because he thought this lacking in dignity Jay contrived the acronym "SHarp And Smooth Transport Algorithm," which is how we presented it to the world.

Initially Jay had been unwilling to reveal the secret of FCT to outsiders, but one day he said to me, "Here comes Super-Klaus!" He suspected that Klaus was writing up his work for publication, and that was what changed his mind. We decided that one paper was not enough. Instead there would be a series, which became the celebrated FCT-1, FCT-2, and FCT-3. Jay was the obvious choice as the senior author of the first paper, most of which he drafted, but he must have felt a little guilty about scooping Klaus, because he suggested that the subsequent papers should bear the names of all three of us. Each of us would write the first draft of one of them; Klaus would be lead author of one and I would be lead author of the other. In the event though, Klaus—true to form—never did write anything, and his name appeared only on the second paper.

Actually, the first publications describing FCT appeared not in 1973, but two years earlier. (Hence the workshop for which I prepared the present paper should properly be called FCT-32, which is a more computational-sounding name than FCT-30 anyway.) I wrote the first journal article, which closely followed the manuscript we were preparing for submission to the Journal of Computational Physics at the time. It appeared in the November issue of NRL Reports of Progress, a house organ read by almost nobody. The only reason I wrote it was because I was then serving on a committee set up in order to make the Reports of Progress more relevant and better known.

But our very first publication was oral and never found its way into print. The application for which FCT was intended was modeling the atmospheric nuclear explosions that would have resulted from the infamous antiballistic missile program. The 1971 Symposium on High-Altitude Nuclear Effects (HANE), held at Stanford in August, was the first exposure of FCT to the computational community at large. The meeting was attended by government and private contractors funded by the Defense Atomic Support Agency, which that year became the Defense Nuclear Agency (DNA), afterward called the Defense Special Weapons Agency, and currently the Defense Threat Reduction Agency.

Several people from NRL were there, presenting results on various aspects of HANE. The attendees from other organizations were of course competing with us for DNA support. We wanted to impress them and our sponsor, but not to tell them too much. I gave the talk on the design of FCT. Jay was very insistent that I not reveal any secrets, and I didn't. I showed the results of the square-wave tests, I described SHASTA, but I didn't explain how the flux limiter worked, nor did I mention an embellishment called a "steepener." A lot of what I said was sheer doubletalk. I have a vivid recollection of a frustrated Greg Canavan from the Air Force Weapons Laboratory standing in the audience, asking question after question, trying to pin me down. Finally he said, "You keep saying, 'Another way to look at it is so-and-so.' Just tell us how it works!"

The meeting was a triumph for FCT and for our group at NRL. Our co-authors on that initial publication, Carl Wagner and Ed McDonald, were among the first users of SHASTA. Carl later worked on controlled fusion in private industry. Ed, who is still at NRL, changed fields and is now widely known for his work modeling sound propagation in the ocean. Jay and I both began applying SHASTA to a variety of problems, as did at least half a dozen of our colleagues in the Plasma Dynamics Branch. Many of these led to plasma and ionospheric physics papers.

At the same time we continued to refine and extend the method while preparing for publication in JCP. Some of the improvements resulted in making the code more efficient. Jay found that the flux limiter, which originally required a nested sequence of IF statements, could be expressed by a one-line formula:

$$\phi^C_{j+1/2} = \sigma_{j+1/2} \max\left[0, \min(\sigma_{j+1/2}\Delta_{j-1/2}, |\phi_{j+1/2}|, \sigma_{j+1/2}\Delta_{j+3/2})\right].$$

Here

$$\Delta_{j+1/2} = \rho^{TD}_{j+1} - \rho^{TD}_j,$$
$$\phi_{j+1/2} = \mu\left(\rho^{TD}_{j+1} - \rho^{TD}_j\right) \quad \text{or} \quad \phi_{j+1/2} = \mu\left(\rho^T_{j+1} - \rho^T_j\right),$$

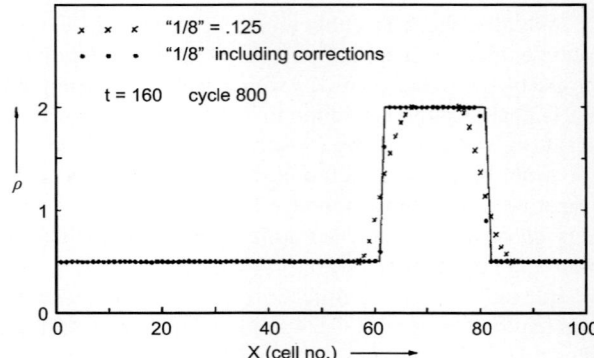

Fig. 9 The effect of introducing an artificial steepener

where ρ_j^T is the transported density, ρ_j^{TD} is the transported diffused density, and

$$\sigma_{j+1/2} = \text{sign}(\phi_{j+1/2}).$$

The two versions given here for the raw antidiffusive flux both use the same dimensionless coefficient μ, but the second version, which we called phoenical, has the advantage that when the flow velocity u vanishes $\rho_j^T \to \rho_j$. This permits the profile, which has been smeared out by diffusion, to be restored "like a phoenix," so that the algorithm reduces to the identity operation.

There was one aspect of the early versions of FCT that I felt uneasy about, the use of steepeners. In FCT-1 the antidiffusion coefficient μ was given as "1/8." We wrote "The quotation marks indicate that more exact cancellation of errors can be achieved if one expends a small amount of computational effort by including at least rough approximations to the velocity- and wavenumber-dependent corrections [11]." Footnote 11 explained just what was meant by wavenumber dependence: The antidiffusion coefficient was bigger than the diffusion coefficient by an amount that depended on the size of the discontinuity. Naturally, this yielded very nice square waves (Fig. 9). The drawback was that it turned every bump into a square wave!

The steepener was Jay's idea. The rationale for it was that our tests produced profiles that were actually less sharp than shocks calculated with FCT because, as I mentioned earlier, the passive advection equation has no mechanism for self-steepening. The steepener was supposed to model this mechanism, but to me it seemed an out-and-out kludge, and people who read footnote 11 apparently agreed. In the event steepeners disappeared after FCT-1 and were never mentioned again.

4 Infancy

As we continued to improve and extend SHASTA we gradually realized that what we had was more widely applicable and more general than a mere algorithm. Jay insisted on using the name "flux-corrected transport," but it was several years before the rest of the community distinguished between FCT and SHASTA. (Jay also tried to reserve the term "scheme" for competing algorithms, while referring to FCT as a "method" or "technique," but that invidious distinction, not surprisingly, never caught on. Neither did the term "Flux-Uncorrected Transport," which he used once in a talk.)

My first contribution to helping transform FCT from an algorithm into a method was to derive a formula for the diffusive transport step of SHASTA, i.e., the result of assigning mass to the trapezoids, transporting them, and reassigning it to the mesh. (The original code simply followed the geometric construction.) This formula can be expressed algebraically as

$$\rho_j^{n+1} = \frac{1}{2}Q_-^2(\rho_j^n - \rho_{j-1}^n) + \frac{1}{2}Q_+^2(\rho_{j+1}^n - \rho_j^n) + (Q_- + Q_+)\rho_j^n,$$

where

$$Q_\sigma = \frac{1/2 - \sigma u_j \Delta t/\Delta x}{1 + \sigma(u_{j+\sigma} - u_j)\Delta t/\Delta x}, \quad \sigma = \pm 1.$$

For a uniform velocity field, $u_j = u$, it reduces to

$$\rho_j^{n+1} = \rho_j^n - \frac{\varepsilon}{2}(\rho_{j+1}^n - \rho_{j-1}^n) + \left(\frac{1}{8} + \varepsilon^2\right)(\rho_{j+1}^n - 2\rho_j^n + \rho_{j-1}^n),$$

which is just Lax–Wendroff plus a zeroth-order (in ε) diffusion with coefficient 1/8. So, Jay thought, why not use Lax–Wendroff as the transport algorithm even when the flow field is nonuniform, adding diffusion to it and then applying antidiffusion? This worked just as well as SHASTA.

We tried flux-correcting leapfrog, and that worked fine too. Initially we used a diffusion/antidiffusion coefficient of 0.125 because that was what SHASTA used. It turned out that the limit on the Courant number in order to ensure positivity, $\varepsilon < 0.5$, is the same as for SHASTA, although the geometric interpretation no longer holds. Why is 1/8 the best choice? Wouldn't it be better to use less diffusion sometimes, especially when the flow velocity is very small and less is needed to ensure positivity? I tried running our standard test using different values of the diffusion/antidiffusion coefficient (Fig. 10). It is evident that the best results come from using 1/8, and that using too much can be as bad as using too little.

We noticed that combining a velocity-dependent antidiffusion with donor cell created a second-order-accurate algorithm formally identical to Lax–Wendroff. If the antidiffusive fluxes are limited with the same prescription that was used in SHASTA the result is a flux-corrected version of donor cell. In this algorithm the diffusion and antidiffusion coefficients vanish when the flow velocity does. In fact, for any value of u flux-corrected donor cell embodies the smallest diffusion coefficient consistent with positivity. But when we tested the new algorithm on square waves we found that the results were inferior to those obtained with flux-corrected Lax–Wendroff or leapfrog. Evidently minimizing the diffusion does not produce the best algorithm.

Jay found the explanation: phase accuracy is more important than amplitude accuracy. This is because the cumulative residual diffusion due to flux limiting results from mashing down the short-wavelength harmonics, which always propagate too

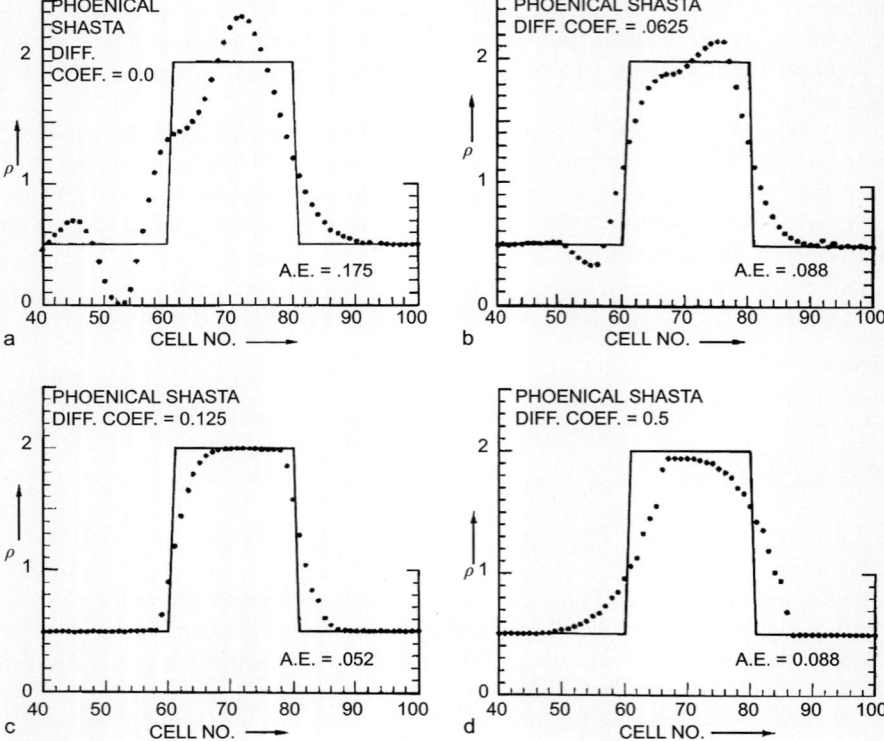

Fig. 10 Result of varying the diffusion/antidiffusion coefficient

slowly or too fast in a finite-difference algorithm. Minimizing the relative phase error works better (because fewer harmonics need mashing) than making the amplification factor as close as possible to unity. In fact, the amplification factor must go to zero for very short wavelengths, or else they wouldn't get mashed. For good results an FCT algorithm should have phase error that is at least second-order (i.e., proportional to k^2 when expanded in powers of wavenumber).

Another lesson we learned from these experiments was that the diffusion and antidiffusion coefficients can be velocity-dependent, provided that the diffusive and antidiffusive fluxes cancel out. This gave us an extra degree of freedom, an additional knob to turn in fine-tuning algorithms.

While musing on an algorithm called hopscotch I had read about in JCP I dreamed up "reversible FCT," which is basically flux-corrected Crank–Nicolson. It applies half the transport step to the old values and half to the new, together with a diffusion/antidiffusion that is also symmetric between the old and new values:

$$\rho_j^T + \frac{\varepsilon}{4}\left(\rho_{j+1}^T - \rho_{j-1}^T\right) + \nu\left(\rho_{j+1}^T - 2\rho_j^T + \rho_{j-1}^T\right)$$
$$= \rho_j^0 - \frac{\varepsilon}{4}\left(\rho_{j+1}^0 - \rho_{j-1}^0\right) + \nu\left(\rho_{j+1}^0 - 2\rho_j^0 + \rho_{j-1}^0\right).$$

Fig. 11 Reversible FCT

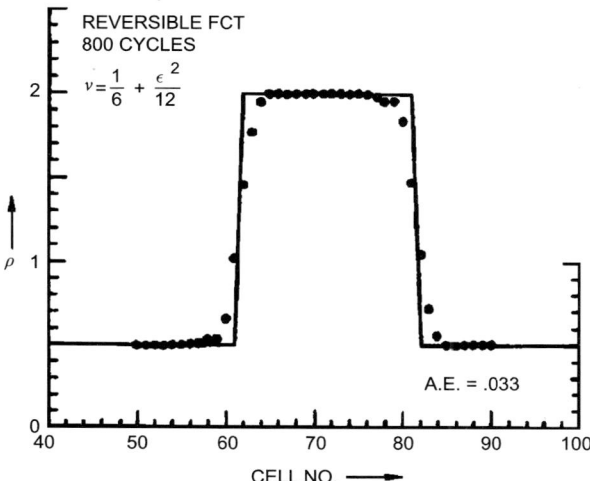

The transported diffused density ρ_j^{TD} is found by adding $\nu(\rho_{j+1}^T - 2\rho_j^T + \rho_{j-1}^T)$ to ρ_j^T. Then the raw antidiffusive flux $\phi_{j+1/2} = \nu(\rho_{j+1}^T - \rho_j^T)$ is corrected with respect to ρ_j^{TD} and reapplied. This algorithm is second-order for any choice of ν because of symmetry. Setting $\nu = 1/6 + \varepsilon^2/12$ makes the phase error fourth-order. With this choice the square-wave test yielded an A.E. of 0.033, the best we had yet found (Fig. 11).

Even without flux correction the underlying transport routine in an FCT algorithm gives better results than conventional algorithms. In a sense it *should*, because conventional algorithms are based on three-point stencils (they involve only the mesh point in question and its two nearest neighbors), while the extra antidiffusion step in FCT introduces information from next-nearest neighbors as well. Can FCT algorithms be made even more accurate by using more complicated stencils?

Thinking about this led me to invent Fourier-transform FCT. Start by Fourier-transforming the density ρ:

$$\rho_j = \sum_{\kappa=1}^{N} \tilde{\rho}_\kappa \exp(2\pi i j \kappa/N).$$

(This is of course implies an N-point stencil, but with fast transforms the computational overhead is acceptable.) Advance each component according to the exact solution of the transformed advection equation:

$$\tilde{\rho}_\kappa(t + \Delta t) = \tilde{\rho}_\kappa(t) \exp(-2\pi i \kappa u \Delta t/\Delta x).$$

Now transform back to x space. The resulting solution has no dissipation and no phase error. On the face of it this algorithm should be error-free, at least for passive advection with a uniform velocity. Indeed, if $\varepsilon = u\Delta t/\Delta x$ is an integer the solution reproduces the analytic solution exactly. There is no need for additional diffusion, antidiffusion, or flux correction.

Fig. 12 Continuous function obtained from the discrete Fourier transform of a jump

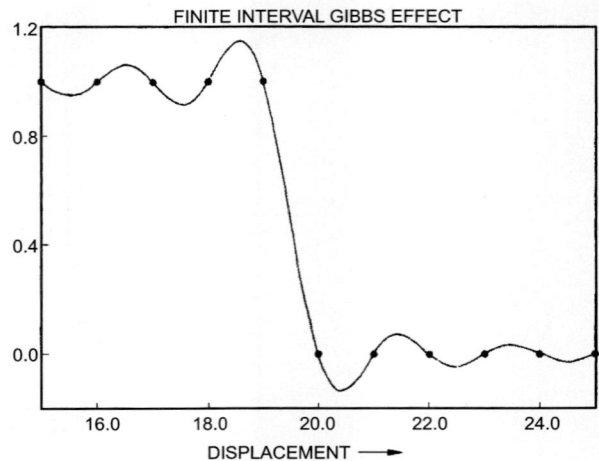

If $u \Delta t / \Delta x$ is not an integer, however, the solution that results from inverting the Fourier transform has ripples. Jay called this the finite-interval Gibbs effect. These ripples are related to the "window problem," i.e., the impossibility of knowing what goes on between the mesh points. Plotting the inverse of the discrete Fourier transform over the entire interval on which the original discrete function is defined yields a continuous curve. This can be shown to be the smoothest function that passes through the values of the function on the mesh. If the original function contains a sharp discontinuity, this plot not only exhibits the usual Gibbs over- and undershoots at the top and bottom of the jump, but also has wiggles between the mesh points (Fig. 12).

Translating the profile over a fraction of a mesh space exposes these wiggles to view. The Fourier transform thinks a function should behave this way in order to be as smooth as possible, whereas the physics favors one that has as few wiggles as possible. But we know how to fix that: flux-correct it. In other words, add some diffusion, then apply an equal amount of antidiffusion with a flux limiter. Nothing dictates the choice of the coefficient, so we used $\mu = \nu = 1/8$. (Why not? It worked in other algorithms.) The resulting value of 0.022 for the A.E. is the smallest one we ever found (Fig. 13). Thus, at the cost of a higher operation count, Fourier-transform FCT emerged as the optimum FCT algorithm, at least for this test problem.

The three JCP papers focused almost entirely on how FCT worked and on the design of FCT algorithms. In 1976 we contributed an article to the series *Methods of Computational Physics*. It appeared as a chapter in volume 16, which surveyed numerical techniques for plasma physics problems. This article (which perhaps deserves to be designated FCT-4) reviewed our previously published work, but also discussed some of the codes incorporating FCT and the problems to which we had applied them.

When it came to real applications we found that FCT is not perfect. One source of error is something than can be called residual diffusion, which results when antidiffusion fails to completely cancel diffusion.

Fig. 13 The optimum FCT algorithm

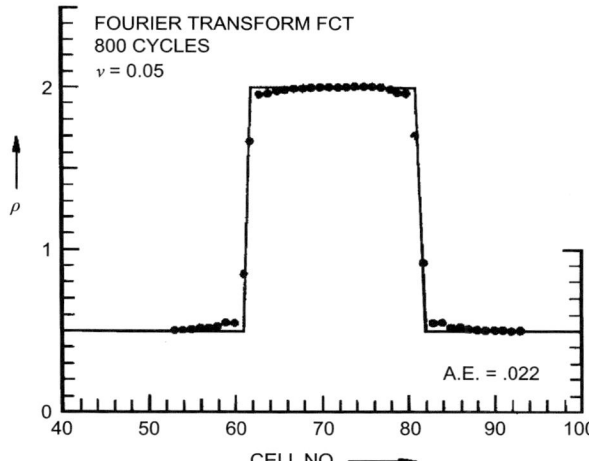

Fig. 14 Result of 20 repetitions to profile (*a*) of diffusion and (*b*) explicit; (*c*) phoenical; (*d*) implicit antidiffusion with strong flux limiting, using coefficients $\mu = \nu = 0.2$

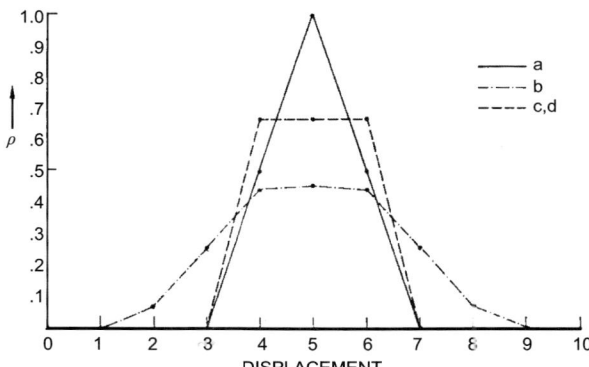

The most dramatic manifestation of this is "clipping." An extremum loses a little bit of amplitude each timestep, even if it isn't being advected across a grid, because diffusion squashes the extremum down and strong flux limiting doesn't allow the antidiffusion to push it back up. Ultimately the peak changes into a characteristic flat-topped structure, which we called a plateau. For example, an initially sharp maximum subjected to repeated diffusion and antidiffusion operations gradually flattens out until it forms a plateau three points across (Fig. 14), which is stable.

Figure 14 shows that this flattening is less severe with phoenical or implicit antidiffusion than with the original explicit form of antidiffusion, in which the raw flux is calculated from the diffused profile. The reason is obvious. If T stands for the transport operation, D stands for a three-point centered second difference with coefficient ν, and A represents the same operation with coefficient $\mu = -\nu$, the three versions of FCT can be represented symbolically as

Explicit: $\quad \rho^1 = (1+A)\rho^{TD} = (1+A)(1+T+D)\rho^0;$

Phoenical: $\quad \rho^1 = \left[(1+A)(1+T) + D\right]\rho^0;$

Implicit: $\quad \rho^1 = (1+D)^{-1}\rho^{TD} = (1+D)^{-1}(1+T+D)\rho^0.$

It is clear that in the limit $T \to 0$ phoenical and implicit FCT reduce to the identity operation, or would except for the action of the flux limiter, but explicit FCT does not. (This is of course why phoenical antidiffusion was invented.)

In an effort to eliminate clipping we tried a number of different flux limiters. One of my early attempts was the one-sided flux limiter. This involved changing the flux limiter so that maxima could grow on positive profiles but minima could not (and vice-versa for negative profiles). The resulting algorithm preserved positivity, but gave rise—unsurprisingly—to one-sided ripples.

Another idea of mine that didn't work out was called "flux-limited diffusion." The innovation here was to apply diffusion only where needed to prevent extrema from growing relative to their original values, rather than put it in everywhere and remove it where it was not needed. The test results, however, were disappointing. Square waves propagated using flux-limited diffusion were badly eroded. The trouble with this approach was that it failed to ensure high phase accuracy in the underlying transport scheme. Ultimately we decided to stick with strong flux limiting. It was easier to live with the symptoms of the disease than with the side effects of the cures.

Another form of residual diffusion is more subtle. If the antidiffusion coefficient is smaller than the diffusion coefficient, some residual diffusion is present even in the absence of flux limiting. The algorithm with fourth-order phase accuracy described in FCT-3 has an amplification factor that just barely—by less than 0.5%—exceeds unity. (We didn't know this until Phil Colella pointed it out years later.) Because of this it was found that codes yielded the best results when the antidiffusion coefficient was reduced slightly by multiplying by a factor called a "mask." (The name arose because the correction was applied using MASK, a Fortran IV bit-manipulating instruction.)

An annoying but fairly innocuous departure from realism arises because the flux limiter stops dispersive ripples from growing on the slopes of hills or valleys only when they are about to form new extrema. This leaves flat "terraces" on the slopes, which I like to think of as the ghosts of departed ripples. Improving the phase accuracy of the algorithm helps reduce terracing, but the only real cure is to use a more elaborate form of flux limiting that takes into account second derivatives of the profile.

When we started using FCT for two-dimensional problems a new and much more serious question arose: How could we generalize the flux limiter to multidimensions? Should all the fluxes be corrected in one sweep? In applying antidiffusive fluxes in, say, the x direction, should we worry about extrema only along that axis, or should we look at all points in the neighborhood? The coding was tortuous, and every prescription we tried seemed to fail in some combination of circumstances.

Finally we gave up and decided to use coordinate splitting. That is, on each timestep we treated each coordinate independently, carrying out 1-D transport and 1-D flux limiting first in the x direction and then in y. Symbolically this can be written, e.g.,

$$\rho^{TD} = (1 + T_x + D_x)(1 + T_y + D_y)\rho^0;$$
$$\rho^1 = (1 + A_x)(1 + A_y)\rho^{TD}.$$

Obviously this introduces spurious terms, e.g., $A_x A_y$. Most of the time they do no great harm because the errors tend to cancel out, but there are some situations where splitting creates unphysical effects. One example is when plateau formation occurs in both coordinate directions at the same time, for example on a hilltop. The result is that contours of constant density (or pressure, etc.) become square. I got very tired of going to meetings where I had to explain why my code generated square fireballs.

5 The Next Generation

Despite all our efforts we never found a way around the problem of creating a multidimensional flux limiter. Eventually I decided that it was insoluble, and I told every colleague who expressed an interest in it not to waste his time. As is well known, the problem turned out not to be insoluble. Likewise, the "pioneering idea of blending high- and low-order discretizations," cited on the FCT-30 web page, was not part of the original concept. It was a later development, due to the same individual who created the multidimensional flux limiter, Steve Zalesak. But that is his story to tell.

References

1. Book, D.L., Boris, J.P., McDonald, B.E., Wagner, C.E.: SHASTA, a transport algorithm that works. In: Proc. Symposium on High-Altitude Nuclear Effects, Stanford, CA, August 10–12 (1971)
2. Book, D.L., Boris, J.P.: A transport algorithm that works. NRL Reports of Progress (Nov. 1971), p. 1
3. Boris, J.P., Book, D.L.: Flux-corrected transport: I. SHASTA, a fluid transport algorithm that works. J. Comput. Phys. **11**, 38–69 (1973) [FCT-1]
4. Book, D.L., Boris, J.P., Hain, K.: Flux-corrected transport: II. Generalizations of the method. J. Comput. Phys. **18**, 248–283 (1975) [FCT-2]
5. Boris, J.P., Book, D.L.: Flux-corrected transport: III. Minimal-error FCT algorithms. J. Comput. Phys. **20**, 397–431 (1976) [FCT-3]
6. Boris, J.P., Book, D.L.: Solution of continuity equations by the method of flux-corrected transport. In: Alder, B., Fernbach, S., Rotenberg, M., Killeen, J. (eds.) Methods in Computational Physics, vol. 16. Academic Press, New York (1976)

The Design of Flux-Corrected Transport (FCT) Algorithms for Structured Grids

Steven T. Zalesak

Abstract A given flux-corrected transport (FCT) algorithm consists of three components: (1) a high order algorithm to which it reduces in smooth parts of the flow; (2) a low order algorithm to which it reduces in parts of the flow devoid of smoothness; and (3) a flux limiter which calculates the weights assigned to the high and low order fluxes in various regions of the flow field. One way of optimizing an FCT algorithm is to optimize each of these three components individually. We present some of the ideas that have been developed over the past 30 years toward this end. These include the use of very high order spatial operators in the design of the high order fluxes, non-clipping flux limiters, the appropriate choice of constraint variables in the critical flux-limiting step, and the implementation of a "failsafe" flux-limiting strategy. This chapter confines itself to the design of FCT algorithms for structured grids, using a finite volume formalism, for this is the area with which the present author is most familiar. The reader will find excellent material on the design of FCT algorithms for unstructured grids, using both finite volume and finite element formalisms, in the chapters by Professors Löhner, Baum, Kuzmin, Turek, and Möller in the present volume.

1 Introduction: Modern Front-Capturing Methods

We are interested in systems of conservation laws of the form

$$\frac{\partial \mathbf{q}(\mathbf{x}, t)}{\partial t} + \nabla \cdot \mathbf{f}(\mathbf{q}, \mathbf{x}, t) = 0 \tag{1}$$

where $\mathbf{q}(\mathbf{x}, t)$ and $\mathbf{f}(\mathbf{q}, \mathbf{x}, t)$ are vector functions of the independent variables $\mathbf{x}$ and t, which we henceforth refer to as space and time respectively. Examples of such equations include the Navier-Stokes equations, the equations of magnetohydrodynamics (MHD), the Vlasov equation, and passively-driven convection.

This work was supported by the U.S. Department of Energy.

S.T. Zalesak (✉)
Plasma Physics Division, Naval Research Laboratory, Washington, DC 20375, USA
e-mail: steven.zalesak.ctr@this.nrl.navy.mil

It is well known that differentiable solutions to (1) may cease to exist after a finite time t_s, even if the initial conditions $\mathbf{q}(\mathbf{x}, 0)$ are smooth. After t_s, only the integral or "weak" form of (1) will have solutions, and these will contain discontinuities in $\mathbf{q}$ and/or one or more of its derivatives. We will term such discontinuities "fronts" for the purpose of this chapter. This situation is addressed by the Lax-Wendroff Theorem, which states that if one's numerical approximation to Eq. (1) is in "flux" or "conservation" form, and the numerical solution converges everywhere but on a set of measure zero to some solution, then that solution is a weak solution to Eq. (1). Thus the great majority of methods designed to treat fronts in the context of Eq. (1) are in conservation form, i.e., a form consisting of numerical fluxes connecting adjacent grid points, these fluxes being used to advance the numerical solution in time.

Using conservation form does not by itself give the desired result however, since one still needs to compute a convergent solution. In general, numerical methods not designed to deal with fronts will not produce the desired convergence in their presence, often producing numerical oscillations that degrade the solution severely. It is precisely this situation that prompted Von Neumann and Richtmyer to add an explicit artificial dissipation term to Eq. (1), the idea being to smooth the fronts to the point where they are resolved on the grid as narrow but smooth features, thereby producing the desired convergence. This is the fundamental idea underlying nearly all conservative numerical methods designed to handle fronts, including the FCT algorithms we address here. (We are excluding, of course, methods that treat fronts as moving internal boundaries, the class of methods known as "front-tracking methods." We are also excluding random choice methods [3, 4, 8].) For the purposes of this chapter, we shall refer to methods which attempt to smooth a front into a narrow but smooth transition as "front-capturing methods."

Over the past 30 years a host of algorithms, known variously as "modern" front-capturing methods or "high resolution methods," have been developed in an attempt to perform calculations more accurately and more efficiently than with the more traditional explicit artificial dissipation approach. The first of these methods was flux-corrected transport (FCT) [1, 2], but there are now a large number of others. What distinguishes the "modern" front-capturing methods from their predecessors is their attempt to constrain the numerical fluxes, grid point by grid point and timestep by timestep, in such a way as to avoid the production of unphysical values in the solution vector $\mathbf{q}$ in and near the fronts, and at the same time treat the regions in space and time in which $\mathbf{q}$ is smooth as accurately as possible. Clearly the success of these methods depends critically on an accurate criterion for determining what constitutes an unphysical value for $\mathbf{q}$, one of the primary topics of this chapter.

One way of stating the design philosophy of these methods, and the one we shall embrace in this chapter, is as follows:

When the numerics fails, substitute the physics.

Clearly the designers of such algorithms must possess a knowledge of the physics being addressed if they are to be successful.

These modern front-capturing methods may be thought of as consisting of three parts:

1. an algorithm to which they reduce in regions of time and space where **q** is smooth;
2. an algorithm to which they reduce at fronts; and
3. a mechanism for weighting each of the above algorithms at each grid point and timestep.

Obviously the accuracy of a given modern front-capturing method may depend strongly on the choices made in each of its three parts. In the FCT algorithms we shall consider here, these three parts correspond to the high-order fluxes, the low-order fluxes, and the flux limiter respectively, terms we shall define shortly. Our experience is that FCT algorithms are capable of solving most problems involving fronts with both robustness and accuracy, as long as certain design principles are adhered to. Toward that end, this chapter shall present to the reader a collection of design principles that we have found to be of value in the creation of an FCT algorithm for a given situation. In general, they involve optimizing one's choice of each of the above three components of the algorithm. The reader will not be surprised to learn that a knowledge of the physics problem being addressed is an essential part of the design criteria.

In Sect. 2, we give a formal definition of FCT, first for the special case of one spatial dimension, and then for multidimensions. In Sect. 3 we give six design criteria that collectively define what we mean by a "properly designed" FCT algorithm. In Sect. 4 we give examples of the kind of performance one can expect from a properly designed FCT algorithm, using the scalar linear advection problem. For this problem, the "physics" that must be incorporated into the algorithm is simple and intuitive, and accurate and robust algorithms are easy to construct. In Sect. 5 we move on to nonlinear systems of equations, using the Euler equations as an example. Here the physics is not trivial as it was in the case of linear advection, and we find that blindly applying the methods that worked well for advection can be disappointing. However, when we transform the problem into a set of variables for which we have a legitimate set of physical constraints that can be imposed, we recover the kind of performance that we saw in the linear advection case. In Sect. 6 we treat both passively-driven convection and compressible gas dynamics in two space dimensions, and again have to face and solve the question of physically appropriate constraints. Finally in Sect. 7 we give our conclusions.

2 Flux-Corrected Transport (FCT) Defined

As we mentioned in the previous section, the great majority of methods designed to treat fronts in the context of Eq. (1), are in conservation form, i.e., a form consisting of numerical fluxes connecting adjacent grid points, these fluxes being used to advance the numerical solution in time. In FCT, at every timestep and at every flux point, these fluxes are computed twice, once using an algorithm guaranteed not to generate unphysical values (the "low order fluxes"), and once using an algorithm that is formally of high accuracy in the smooth portions of the solution (the "high order fluxes"). FCT then constructs the net fluxes for the timestep as weighted averages of these two candidate fluxes. The weighting is performed in a manner which ensures that the high order fluxes are used to the greatest extent possible without introducing unphysical values into the solution. The procedure is referred to as "flux-correction" or "flux limiting" for reasons which will become clear shortly.

From the above description, it should be clear that one may easily define an FCT algorithm on any structured or unstructured grid in any number of spatial dimensions, as long as one can define a numerical technique for which the difference between a low order time advancement operator and its higher order counterpart can be written as an array of fluxes between adjacent grid points. Rather than attempt to give a definition at that level of generality, we will give formal definitions for the cases of one spatial dimension, and for two spatial dimensions on a structured grid. From these two examples it should be clear how to construct an FCT algorithm in any number of dimensions, and on any grid, structured or unstructured.

2.1 FCT in One Spatial Dimension

In one spatial dimension, Eq. (1) takes the simpler form

$$\frac{\partial q(x,t)}{\partial t} + \frac{\partial f(q,x,t)}{\partial x} = 0. \tag{2}$$

A simple example of such a system of equations is the system describing one-dimensional ideal inviscid fluid flow, also known as the Euler equations:

$$q = \begin{pmatrix} \rho \\ \rho u \\ \rho E \end{pmatrix}; \quad f = \begin{pmatrix} \rho u \\ \rho u u + P \\ \rho u E + P u \end{pmatrix} \tag{3}$$

where ρ, u, P, and E are the fluid density, velocity, pressure, and specific total energy respectively.

We say that a discrete approximation to Eq. (2) is in conservation or "flux" form when it can be written in the form

$$q_i^{n+1} = q_i^n - \Delta x_i^{-1}[F_{i+(1/2)} - F_{i-(1/2)}]. \tag{4}$$

Here q and f are defined on the spatial grid points x_i and temporal grid points t^n, and Δx_i is the cell width associated with cell i. The $F_{i+(1/2)}$ are called numerical fluxes, and are functions of f and q at one or more of the time levels t^n. The functional dependence of F on f and q *defines* the particular discrete approximation.

As mentioned above, FCT constructs the net flux $F_{i+(1/2)}$ point by point and timestep by timestep (nonlinearly) as the weighted average of two fluxes, one produced by a "high order" method and the other by a "low order" method. The formal procedure introduced in [13] is as follows:

1. Compute $F_{i+(1/2)}^L$, the "low order fluxes," using a method guaranteed not to generate unphysical values in the solution for the problem at hand.

2. Compute $F^H_{i+(1/2)}$, the "high order fluxes" using a method chosen to be accurate in smooth regions for the problem at hand.
3. *Define* the "antidiffusive fluxes" [2]

$$A_{i+(1/2)} \equiv F^H_{i+(1/2)} - F^L_{i+(1/2)}. \tag{5}$$

4. Compute the time advanced low order ("transported and diffused" [2]) solution:

$$q^{td}_i = q^n_i - \Delta x_i^{-1}\left[F^L_{i+(1/2)} - F^L_{i-(1/2)}\right]. \tag{6}$$

5. *Limit* the antidiffusive fluxes in a manner such that q^{n+1}_i as computed in step 6 below does not take on nonphysical values:

$$A^C_{i+(1/2)} = C_{i+(1/2)} A_{i+(1/2)}, \quad 0 \le C_{i+(1/2)} \le 1. \tag{7}$$

6. Apply the limited antidiffusive fluxes:

$$q^{n+1}_i = q^{td}_i - \Delta x_i^{-1}\left[A^C_{i+(1/2)} - A^C_{i-(1/2)}\right].$$

The critical step in the above is step 5, the flux limiting step. In the absence of step 5 (i.e., $A^C_{i+(1/2)} = A_{i+(1/2)}$), q^{n+1}_i would simply be the time-advanced high order solution.

2.2 Multidimensional Flux-Corrected Transport

Let us see how the procedure above might be implemented in multidimensions. An obvious choice would be to use an operator-splitting technique, splitting along spatial dimensions, when it can be shown that the equations allow such a technique to be used without serious error. Indeed, such a procedure may even be preferable from programming and time-step considerations. However, there are many problems for which such splitting produces unacceptable numerical results, among which are incompressible or nearly incompressible flow fields. The technique is straightforward and shall not be discussed here. Instead, let us now consider the two-dimensional system of conservation laws

$$q(\mathbf{x}, t)_t + f(q, \mathbf{x}, t)_x + g(q, \mathbf{x}, t)_y = 0. \tag{8}$$

A simple example of such a system of equations, and one we consider later, is the system describing two-dimensional ideal inviscid fluid flow, also known as the two-dimensional Euler equations:

$$q = \begin{pmatrix} \rho \\ \rho u \\ \rho v \\ \rho E \end{pmatrix}; \quad f = \begin{pmatrix} \rho u \\ \rho u u + P \\ \rho u v \\ \rho u E + P u \end{pmatrix}; \quad g = \begin{pmatrix} \rho v \\ \rho v u \\ \rho v v + P \\ \rho v E + P v \end{pmatrix} \tag{9}$$

where ρ, u, v, P, and E are the fluid density, x velocity, y velocity, pressure, and specific total energy respectively.

If we work on a finite volume coordinate-aligned mesh, we can define our two-dimensional FCT algorithm thus:

$$q_{ij}^{n+1} = q_{ij}^n - \Delta V_{ij}^{-1}[F_{i+(1/2),j} - F_{i-(1/2),j} + G_{i,j+(1/2)} - G_{i,j-(1/2)}] \quad (10)$$

where ΔV_{ij} is the volume of cell ij.

Now there are two sets of transportive fluxes F and G, and the FCT algorithm proceeds as before:

1. Compute $F_{i+(1/2),j}^L$ and $G_{i,j+(1/2)}^L$, the "low order fluxes," using a method guaranteed not to generate unphysical values in the solution for the problem at hand.
2. Compute $F_{i+(1/2),j}^H$ and $G_{i,j+(1/2)}^H$, the "high order fluxes," using a method chosen to be accurate in smooth regions for the problem at hand.
3. *Define* the "antidiffusive fluxes" [2]

$$A_{i+(1/2),j} \equiv F_{i+(1/2),j}^H - F_{i+(1/2),j}^L,$$
$$A_{i,j+(1/2)} \equiv G_{i,j+(1/2)}^H - G_{i,j+(1/2)}^L.$$

4. Compute the time advanced low order ("transported and diffused" [2]) solution:

$$q_{ij}^{td} = q_{ij}^n - \Delta V_{ij}^{-1}[F_{i+(1/2),j}^L - F_{i-(1/2),j}^L + G_{i,j+(1/2)}^L - G_{i,j-(1/2)}^L].$$

5. *Limit* the antidiffusive fluxes in a manner such that q_{ij}^{n+1} as computed in step 6 below does not take on nonphysical values:

$$A_{i+(1/2),j}^C = C_{i+(1/2),j} A_{i+(1/2),j}, \quad 0 \le C_{i+(1/2),j} \le 1,$$
$$A_{i,j+(1/2)}^C = C_{i,j+(1/2)} A_{i,j+(1/2)}, \quad 0 \le C_{i,j+(1/2)} \le 1.$$

6. Apply the limited antidiffusive fluxes:

$$q_{ij}^{n+1} = q_{ij}^{td} - \Delta V_{ij}^{-1}[A_{i+(1/2),j}^C - A_{i-(1/2),j}^C + A_{i,j+(1/2)}^C - A_{i,j-(1/2)}^C].$$

As can be easily seen, implementation of FCT in multidimensions is straightforward, with the possible exception of Step 5, the flux limiter, which will be addressed in a later section.

3 Design Criteria for FCT Algorithms

Here we give, with only modest detail, six criteria that we believe are necessary for the construction of properly designed (robust but accurate) FCT algorithms. They are:

1. The resolving power of the high order fluxes should be as high as is practical. The term "resolving power" will be defined precisely below.
2. The high order fluxes should have a dissipative component which adapts itself to the resolving power of the nondissipative component.
3. The high order fluxes should be "pre-constrained" with respect to physically appropriate bounds before being input to the flux limiter.

4. The low order flux must be dissipative enough to guarantee that unphysical values in the solution cannot be generated, but should otherwise be as accurate as is practical.
5. The flux limiter should accommodate as flexible a specification of solution bounds as possible, and should utilize constraints that have a strong physical basis.
6. The flux limiter should have a simple fail-safe feature that reduces all fluxes into a grid point to their low order values when the normal flux limiter machinery fails.

We will treat each one of these in turn.

3.1 High Order Fluxes with Very High Resolving Power

The primary result of [14] was that there is a significant advantage in using fluxes derived from very high order, spatially-centered finite difference operators (fourth order or higher) for the "high order fluxes" in FCT algorithms. That conclusion was based on some analysis showing a strong empirical relationship between order and resolving power (a term which we define below) for centered finite difference operators, some heuristic reasoning as to how FCT works in practice, and several one-dimensional test calculations dominated by linear advection test problems. We review that work in this section.

For analysis, Eq. (2) is usually reduced to a scalar conservation law and linearized:

$$\frac{\partial q}{\partial t} + u \frac{\partial q}{\partial x} = 0 \qquad (11)$$

where u is a constant. This is the linear advection equation with advection speed u. Its solution is simply

$$q(x, t) = q(x - ut, 0). \qquad (12)$$

That is, the profile is simply translated right or left with velocity u and no change in shape. In Fourier space, this takes the form of each Fourier mode moving with a phase velocity u, with no change in amplitude. Thus all numerical errors associated with a given numerical algorithm can be quantified by a specification of phase velocity error per timestep and amplitude error per timestep as a function of the wavenumber k, and as a function of the discretization step in space and time Δx and Δt respectively.

In [14] we examined a particular algorithm for solving Eq. (11) on a uniform mesh, that of using a leapfrog discretization in time and centered finite differences of arbitrary order in space. This choice allowed us to ignore the amplitude errors entirely, since for the leapfrog discretization these errors vanish for all k and for all Δx and Δt satisfying the Courant condition $\epsilon \equiv |u|\Delta t/\Delta x < 1$. Further noting that the total phase error was the algebraic sum of that induced by the temporal and

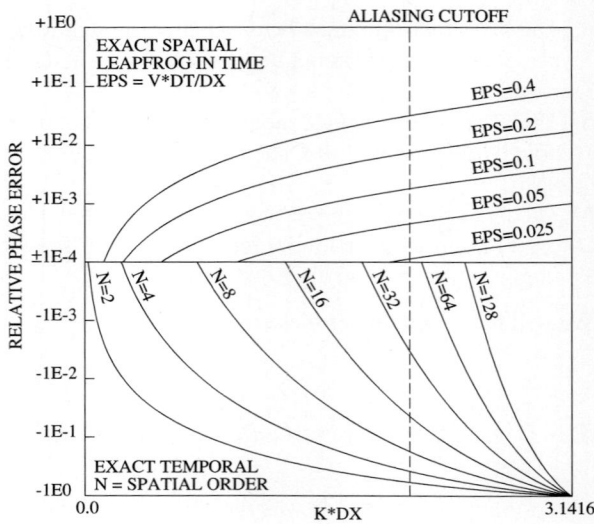

Fig. 1 Plot of relative phase error versus the normalized wavenumber $k\Delta x$ for the leapfrog time-marching scheme and analytic spatial derivatives (*top*), and for centered finite difference schemes of order N and analytic temporal derivatives (*bottom*). Figure taken from Ref. [14]

spatial discretizations separately, as long as both were small, we were able to reduce our algorithm analysis to a single plot, reproduced here in Fig. 1.

The striking aspect of Fig. 1 is the marked effect of the order N of the centered spatial finite difference operator, as well as the timestep Δt, on the resolving power of the algorithm. By "resolving power" we mean, loosely, the ability of an algorithm to maintain low phase errors over a large part of k-space. In more precise terms, we define the resolving power $k_r(E)$ of a given combination of N and Δt to be the largest wavenumber k for which all phase errors are smaller than a pre-specified value E. Thus Fig. 1 tells us that for a given E, we can resolve more and more of k-space if, on a given grid, we simply increase the spatial order N of our centered finite difference operator, decreasing Δt appropriately as we do so. The primary conclusion of [14] was that not only is this statement true for smooth functions, but that it is true in the presence of fronts also, as long as one is treating the fronts with a front-capturing algorithm such as FCT. Nothing since 1981 has dissuaded us from that view, and thus we present it as the first of our FCT design criteria. Computational examples presented later in this chapter will hopefully provide the reader more evidence of its correctness.

3.2 High Order Fluxes with an Adaptive Dissipation Component

Looking at the bottom portion of Fig. 1, we see that the arguments of the last section become less and less convincing as one moves to the extreme right portion of the plot. Phase error in this portion of the plot becomes increasing resistant to reduction by simply increasing N. Indeed, at the Nyquist frequency $k\Delta x = \pi$, the phase error is -100% regardless of how large we make N or how small we make Δt. Thus for

Fig. 2 *Top*: Plot of the damping induced by a centered finite difference approximation to the N_Dth derivative versus the normalized wavenumber $k\Delta x$, normalized so that the Nyquist mode is completely eliminated. *Bottom*: Same as bottom of Fig. 1: Plot of relative phase error versus the normalized wavenumber $k\Delta x$ for centered finite difference schemes of order N and analytic temporal derivatives

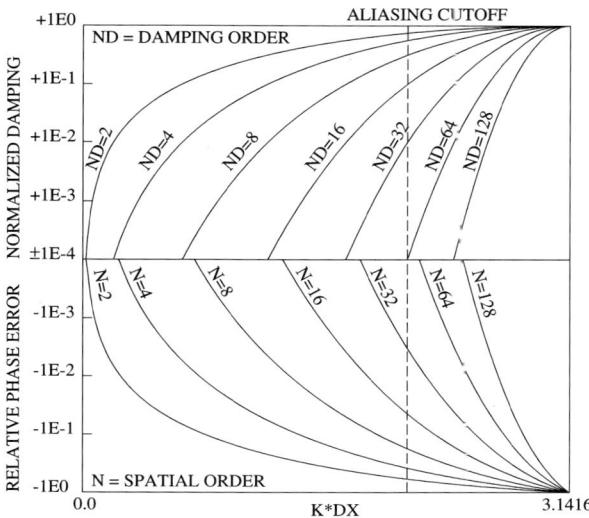

any given N, there will be some portion of k-space which is resolved poorly. Logic dictates that these Fourier modes be damped rather than carried with an erroneous speed, and that any given Fourier mode be damped nearly entirely by the time it is 180 degrees out of phase with the analytic result. Clearly, from Fig. 1, the functional dependence of this dissipation on k must itself depend on N if we are to achieve this result without damping the modes which are actually being carried accurately.

We have found that one can form such dissipative fluxes from the centered finite difference forms of $d^{N_D}q/dx^{N_D}$ where $N_D \leq N+2$. One then simply adds these fluxes to FCT's "high order" fluxes. Indeed, the early work of Kreiss and Oliger [9] contained calculations of the linear advection of a triangular wave using $N_D = N = 4$ and leapfrog time differencing, with results that, although not oscillation free, were considerably better than without the dissipation. These results are what prompted us to use such forms to construct FCT algorithms, and we have found them to be of sufficient value to include them among our design criteria.

In the top portion of Fig. 2 we plot the damping induced by centered finite difference approximations to $d^{N_D}q/dx^{N_D}$ versus $k\Delta x$, normalized so that the Nyquist mode is completely eliminated. In the bottom portion, we simply reproduce the bottom portion of Fig. 1. Focusing our attention on pairs of lines for which $N = N_D$, we see a rather remarkable match between phase error amplitude and dissipation amplitude as a function of $k\Delta x$. That is, for such pairs, as the relative phase error increases, so does the damping, with approximately the same functional dependence on $k\Delta x$. This is in accordance with our expressed desire to induce damping in proportion to the relative phase error. Thus, for the computational examples in this chapter, we have chosen to use $N_D = N$. An equally good case can be made for the choice $N_D = N + 2$, since this will leave the overall order of the algorithm intact. Indeed we have made that choice ourselves in some contexts. The specific construction of these operators in flux form is given in the appropriate later sections.

3.3 Imposition of Physically-Motivated Constraints on the High Order Fluxes Before the Flux Limiting Step

In general, the numerical algorithms used to construct high order numerical fluxes from the cell-centered values of q assume by necessity a degree of smoothness to q. Thus near fronts it is not unusual, especially for spatial orders higher than 2, to find that these fluxes violate physically-motivated bounds on their values. One could, of course, take the position that the flux limiter itself will ensure that these fluxes are prevented from producing values of q in the next time step that violate appropriate bounds for q, and thus that these unphysical values for the high order fluxes should be allowed to stand when computing the antidiffusive fluxes A. Nonetheless, it is wise to attempt to address this problem at its source, rather than shift the burden to the flux limiter at a later stage, and simply not allow the high order fluxes to take on values that are clearly outside the bounds of possibility. This design principle is not truly new, although not generally expressed as we have above. The prime example is the default behavior of the original Boris-Book flux limiter [2], which the reader will meet shortly. As explained in [13], this flux limiter sets the antidiffusive flux A to zero in virtually all cases where the antidiffusive flux has the same direction as the gradient of q^{td}, i.e., where A is actually diffusive, and in most cases would not actually cause the adjacent values of q to take on unphysical values. Although it is difficult to make rigorous statements in this context, in the great majority of the cases for which this flux-canceling machinery is active, and for which one can place physically-motivated upper and lower bounds on the value of the flux, the high order flux itself can be shown to be outside those bounds, without resort to arguments about its effect of the subsequent values of q. Another example of constraints imposed on the high order fluxes prior to the primary monotonicity machinery is to be found in the PPM algorithm [5], wherein candidate point values of q at cell interfaces are computed, given its cell averages. Rather than simply calculate these "high order" point values of q in a straightforward way, Colella and Woodward use a multistep algorithm that utilizes MUSCL slope limiting in a way that guarantees that the candidate "high order" interface value of q is bounded by the corresponding cell averages at the adjacent grid points. The motivation and the effects of this "pre-limiting" is similar to that of the Boris-Book limiter.

If one can place rigorous physically-motivated bounds on the high order flux, this step can be quite simple, as well as quite effective. We will see an example of such a situation when we construct a non-clipping flux limiter for advection in the next section. However, in general it is difficult to find such rigorous bounds, for at least two reasons. The first is that the flux **f** lives in a space one dimension lower than the corresponding q. For example, in three spatial dimensions in a finite volume context, q represents a volume average over the cell while the fluxes **f** are area averages at cell faces. The second is that in general **f** is a nonlinear function of q. These two combine to make it quite difficult to reliably place bounds on the high order fluxes. Thus we often use the reasoning implicit in the Boris-Book limiter: if the antidiffusive flux is actually diffusive, then there must be something wrong with the high order flux, and we set that antidiffusive flux to zero before limiting. We are

3.4 Low Order Fluxes That Guarantee That Physical Bounds Are Not Violated

The prime requirement of FCT's low order flux is that it be guaranteed to produce a solution free of unphysical grid point values. If this cannot be guaranteed, then there is no hope of guaranteeing that property in the final FCT solution, even with fail-safe limiting (item 6). One way to satisfy this requirement is to simply incorporate enough numerical dissipation in the algorithm, but overkill here is to be avoided (see below). Another way is to use first-order upwind methods or kinetics-based methods such as the beam scheme of Sanders and Prendergast. When properly chosen, these are probably the best choice, for their inherent dissipation is usually close to the minimum required to meet the prime requirement. However, be aware that many such schemes using "approximate" Riemann solvers cannot meet the prime requirement. It should be obvious that, within the prime requirement, the low order algorithm should be as accurate as is practical. In particular, any low order flux with more dissipation than is necessary to meet the prime requirement is harmful, putting an extra burden on the flux limiter, which is arguably the weakest link of the algorithm. Thus, for example, one would not use a Lax-Friedrichs flux for an advection problem, since the less diffusive donor cell flux already satisfies the prime requirement. As another example, if we are solving a two-dimensional advection problem, and we must choose between two donor cell algorithms, the first of which allows corner transport, and the second of which does not, we would choose the former as long as it satisfied the prime requirement.

3.5 Flexible Flux Limiters That Utilize Constraints with a Strong Physical Basis

It is useful to consider a flux limiter as being comprised of two components:

1. A physics component which specifies physically-motivated upper and lower bounds on grid point values in the next time step; and
2. An algorithmic machinery component for enforcing the above bounds.

It is clear that the algorithmic machinery component must have sufficient flexibility to accommodate the needs of the physics component. Thus, in our view, a good flux limiter must possess both a robust and accurate physics component and a flexible algorithmic machinery component. There exist several flux limiters which are of extremely simple form, expressible in as little as one line of Fortran. The original flux limiter of Boris and Book, and the ones typically used in algorithms that describe themselves as "TVD" are prime examples. The simplicity of these flux limiters is due to an extremely simple physics component and a very inflexible algorithmic machinery component: they make rather strong assumptions about what constitutes proper upper and lower bounds on the solution at a given grid point at a given timestep, and their algorithmic machinery is capable of accommodating those strong assumptions and little more. These assumptions may be overly restrictive, as they are in the case of the "clipping" phenomenon, or overly loose, as in the case of the "terracing" phenomenon. Thus we strongly encourage the reader to derive the above flux limiters for himself or herself. This exercise will reveal the assumptions implicit in these limiters, and allow an assessment as to their appropriateness for the problem at hand. If they are not appropriate, the reader may wish to consider a more flexible flux limiter, albeit one probably more complex and longer that one line of Fortran. We give an example of a more flexible flux limiter later. In our view the most difficult issue in the design of flux limiters is the physics component. Most commonly the antidiffusive fluxes are those which update conservative variables directly, and hence the default choice for most FCT algorithms has been to constrain the conservative variables using the default bounds built into the Boris-Book flux limiter. But this can often be a bad choice. Even if we were to circumvent the default bounds by using more flexible algorithmic machinery, it can often be extremely difficult to determine the appropriate bounds for the conserved variables. Using gas dynamics as an example, one would be hard pressed to specify the physically appropriate bounds on mass, momentum, and energy per unit volume at a given grid point at a given timestep, even if he or she were given the complete time history of the computed solution up to that point. We know that the physics allows the formation of new extrema in all three of these quantities. Thus looking at the adjacent grid point values at the previous time step, or even at the values of the time-advanced low order solution (FCT's default), can lead to bounds on the solution that are not physically appropriate. When we discuss the construction of FCT algorithms for gas dynamics in Sect. 5, we will put forth the hypothesis that much more reliable constraints are to be obtained by performing the flux limiting step in characteristic variables rather than conservative variables.

3.6 Failsafe Flux Limiters

We suppose that this topic comes under the general heading of "dirty laundry," but it cannot be ignored in any objective discussion of front-capturing algorithms. If one is attempting to solve difficult problems, our experience is that, no matter how carefully one tries to design algorithms that are consistent both with numerical analysis and with the physics problem one is attempting to solve, there will be situations in which at least one grid point at least one time step takes on values that are outside the bounds of physical possibility. For FCT and similar algorithms, these will usually be variables that one is not directly constraining. For example, if one is performing flux limiting on the conserved variables (not recommended here, but done often by many users of FCT), the density by construction can never become negative, but the internal energy can do so, since it is not directly constrained by the limiting process. What should one do in this case? As another example that we ourselves will face later in this chapter in the Woodward-Colella double shock tube problem, even when we limit with respect to what in our view are the most physically appropriate variables, the characteristic variables, we can occasionally generate unphysical grid point values. The characteristic variables are, after all, a linearization, and a linearization can be inaccurate at large jumps. Thus we could generate negative internal energies in that case also (or, in theory, even negative densities!). What action should we take when this happens?

Prudence, if nothing else, would dictate that the algorithm make provisions for such a case. Assuming we do not wish to simply terminate the calculation, we desire a solution which is as consistent with the design philosophy of the algorithm as possible and, most important of all, is explicitly stated.

For FCT, we believe that there is an obvious solution consistent with FCT's design, and with the numerical analysis goal of being at least first order accurate, and that is the one we choose here: For any such offending grid point, we iteratively drive all the fluxes into or out of that grid point toward their low order values, until the offense is eliminated. Thus it is especially important that FCT's low order scheme be guaranteed to be free of unphysical values! We use an especially simple algorithm here, which we describe in a later section.

4 FCT Algorithms for One Dimensional Linear Advection

We wish to give the reader an idea of the kind of performance one can expect of an FCT algorithm for the simplest of scalar conservation laws, linear advection. That is, we have Eq. (2) with $f = qu$ and u a constant. We use a uniform spatial mesh of cell size Δx. We utilize a method of lines approach, choosing our spatial and temporal discretization independently. All temporal discretizations we shall use (e.g., modified Euler, explicit Runge-Kutta, leapfrog, leapfrog-trapezoidal) involve one or more leapfrog-like substeps of the following form:

$$q_i^{n+1} = q_i^n - \Delta x_i^{-1}[F_{i+(1/2)} - F_{i-(1/2)}]. \tag{13}$$

Here t^{n+1} and t^n are substep time levels associated with a particular substep, with associated timestep $\Delta t^{n+1/2}$. The fluxes F are functions of f at one or more of the time levels, not necessarily t^{n+1} and t^n. The timestep $\Delta t^{n+1/2}$ has been absorbed into the definition of the fluxes. This leapfrog-like substep will be used as the fundamental building block for any time discretization we use. Thus we can describe our treatment for all temporal discretizations by describing our treatment of this substep.

Our low order flux for advection is given by the first order upwind scheme:

$$F_{i+(1/2)}^L = \left[\frac{1}{2}(f_{i+1}^n + f_i^n) - \frac{1}{2}|u|(q_{i+1}^n - q_i^n)\right]\Delta t^{n+1/2}. \tag{14}$$

The high order fluxes are given by the formulae in the Appendix of [13]. As an example, the fourth order flux is given by:

$$F^{H4}_{i+(1/2)} = \left[\frac{7}{12}(f^a_{i+1} + f^a_i) - \frac{1}{12}(f^a_{i+2} + f^a_{i-1})\right]\Delta t^{n+1/2} \quad (15)$$

where the time level t^a is meant to denote whatever time level or average of time levels is required by the particular substep of the particular time discretization chosen.

The high order dissipative fluxes of order N_D, which are added to the above high order fluxes, are simply the flux form representation of $\partial^{N_D} q/\partial x^{N_D}$, normalized to damp the Nyquist mode completely in one timestep at a Courant number of unity. As an example, the order 4 dissipative flux is given by:

$$F^{D4}_{i+(1/2)} = -|u|\left[\frac{3}{16}(q^n_{i+1} - q^n_i) - \frac{1}{16}(q^n_{i+2} - q^n_{i-1})\right]\Delta t^{n+1/2}. \quad (16)$$

Thus far we have dealt with only three of our six FCT design criteria, the design of the high and low order fluxes. The other three are the pre-constraint of the high order fluxes, the construction of the flux limiter, and the failsafe limiter. A failsafe limiter is not needed here, since we are directly constraining the only variable of interest. For the moment, we will choose a simple default for the remaining two criteria, the original Boris-Book limiter:

$$A^C_{i+(1/2)} = S\max\left(0, \min\left(|A_{i+(1/2)}|, S(q^{td}_{i+2} - q^{td}_{i+1})\Delta x, S(q^{td}_i - q^{td}_{i-1})\Delta x\right)\right)$$
$$\text{where } S \equiv \text{sign}(1, A_{i+(1/2)}). \quad (17)$$

This simple formula implicitly determines our choices for the remaining two design criteria. These choices turn out to be reasonable for this advection problem, at least away from extrema. However, as we will shortly see, we can improve on FCT's performance at extrema by addressing these remaining two criteria explicitly.

All of the tests in this section use the classic explicit fourth order Runge-Kutta time discretization, each substep of which is treated in the manner described above.

4.1 Tests of FCT Advection Algorithms on Three Classic Test Problems

In [15] we compared a number of advection algorithms on three test problems chosen from the open literature: the square wave test of Boris and Book [2], the Gaussian of Forester [7], and the semi-ellipse of McDonald [10]. The first test consists of a square wave 20 cells wide to be advected 800 time steps at a Courant number of 0.2. The second test consists of a Gaussian of half width 2 cells to be advected 600 time steps at a Courant number of 0.1. The third and final test consists of a semi-ellipse of radius 15 cells to be advected 600 time steps at a Courant number

The Design of Flux-Corrected Transport (FCT) Algorithms

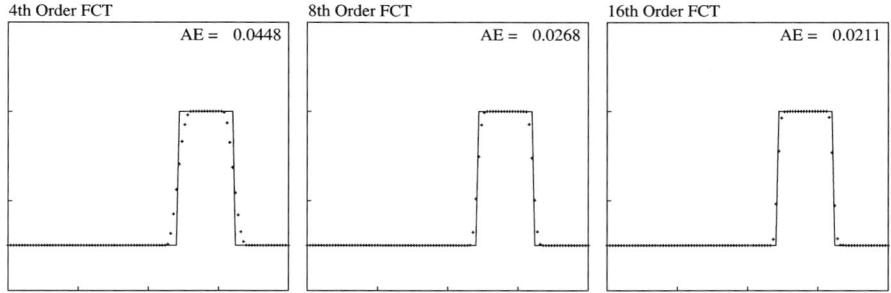

Fig. 3 Results for the Boris-Book square wave using FCT algorithms with high order fluxes of fourth, eighth, and sixteenth order. The analytic solution is shown as a *solid line*, while the computed solution is shown as *discrete data points*. The L_1 error is denoted "AE" in the plots, for consistency with the original plots of Boris and Book. Note the marked improvement with resolving power

of 0.1. We will use those same test problems here to demonstrate various aspects of the FCT algorithms we have just described.

In Fig. 3, we examine the effect that we had observed in our earlier work [14]. Running the square wave problem, we vary only the order of the high order flux from fourth to eighth to sixteenth, and see a marked increase in the resolution of the discontinuities. Our interpretation of this effect was, and continues to be, that since the discontinuity is the result of the superposition of a large number of Fourier modes, with precise phase relationships being critical, increasing the resolving power, i.e., the percentage of k-space for which the phase speed is accurate, makes it possible for the flux limiter to introduce less and less dissipation to prevent unphysical values, thus yielding more accurate results.

In Fig. 4, we show the same sequence of algorithms, but for the semi-ellipse of McDonald. Again we see an increase of performance with resolving power. However, this problem is prone to the "terracing" phenomenon, some hints of which can be seen at the right edge of the semi-ellipse. To show the value of the dissipative component of the high order flux, we show the same problem with the same set of algorithms in Fig. 5, but with the dissipative component eliminated. Although not as dramatic as the effect of increasing the resolving power of the high order flux, the dissipative flux clearly is of value in preventing the occurrence of errors that are not detected by the flux limiter. What is happening here is that dispersive oscillations are being shed by the leading (right) edge of the semi-ellipse. As they propagate into the semi-ellipse they are not detected as oscillations because they are hidden by the large gradient in the right side of the ellipse. As they get closer to the center of the ellipse, they try to take the form of true extrema, at which point they are prevented from doing so by the flux limiter. The damage is already done, however. The effect of the dissipative component in the high order flux is to damp the modes moving with the wrong phase velocity before the fact. Calculations like these, as well as analytic arguments, are the reason we believe that some high order dissipation should be present in most calculations, whether one is using front capturing techniques or not.

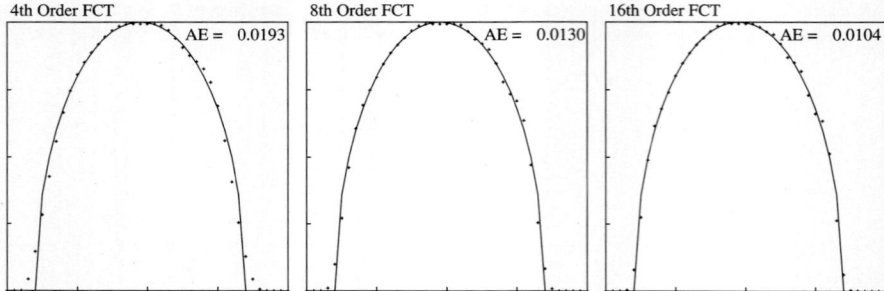

Fig. 4 Results for the semi-ellipse of McDonald using FCT algorithms with high order fluxes of fourth, eighth, and sixteenth order. The analytic solution is shown as a *solid line*, while the computed solution is shown as *discrete data points*. The L_1 error is denoted "AE" in the plots. Note the improvement, albeit modest, with increased resolving power. Although mitigated significantly by the high order dissipation, clear hints of the terracing phenomenon are still visible. Compare to Fig. 5

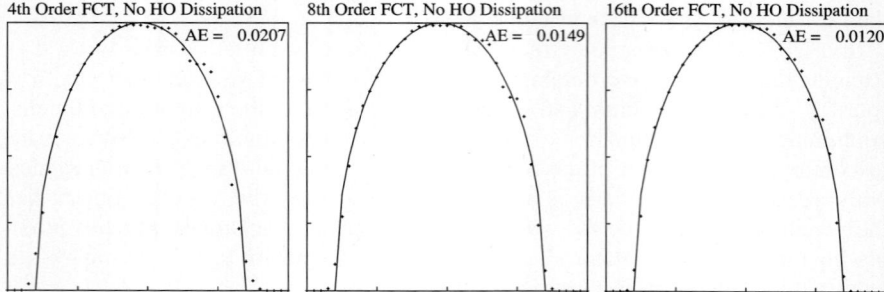

Fig. 5 Same as Fig. 4, but with the dissipative component of the high order fluxes removed. Note the "terracing" phenomenon on the right edge of the semi-ellipse, which is the result of dispersive waves being ignored by the flux limiter until they attempt to become extrema. Compare to Fig. 4

The previous set of calculations was an example of one way a flux limiter can fail. In that case, the flux limiter failed to perceive and prevent an error because its definition of an error was the creation of new extrema in q. Note that the algorithmic machinery component of the flux limiter did not fail, but rather its physics component. In this case its "physics" criterion for what constituted an error was too weak. In the next set of calculations, we see an example where exactly the same criterion is too strong, preventing the formation of an extremum when it is physically allowable. In Fig. 6 we show the same sequence of algorithms, but for the Gaussian of Forester. Although we again see the same pattern of increased performance with increased resolving power, we also see the well-known "clipping" problem. Here, as the true peak of the Gaussian passes between grid point centers, the true grid point extrema value should increase and decrease in an oscillatory fashion. However, the flux limiter used here does not allow for that possibility, treating all attempts to accentuate an extremum during a time step as an error to be prevented. The problem

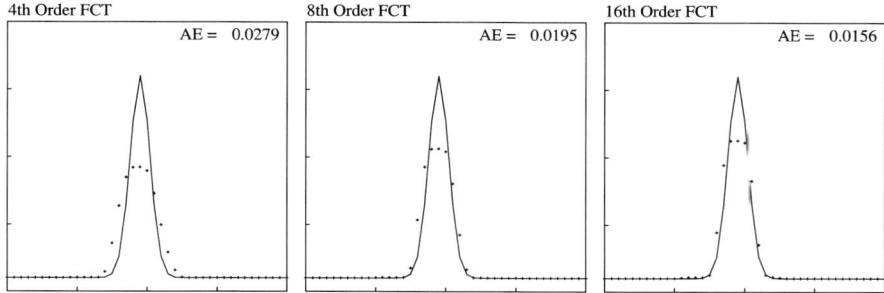

Fig. 6 Results for the Gaussian of Forester using FCT algorithms with high order fluxes of fourth, eighth, and sixteenth order. The analytic solution is shown as a *solid line*, while the computed solution is shown as *discrete data points*. The L_1 error is denoted "AE" in the plots. Again we see improvement with resolving power, but the errors are dominated by the "clipping" phenomenon

can be addressed by using a more flexible limiter and a better estimate of the allowable upper and lower bounds on the solution, as we show in the next subsection.

4.2 An Alternative to the Boris-Book Flux Limiter

In [13], we described a new flux-limiting algorithm for FCT. Although developed primarily to allow the construction of fully multidimensional FCT algorithms, that flux limiter also allowed a much more flexible specification of upper and lower bounds on the solution than did the original Boris-Book limiter Eq. (17). In particular, it allowed the construction of flux limiters which do not clip physical extrema. We describe that algorithm in one spatial dimension in this section, and then use it to construct a non-clipping flux limiter for one dimensional advection. In Sect. 5 we describe and use the algorithm in two spatial dimensions.

In words, the alternative flux limiter constrains the solution by first computing two independent sets of provisional coefficients $C_{i+(1/2)}$ for each antidiffusive flux, one to enforce the user-supplied upper bounds on the solution, and the other to enforce the user-supplied lower bounds. Both bounds are satisfied simply by choosing the final coefficients to be the minimum of the two provisional coefficients.

The upper bounds constraint is computed by dividing Q_i^+, the maximum allowable net flux into a cell, by P_i^+, the sum of all those fluxes whose effect is to increase the value of q_i. That fraction, bounded by 0 and 1, is provisionally assigned to the $C_{i+(1/2)}$ of each of those fluxes. A similar procedure is undertaken for the lower bounds constraint, and still another provisional value of $C_{i+(1/2)}$ assigned to each of the fluxes whose effect is to decrease the value of q_i. The net $C_{i+(1/2)}$ is simply the minimum of the two temporary values. From the above description it should be clear that this limiter is unambiguously defined for any number of spatial dimensions and for both structured and unstructured meshes, as long as the difference between the low and high order components can be written as fluxes flowing between adjacent cells. In one spatial dimension, the procedure is as follows:

1. Compute, for each grid point i, physically-motivated upper and lower bounds on the solution in the next timestep, q_i^{max} and q_i^{min} respectively. This step is both flexible and critical, requiring intimate knowledge of the science underlying one's equation. It is important that q_i^{td} already satisfy these bounds.
2. For the upper bound, compute P, Q, and their ratio R at each grid point:

$$P_i^+ = \max(A_{i-(1/2)}, 0) - \min(A_{i+(1/2)}, 0), \tag{18}$$

$$Q_i^+ = \left(q_i^{max} - q_i^{td}\right)\Delta x_i, \tag{19}$$

$$R_i^+ = \min\left(1, Q_i^+/P_i^+\right), \quad P_i^+ > 0, \quad 0 \text{ otherwise.} \tag{20}$$

3. For the lower bound, compute P, Q, and their ratio R at each grid point:

$$P_i^- = \max(A_{i+(1/2)}, 0) - \min(A_{i-(1/2)}, 0), \tag{21}$$

$$Q_i^- = \left(q_i^{td} - q_i^{min}\right)\Delta x_i, \tag{22}$$

$$R_i^- = \min\left(1, Q_i^-/P_i^-\right), \quad P_i^- > 0, \quad 0 \text{ otherwise.} \tag{23}$$

4. Compute $C_{i+(1/2)}$ by taking a minimum:

$$C_{i+(1/2)} = \begin{cases} \min(R_{i+1}^+, R_i^-) & \text{when } A_{i+(1/2)} > 0, \\ \min(R_i^+, R_{i+1}^-) & \text{when } A_{i+(1/2)} \leq 0. \end{cases} \tag{24}$$

Note that in the above we do not specify the equivalent of Eq. (14) in [13], which, as we explained in the previous section, can be thought of as a method for pre-constraining the high order fluxes prior to the flux limiting step. In the case of linear advection in one dimension, we have a much more robust way of pre-limiting those fluxes, as we will see below.

4.3 A Non-clipping Version of the Alternative Flux Limiter

Let us now specify our non-clipping flux limiter for one-dimensional linear advection. To do so we need to define an algorithm for computing q_i^{min} and q_i^{max} above. We also need to address our third criterion and specify an algorithm for pre-constraining our high order fluxes. We shall use a similar approach for both. In Fig. 7, we show a technique we shall use to reconstruct extrema between grid points, for use both in specifying q_i^{min} and q_i^{max} and in pre-constraining our high order fluxes. On each interval $[x_i, x_{i+1}]$ we define $q_{i+(1/2)}^{peak}$ to be the value of q at the intersection of the lines formed by connecting the point (x_{i-1}, q_{i-1}) with (x_i, q_i) and the point (x_{i+1}, q_{i+1}) with (x_{i+2}, q_{i+2}). If the x coordinate of this intersection lies between x_i and x_{i+1}, then we consider this $q_{i+(1/2)}^{peak}$ to be a physically legitimate value for q on the interval $[x_i, x_{i+1}]$.

Let us now define the upper and lower bounds for q on the interval $[x_i, x_{i+1}]$ to be

$$q_{i+(1/2)}^{max} = \max\left(q_i, q_{i+1}, q_{i+(1/2)}^{peak}\right), \tag{25}$$

$$q_{i+(1/2)}^{min} = \min\left(q_i, q_{i+1}, q_{i+(1/2)}^{peak}\right) \tag{26}$$

Fig. 7 A possible scheme for extracting information about extrema which exist between grid points at a given point in time. An extremum is assumed to exist between grid points i and $i+1$ if the intersection of the right and left extrapolation of q has an x coordinate between x_i and x_{i+1}. The q coordinate of this intersection in then used both to compute q^{max} and q^{min}, and to pre-constrain the high order flux $F^H_{i+(1/2)}$ (see text)

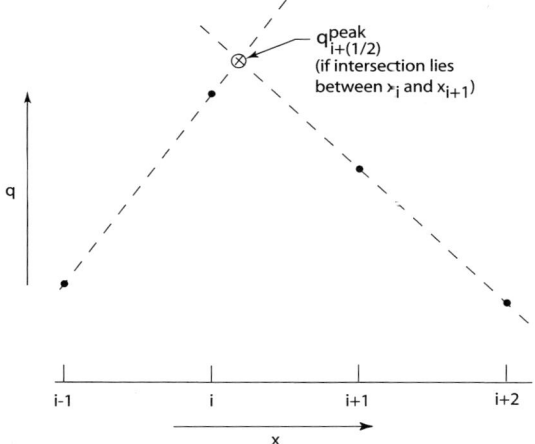

where all quantities are evaluated at time level n.

We are now in a position to introduce the physics of the problem into the flux limiter. Given that this is an advection problem, and that we choose our Courant number to be less than unity, we know that q_i^{n+1} must be bounded by $q^{min}_{i-(1/2)}$ and $q^{max}_{i-(1/2)}$ if $u > 0$, and by $q^{min}_{i+(1/2)}$ and $q^{max}_{i+(1/2)}$ if $u < 0$. Thus we take

$$q_i^{min} = \begin{cases} \min(q_i^{td}, q^{min}_{i-(1/2)}) & \text{when } u > 0, \\ \min(q_i^{td}, q^{min}_{i+(1/2)}) & \text{when } u \leq 0, \end{cases} \quad (27)$$

$$q_i^{max} = \begin{cases} \max(q_i^{td}, q^{max}_{i-(1/2)}) & \text{when } u > 0, \\ \max(q_i^{td}, q^{max}_{i+(1/2)}) & \text{when } u \leq 0. \end{cases} \quad (28)$$

Finally we specify our pre-constraint condition on the high order fluxes. Again we use the physics of the problem. Since this is advection, the physical fluxes $F_{i+(1/2)}$ must be bounded by $uq^{max}_{i+(1/2)}$ and $uq^{min}_{i+(1/2)}$. Thus we define $F^{max}_{i+(1/2)} \equiv \max(uq^{max}_{i+(1/2)}, uq^{min}_{i+(1/2)})$ and $F^{min}_{i+(1/2)} \equiv \min(uq^{max}_{i+(1/2)}, uq^{min}_{i+(1/2)})$, and after computing the unconstrained high order fluxes $F^H_{i+(1/2)}$, we constrain them thus:

$$F^H_{i+(1/2)} = \min\left(F^{max}_{i+(1/2)}, \max\left(F^{min}_{i+(1/2)}, F^H_{i+(1/2)}\right)\right). \quad (29)$$

The results of using the above pre-constraint condition and flux limiter are shown in Fig. 8. For sufficiently high resolving power, the clipping phenomenon has been virtually eliminated. We believe that this demonstrates the advantage of using one's knowledge of the physics of the problem to design FCT and other front-capturing algorithms, rather than accepting their default behavior.

Before leaving this section, let us try to design the "ultimate" high order FCT scheme, and see how it performs on the three test problems we have examined in this section. It was shown by Fornberg that the asymptotic limit of an Nth order finite difference scheme on a periodic domain as N goes to infinity is in fact just the

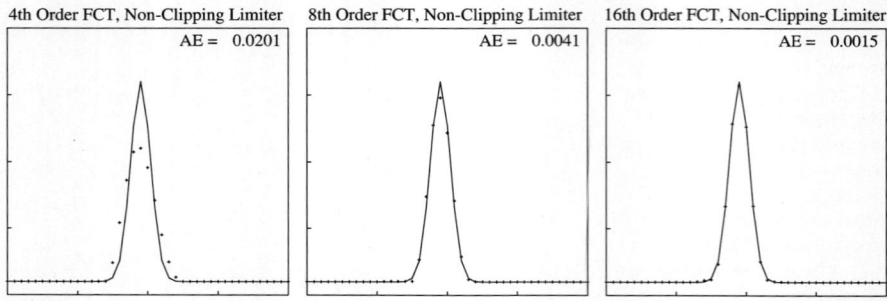

Fig. 8 Same as Fig. 6, but using the non-clipping flux limiter and the pre-constraint of the high order fluxes described in the text. Note the marked increase of accuracy with increased resolving power. Also note that the clipping can be virtually eliminated as long as one has sufficient resolving power

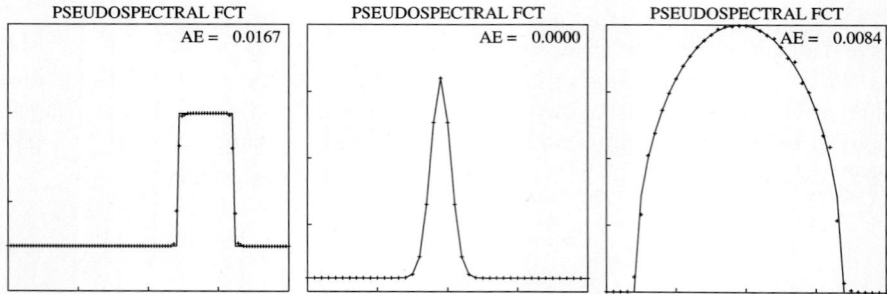

Fig. 9 Performance of a pseudospectral FCT algorithm on all three of the test problems used in this section. We have used the non-clipping flux limiter and the pre-constraint of the high order fluxes described in the text. These are the best results we have been able to produce for these problems using front-capturing algorithms whose high-order components are stable

pseudospectral approximation using Fourier modes as basis functions. Thus we will take our high order fluxes to be those which reproduce the pseudospectral discretization. The results of using these pseudospectral fluxes and the non-clipping limiter and pre-constraint algorithm described above are shown in Fig. 9. These are the best results we have been able to produce for these problems using front-capturing algorithms whose high-order components are stable. (Using unstable schemes as the high order component of front-capturing algorithms can produce extremely sharp square waves, but can severely distort the Gaussian and semi-ellipse. Examples are Superbee, Ultrabee, ACM, and the contact detection algorithm in PPM.)

The above examples provide a springboard for the next section, where we address a nonlinear system of equations, and the construction of our FCT algorithms will become more complex. The path we will choose to success will be the same, however: We will incorporate as much knowledge of the physics as possible into the design of the algorithm.

5 FCT Algorithms for One Dimensional Nonlinear Systems of Conservation Laws

We now consider Eq. (2), where f is a fully nonlinear function of q. The example we shall use is that of the Euler equations (3). If we go down our list of FCT design criteria, we find that many of the optimal choices are the same as those for linear advection, or are obvious generalizations thereof. However, the proper construction of the flux limiter, both with regard to its physics component and with regard to its algorithmic machinery, is less than obvious. Hence we will focus primarily on the construction of the flux limiter in this section.

We saw in the last section that the results of an FCT calculation can be sensitive to the choice of the flux limiter. But we also learned that modest modifications of the basic FCT machinery allowed us to produce results which are quite good. Thus it is worth looking at both the advection equation and at the advection flux limiters for the purpose of determining how we wish to proceed for systems of hyperbolic conservation laws.

The simplicity of the advection equation allows us to make some rather strong statements about the allowable bounds for q in the next timestep. In particular, we know that for any Courant number less than unity, the value of q_i^{n+1} is bounded by the values of q^n on the interval $[x_{i-1}, x_{i+1}]$. This fact allows us to construct reasonably precise upper and lower bounds on the solution. The bounds for q_i^{n+1} used by the Boris and Book flux limiter Eq. (17) are simply the maximum and minimum of $(q_{i-1}^{td}, q_i^{td}, q_{i+1}^{td})$ respectively. While this could certainly be refined, we saw in the last section that this algorithm produces reasonable results if one is willing to tolerate the clipping of extrema. The reason for this, we believe, is that the built-it physics component of the limiter is reasonably close to one physically appropriate to the advection problem, again except near extrema.

5.1 Hyperbolic Systems of Conservation Laws: The Case for Characteristic Variables

Let us now consider systems of hyperbolic conservation laws, using the Euler equations as an example. If we choose to deal completely with the conserved variables q, what sort of statements can we make about upper and lower bounds on q in the next timestep? In contrast to the case for advection, we are at a loss. In fact we know for certain that q_i^{n+1} is not necessarily bounded by q^n anywhere in the vicinity of grid point i. The default route taken by most FCT algorithms is to simply use what worked for advection, and take the upper and lower bounds for q^{n+1} to be the maximum and minimum of $(q_{i-1}^{td}, q_i^{td}, q_{i+1}^{td})$ respectively. While certainly better than the disastrous choice of using the maximum and minimum of $(q_{i-1}^n, q_i^n, q_{i+1}^n)$, in contrast to the case for advection, brand new extrema will be a common occurrence. The default choice would allow strong suppression of these new extrema by the combination of the flux limiter and the dissipation in the low order fluxes. Using the non-clipping flux limiter described in the last section would not help either, again because we have no basis by which to determine the magnitude that a new extremum could attain before being declared unphysical. These difficulties are reflected in the difficulty that FCT has, in our experience, in attaining the same kind of clean results for systems of equations that are relatively easy to come by for scalar equations, when using the default strategy of flux limiting with respect to the conserved variables directly. This is because we have strong and simple statements that we can make about the upper and lower bounds in q in the next time step for the scalar case that we just don't have in the case of systems.

However, there is a set of variables in which a one-dimensional hyperbolic system looks exactly like an advection equation, a set of uncoupled advection equations to be precise, and that is the set of characteristic variables. These variables are not global variables, but rather the result of locally linearizing the equations. Briefly, what we will do in what follows is take the entire flux limiting problem, consisting of the low order solution q^{td} and the set of "antidiffusive fluxes" A, and transform them both into a set of variables in which the same flux limiting problem looks like a set of uncoupled linear advection flux limiting problems. We then limit the fluxes using constraints physically and mathematically appropriate to an advection problem, and then transform the limited fluxes back into conserved variables, where they will be applied to q^{td} to produce the new solution q^{n+1}. This will produce results that in our view are far superior to those produced using the conserved variables directly.

Let us specialize our one-dimensional system of conservation laws Eq. (2) to the case of a flux function f which is solely a function of q (the usual case), and write it in the more compact notation

$$q_t + f(q)_x = 0 \tag{30}$$

where the subscript denotes partial differentiation. We can then further rewrite Eq. (30) in the following form:

$$q_t + A(q)q_x = 0 \tag{31}$$

where A is the $m \times m$ Jacobian matrix $\partial f/\partial q$, and m is the number of conservation laws in Eq. (30). It is not clear that Eq. (31) is any improvement over Eq. (30), since we have lost our explicit conservation form, and in general most entries of A are nonzero. If our goal is to find an appropriate set of constraints on the values of q for the purpose of flux limiting, we apparently have made no progress. But for most hyperbolic systems it is possible to find a new set of variables q' defined by a transformation matrix $T^{-1} = T^{-1}(q)$ such that Eq. (31) takes the form of a series of m decoupled advection equations:

$$T^{-1}q_t + T^{-1}A(q)TT^{-1}q_x = 0, \tag{32}$$

$$q'_t + \Lambda q'_x = 0 \tag{33}$$

where $q' = T^{-1}q$ and $\Lambda = T^{-1}A(q)T$ is a diagonal matrix with diagonal elements λ_j, $1 \le j \le m$.

Specifically, we will have m scalar equations of the form

$$\frac{\partial q'_j}{\partial t} + \lambda_j \frac{\partial q'_j}{\partial x} = 0. \tag{34}$$

Our problem has now been reduced to performing the flux limiting step for m independent advection equations, something that we know how to do well, precisely because we have very good information on how to constrain the solution, as we demonstrated in the last section. We note that the characteristic variables q' are not the conserved quantities, and that we wish to construct our FCT algorithm such that the conserved variables q are updated in flux form. Thus it will be important to transform the fluxes themselves between the two spaces, not just the solution vectors.

To construct our characteristic variable (CV) flux limiter, we first look at the basic Boris-Book limiter, Eq. (17)

$$A^C_{i+(1/2)} = S \max\left(0, \min\left(|A_{i+(1/2)}|, S(q^{td}_{i+2} - q^{td}_{i+1})\Delta x, S(q^{td}_i - q^{td}_{i-1})\Delta x\right)\right). \tag{35}$$

This one-line formula provides one possible answer to the following question, which we term the "Flux Limiting Problem" (FLP) for advection: Given the time-advanced low order solution q^{td} and perhaps other auxiliary solution vectors, and given a set of antidiffusive fluxes A, what is a set of corrected antidiffusive fluxes A^C that are as close to A as possible, and that will constrain q^{n+1} to lie within the bounds appropriate to the advection problem? The FLP requires at least two inputs q^{td} and A, and asks for one output A^C. A look at Eq. (35), however, will convince the reader that q^{td} itself is not really needed, but rather only its first differences at flux evaluation points $i + (1/2)$. This is not atypical. All flux limiting algorithms of which we are aware have the property of depending only on local variations in q, not on the values of q themselves. This observation makes the construction of a CV limiter particularly simple.

5.2 A Characteristic Variable Implementation of the Boris-Book Flux Limiter

To be concrete here, we present a version of the CV flux limiter using the Boris-Book limiter as a building block. Using other limiters as building blocks may require some modification which will hopefully be obvious to the reader.

Given a hyperbolic system of conservation laws of length m with components j, $1 \leq j \leq m$ in one spatial dimension, a low-order solution vector q^{td} with components denoted $q^{td(j)}$, $1 \leq j \leq m$ defined on grid points x_i, and a vector of antidiffusive fluxes A with components denoted $A^{(j)}$, $1 \leq j \leq m$ defined at flux points $x_{i+(1/2)}$, the following steps define a characteristic variable-based implementation of the Boris-Book flux limiter.

1. Calculate some appropriate average $q^{td(j)}_{i+(1/2)}$, $\forall j, i$.

2. From $q^{td}_{i+(1/2)}$ calculate $T^{-1}_{i+(1/2)}$ and $T_{i+(1/2)}$, $\forall i$.
3. Set $D_{i+(1/2)} = T^{-1}_{i+(1/2)}(q^{td}_{i+1} - q^{td}_i)$, $\forall i$.
4. Set $B_{i+(1/2)} = T^{-1}_{i+(1/2)} A_{i+(1/2)}$, $\forall i$.
5. Set $B^{C(j)}_{i-(1/2)} = S \max(0, \min(|B^{(j)}_{i+(1/2)}|, SD^{(j)}_{i+(3/2)}\Delta x, SD^{(j)}_{i-(1/2)}\Delta x))$, $\forall j, i$, where $S = \text{sign}(1, B^{(j)}_{i+(1/2)})$.
6. Set $A^C_{i+(1/2)} = T_{i+(1/2)} B^C_{i+(1/2)}$, $\forall i$.

The notation above uses the superscript (j) on quantities only when it is necessary to emphasize that each component of the vector is being manipulated separately. Otherwise when the superscript is not present, a vector operation of length m is assumed.

5.3 Computational Examples: The One Dimensional Euler Equations

The equations of interest are

$$w = \begin{pmatrix} \rho \\ \rho u \\ \rho E \end{pmatrix}; \quad f = \begin{pmatrix} \rho u \\ \rho u u + P \\ \rho u E + P u \end{pmatrix} \quad (36)$$

where ρ, u, P, and E are the fluid density, velocity, pressure, and specific total energy respectively. We will assume an ideal gas equation of state

$$P = (\gamma - 1)\left(\rho E - \frac{1}{2}\rho u^2\right). \quad (37)$$

The matrices T and T^{-1} that we shall need are found by first setting

$$|A - \lambda I| = 0$$

and solving for the eigenvalues λ_j of A. Then for each of these eigenvalues the right and left eigenvectors are found. T is the matrix whose columns are the right eigenvectors of A. T^{-1} is the matrix whose rows are the corresponding left eigenvectors of A. These matrices are well known for this system. They are

$$T = \begin{bmatrix} 1 & 1 & 1 \\ u-c & u & u+c \\ H-uc & \frac{1}{2}u^2 & H+uc \end{bmatrix},$$

$$T^{-1} = \begin{bmatrix} \frac{1}{2}(\frac{\gamma-1}{2}M^2 + \frac{u}{c}) & -\frac{1}{2c} - \frac{(\gamma-1)u}{2c^2} & \frac{\gamma-1}{2c^2} \\ 1 - \frac{\gamma-1}{2}M^2 & \frac{(\gamma-1)u}{c^2} & -\frac{\gamma-1}{c^2} \\ \frac{1}{2}(\frac{\gamma-1}{2}M^2 - \frac{u}{c}) & \frac{1}{2c} - \frac{(\gamma-1)u}{2c^2} & \frac{\gamma-1}{2c^2} \end{bmatrix}$$

where $M^2 \equiv u^2/c^2$, $c^2 = \gamma P/\rho$, and $H = c^2/(\gamma - 1) + u^2/2$ is the stagnation enthalpy.

For our low order flux for the Euler equations we choose the Rusanov scheme:

$$F^L_{i+(1/2)} = \left[\frac{1}{2}(f^n_{i+1} + f^n_i) - \frac{1}{4}(Q_i + Q_{i+1})(q^n_{i+1} - q^n_i)\right]\Delta t^{n+1/2} \qquad (38)$$

where Q_i is the maximum characteristic speed at i:

$$Q_i = |u_i| + c_i. \qquad (39)$$

The high order fluxes are as given before for advection. As an example, the fourth order flux is given by:

$$F^{H4}_{i+(1/2)} = \left[\frac{7}{12}(f^a_{i+1} + f^a_i) - \frac{1}{12}(f^a_{i+2} + f^a_{i-1})\right]\Delta t^{n+1/2} \qquad (40)$$

where the time level t^a is meant to denote whatever time level or average of time levels is required by the particular substep of the particular time discretization chosen.

The high order dissipative fluxes are modified versions of those used for advection, with the advection speed u replaced by the maximum characteristic speed Q. As an example, the order 4 dissipative flux is given by:

$$F^{D4}_{i+(1/2)} = -\frac{1}{2}(Q_i + Q_{i+1})\left[\frac{3}{16}(q^n_{i+1} - q^n_i) - \frac{1}{16}(q^n_{i+2} - q^n_{i-1})\right]\Delta t^{n+1/2}. \qquad (41)$$

The flux limiter, when we are not using the CV limiter described above, is again given by the original Boris-Book limiter:

$$A^C_{i+(1/2)} = S\max\left(0, \min\left(|A_{i+(1/2)}|, S(q^{td}_{i+2} - q^{td}_{i+1})\Delta x, S(q^{td}_i - q^{td}_{i-1})\Delta x\right)\right)$$
where $S \equiv \text{sign}(1, A_{i+(1/2)})$. \qquad (42)

All of the tests in this section use a modified Euler time discretization, each substep of which is treated in the manner described in Sect. 4.

Our failsafe limiter is the simplest imaginable: If, after flux limiting, either the density or the pressure in a cell is negative, all the fluxes into that cell are set to their low order values, and the grid point values recalculated. Clearly there is much room for a more precise failsafe mechanism, but this one has proved adequate for the problems presented here.

Now that we have described the algorithms we will be using in this section, we show some results using standard test problems. The first is the shock tube problem due to Sod [11]. The initial conditions consist of a single discontinuity in density (8 : 1) and pressure (10 : 1), with both $\gamma = 1.4$ gases at rest. All of our results plot the analytic solution as a solid line, and the computed grid point values as data points, using the temperature field, which we have found to be the field most sensitive to numerical error. In Fig. 10 we show the results of our CV FCT algorithms for $N = N_D = 4, 8,$ and 16. From left to right in each of the three plots, the reader will recognize the shock wave, the contact discontinuity, and the rarefaction fan associated with this problem. Note the marked increase in the accuracy of the contact discontinuity as the resolving power of the high order fluxes increases, similar to our experiences with advection. For comparison, in Fig. 11 we show the same

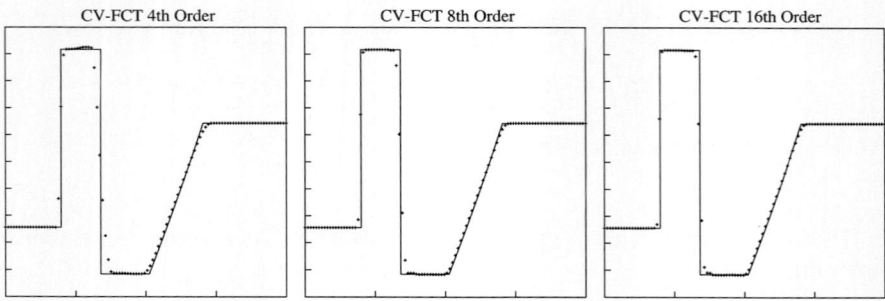

Fig. 10 Results for the temperature field, Sod shock tube problem using CV FCT algorithms with $N = N_D = 4, 8$, and 16

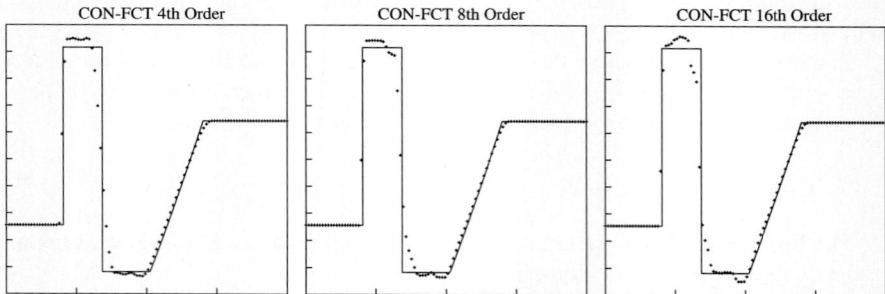

Fig. 11 Same as Fig. 10, but using a flux limiter which limits only with respect to conserved variables

three calculations, but using the more conventional FCT flux limiter which limits the fluxes based solely on the conserved variables. Note the marked superiority of the characteristic variable-based CV flux limiter.

The next test problem is the double shock tube of Woodward and Colella [12]. This problem involves the complex interaction of very strong waves of all types, and is considerably more difficult than the Sod problem. Here we show the performance of our CV FCT algorithms on three grids, of size 200, 400, and 800 grid points, testing both absolute performance and convergence. In Fig. 12, we show the density field at $t = 0.2$ using a grid of 200 points, and using our CV FCT algorithms with $N = N_D = 4, 8$, and 16. As with the advection tests, and with the Sod test problem, we see increased accuracy with resolving power, but all calculations suffer from lack of resolution.

Figures 13 and 14 show the same calculations with 400 and 800 grid points respectively. We see increased accuracy with resolving power, as well as with grid refinement. Note that the two shock waves are resolved over 1–2 grid points regardless of the resolving power, but that the accuracy (sharpness in this case) of the three contact discontinuities increases markedly with increased resolving power at all refinement levels.

The Design of Flux-Corrected Transport (FCT) Algorithms

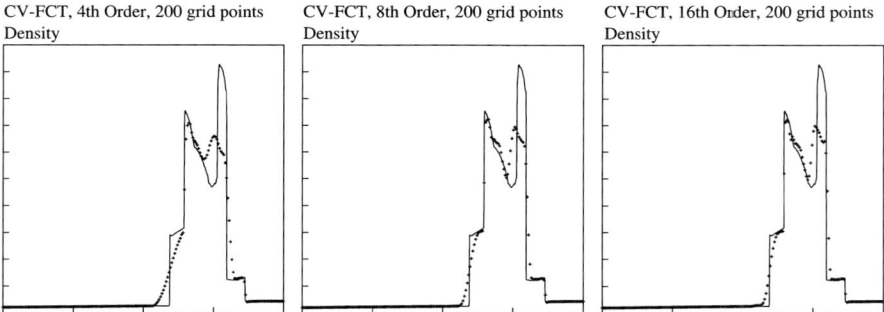

Fig. 12 Results for the density field for the 200-point Woodward-Colella double shock tube problem using CV FCT algorithms with $N = N_D = 4$, 8, and 16

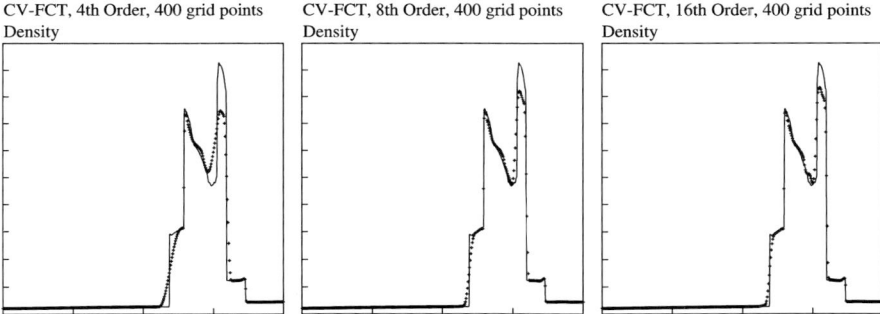

Fig. 13 Same as Fig. 12, but using a grid of 400 points

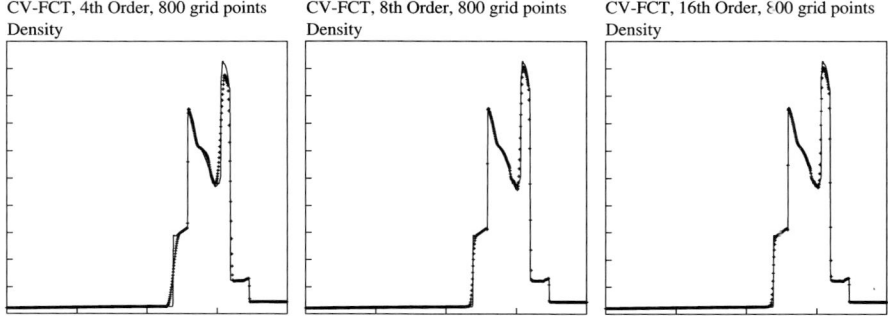

Fig. 14 Same as Fig. 12, but using a grid of 800 points

5.4 Using Characteristic Variables in Other FCT Components

Thus far, we have dealt with the use of characteristic variables only in the flux limiter, but there are two other FCT components that could conceivably benefit from their use: the low order fluxes, and the dissipative component of the high order fluxes. The treatment of both is quite similar, so we will discuss them together. Recall that our low order flux for advection was given by the first order upwind method:

$$F^L_{i+(1/2)} = \left[\frac{1}{2}(f^n_{i+1} + f^n_i) - \frac{1}{2}|u|(q^n_{i+1} - q^n_i)\right]\Delta t^{n+1/2}. \tag{43}$$

It can be proven that for any flux of the above form, the coefficient $|u|/2$ used above is the smallest that will guarantee that the flux will maintain the monotonicity of a monotone profile. Thus this flux is in some sense the optimum low order flux for advection.

By contrast, our low order flux for the Euler equations was the Rusanov flux:

$$F^L_{i+(1/2)} = \left[\frac{1}{2}(f^n_{i+1} + f^n_i) - \frac{1}{4}(Q_i + Q_{i+1})(q^n_{i+1} - q^n_i)\right]\Delta t^{n+1/2} \tag{44}$$

where Q_i is the maximum characteristic speed at i:

$$Q_i = |u_i| + c_i. \tag{45}$$

This flux is *not* the optimum low order flux for the Euler equations, and is in general considerably more dissipative than necessary to guarantee that unphysical solutions cannot be generated. Rather, the Godunov flux is the optimal choice. This flux requires the solution of the full nonlinear Riemann problem at each flux point. However, a good approximation to the Godunov flux $F^G_{i+(1/2)}$ is obtained by doing exactly what we did to limit fluxes: We transform the entire "low order flux" problem into characteristic variables, where the system is of the form of m uncoupled advection problems, compute first order upwind fluxes in those variables, and then transform the fluxes back to conserved variables:

1. Calculate some appropriate average $q^{n(j)}_{i+(1/2)}$, $\forall j, i$.
2. From $q^n_{i+(1/2)}$ calculate $T^{-1}_{i+(1/2)}$ and $T_{i+(1/2)}$, $\forall i$.
3. Set $D_{i+(1/2)} = T^{-1}_{i+(1/2)}(q^n_{i+1} - q^n_i)$, $\forall i$.
4. Set $D^{(j)}_{i+(1/2)} = -\frac{|\lambda^{(j)}_{i+(1/2)}|}{2} D^{(j)}_{i+(1/2)}$, $\forall i, j$.
5. Set $F^L_{i+(1/2)} = [\frac{1}{2}(f^n_{i+1} + f^n_i) + T_{i+(1/2)} D_{i+(1/2)}]\Delta t^{n+1/2}$.

In principle, this would be a much better choice for our low order flux than the Rusanov flux we have chosen, because it would be less dissipative. However, we cannot forget that the primary property that we want of our low order flux is its guaranteed freedom from unphysical behavior. Anyone familiar with the modern literature on approximate Riemann solvers will recognize the above as one of the popular ways of constructing them. He or she will also know that such approximate solvers are in general devoid of the guarantees that we need. Thus we leave this

The Design of Flux-Corrected Transport (FCT) Algorithms

promising topic for future exploration, and move on to the related topic of optimizing the adaptive dissipation in the high order flux.

Recall that our order 4 dissipative flux for advection was given by:

$$F^{D4}_{i+(1/2)} = -|u|\left[\frac{3}{16}(q^n_{i+1} - q^n_i) - \frac{1}{16}(q^n_{i+2} - q^n_{i-1})\right]\Delta t^{n+1/2} \quad (46)$$

while the corresponding flux for the Euler equations was

$$F^{D4}_{i+(1/2)} = -\frac{1}{2}(Q_i + Q_{i+1})\left[\frac{3}{16}(q^n_{i+1} - q^n_i) - \frac{1}{16}(q^n_{i+2} - q^n_{i-1})\right]\Delta t^{n+1/2} \quad (47)$$

where Q_i is again the maximum characteristic speed at i:

$$Q_i = |u_i| + c_i. \quad (48)$$

Comparing the above pair of equations with the preceding pair, we see that the order 4 dissipative flux for the Euler equations suffers from the same flaw as does the Rusanov flux: in general it will provide more dissipation than is needed. The most extreme example of this is that of very low Mach number flow in which advected waves would be subject to a dissipation proportional to c, while the waves themselves were moving with a velocity of $v \ll c$. A way of addressing this is, again, to use the characteristic variables, dissipating each of the component waves in proportion to its own wave speed.

To give a concrete example of the procedure, let us first rewrite Eq. (47):

$$F^{D4}_{i+(1/2)} = \frac{1}{32}(Q_i + Q_{i+1})[\Delta q_{i-(1/2)} - 2\Delta q_{i+(1/2)} + \Delta q_{i+(3/2)}]\Delta t^{n+1/2}$$
where $\Delta q_{i+(1/2)} \equiv q^n_{i+1} - q^n_i$. \quad (49)

Our CV dissipative flux of order 4 would then be computed as follows:

1. Calculate some appropriate average $q^{n(j)}_{i+(1/2)}$, $\forall j, i$.
2. From $q^n_{i+(1/2)}$ calculate $T^{-1}_{i+(1/2)}$ and $T_{i+(1/2)}$, $\forall i$.
3. Set $\Delta_{i+(1/2)} = T^{-1}_{i+(1/2)}(q^n_{i+1} - q^n_i)$, $\forall i$.
4. Set $D^{(j)}_{i+(1/2)} = \frac{|\lambda^{(j)}_{i+(1/2)}|}{16}(\Delta^{(j)}_{i-(1/2)} - 2\Delta^{(j)}_{i+(1/2)} + \Delta^{(j)}_{i+(3/2)})$, $\forall j, i$.
5. Set $F^{D4}_{i+(1/2)} = [T_{i+(1/2)} D_{i+(1/2)}]\Delta t^{n+1/2}$.

Other adaptive dissipative fluxes can be computed in a similar manner. Rerunning all of our previous CV limiter calculations with this CV adaptive dissipation, we find that only the $N = 4$ calculations display any significant differences. We show only those here. In Fig. 15 we compare two calculations, both using CV limiting, for the Sod shock tube problem. The left panel is the same as that of the left panel in Fig. 10, using the adaptive dissipation given by Eq. (47), while the right panel instead uses the CV adaptive dissipation given by the above algorithm. Note a significant increase in the sharpness of the contact discontinuity, without any other adverse effects. This is, of course, what we hoped we would achieve.

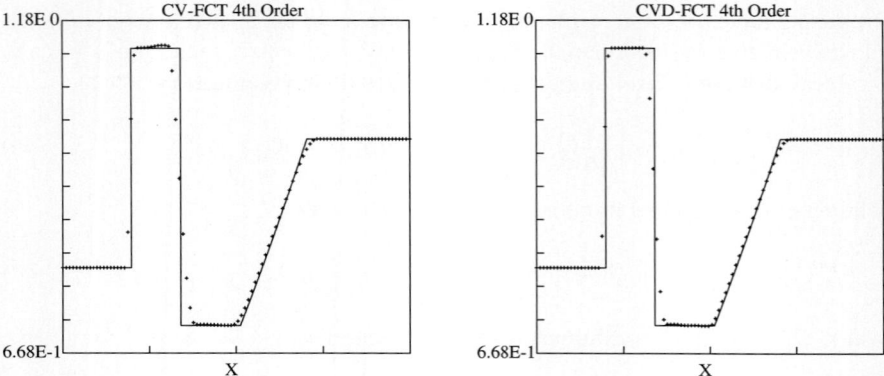

Fig. 15 Sod shock tube problem: Comparison of the $N = N_d = 4$ CV FCT algorithm using the conventional adaptive dissipation given by Eq. (47) (*left*), and the CV-based adaptive dissipation described in the text (*right*)

6 Flux-Corrected Transport in Multidimensions

As we have stated before, there is a large class of problems for which an operator splitting strategy, using sequences of one-dimensional time-advancement operators, can be successful. We are assuming here, however, that we are interested in pursuing a more fully multidimensional approach wherein the results are independent, or as independent as possible, of any apparent ordering of one-dimensional operators. Of the three components of an FCT algorithm, only the flux limiter normally presents any difficulty in this regard. Indeed, much of [13] was devoted to defining a fully multidimensional flux limiter, which we present below.

6.1 A Fully Multidimensional Flux Limiter

The alternative flux limiting algorithm presented in Sect. 4.2 generalizes trivially to any number of spatial dimensions, and in fact to unstructured as well as the structured meshes we consider here. For the sake of completeness we present the algorithm for the structured coordinate-aligned two dimensional mesh referred to in Eq. (10).

Referring to Fig. 16, we seek to limit the antidiffusive fluxes $A_{i+(1/2),j}$ and $A_{i,j+(1/2)}$ by finding coefficients $C_{i+(1/2),j}$ and $C_{i,j+(1/2)}$ such that

$$A^C_{i+(1/2),j} = C_{i+(1/2),j} A_{i+(1/2),j}, \quad 0 \le C_{i+(1/2),j} \le 1,$$
$$A^C_{i,j+(1/2)} = C_{i,j+(1/2)} A_{i,j+(1/2)}, \quad 0 \le C_{i,j+(1/2)} \le 1$$

and such that $A^C_{i+(1/2),j}$, $A^C_{i-(1/2),j}$, $A^C_{i,j+(1/2)}$, and $A^C_{i,j-(1/2)}$ acting in concert shall not cause

$$q^{n+1}_{ij} = q^{td}_{ij} - \Delta V^{-1}_{ij} \left[A^C_{i+(1/2),j} - A^C_{i-(1/2),j} + A^C_{i,j+(1/2)} - A^C_{i,j-(1/2)} \right]$$

Fig. 16 Schematic of the flux limiting problem in two dimensions

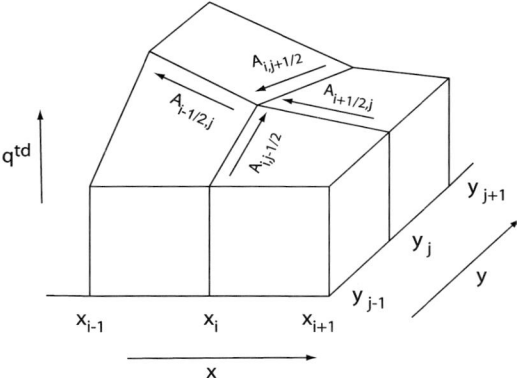

to exceed some maximum value q_{ij}^{max} or fall below some minimum value q_{ij}^{min}. The procedure is completely analogous to that given in Sect. 2:

1. Compute, for each grid point ij, physically-motivated upper and lower bounds on the solution in the next timestep, q_{ij}^{max} and q_{ij}^{min} respectively.
2. For the upper bound, compute P, Q, and their ratio R at each grid point:

$$P_{ij}^+ = \max(A_{i-(1/2),j}, 0) - \min(A_{i+(1/2),j}, 0) \tag{50}$$
$$+ \max(A_{i,j-(1/2)}, 0) - \min(A_{i,j+(1/2)}, 0), \tag{51}$$
$$Q_{ij}^+ = \left(q_{ij}^{max} - q_{ij}^{td}\right)\Delta V_{ij}, \tag{52}$$
$$R_{ij}^+ = \min\left(1, Q_{ij}^+/P_{ij}^+\right), \quad P_{ij}^+ > 0, \quad 0 \text{ otherwise.} \tag{53}$$

3. For the lower bound, compute P, Q, and their ratio R at each grid point:

$$P_{ij}^- = \max(A_{i+(1/2),j}, 0) - \min(A_{i-(1/2),j}, 0) \tag{54}$$
$$+ \max(A_{i,j+(1/2)}, 0) - \min(A_{i,j-(1/2)}, 0), \tag{55}$$
$$Q_{ij}^- = \left(q_{ij}^{td} - q_{ij}^{min}\right)\Delta V_{ij}, \tag{56}$$
$$R_{ij}^- = \min\left(1, Q_{ij}^-/P_{ij}^-\right), \quad P_{ij}^- > 0, \quad 0 \text{ otherwise.} \tag{57}$$

4. Compute $C_{i+(1/2),j}$ and $C_{i,j+(1/2)}$ by taking a minimum:

$$C_{i+(1/2),j} = \begin{cases} \min(R_{i+1,j}^+, R_{ij}^-) & \text{when } A_{i+(1/2),j} > 0, \\ \min(R_{ij}^+, R_{i+1,j}^-) & \text{when } A_{i+(1/2),j} \leq 0, \end{cases} \tag{58}$$

$$C_{i,j+(1/2)} = \begin{cases} \min(R_{i,j+1}^+, R_{ij}^-) & \text{when } A_{i,j+(1/2)} > 0, \\ \min(R_{ij}^+, R_{i,j+1}^-) & \text{when } A_{i,j+(1/2)} \leq 0. \end{cases} \tag{59}$$

Again note that in the above we do not specify the equivalent of Eq. (14) in [13]. As we have stated, we do not consider that equation to be part of the flux limiter proper, but rather an algorithm for pre-constraining the high order fluxes. Nonetheless, for fully multidimensional problems we have yet to find anything better, and we use an abbreviated form of that equation to pre-constrain the high order fluxes in the multidimensional advection problems that follow:

$$A_{i+(1/2),j} = 0 \quad \text{if } A_{i+(1/2),j}\left(q^{td}_{i+1,j} - q^{td}_{ij}\right) \leq 0,$$
$$A_{i,j+(1/2)} = 0 \quad \text{if } A_{i,j+(1/2)}\left(q^{td}_{i,j+1} - q^{td}_{ij}\right) \leq 0. \tag{60}$$

With our fully multidimensional flux limiter in hand, along with our algorithm for pre-constraining the high order fluxes Eq. (60), let us consider two multidimensional problems: passively-driven convection in two dimensions, and compressible gas dynamics in two dimensions.

6.2 FCT Algorithms for Two-Dimensional Passively-Driven Convection

We shall be interested in solving Eq. (8) for the special case where $q(x, y)$ is a scalar and where

$$f = qu, \tag{61}$$
$$g = qv. \tag{62}$$

Here $u(x, y)$ and $v(x, y)$ are convection velocity components in the x and y directions respectively. They are assumed to be specified either globally or at the very least at cell boundaries. Thus our equation is

$$q_t + (qu)_x + (qv)_y = 0. \tag{63}$$

Our first order of business is to specify high and low order fluxes. Since $u_{i+(1/2),j}$ and $v_{i,j+(1/2)}$ are specified at cell faces, our job reduces to specifying $q_{i+(1/2),j}$ and $q_{i,j+(1/2)}$ at cell faces, and then multiplying them by the appropriate cell face velocity. The low order fluxes, the high order fluxes, and the high order dissipation components are all straightforward generalizations of the fluxes we used in one dimensional linear advection.

For our low order fluxes, we choose a two-dimensional donor cell algorithm:

$$q^L_{i+(1/2),j} = \begin{cases} q_{ij} & \text{when } u_{i+(1/2),j} > 0 \\ q_{i+1,j} & \text{when } u_{i+(1/2),j} \leq 0 \end{cases}, \tag{64}$$

$$q^L_{i,j+(1/2)} = \begin{cases} q_{ij} & \text{when } v_{i,j+(1/2)} > 0 \\ q_{i,j+1} & \text{when } v_{i,j+(1/2)} \leq 0 \end{cases}, \tag{65}$$

$$F^L_{i+(1/2),j} = q^L_{i+(1/2),j} u_{i+(1/2),j} S_{i+(1/2),j} \Delta t^{n+1/2}, \tag{66}$$

$$G^L_{i,j+(1/2)} = q^L_{i,j+(1/2)} v_{i,j+(1/2)} S_{i,j+(1/2)} \Delta t^{n+1/2} \tag{67}$$

where $S_{i+(1/2),j}$ and $S_{i,j+(1/2)}$ are the areas of the x and y cell faces respectively. We note that the above "four-flux" donor cell algorithm does not account for corner transport in a single step. While we do not describe them here, variants of the above do allow corner transport and at the same time satisfy the prime requirement of preventing unphysical values of q. Thus these variants are the preferred low order fluxes for this problem, and are the ones we use here.

The high order fluxes are again computed using the formulae in the Appendix of [13]. As an example, the fourth order fluxes are given by:

$$q^{H4}_{i+(1/2),j} = \frac{7}{12}(q^a_{i+1,j} + q^a_{ij}) - \frac{1}{12}(q^a_{i+2,j} + q^a_{i-1,j}), \tag{68}$$

$$q^{H4}_{i,j+(1/2)} = \frac{7}{12}(q^a_{i,j+1} + q^a_{ij}) - \frac{1}{12}(q^a_{i,j+2} + q^a_{i,j-1}), \tag{69}$$

$$F^{H4}_{i+(1/2),j} = q^{H4}_{i+(1/2),j} u_{i+(1/2),j} S_{i+(1/2),j} \Delta t^{n+1/2}, \tag{70}$$

$$G^{H4}_{i,j+(1/2)} = q^{H4}_{i,j+(1/2)} v_{i,j+(1/2)} S_{i,j+(1/2)} \Delta t^{n+1/2} \tag{71}$$

where the time level t^a is meant to denote whatever time level or average of time levels is required by the particular substep of the particular time discretization chosen.

The high order dissipative fluxes of order N_D, which are added to the above high order fluxes, again follow very closely to their one-dimensional counterparts. As an example, the order 4 dissipative fluxes are given by:

$$F^{D4}_{i+(1/2),j} = -|u_{i+(1/2),j}|\left[\frac{3}{16}(q^n_{i+1,j} - q^n_{ij}) - \frac{1}{16}(q^n_{i+2,j} - q^n_{i-1,j})\right]$$
$$\times S_{i+(1/2),j} \Delta t^{n+1/2}, \tag{72}$$

$$F^{D4}_{i,j+(1/2)} = -|v_{i,j+(1/2)}|\left[\frac{3}{16}(q^n_{i,j+1} - q^n_{ij}) - \frac{1}{16}(q^n_{i,j+2} - q^n_{i,j-1})\right]$$
$$\times S_{i,j+(1/2)} \Delta t^{n+1/2}. \tag{73}$$

The pre-constraint of the high order fluxes is given by Eq. (60). Since we will be limiting directly on the variable q, there is no need for a fail-safe procedure.

For our flux limiter, we choose the multidimensional limiter given above, with q^{max}_{ij} and q^{min}_{ij} specified thus:

$$\begin{aligned} q^+_{ij} &= \max(q^n_{ij}, q^{td}_{ij}), \\ q^{max}_{ij} &= \max(q^+_{i-1,j}, q^+_{i,j}, q^+_{i+1,j}, q^+_{i,j-1}, q^+_{i,j+1}), \\ q^-_{ij} &= \min(q^n_{ij}, q^{td}_{ij}), \\ q^{min}_{ij} &= \min(q^-_{i-1,j}, q^-_{i,j}, q^-_{i+1,j}, q^-_{i,j-1}, q^-_{i,j+1}). \end{aligned} \tag{74}$$

For our test problem we choose the solid body rotation problem given in [13]. We have Eq. (63) with $u = -\Omega(y - y_0)$ and $v = \Omega(x - x_0)$, where Ω is a constant angular velocity, and (x_0, y_0) is the axis of rotation. The computational grid is 100×100 cells, $\Delta x = \Delta y$, with counterclockwise rotation taking place about grid point $(50, 50)$. Centered at grid point $(50, 75)$ is a cylinder of radius 15 grid points, through which a slot has been cut of width 5 grid points. The time step and rotation

Fig. 17 Initial condition. Grid points inside the slotted cylinder have $q = 3.0$. All others have $q = 1.0$. Only the central 50×50 array of grid points around the analytic center of the distribution is shown

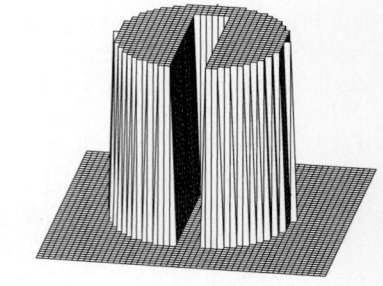

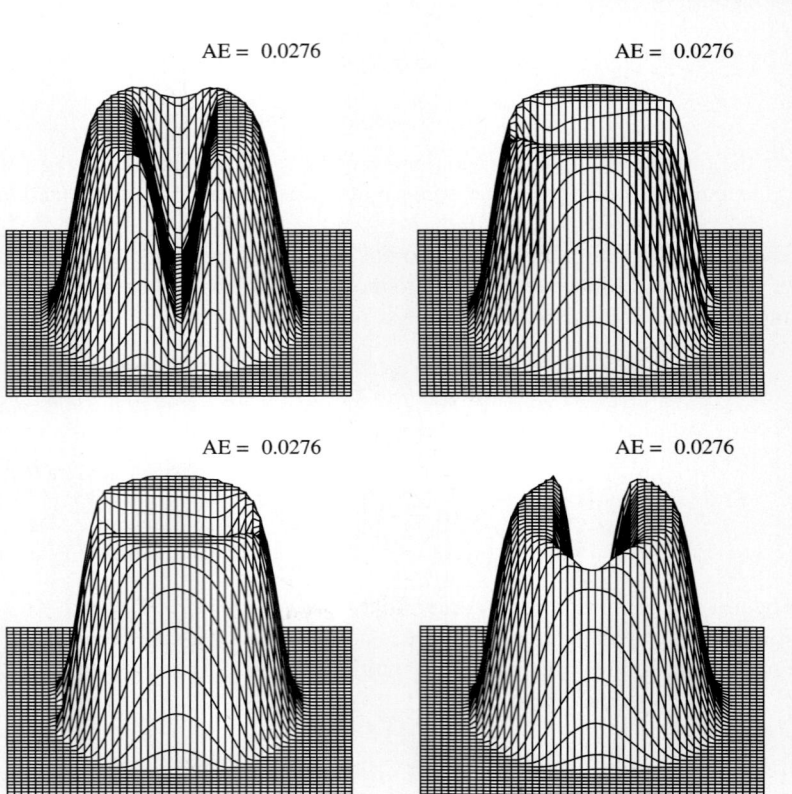

Fig. 18 Results after one revolution with $N = N_D = 4$

speed are such that 1256 time steps will effect one complete revolution of the cylinder about the central point. A perspective view of the initial conditions is shown in Fig. 17. In this and following figures, only the central 50×50 array of grid points around the analytic center of the distribution is shown.

In Fig. 18 we show the results after one revolution of the cylinder about the axis, using $N = N_D = 4$. We show the profile from four different angles, with the L_1 error denoted by "AE." Overall, the FCT algorithm has performed well. Nowhere on the

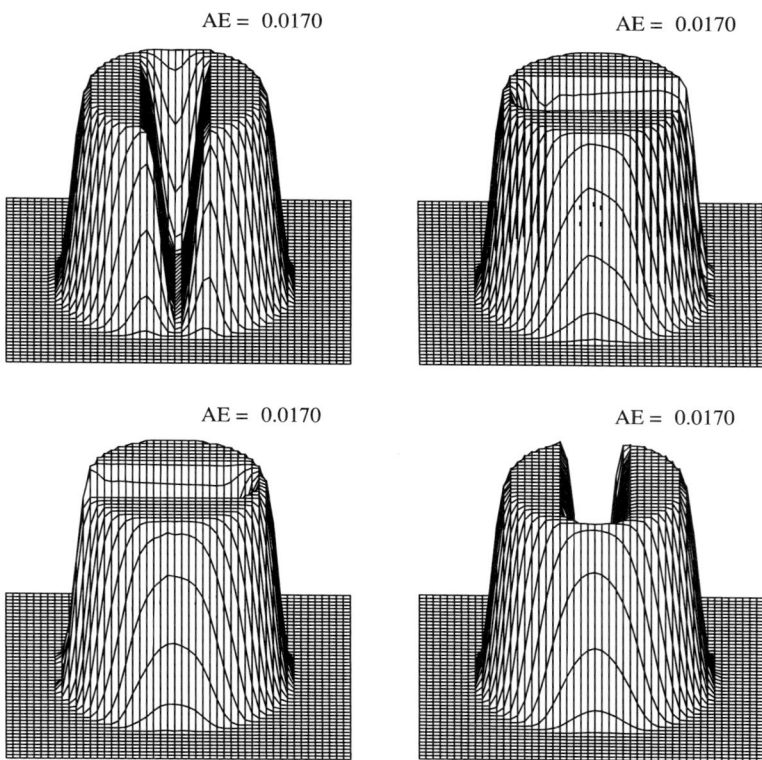

Fig. 19 Results after one revolution with $N = N_D = 8$

grid are there values of q outside the bounds of the analytic result, bounds the high order algorithm would have violated in the very first timestep. Yet the numerical diffusion, although certainly present, is far less than that which would have been generated by the low order algorithm. The top of the cylinder has remained flat and free of oscillations, and kept its original value of 3.0. The flat area outside the cylinder has also remained flat and free of oscillations, and kept its original value of 1.0. The L_1 error is 0.0276. The profile is a bit more diffuse than the fourth order calculation shown in [13]. This can be attributed to the fact that we include a fourth order dissipation term in the high order flux in the present calculation, and did not do so in [13].

In Sect. 4, we found that by increasing the resolving power of the high order fluxes, we could improve the performance of the corresponding FCT algorithm for one-dimensional advection. Let us see if that pattern plays out in multidimensional advection as well. In Fig. 19 we show the results after one revolution of the cylinder about the axis, using $N = N_D = 8$. The L_1 error is 0.0170. The results are clearly quite a bit better than the $N = N_D = 4$ calculation. There is far less erosion in the slot, and the bridge connecting the two halves of the cylinder has maintained its integrity. In Fig. 20 we show the results after one revolution of the cylinder about

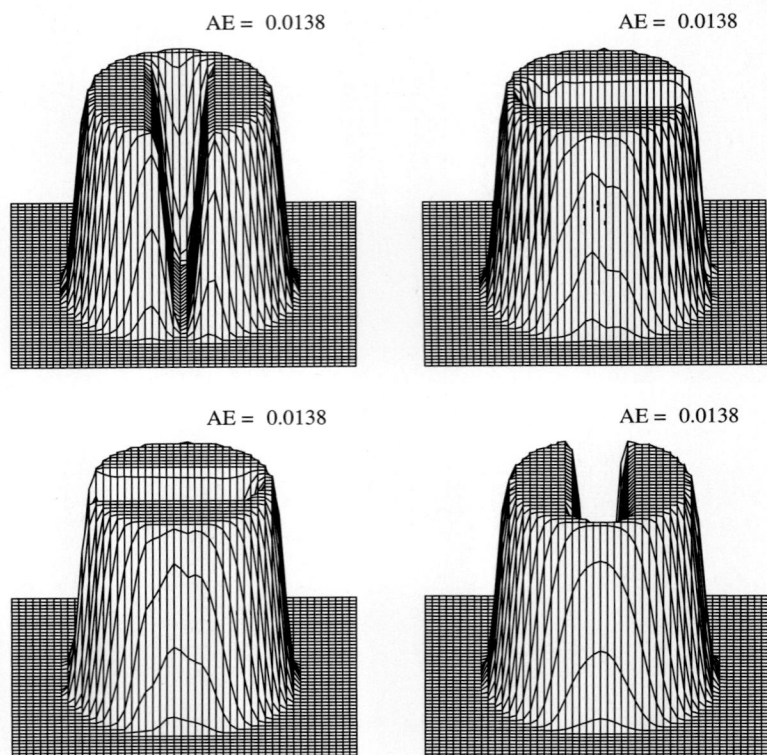

Fig. 20 Results after one revolution with $N = N_D = 16$

the axis, using $N = N_D = 16$. The L_1 error is 0.0138. Again we see a marked improvement with increasing resolving power in the high order flux.

A careful look at Fig. 20 will reveal an aspect of the multidimensional limiter used with the bounds given in Eq. (74) that has been noted both in [13] and more recently by DeVore [6]: Even though this combination of limiter and upper and lower bounds does prevent the occurrence of maxima and minima beyond those bounds, this property is not synonymous with the enforcement of monotonicity in any given coordinate direction. Note in particular the breaking of one-dimensional monotonicity along the front upper portion of the cylinder in the lower left panel of Fig. 20. Such breaking of monotonicity is often, but not always, caused by the development of dispersive ripples due to high order fluxes in one direction which are not seen as errors by the multidimensional limiter due to the presence of a steep gradient in a transverse direction. To address this issue, both [13] and [6] recommended adding a "pre-limiting" step before the multidimensional flux limiter, consisting of a call to the Boris-Book flux limiter for each of the one-dimensional fluxes. That is, prior to the multidimensional flux limiter, $A_{i+(1/2),j}$ is limited with respect to q^{td} in the x-direction, and $A_{i,j+(1/2)}$ is limited with respect to q^{td} in the y-direction, using the Boris-Book limiter. In Fig. 21 we show the results of applying that technique

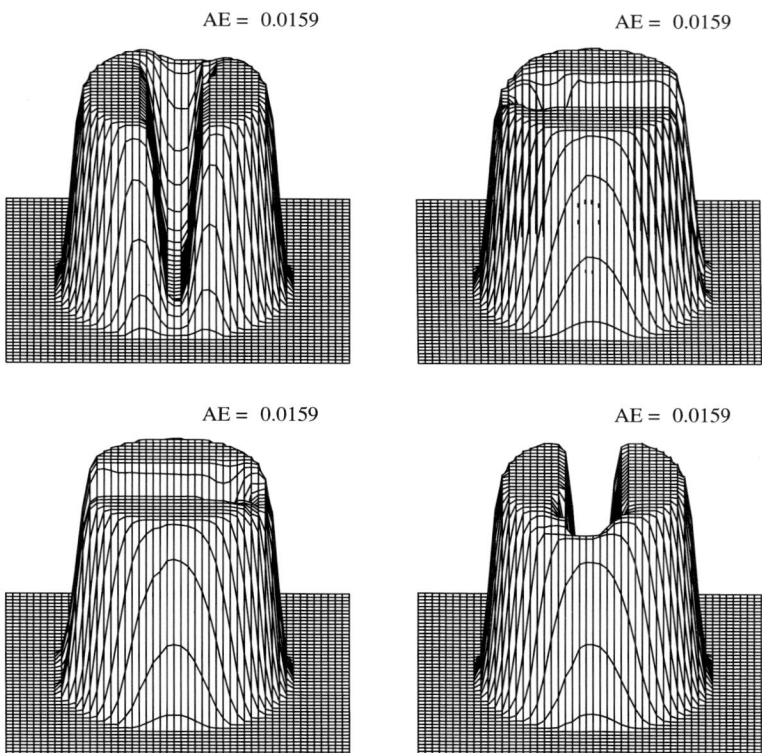

Fig. 21 $N = N_D = 16$ with the Boris-Book pre-limiter

to the $N = N_D = 16$ calculation shown previously in Fig. 20. Although many of the regions of broken monotonicity have been eliminated, the overall solution has been degraded. Significant erosion of the slot and the bridge have taken place, and we now have an L_1 error of 0.0159. This degradation is due primarily to the fact that peaked profiles naturally occur along the outer portions of the cylinder, both initially and as the profile moves and diffuses slightly. These peaked profiles are subsequently "clipped" by the Boris-Book limiter, giving us worse results than if we had not invoked the pre-limiter at all, at least for this problem.

A solution to the above dilemma is to pre-limit using a one-dimensional limiter as above, but to do so using a limiter which does not clip extrema, rather than the Boris-Book limiter. In Fig. 22 we show the results of using a slightly modified version of the non-clipping one-dimensional flux limiter described in Sect. 4.3 to "pre-limit" the $N = N_D = 16$ calculation shown previously in Fig. 20. We see that not only have many of the regions of broken monotonicity vanished, but the overall solution has improved, with an L_1 error of 0.0137. Thus if pre-limiting is deemed advisable, our recommendation is to use non-clipping limiters rather than the Boris-Book limiter to accomplish that task.

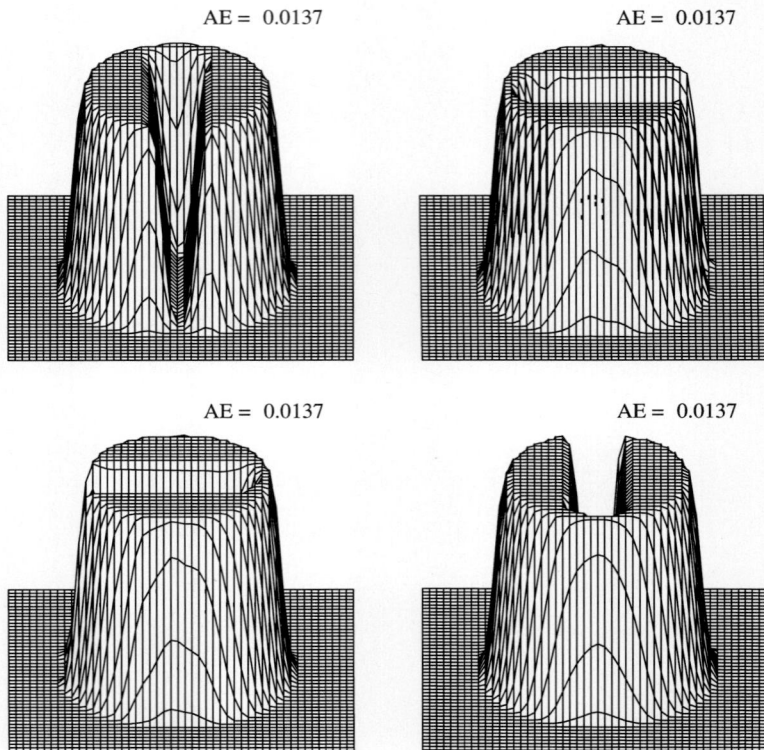

Fig. 22 $N = N_D = 16$ with a non-clipping pre-limiter

6.3 FCT Algorithms for Two-Dimensional Compressible Gas Dynamics

We are interested in solving the equations of two-dimensional compressible inviscid fluid flow Eq. (9). Recall that when we studied the corresponding one-dimensional system, we found a distinct advantage to limiting with respect to the characteristic variables rather than the conserved variables. We also found that we could use the Boris-Book limiter with fairly good success, indicating that clipping was not a serious problem, at least when one uses characteristic variables. Here we will try to build on that success.

We immediately face an apparent problem, however. The characteristic variables are only rigorously defined for one spatial dimension, i.e., it is not possible to simultaneously diagonalize both f and g with the same similarity transformation (for gas dynamics). It is clear, then, that if we wish to limit with respect to characteristic variables, we can only perform flux limiting in one direction at a time. We shall use the characteristic form of the Boris-Book limiter that we developed in Sect. 5.2 for

The Design of Flux-Corrected Transport (FCT) Algorithms

this task, using it in such a manner as to preserve as much full multidimensionality as possible in the algorithm.

The flux limiter we shall use is exactly as described in Sect. 5.2, except that we shall require similarity transformations appropriate for the full set of four conserved variables. The ones we actually use here are those appropriate for three-dimensional gas dynamics, with five conserved variables, with the third component of momentum set to zero. The matrices T and T^{-1} in the x direction are given by

$$T = \begin{bmatrix} 1 & 0 & 0 & 1 & 1 \\ u-c & 0 & 0 & u & u+c \\ v & 1 & 0 & v & v \\ w & 0 & 1 & w & w \\ H-uc & v & w & \frac{1}{2}q^2 & H+uc \end{bmatrix}, \tag{75}$$

$$T^{-1} = \begin{bmatrix} \frac{1}{2}(\frac{\gamma-1}{2}M^2 + \frac{u}{c}) & -\frac{1}{2c} - \frac{(\gamma-1)u}{2c^2} & -\frac{(\gamma-1)v}{2c^2} & -\frac{(\gamma-1)w}{2c^2} & \frac{\gamma-1}{2c^2} \\ -v & 0 & 1 & 0 & 0 \\ -w & 0 & 0 & 1 & 0 \\ 1-\frac{\gamma-1}{2}M^2 & \frac{(\gamma-1)u}{c^2} & \frac{(\gamma-1)v}{c^2} & \frac{(\gamma-1)w}{c^2} & -\frac{\gamma-1}{c^2} \\ \frac{1}{2}(\frac{\gamma-1}{2}M^2 - \frac{u}{c}) & \frac{1}{2c} - \frac{(\gamma-1)u}{2c^2} & -\frac{(\gamma-1)v}{2c^2} & -\frac{(\gamma-1)w}{2c^2} & \frac{\gamma-1}{2c^2} \end{bmatrix} \tag{76}$$

where $q^2 \equiv u^2 + v^2 + w^2$, $M^2 \equiv q^2/c^2$, H is the stagnation enthalpy, and u, v, and w are the x, y, and z components of velocity respectively. For the y direction, a corresponding set of transformation matrices is used.

The easiest solution is, of course, to simply use directional operator splitting. To demonstrate that such a technique is viable, in Fig. 23 we show a calculation using a directionally split version of the $N = 8$, $N_D = 8$ CV FCT algorithm given here to solve the Mach reflection problem given by Woodward and Colella [12]. The problem consists of a Mach 10 shock reflecting from a 30 degree wedge. The three resolutions used in [12] are shown, corresponding to meshes of 120×30, 240×60, and 480×120 from top to bottom. We invite the reader to compare the results to those obtained elsewhere. In Fig. 24 we show the same calculation, but using the conventional non-CV limiter which limits only on the conserved variables. The morphology of the jet along the bottom wall disagrees both with experimental data and with other published numerical calculations. We conclude that the CV limiter used in Fig. 23 is by far the better choice. We also conclude that, for this particular test problem, directional splitting is satisfactory.

Of course one would prefer not to use directional splitting, since one cannot be sure in advance that the particular physics problem of interest will yield satisfactory results when such splitting is employed. Thus we would prefer not to use full-blown directional splitting, and yet the variables which we desire to use for flux limiting would seem to require that the limiting step itself be directionally split. Is there a way to satisfy both requirements? Is there some way to be "fully multidimensional" and also use characteristic variables?

One way to define the term "fully multidimensional algorithm" is to demand that the results be independent of any choice of ordering that may be present in an algorithm. Another, perhaps just as satisfactory, is to demand that same independence,

Fig. 23 Isodensity contours for the Woodward-Colella ramp problem at $t = 0.2$ using a CV limiter with $N = N_D = 8$, and directional splitting. From top to bottom, the displayed grids are 120×30, 240×60, and 480×120

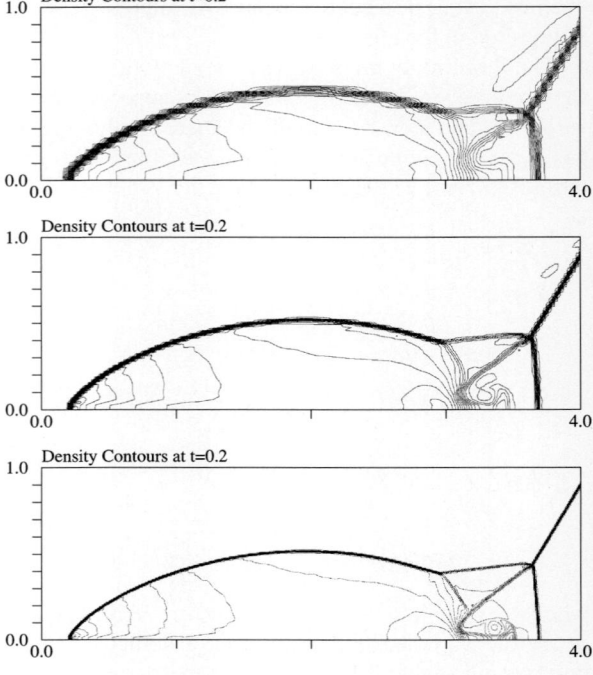

Fig. 24 Same as Fig. 23, but using the conventional flux limiter which limits only on the conserved variables. The morphology of the jet along the bottom wall disagrees both with experimental data and with other published numerical calculations. We conclude that the CV limiter used in Fig. 23 is by far the better choice

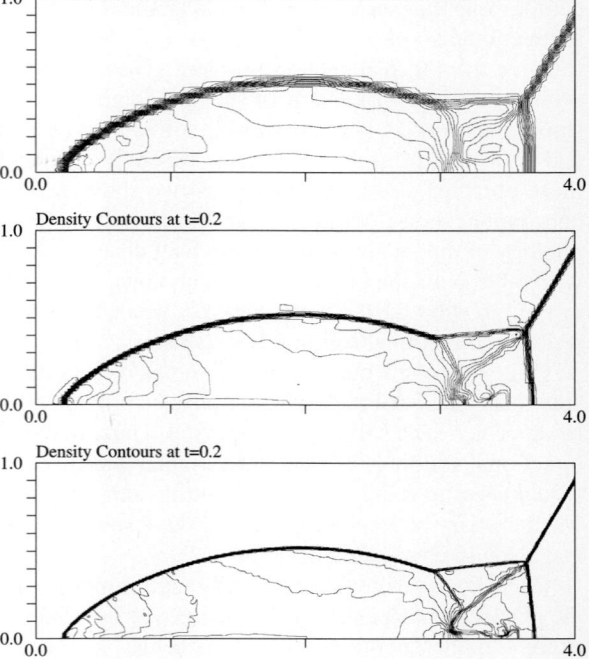

The Design of Flux-Corrected Transport (FCT) Algorithms

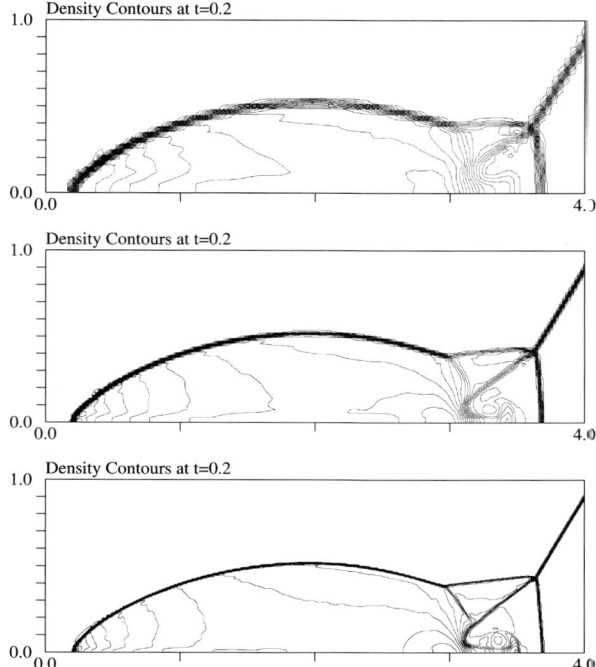

Fig. 25 Same as Fig. 23, but using a CV limiter independently on fully multidimensional antidiffusive fluxes

except for the flux limiting step itself. We use one variant of each here, which we describe in compact form:

1. Compute all high and low order fluxes fully multidimensionally.
2. Either
   - limit the x-, y-, and z-directed fluxes independently; or
   - limit the x-, y-, and z-directed fluxes sequentially, updating solution values between steps.

The first choice increases the risk that the failsafe limiter will be brought into play, but is truly multidimensional. The second is less likely to generate the need for the failsafe limiter, but is multidimensional only in the second sense above. In Fig. 25, we show the results of the Woodward-Colella Mach reflection problem using a CV limiter and limiting the x- and y-directed fluxes independently. In Fig. 26, we show the results using a CV limiter and limiting the x- and y-directed fluxes sequentially. Both are seen to perform quite well, albeit with results that are virtually indistinguishable from those in Fig. 23. Clearly this test problem, although it is the standard test problem for multidimensional compressible flow, is not one for which one needs a fully multidimensional algorithm to achieve accurate solutions. Nonetheless, we would recommend either of the two multidimensional approaches over the fully split one, as a means of avoiding the splitting errors that may occur when simulating more general flows.

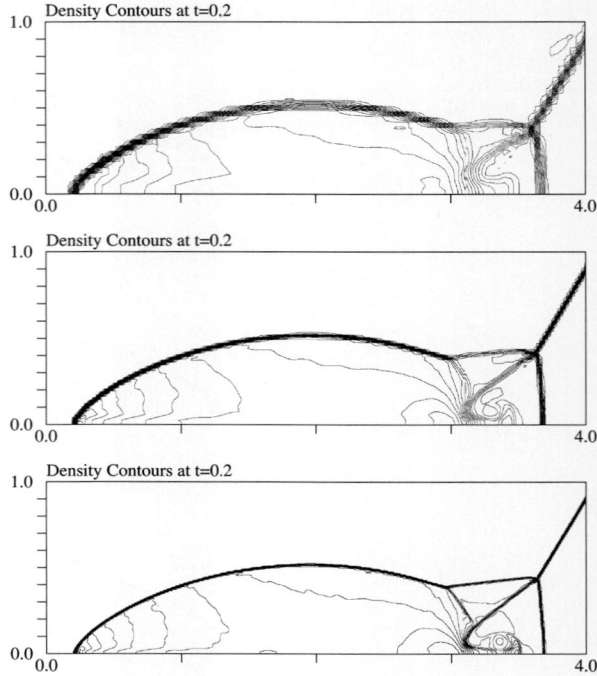

Fig. 26 Same as Fig. 23, but using a CV limiter sequentially on fully multidimensional antidiffusive fluxes

7 Conclusions

We have tried to give the reader a distillation of the design principles for building FCT algorithms, and front-capturing algorithms in general, that we have gleaned from experience over the past several decades. If there is a common thread to all of them it is this: the scientists who use such algorithms must have both input to and knowledge of their design. There may come a day when we no longer hold to this view, when the design of such algorithms can be left to expert numerical analysts alone, but that day has not yet arrived.

References

1. Boris, J.P.: A fluid transport algorithm that works. In: Computing as a Language of Physics, International Atomic Energy Agency, pp. 171–189 (1971)
2. Boris, J.P., Book, D.L.: Flux-Corrected Transport I: SHASTA, a fluid-transport algorithm that works. J. Comput. Phys. **11**, 38–69 (1973)
3. Chorin, A.J.: Random choice solution of hyperbolic systems. J. Comput. Phys. **22**, 517–536 (1976)
4. Chorin, A.J.: Random choice methods with application to reacting gas flow. J. Comput. Phys. **25**, 252–272 (1977)
5. Colella, P., Woodward, P.R.: The Piecewise-Parabolic Method (PPM) for gas-dynamical simulations. J. Comput. Phys. **54**, 174–201 (1984)

6. DeVore, C.R.: An improved limiter for multidimensional flux-corrected transport. NRL Memorandum Report 6440-98-8330, Naval Research Laboratory, Washington, DC (1998)
7. Forester, C.K.: Higher order monotonic convective difference schemes. J. Comput. Phys. **23**, 1–22 (1977)
8. Glimm, J.: Solution in the large for nonlinear hyperbolic systems of equations. Commun. Pure Appl. Math. **18**, 697–715 (1955)
9. Kreiss, H.-O., Oliger, J.: Comparison of accurate methods for the integration of hyperbolic equations. Tellus **24**, 199 (1972)
10. McDonald, B.E.: Flux-corrected pseudospectral method for scalar hyperbolic conservation laws. J. Comput. Phys. **82**, 413 (1989)
11. Sod, G.A.: A survey of several finite difference methods for systems of nonlinear hyperbolic conservation laws. J. Comput. Phys. **27**, 1–31 (1978)
12. Woodward, P.R., Colella, P.: The numerical simulation of two-dimensional flow with strong shocks. J. Comput. Phys. **54**, 115–173 (1984)
13. Zalesak, S.T.: Fully multidimensional Flux-Corrected Transport algorithms for fluids. J. Comput. Phys. **31**, 335–362 (1979)
14. Zalesak, S.T.: Very high order and pseudospectral Flux-Corrected Transport (FCT) algorithms for conservation laws. In: Vichnevetsky, R., Stepleman, R.S. (eds.) Advances in Computer Methods for Partial Differential Equations IV, IMACS, Rutgers University, pp. 126–134 (1981)
15. Zalesak, S.T.: A preliminary comparison of modern shock-capturing schemes: Linear advection. In: Vichnevetsky, R., Stepleman, R.S. (eds.) Advances in Computer Methods for Partial Differential Equations VI, IMACS, Rutgers University, pp. 15–22 (1987)

On Monotonically Integrated Large Eddy Simulation of Turbulent Flows Based on FCT Algorithms

Fernando F. Grinstein and Christer Fureby

Abstract Non-classical Large Eddy Simulation (LES) approaches based on using the unfiltered flow equations instead of the filtered ones have been the subject of considerable interest during the last decade. In the Monotonically Integrated LES (MILES) approach, flux-limiting schemes are used to emulate the characteristic turbulent flow features in the high-wavenumber end of the inertial subrange region. Mathematical and physical aspects of implicit SGS modeling using non-linear flux-limiters are addressed using the modified LES-equation formalism. FCT based MILES performance is demonstrated in selected case studies including (1) canonical flows (homogeneous isotropic turbulence and turbulent channel flows), (2) complex free and wall-bounded flows (rectangular jets and flow past a prolate spheroid), (3) very-complex flows at the frontiers of current unsteady flow simulation capabilities (submarine hydrodynamics).

1 Background

High Reynolds (Re) number turbulent flows are of considerable importance in many fields of engineering, geophysics, and astrophysics. Turbulent flows involve multiscale space/time developing flow physics largely governed by large-scale vortical Coherent Structures (CS's). Typical turbulent energy spectra exhibit a large-wavelength portion dependent on the flow features imposed by geometry and boundary conditions, followed by an intermediate inertial subrange—which becomes longer for higher Re and characterizes the virtually inviscid cascade processes, and then by much-faster decaying portion in the dissipation region (e.g., Sect. 5.1.1 below).

F.F. Grinstein (✉)
X-Computational Physics Division, Los Alamos National Laboratory, Los Alamos, NM 87545, USA
e-mail: fgrinstein@lanl.gov

C. Fureby
Dept. of Weapons and Protection, The Swedish Defence Research Agency—FOI, 172 90, Stockholm, Sweden
e-mail: fureby@foi.se

Capturing the dynamics of all relevant scales based on the numerical solution of the Navier-Stokes Equations (NSE) constitutes Direct Numerical Simulation (DNS), which is prohibitively expensive for practical flows at moderate-to-high Re. On the other end of computer simulation possibilities, the industrial standard is Reynolds-Averaged Navier-Stokes (RANS) modeling, which involves simulating only the mean flow and modeling the effects of the turbulent scales.

Large Eddy Simulation (LES) is an effective intermediate approach between DNS and RANS, capable of simulating flow features which cannot be handled with RANS such as flow unsteadiness and strong vortex-acoustic couplings. Furthermore, LES provides higher accuracy than RANS at reasonable cost but still typically an order of magnitude more expensive. Desirable modeling choices involve selecting an appropriate discretization of the flow problem at hand, such that the LES cutoff lies within the inertial subrange, and ensuring that a smooth transition can be enforced at the cutoff. The main assumptions of LES are that: (i) transport is largely governed by large-scale unsteady features and that such dominant features of the flow can be resolved, (ii) the less-demanding accounting of the small-scale flow features can be undertaken by using suitable Sub Grid Scale (SGS) models.

In the absence of an accepted universal theory of turbulence to solve the problem of SGS modeling, the development and improvement of such models must include the rational use of empirical information. Several strategies to the problem of SGS modeling are being attempted, see e.g., [1], for a recent survey. After more than thirty years of intense research on LES of turbulent flows based on eddy-viscosity models there is now consensus that such approach is subject to fundamental limitations [2]. It has been demonstrated, for a number of flows, that the eigenvectors of the SGS stress and rate-of-strain tensors involved in SGS eddy-viscosity models are not parallel, rendering eddy-viscosity models to be inaccurate.

There have been other proposals that do not employ the assumption of co-linearity of SGS stress and rate-of-strain embedded in the eddy-viscosity models, e.g. the scale-similarity model (SSM) [3] and the Approximate Deconvolution Method (ADM) [4]. Such models may however be numerically unstable, and the more recent efforts have focused on developing mixed models, combining in essence the dissipative eddy-viscosity models with the more accurate but unstable SSM's. The results from such mixed models have been mostly satisfactory but the implementation and computational complexity of these improved combined approaches have limited their popularity. In fact, because of the need to distinctly separate (i.e. resolve) the effects of explicit filtering and SGS reconstruction models from those due to discretization, carrying out such well-resolved LES can typically amount in practice to performing a coarse DNS. As a consequence, it has been argued that the use of hybrid RANS/LES models for realistic whole-domain complex configurations might be unavoidable in the foreseeable future, e.g., [5].

Recognizing the aforementioned difficulties but also motivated by new ideas pioneered at NRL by Boris and collaborators [6, 7], several researchers have abandoned the classical LES formulations and started employing the unfiltered flow equations instead of the filtered ones. Major focus of the new approaches [8, 9] has been on the inviscid inertial-range dynamics and regularization of the under-resolved flow, based on *ab initio* scale separation with additional assumptions for stabilization, or applying monotonicity via non-linear limiters that implicitly act as a filtering mechanism for the small scales—the original proposal of Boris *et al.* [7]. The latter concept goes back to the 50's to von Neumann and Richtmyer [10], who used artificial dissipation to stabilize finite-difference simulations of flows involving shocks. This artificial dissipation concept also motivated Smagorinsky [11] in developing his scalar viscosity concept based upon the principles of similarity in the inertial range of 3D isotropic turbulence. However, the recognition of the more broadly defined implicit LES (ILES) framework is more recent [12]. In ILES, the effects of the SGS physics on the resolved scales are incorporated through functional reconstruction of the convective fluxes using non-oscillatory—but not necessarily monotonic—finite-volume (NFV) algorithms.

In what follows, we use the modified LES equation formalism to carry out a formal comparative analysis of conventional LES and MILES. The performance of MILES is demonstrated for selected representative case studies including canonical flows, moderately complex free and wall-bounded flows, and extremely complex flows at the frontiers of current flow simulation capabilities. We conclude our presentation by addressing fundamental challenges for further development of the concept of nonlinear Implicit LES (ILES).

2 Conventional LES

For simplicity, we restrict the discussion to incompressible flows described by the Navier-Stokes momentum balance equation,

$$\partial_t(\mathbf{v}) + \nabla \cdot (\mathbf{v} \otimes \mathbf{v}) = -\nabla p + \nabla \cdot \mathbf{S}, \tag{1}$$

in conjunction with the incompressibility (or divergence) constraint $\nabla \cdot \mathbf{v} = 0$, where $\otimes$ denotes the tensorial product, and $\mathbf{S} = 2\nu \mathbf{D}$ and $\mathbf{D} = \frac{1}{2}(\nabla \mathbf{v} + \nabla \mathbf{v}^T)$ are the viscous-stress and strain-rate tensors. The conventional LES procedure [1] involves three basic ingredients:

(i) low-pass filtering by the convolution

$$\bar{f}(\mathbf{x},t) = G * f(\mathbf{x},t) = \int_D G(\mathbf{x} - \mathbf{x}', \Delta) f(\mathbf{x}',t) d^3\mathbf{x}',$$

   with a prescribed kernel $G = G(\mathbf{x}, \Delta)$ of width Δ,
(ii) finite volume, element or difference discretization,
(iii) *explicit* SGS modeling to close the low-pass filtered equations.

Applying (i) and (ii), using a second order accurate finite volume algorithm, to (1), and rewriting the results in terms of the *modified equations approach*, i.e., the equation satisfied by the numerical solutions being actually calculated yields [13, 14],

$$\partial_t(\bar{\mathbf{v}}) + \nabla \cdot (\bar{\mathbf{v}} \otimes \bar{\mathbf{v}}) = -\nabla \bar{p} + \nabla \cdot \bar{\mathbf{S}} - \nabla \cdot \mathbf{B} + \mathbf{m}^v + \tau, \tag{2}$$

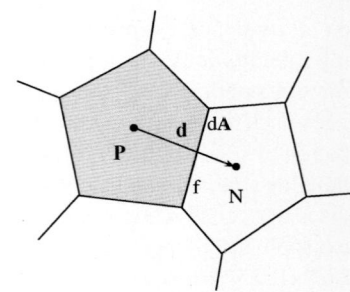

Fig. 1 Grid schematic. P and N denote typical computational cell centers and f an interface; **n** denotes a unit vector normal to the interface, and **A** its area; **d** is the topological vector connecting neighboring cells

where,

$$\mathbf{B} = \overline{\mathbf{v} \otimes \mathbf{v}} - \bar{\mathbf{v}} \otimes \bar{\mathbf{v}}, \qquad \mathbf{m}^v = [G*, \nabla](\mathbf{v} \otimes \mathbf{v} + p\mathbf{I} - \mathbf{S}),$$

$$\tau = \nabla \cdot \left[\left[\frac{1}{6} \nu \nabla^3 \mathbf{v} - \frac{1}{8} \nabla^2 \mathbf{v} \right] (\mathbf{d} \otimes \mathbf{d}) + \cdots \right] \tag{3}$$

are the SGS stress tensor, commutation error term, and the total (convective, temporal and viscous) truncation error, respectively, **I** is the unit tensor, and **d** is the topological vector connecting neighboring control volumes (see Fig. 1), and, $[G*, \nabla]f = \overline{\nabla f} - \nabla \bar{f}$. The commutation error term is often lumped together with the SGS force $\nabla \cdot \mathbf{B}$, prior to modeling, and hence a generalized SGS stress tensor **B** needs to be prescribed in terms of discretized filtered fields for closure of the new equations—which constitutes (iii) above.

Functional modeling consists of the modeling action of the SGS's on the resolved scales. It involves modeling of energetic nature, by which balances of energy are transferred between resolved and subgrid scale ranges, thus accounting for the SGS effects. The energy transfer mechanism from resolved to SGS's is assumed analogous to a Brownian motion superimposed on the large-scale motion. An example of this is the eddy-viscosity approach, in which $\mathbf{B} = -2\nu_k \bar{\mathbf{D}}$ where ν_k is the SGS viscosity—for example, using the Smagorinsky model [11] or the one equation eddy-viscosity model (OEEVM) [15], its principal drawback is the well-established lack of collinearity between **B** and $\bar{\mathbf{D}}$. Natural improvements to these models use anisotropic counterparts based on tensor forms of the SGS turbulent viscosity [16]. These more sophisticated closures involve structural modeling, which attempts to model **B** without incorporating the interactions between SGS and resolved scales. By relying on actual SGS's in the upper resolved subrange—rather than on those modeled through dissipative eddy viscosity—we can better emulate scatter and backscatter, and the modeling procedures won't require assumptions on local isotropy and inertial range. Potential drawbacks arise, however, because structural models are computationally more expensive and typically not dissipative enough; accordingly, mixed models, combined with an eddy-viscosity model, are often used instead.

3 Implicit LES

A key self-consistency issue required in the conventional LES approach involves separating the computing effects of its three basic elements: filtering, discretization, and reconstruction. Filtering and reconstruction contributions must be resolved, i.e., their effective contributions in (2) must be larger than the total truncation error τ. Also, their upper range of represented (but inaccurate) scales interactions must be addressed—in addition to those between resolved and SGS's. Thus, it is useful to examine **B** written in the following way,

$$\mathbf{B} = \overline{\mathbf{v} \otimes \mathbf{v}} - \bar{\mathbf{v}} \otimes \bar{\mathbf{v}} = (\overline{\mathbf{v} \otimes \mathbf{v}} - \overline{\mathbf{v}_P \otimes \mathbf{v}_P}) + (\overline{\mathbf{v}_P \otimes \mathbf{v}_P} - \bar{\mathbf{v}} \otimes \bar{\mathbf{v}}) = \mathbf{B}_1 + \mathbf{B}_2, \quad (4)$$

where $\bar{\mathbf{v}}_P$ denotes the (grid) represented velocity scales, $\mathbf{B}_1$ the interaction between represented and nonrepresented scales—which is not known a priori, and therefore must be modeled—whereas $\mathbf{B}_2$ relates to the interaction between filtered and discretized represented scales, and it can be approximated by prescribing an estimated $\mathbf{v}_P$ in the represented-velocity space (i.e., the solution to the so-called *soft deconvolution problem*) [4]. In this framework, a basic structural SGS model, such as the scale-similarity model, provides $\mathbf{B}_2$, and the eventual need of mixed models results from the recognition that $\mathbf{B}_2$ is not dissipative enough so a secondary regularization through $\mathbf{B}_1$ is needed—i.e., an approximation to **v** in physical-velocity space must be prescribed (the *hard deconvolution problem*).

Traditional approaches, motivated by physical considerations on the energy transfer mechanism from *resolved* to SGS's, express $\mathbf{B}_1$ with an appropriately functional model (for example, an eddy-viscosity SGS model), and seek sufficiently high-order discretization and grid resolution to ensure that effects due to τ are sufficiently small. However, we could argue that discretization could implicitly provide $\mathbf{B}_1$ if nonlinear stabilization can be achieved algorithmically via a particular class of numerical algorithms or based on regularizing the discretization of the conservation laws. In fact, (2) suggests that most schemes can potentially provide built-in or implicit SGS models enforced by the discretization errors τ, provided that their leading order terms are dissipative. We are thus led to the natural question: To what extent can we avoid the (explicit) filtering and modeling phases of LES (i.e., $\mathbf{B}_2 \equiv \mathbf{0}$ and $\mathbf{m}^v \equiv \mathbf{0}$) and focus on the implicit $\mathbf{B}_1$ provided by a suitably chosen discretization scheme?

Not all implicitly implemented SGS models are expected to work: good or bad SGS physics can be built into the simulation model depending on the choice of numerics and its particular implementation. Moreover, the numerical scheme has to be constructed such that the leading order truncation errors satisfy physically required SGS properties, and hence non-linear discretization procedures will here be required. The analogy to be recalled is that of shock-capturing schemes designed under the requirements of convergence to weak solution while satisfying the entropy condition [17].

4 Monotonically Integrated LES (MILES)

The relevancy of NFV algorithms for ILES of turbulent flows have been motivated [14, 18] by proposing to focus on two distinct inherent physical SGS features to be emulated:

- the anisotropy of high-Re turbulent flows in the high-wave-number end of the inertial subrange region, characterized by very thin filaments of intense vorticity and largely irrelevant internal structure, embedded in a background of weak vorticity, e.g., [19],
- the particular (discrete) nature of laboratory observables (only finite fluid portions transported over finite periods of time can be measured) [18].

We thus require that ILES be based on NFV numerics having a *sharp velocity-gradient capturing capability* operating at the smallest resolved scales. By focusing on the inviscid inertial-range dynamics and on adaptive regularization of the under-resolved flow, ILES thus follows very naturally on the historical precedent of using this kind of schemes for shock capturing—in the sense that requiring emulation (near the cutoff) of the high wavenumber-end features of the inertial subrange region of turbulent flows is analogous to spreading the shock width to the point that it can be resolved by the grid.

Although the history of ILES draws on the development of shock-capturing schemes, the MILES concept—as originally introduced by Boris and his colleagues [7] and further developed in our previous work [13, 14]—embodies a computational procedure for solving the NSE as accurately as possible by using a particular class of flux-limiting schemes and their associated built-in (or implicit) SGS models. An intriguing MILES feature is the convection discretization that implicitly generates a nonlinear tensor-valued eddy-viscosity, which acts predominantly to stabilize the flow and suppress unphysical oscillations.

MILES draws on the fact that FV methods filter the NSE over nonoverlapping computational cells Ω_P, with the typical dimension $|\mathbf{d}|$—using a top-hat-shaped kernel, $f_P = \frac{1}{\delta V_P} \int_{\Omega_P} f dV$. In the finite-volume context, discretized equations are obtained from the NSE using Gauss's theorem and by integrating over time with a multistep method parametrized by m, α_i, and β_i,

$$\begin{cases} \dfrac{\beta_i \Delta t}{\delta V_P} \sum_f [F_f^{C,\rho}]^{n+i} = 0, \\ \sum_{i=0}^{m} \left(\alpha_i (\mathbf{v})_P^{n+i} + \dfrac{\beta_i \Delta t}{\delta V_P} \sum_f [F_f^{C,\rho} \mathbf{v}_f + \mathbf{F}_f^{D,v}]^{n+i} + \beta_i (\nabla p)_P^{n+i} \Delta t \right) = \mathbf{0}, \end{cases} \quad (5)$$

where α, β and m are parameters of the scheme, and $F_f^{C,\rho} = (\mathbf{v} \cdot d\mathbf{A})_f$ and $\mathbf{F}_f^{C,v} = F_f^{C,\rho} \mathbf{v}_f$ are the convective and $\mathbf{F}_f^{D,v} = (\nu \nabla \mathbf{v})_f d\mathbf{A}$ the viscous fluxes. To complete the discretization, all fluxes at face 'f' need to be reconstructed from the dependent variables at adjacent cells. This requires flux interpolation for the convective fluxes and difference approximations for the inner derivatives in the viscous fluxes.

For conventional LES, it is appropriate to use linear (or cubic) interpolation for the convective fluxes and central difference approximations for the inner gradients in the viscous fluxes. This then results in a cell-centered second- or fourth-order accurate scheme. Scheme stability can be enforced not only by conserving momentum, but also kinetic energy, which ensures robustness without numerical dissipation (which compromises accuracy).

Given (5), the methods available for constructing implicit SGS models by means of the leading order truncation errors are generally restricted to nonlinear high-resolution methods for the convective flux $\mathbf{F}_f^{C,v}$ to maintain second-order accuracy in smooth regions of the flow (such high-resolution methods are at least second-order accurate on smooth solutions while giving well-resolved, non-oscillatory discontinuities) [17]. In addition, these schemes are required to provide a leading order truncation error that vanishes as $\mathbf{d} \to \mathbf{0}$ so that it remains consistent with the NSE and the conventional LES model. We focus here on the certain flux-limiting and correcting methods.

To this end, we introduce a flux-limiter Γ that combines a high-order convective flux-function $\mathbf{v}_f^H$ which is well-behaved in smooth flow regions, with a low-order dispersion-free flux-function $\mathbf{v}_f^L$, being well-behaved near sharp gradients, so that the total flux-function becomes $\mathbf{v}_f = \mathbf{v}_f^H - (1 - \Gamma)[\mathbf{v}_f^H - \mathbf{v}_f^L]$. Choosing the particular flux limiting scheme also involves specific selections for $\mathbf{v}_f^L$ and $\mathbf{v}_f^H$. In the analysis that follows, $\mathbf{v}_f^H$ and $\mathbf{v}_f^L$ are assumed to be based on linear interpolation, and upwind-biased piecewise constant approximation, respectively, e.g.,

$$\begin{cases} \mathbf{F}_f^{C,v,H} = F_f^{C,\rho}\left[\ell \mathbf{v}_P + (1-\ell)\mathbf{v}_N - \frac{1}{8}(\mathbf{d} \otimes \mathbf{d})\nabla^2 \mathbf{v} + O(|\mathbf{d}|^3)\right], \\ \mathbf{F}_f^{C,v,L} = F_f^{C,\rho}\left[\beta^+ \mathbf{v}_P + \beta^- \mathbf{v}_N + (\beta^+ - \beta^-)(\nabla \mathbf{v})\mathbf{d} + O(|\mathbf{d}|^2)\right], \end{cases} \quad (6)$$

where $\beta^\pm = \frac{1}{2}(\mathbf{v}_f \cdot d\mathbf{A} \pm |\mathbf{v}_f \cdot d\mathbf{A}|)/|\mathbf{v}_f \cdot d\mathbf{A}|$, and $-\frac{1}{8}(\mathbf{d} \otimes \mathbf{d})\nabla^2 \mathbf{v}$ and $(\beta^+ - \beta^-)(\nabla \mathbf{v})\mathbf{d}$ are the leading order truncation errors. The flux limiter Γ is to be formulated as to allow as much as possible of the correction $[\mathbf{v}_f^H - \mathbf{v}_f^L]$ to be included without increasing the variation of the solution—e.g., to comply with the physical principles of causality, monotonicity and positivity [7] (when applicable) and thus to preserve the properties of the NSE. To see the effects of this particular convection discretization we consider the modified equations corresponding to the semi-discretized equations (5) with the flux-limiting functions in (6) being used for the convective fluxes,

$$\partial_t(\mathbf{v}) + \nabla \cdot (\mathbf{v} \otimes \mathbf{v}) = -\nabla p + \nabla \cdot \mathbf{S} + \nabla \cdot \left[\mathbf{C}(\nabla \mathbf{v})^T + (\nabla \mathbf{v})\mathbf{C}^T \right.$$
$$\left. + \chi^2 (\nabla \mathbf{v})\mathbf{d} \otimes (\nabla \mathbf{v})\mathbf{d} + \left[\frac{1}{6}v\nabla^3 \mathbf{v} - \frac{1}{8}\nabla^2 \mathbf{v}\right](\mathbf{d} \otimes \mathbf{d}) + \cdots \right], \quad (7)$$

with $\nabla \cdot \mathbf{v} = 0$, and where $\mathbf{C} = \chi(\mathbf{v} \otimes \mathbf{d})$ and $\chi = \frac{1}{2}(1 - \Gamma)(\beta^- - \beta^+)$. In particular, we note that in smooth regions, $\Gamma = 1$ implies that $\chi = 0$ and $\mathbf{C} = \mathbf{0}$, and the leading order truncation error becomes $\tau = \nabla \cdot [[\frac{1}{6}v\nabla^3 \mathbf{v} - \frac{1}{8}\nabla^2 \mathbf{v}](\mathbf{d} \otimes \mathbf{d})]$. Comparing with the analysis of the momentum equation in the framework of the conventional LES approach (equation (2)) suggests that the MILES modified equation incorporates additional dissipative and dispersive terms, and we can consistently identify the implicit SGS stress term,

$$\mathbf{B} = \mathbf{C}(\nabla \mathbf{v})^T + (\nabla \mathbf{v})\mathbf{C}^T + \chi^2 (\nabla \mathbf{v})\mathbf{d} \otimes (\nabla \mathbf{v})\mathbf{d}. \tag{8}$$

The implicit SGS stress tensor can according to (8) be decomposed into $\mathbf{B}^{(1)} = \mathbf{C}(\nabla \mathbf{v})^T + (\nabla \mathbf{v})\mathbf{C}^T$ and $\mathbf{B}^{(2)} = \chi^2 (\nabla \mathbf{v})\mathbf{d} \otimes (\nabla \mathbf{v})\mathbf{d}$, in which the former is a tensor-valued eddy-viscosity model, while the latter is of a form similar to the scale similarity model. The decomposition in (8) can also be interpreted as breaking $\mathbf{B}$ into its slow and rapid varying parts—relative to the time scale of its response to variations in the mean flow [20]. In MILES, the rapid part that cannot be captured by isotropic models relates to $\mathbf{B}^{(2)}$, while the slow part relates to $\mathbf{B}^{(1)}$. Borue and Orszag [21] have shown that a $\mathbf{B}^{(2)}$ type term improves the correlations between the exact and modeled SGS stress tensor. A closely-related view further explaining the effectiveness of ILES formulations based on local monotonicity (or sign) preservation concepts has been given by Margolin and Rider [18]; they argued that the leading order truncation error introduced by NFV algorithms represents a physical flow regularization term, providing necessary modifications to the governing equations that arise when the motion of *observables*—finite volumes of fluid convected over finite intervals of time—is considered.

Detailed properties of the implicit SGS model are related to the flux limiter Γ and to the choice of low- and high-order schemes; they also relate as well to other specific features of the scheme—e.g., such as monotonicity, l_1-contraction, local monotonicity preservation, and griding. We have illustrated above in (8) and discussed elsewhere [13, 14] how some of these properties can directly affect the implicit SGS modeling effectiveness in the MILES context. MILES performance as a function of flux limiter is discussed further below; dependence on the choice of low order scheme has been examined in Ref. [22].

In what follows we address effects of variations in the flux-limiter Γ. To this end we consider first high-resolution schemes that can be formulated using the ratio of consecutive gradients,

$$r = \frac{\delta \mathbf{v}_{P-1/2}^n}{\delta \mathbf{v}_{P+1/2}^n} = \frac{(\mathbf{v}_P^n - \mathbf{v}_{P-1}^n)}{(\mathbf{v}_{P+1}^n - \mathbf{v}_P^n)}.$$

Examples of well-known flux-limiters that fit into this category are:

1. the minmod flux-limiter of Roe, e.g. [23], with

$$\Gamma = \max(0, \min(1, r)),$$

2. the van-Leer flux-limiter, e.g. [23], with

$$\Gamma = \frac{r + |r|}{1 + |r|},$$

3. the superbee flux-limiter, Roe, e.g. [23], with

$$\Gamma = \max(0, \max(\min(2r, 1), \min(r, 2))),$$

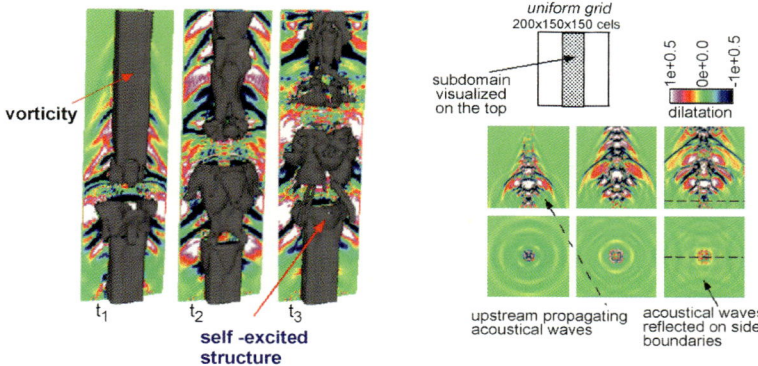

Fig. 6 MILES studies of global instabilities in a countercurrent supersonic cold square jet in terms of instantaneous visualizations [40]

- The numerical algorithm's dispersiveness should be minimized to ensure good modeling of the acoustical propagation properties of the small wavelengths.
- Because of the very small energy of the acoustic field compared to that of the flow field, there is a potential for spurious sound sources due to numerical discretization.

Because of the tensorial nature of its implicit SGS model, and the inherently low numerical diffusion involved, the use of flux limiting in MILES offers an overall effective computational alternative to conventional SGS models in this context. MILES was used to extensively investigate the natural mechanisms of transition to turbulence in rectangular jets evolving from laminar conditions [41], in compressible (subsonic) jet regimes with aspect ratio $AR = 1$ to 4 and moderately high Re. The studies demonstrated qualitatively different dynamical vorticity geometries characterizing the near jet, involving

- self-deforming and splitting vortex rings,
- interacting ring and braid (rib) vortices, including single ribs aligned with corner regions ($AR \geq 2$), and rib pairs (hairpins) aligned with the corners ($AR = 1$), and,
- a more disorganized flow regime in the far jet downstream, where the rotational fluid volume is occupied by a relatively weak vorticity background with strong, slender tube-like filament vortices filling a small fraction of the domain.

Figure 7(a) illustrates characteristic axis-switching and bifurcation phenomena from visualizations of laboratory elliptical jets subject to strong excitation at the preferred mode [42, 43]. We compare it to the carefully developed simulation results (see Figs. 7(b) and 7(c)) designed to address unresolved issues in vortex dynamics. Detailed key aspects—namely, reconnection, bridging, and threading (see Fig. 7(b))—could not be captured in the laboratory studies and were first demonstrated by the simulations.

Jet flows develop in different possible ways, depending on

- their particular initial conditions,

Fig. 7 Vortex dynamics and transition to turbulence in subsonic noncircular jets; (**a**) laboratory studies [42, 43], (**b**) and (**c**) detailed vortex dynamics elucidated by simulations [41]

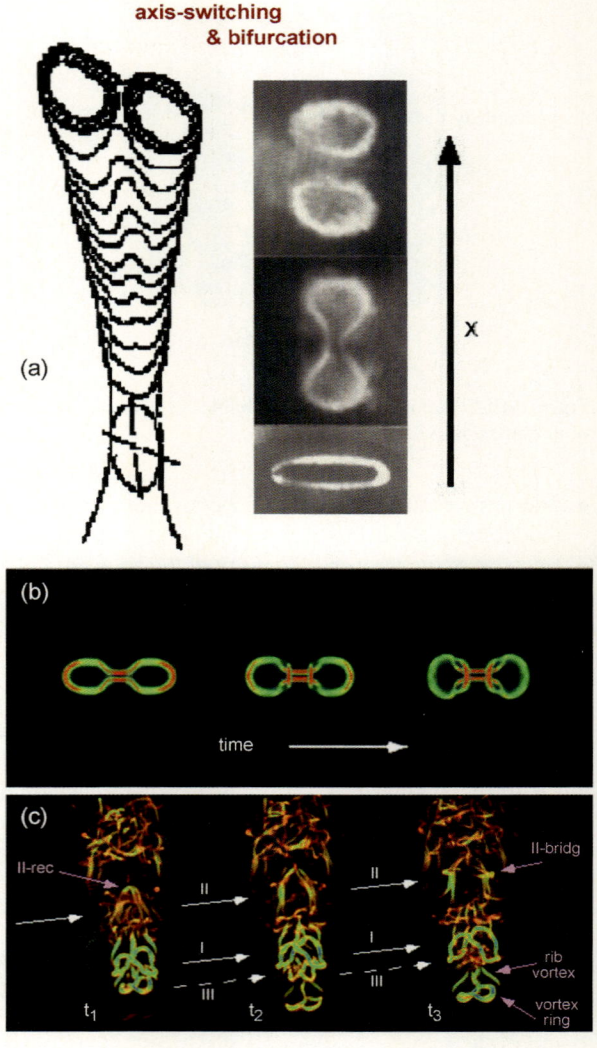

- nozzle geometry and modifications introduced at the jet exit,
- the types of unsteady vortex interactions initiated, and
- local transitions from convectively to absolutely unstable flow.

Taking advantage of these flow control possibilities is of interest to improve the mixing of a jet, or plume, with its surroundings in practical applications demanding:

- enhanced combustion between injected fuel and background oxidizer,
- rapid initial mixing and submergence of effluent fluid,
- less intense jet noise radiation,
- reduced infrared plume signature.

PROPANE TURBULENT DIFFUSION FLAMES
temperature distributions superimposed on vorticity isosurfaces

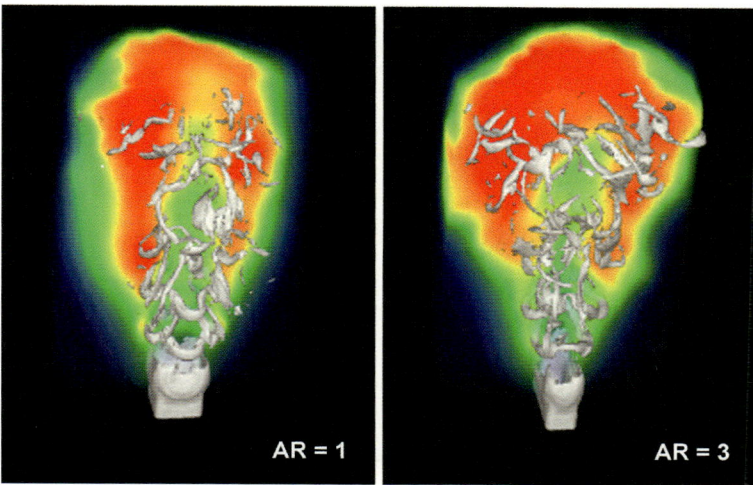

Fig. 8 Visualizations of non-premixed combustion regions as a function of aspect ratio [41]. Temperature distributions (*color*) in the back half of the visualized subvolume are superimposed to isosurfaces of the vorticity magnitude (*gray*)

For example, the jet entrainment rate—the rate at which fluid from the jet becomes entangled or mixed with that from its surroundings—can be largely determined by the characteristic rib-ring coupling geometry and the vortex-ring axis-switching times (see Fig. 8) [40].

5.3 Moderately Complex Geometry: Flow Over a Prolate Spheroid

Crucial additional issues of LES of inhomogeneous high-Re flows to be addressed relate to boundary condition (supergrid) modeling and overall computational model validation [42, 43]. From the practical point of view, it is of utmost importance to consider how the non-linear combination of all—algorithmic, physics-based, SGS, and supergrid—aspects of the model affect the simulation of complex systems for which detailed DNS-type approaches are not possible and for which only limited experimental data might be available at best.

Despite its simple geometry, the flow around a prolate spheroid at an incidence (see Fig. 6(a)) contains a rich gallery of complex 3D flow features. These include:

- stagnation flow,
- 3D boundary layers under influence of pressure gradients and streamline curvature
- cross-flow separation, and
- the formation of free vortex sheets producing streamwise vortices.

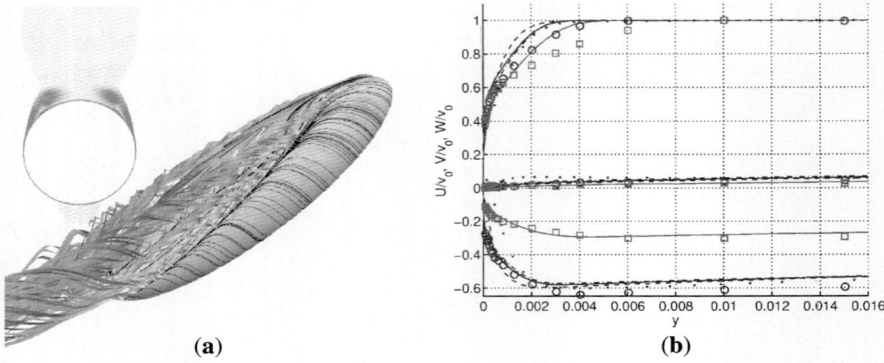

Fig. 9 Flow around a prolate spheroid: (**a**) perspective view and contours of the vorticity magnitude $|\omega|$ at $x/L = 0.772$; (**b**) velocity comparison at $x/L = 0.772$ and $\varphi = 60°$ between experimental ($\bigcirc$ and $\square$) and predicted data at $\alpha = 10°$ (*black*) and $\alpha = 20°$ (*gray*). (—) OEEVM+WM on grid A, (- - -) MILES+WM on grid A, (-.-.) LDKM on grid A, (· · · ·) LDKM on grid B and (· · ·) DES on grid A

These features are archetypes of flows over more complicated airborne and underwater vehicles warranting in-depth study. Previously [42, 43] we studied the flow around a 6 : 1 prolate spheroid mounted in a wind tunnel with a rectangular cross-section [44] at $\alpha = 10°$ and $20°$ angles of attack. Based on the free-stream velocity v_0 and the body length L, the Re number is $Re_L = 4.2 \cdot 10^6$. The domain is discretized with a block-structured mesh, supported by a double O-shaped block structure. Two meshes are used in order to parameterize effects of the grid on the boundary layer resolution. Mesh A has $0.75 \cdot 10^6$ cells and $y^+ \approx 25$ and mesh B has $1.50 \cdot 10^6$ cells with $y^+ \approx 5$. At the inlet, $\bar{\mathbf{v}} = v_0 \mathbf{n}$ and $\partial \bar{p}/\partial n = 0$, where $\mathbf{n}$ is the outward pointing unit normal, and at the outlet $\bar{p} = p_\infty$ and $\partial(\bar{\mathbf{v}} \cdot \mathbf{n})/\partial n = 0$. On the body, no-slip conditions are used.

Figure 9(a) shows perspective views from the port side of the prolate spheroid at $\alpha = 20°$. The flow is represented by surface streamlines, stream-ribbons, and contours of the vorticity magnitude $|\bar{\omega}|$ at $x/L = 0.772$, where $\bar{\omega} = \frac{1}{2}\nabla \times \bar{\mathbf{v}}$ is the vorticity. The stream-ribbons show the complexity of the flow. On the windward side, an attached 3D boundary layer is formed, while on the leeward side, the flow detaches from the hull—because of the circumferentially adverse pressure gradient, and rolls up into a counterrotating pair of longitudinal spiraling vortices on the back of the body. Furthermore, fluid from the windward side is advected across the spheroid, engulfed into the primary vortices and subsequently ejected into the wake.

Figure 9(b) shows the time-averaged velocity components (U, V, W) at $x/L = 0.772$ and at $\varphi = 90°$. The velocity components are presented in the body-surface coordinate system [44]. For V and W, we see good agreement between predictions and measurements for all models—with DES providing the least accurate comparison. We obtained the best agreements with OEEVM and MILES with a wall-model [38] on grid A (OEEVM+WM and MILES+WM). Concerning U, we found significant differences as a function of the various models and grid resolutions. We found best agreements for MILES+WM and OEEVM+WM, whereas the LDKM and DES predictions show larger deviations from the experimental data. The LDKM appears to require better resolution than what we have provided because it underpredicts the boundary layer thickness. The results from MILES+WM and OEEVM+WM appear virtually unaffected by resolution, which is expected because the wall model is designed to take care of the errors introduced by poor resolution in the boundary layer. Also interesting is that the effects of changing the angle of attack α—very important when studying, for example, maneuvering—are very well reproduced in the simulations.

5.4 Challenging New Role of Simulations

For the studies of submarine hydrodynamics and flows in urban areas discussed separately in this volume [22], it is unlikely that we will ever have a deterministic predictive framework based on computational fluid dynamics. This is due to the inherent difficulty in modeling and validating all the relevant physical sub-processes and acquiring all the necessary and relevant boundary condition information. On the other hand, these cases are representative of very fundamental ones for which whole-domain scalable laboratory studies are impossible or very difficult, but for which it is also crucial to develop predictability.

5.4.1 Submarine Hydrodynamics

The flow around a submarine is extremely complicated and characterized by very high Re, $O(10^9)$. Full-scale experiments are complicated and very expensive and are of limited value due to the difficult measurement settings. RANS of full-scale submarine hydrodynamics are barely within reach, whereas LES is currently out of reach due to the wide range of scales present. For model-scale situations ($Re \approx 10^7$), it might be possible to conduct LES and DES [45]. In particular, if we're interested in vortex dynamics, flow noise, and the coupling between the propeller dynamics and the flow around the hull, LES and DES are our only alternatives for the foreseeable future.

As Fig. 7(a) shows, each appendage generates a wake and several vortex systems. A horseshoe-vortex pair is formed in the junction between the hull and the appendage, whereas a tip-vortex pair is formed at the tip of the appendage. Additional vortex systems can be formed, e.g., on the side of the sail towards the trailing edge or in the boundary layer of the tapered sections of the hull. These vortex systems can interact with each other and with the (unsteady) boundary layer to form a very complex flow entering the propeller, thus causing vibrations and noise. In addition, the ocean water is usually stratified, with density variations caused by differences in temperature and salinity between ocean currents, or between the surface water and deeper water. Stratification influences the turbulence and the large flow

Fig. 10 Submarine hydrodynamics: the main flow features represented by stream-ribbons and contour plots of the vorticity magnitude in three cross-sections

structures in the wake, typically resulting in horizontally flattened flow structures (so-called pancake vortices), which would not occur in nonstratified waters.

The case discussed here is the fully appended DARPA Suboff configuration [46] constructed from analytical surfaces, and shown in Fig. 10. Experimental data, using hot-film techniques, are provided at $Re = 12 \cdot 10^6$ based on the overall hull length L, the free-stream velocity u_0 and ν [47]. The total measurement uncertainty in the velocity data—i.e., the geometrical mean of the bias and precision errors, is estimated to be about 2.5% of u_0. The computational domain consists of the submarine model mounted in a cylinder having the same hydraulic diameter as the wind tunnel used in the scale model experiments. The cylinder extends one hull-length upstream and two hull-lengths downstream, thus being $4L$ in overall length. For the hull an O-O topology is used, while for the sail and stern appendages C-O topologies are used and care is taken to ensure that the cell spacings and aspect ratios are suitable for capturing the boundary layers along the hull.

Typically, about 20 cells are contained within the thickness of the boundary layer on the parallel midsection of the hull, having a typical wall distance for the first cell $y^+ \approx 8$. Two grids of about $3 \cdot 10^6$ and $6 \cdot 10^6$ nodes were used. At the inlet boundary, $\bar{\mathbf{v}} = u_0\mathbf{n}$ and $(\nabla \bar{p} \cdot \mathbf{n}) = 0$, at the outlet $\bar{p} = p_0$ and $(\nabla \bar{\mathbf{v}}) \cdot \mathbf{n} = \mathbf{0}$, whereas free-slip conditions are used at the wind-tunnel walls, and no-slip conditions are used on the hull. All LES are initiated with quiescent conditions and the unsteady flow in LES is allowed to evolve naturally (i.e., without any external forcing).

In Figs. 11(a) and 11(b) we show typical comparisons between predictions of towed and self-propelled cases and experimental data [47] of the distribution of the time-averaged static pressure coefficient $C_P = 2(\langle \bar{p} \rangle - p_0)/u_0^2$ along the meridian line of hull and of the circumferentially averaged velocity in the propeller plane. Very good agreement between the measurement data and the computations is ob-

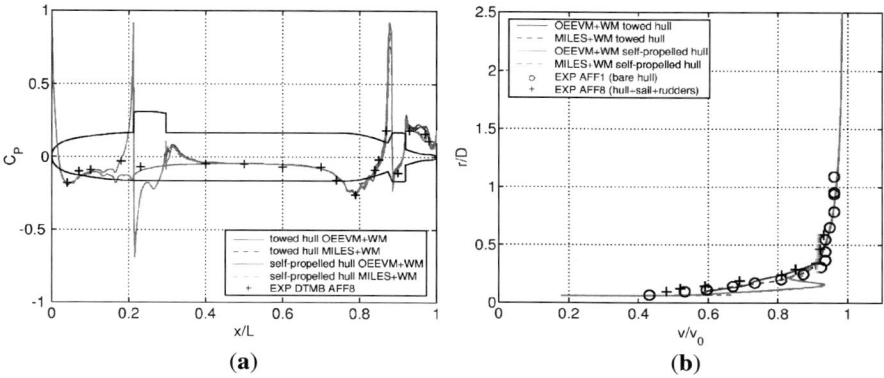

Fig. 11 Submarine hydrodynamics: comparison of the mean pressure and mean velocity: (**a**) along the meridian line of the hull; (**b**) in the propeller plane

served along the entire hull section for the towed case. Virtually no differences in the C_P distribution can be observed between the towed and the self-propelled cases—with the exception of the far-end of the tapered section of the stern, nor do we see significant differences between the MILES+WM and LDKM predictions. Concerning the velocity distributions, the differences are attributed to the presence of the propeller (or rather the actuator-disc used to model the effects of the propeller), and show the effects of the axial pressure gradient, as implicitly imposed by the propeller causing a suction effect along the stern part of the hull. Based on the secondary velocity vector field (not shown) the location of the horseshoe-vortex pair is estimated in the case of the towed case from predictions (measurements) to be at $r/R \approx 0.41$ (0.38) and $\varphi \approx \pm 23°$ ($\pm 22°$), respectively.

6 Outlook

In the absence of an accepted universal theory of turbulence, the development and improvement of SGS models are unavoidably pragmatic and based on the rational use of empirical information. Classical approaches have included many proposals ranging from inherently limited eddy-viscosity formulations to more sophisticated and accurate mixed models. The main drawback of mixed models relates to their computational complexity, and ultimately, to the fact that well-resolved (discretization-independent) LES is prohibitively expensive for the practical flows of interest at moderate-to-high Re. This has recently led many researchers to abandon the classical LES formulations, shifting their focus directly to the SGS modeling implicitly provided by nonlinear stabilization achieved algorithmically, through the use of a particular class of numerical schemes, or based on regularization of the discretization of conservation laws.

In ILES (MILES), the effects of SGS physics on the resolved scales are incorporated in the functional reconstruction of the convective fluxes using NFV methods. Analysis based on the modified equations shows that ILES provides implicitly implemented anisotropic SGS models dependent on the specifics of the particular numerical scheme—i.e., on the flux limiter, on the choice of low- and high-order schemes, and on the griding. By focusing on the inviscid inertial-range dynamics and on regularization of the underresolved flow, ILES follows up very naturally on the historical precedent of using this kind of numerical schemes for shock capturing. Challenges for ILES include constructing a common appropriate mathematical and physical framework for its analysis and development, further understanding the connections between implicit SGS model and numerical schemes, and, in particular, how to address building physics into the numerical scheme to improve global ILES performance, i.e., on the implicitly-implemented SGS dissipation & backscatter features. Moreover, additional (explicit) SGS modeling might be needed to address inherently small-scale physical phenomena such as scalar mixing and combustion—which are actually outside the realm of any LES approach: how do we exploit the implicit SGS modeling provided by the numerics, to build "efficient mixed" (explicit/implicit) SGS models?

Acknowledgements This work was completed while one of us (FFG) was the 2003–2004 Orson Anderson Distinguished Visiting Scholar at the Los Alamos National Laboratory, on Sabbatical leave from the Naval Research Laboratory in Washington DC. Support from the Office of Naval Research through the Naval Research Laboratory 6.1 Computational Physics task area is also greatly appreciated.

References

1. Sagaut, P.: Large Eddy Simulation for Incompressible Flows. Springer, New York (2002)
2. Liu, S., Meneveau, C., Katz, J.: On the properties of similarity subgridscale models as deduced from measurements in a turbulent jet. J. Fluid Mech. **275**, 83 (1994)
3. Bardina, J.: Improved turbulence models based on large eddy simulation of homogeneous incompressible turbulent flows. Ph.D. Thesis, Stanford University (1983)
4. Adams, N.A., Stolz, S.: Deconvolution methods for subgrid-scale approximation in LES. In: Geurts, B.J. (ed.) Modern Simulation Strategies for Turbulent Flows, p. 21. Edwards, Philadelphia (2001)
5. Spalart, P.R., Jou, W.H., Strelets, M., Allmaras, S.R.: Comments on the feasibility of LES for wings, and on hybrid RANS/LES approach. In: Advances in DNS/LES, First AFOSR International Conference in DNS/LES. Greyden Press, Columbus (1997)
6. Boris, J.P.: On large eddy simulation using subgrid turbulence models. In: Lumley, J.L. (ed.) Whither Turbulence? Turbulence at the Crossroads, p. 344. Springer, New York (1989)
7. Boris, J.P., Grinstein, F.F., Oran, E.S., Kolbe, R.J.: New insights into large eddy simulations. Fluid Dyn. Res. **10**, 199 (1992)
8. J. Fluids Eng. **124**(4) (2002). Alternative LES and Hybrid RANS/LES, edited by F.F. Grinstein and G.E. Karniadakis, pp. 821–942
9. Int. J. Numer. Methods Fluids **39**(9) (2002). Special Issue edited by D. Drikakis, pp. 763–864
10. von Neumann, J., Richtmyer, R.D.: A method for the numerical calculation of hydrodynamic shocks. J. Appl. Phys. **21**, 232 (1950)
11. Smagorinsky, J.: The beginnings of numerical weather prediction and general circulation modeling: early recollections. Adv. Geophys. **25**, 3 (1983)
12. Grinstein, F.F., Margolin, L.G., Rider, W.J. (eds.): Implicit Large Eddy Simulation: Computing Turbulent Flow Dynamics, 2nd edn. Cambridge University Press, New York (2010)

13. Fureby, C., Grinstein, F.F.: Monotonically integrated large eddy simulation of free shear flows. AIAA J. **37**, 544 (1999)
14. Fureby, C., Grinstein, F.F.: Large eddy simulation of high Reynolds number free and wall bounded flows. J. Comput. Phys. **181**, 68 (2002)
15. Schumann, U.: Subgrid scale model for finite difference simulation of turbulent flows in plane channels and annuli. J. Comput. Phys. **18**, 376 (1975)
16. Carati, D., Winckelmans, G.S., Jeanmart, H.: Exact expansions for filtered scales modeling with a wide class of LES filters. In: Voke, P.R., Sandham, N.D., Kleiser, L. (eds.) Direct and Large Eddy Simulation III, pp. 213–224. Kluwer Academic, Dordrecht (1999)
17. Godunov, S.K.: Reminiscences about difference schemes. J. Comput. Phys. **153**, 6–25 (1999)
18. Margolin, L.G., Rider, W.J.: A rationale for implicit turbulence modeling. Int. J. Numer. Methods Fluids **39**, 821 (2002)
19. Jimenez, J., Wray, A., Saffman, P., Rogallo, R.: The structure of intense vorticity in isotropic turbulence. J. Fluid Mech. **255**, 65 (1993)
20. Shao, L., Sarkar, S., Pantano, C.: On the relationship between the mean flow and subgrid stresses in large eddy simulation of turbulent shear flows. Phys. Fluids **11**, 1229 (1999)
21. Borue, V., Orszag, S.A.: Local energy flux and subgrid-scale statistics in three dimensional turbulence. J. Fluid Mech. **366**, 1 (1998)
22. Patnaik, G., Boris, J.P., Grinstein, F.F., Iselin, J.P.: Large scale urban simulations with FCT, Chap. 4 in this volume; see also Chap. 17 in Ref. [12] (2004)
23. Hirsch, C.: Numerical Computation of Internal and External Flows. Wiley, New York (1999)
24. Albada, G.D., van Leer, B., van Roberts, W.W.: A comparative study of computational methods in cosmic gas dynamics. Astron. Astrophys. **108**, 76 (1982)
25. Jasak, H., Weller, H.G., Gosman, A.D.: High resolution NVD differencing scheme for arbitrarily unstructured meshes. Int. J. Numer. Methods Fluids **31**, 431 (1999)
26. Boris, J.P., Book, D.L.: Flux corrected transport I, SHASTA, a fluid transport algorithm that works. J. Comput. Phys. **11**, 38 (1973)
27. Colella, P., Woodward, P.: The piecewise parabolic method (PPM) for gas dynamic simulations. J. Comput. Phys. **54**, 174 (1984)
28. Moser, R.D., Kim, J., Mansour, N.N.: Direct numerical simulation of turbulent channel flow up to $Re_\tau = 590$. Phys. Fluids **11**, 943 (1999)
29. Porter, D.H., Pouquet, A., Woodward, P.R.: Kolmogorov-like spectra in decaying three-dimensional supersonic flows. Phys. Fluids **6**, 2133 (1994)
30. Garnier, E., Mossi, M., Sagaut, P., Comte, P., Deville, M.: On the use of shock-capturing schemes for large eddy simulation. J. Comput. Phys. **153**, 273 (2000)
31. Okong'o, N., Knight, D.D., Zhou, G.: Large eddy simulations using an unstructured grid compressible Navier-Stokes algorithm. Int. J. Comput. Fluid Dyn. **13**, 303 (2000)
32. Eswaran, V., Pope, S.B.: An examination of forcing in direct numerical simulation of turbulence. Comput. Fluids **16**, 257 (1988)
33. Fureby, C., Tabor, G., Weller, H., Gosman, D.: A comparative study of sub grid scale models in isotropic homogeneous turbulence. Phys. Fluids **9**, 1416 (1997)
34. Kim, W.-W., Menon, S.: A new incompressible solver for large-eddy simulations. Int. J. Numer. Methods Fluids **31**, 983 (1999)
35. Driscoll, R.J., Kennedy, L.A.: A model for the turbulent energy spectrum. Phys. Fluids **26**, 1228 (1983)
36. Wei, T., Willmarth, W.W.: Reynolds number effects on the structure of a turbulent channel flow. J. Fluid Mech. **204**, 57 (1989)
37. Nikitin, N.V., Nicoud, F., Wasistho, B., Squires, K.D., Spalart, P.R.: An approach to wall modeling in large eddy simulations. Phys. Fluids **12**, 1629 (2000)
38. Fureby, C., Alin, N., Wikström, N., Menon, S., Persson, L., Svanstedt, N.: On large eddy simulations of high Re-number wall bounded flows. AIAA J. **42**, 457–468 (2004)
39. Grinstein, F.F., Oran, E.S., Boris, J.P.: Pressure field, feedback and global instabilities of subsonic spatially developing mixing layers. Phys. Fluids **3**(10), 2401–2409 (1991)
40. Grinstein, F.F., DeVore, C.R.: On global instabilities in countercurrent jets. Phys. Fluids **14**(3), 1095–1100 (2002)

41. Grinstein, F.F.: Vortex dynamics and entrainment in regular free jets. J. Fluid Mech. **437**, 69–101 (2001)
42. Grinstein, F.F.: On integrating large eddy simulation and laboratory turbulent flow experiments. Philos. Trans. R. Soc. Lond. A **367**(1899), 2931–2945 (2009)
43. Hussain, F., Husain, H.S.: Elliptic jets. Part I. Characteristics of unexcited and excited jets. J. Fluid Mech. **208**, 257–320 (1989)
44. Wetzel, T.G., Simpson, R.L., Chesnakas, C.J.: Measurement of three-dimensional crossflow separation. AIAA J. **36**, 557–564 (1998)
45. Alin, N., Svennberg, U., Fureby, C.: Large eddy simulation of flows past simplified submarine hulls. In: Proc. 8th Int'l Conf. Numerical Ship Hydrodynamics, Busan, Korea, pp. 208–222 (2003)
46. Groves, N.C., Huang, T.T., Chang, M.S.: Geometric characteristics of DARPA SUBOFF models. Report DTRC/SHD-1298-01, David Taylor Research Ctr. (1989)
47. Huang, T.T., et al.: Measurements of flows over an axisymmetric body with various appendages (DARPA SUBOFF experiments). In: Proc. 19th Symp. Naval Hydrodynamics, Seoul, Korea (1992)

Large Scale Urban Simulations with FCT

Gopal Patnaik, Jay P. Boris, Fernando F. Grinstein, John P. Iselin, and Denise Hertwig

Abstract Airborne contaminant transport in cities presents challenging new requirements for CFD. The unsteady flow physics is complicated by very complex geometry, multi-phase particle and droplet effects, radiation, latent and sensible heating effects, and buoyancy effects. Turbulence is one of the most important of these phenomena and yet the overall problem is sufficiently difficult that the turbulence must be included efficiently with an absolute minimum of extra memory and computing time. This paper describes the Monotone Integrated Large Eddy Simulation (MILES) methodology used in NRL's FAST3D-CT simulation model for urban contaminant transport (CT) (see Boris in Comput. Sci. Eng. 4:22–32, 2002 and references therein). We also describe important extensions of the underlying Flux-Corrected Transport (FCT) convection algorithms to further reduce numerical dissipation in narrow channels (streets).

1 Background

Urban airflow accompanied by contaminant transport presents new, extremely challenging modeling requirements. Configurations with very complex geometries and

---

G. Patnaik (✉) · J.P. Boris
LCP&FD, Naval Research Laboratory, Washington, DC 20375, USA
e-mail: patnaik@lcp.nrl.navy.mil

J.P. Boris
e-mail: boris@lcp.nrl.navy.mil

F.F. Grinstein
X-Computational Physics Division, Los Alamos National Laboratory, Los Alamos, NM 87545, USA
e-mail: fgrinstein@lanl.gov

J.P. Iselin
University of Wisconsin Platteville, Platteville, WI 53818, USA

D. Hertwig
Meteorological Institute, University of Hamburg, 20146 Hamburg, Germany

unsteady buoyant flow physics are involved. The widely varying temporal and spatial scales exhaust current modeling capacities. Simulations of dispersion of airborne pollutants in urban scale scenarios must predict both the detailed airflow conditions as well as the associated behavior of the gaseous and multiphase pollutants. Reducing health risks from the accidental or deliberate release of Chemical, Biological, or Radioactive (CBR) agents and pollutants from industrial leaks, spills, and fires motivates this work. Crucial technical issues include transport model specifics, boundary condition modeling, and post-processing of the simulation results for practical use by responders to actual real-time emergencies.

Relevant physical processes to be modeled include resolving complex building vortex shedding and recirculation zones. The model must also incorporate a consistent stratified urban boundary layer with realistic wind fluctuations, solar heating including shadows from buildings and trees, aerodynamic drag, turbulence generation, and heat losses due to the presence of trees, surface heat sorption variations and turbulent heat transport. Because of the short time spans and large air volumes involved, modeling a pollutant as well mixed globally is typically not appropriate. It is important to capture the effects of unsteady, non-isothermal, buoyant flow conditions on the evolving pollutant concentration distributions. In fairly typical urban scenarios, both particulate and gaseous contaminants behave similarly insofar as transport and dispersion are concerned. Thus the contaminant spread can be simulated effectively based on appropriate pollutant tracers with suitable sources and sinks. In other cases the full details of multigroup particle distributions are required.

1.1 Established Approach: Gaussian Plume Models

Contaminant plume prediction technology currently in wide use around the world is based on Gaussian similarity solutions ("puffs"). This is a class of extended Lagrangian approximations that only really apply for large scales and flat terrain where separated-flow vortex shedding from buildings, cliffs, or mountains is absent. Diffusion is used in plume/puff models to mimic the effects of turbulent dispersion caused by the complex building geometry and wind gusts of comparable and larger size (e.g., [2–5]). These current aerosol hazard prediction tools for CBR scenarios are relatively fast running models using limited topography, weather and wind data. They give only approximate solutions that ignore the effects of flow encountering 3D structures. The air flowing over and around buildings in urban settings is fully separated. It is characterized by vortex shedding and turbulent fluctuations throughout the fluid volume. In this regime, the usual timesaving approximations such as steady-state flow, potential flow, similarity solutions, and diffusive turbulence models are largely inapplicable. Therefore, a clear need exists for high-resolution numerical models that can compute accurately the flow of contaminant gases and the deposition of contaminant droplets and particles within and around real buildings under a variety of dynamic wind and weather conditions.

1.2 Computational Fluid Dynamics Approach

Since fluid dynamic convection is the most important physical process involved in CBR transport and dispersion, the greatest care and effort should be invested in its modeling. The advantages of the Computational Fluid Dynamics (CFD) approach and representation include the ability to quantify complex geometry effects, to predict dynamic nonlinear processes faithfully, and to handle problems reliably in regimes where experiments, and therefore model validations, are impossible or impractical.

1.2.1 Standard CFD Simulations

Some "time-accurate" flow simulations that attempt to capture the urban geometry and fluid dynamic details are a direct application of standard (aerodynamic) CFD methodology to the urban-scale problem. An example is the work at Clark Atlanta University where researchers conduct finite element CFD simulations of the dispersion of a contaminant in the Atlanta, Georgia metropolitan area. The finite element model includes topology and terrain data and a typical mesh contains approximately 200 million nodes and 55 million tetrahedral elements [6]. These are grand-challenge size calculations were run on 1024 processors of a CRAY T3E. Other research groups have used similar approaches (e.g., [7, 8]). The chief difficulty with this approach for large regions is that they are computer intensive and involve severe overhead associated with mesh generation.

1.2.2 The Large-Eddy Simulation Approach

Capturing the dynamics of all relevant scales of motion, based on the numerical solution of the Navier-Stokes Equations (NSE), constitutes Direct Numerical Simulation (DNS), which is prohibitively expensive for most practical flows at moderate-to-high Reynolds Number (Re). On the other end of the CFD spectrum are the industrial standard methods such as the Reynolds-Averaged Navier-Stokes (RANS) approach, e.g., involving $k-\varepsilon$ models, and other first- and second-order closure methods, which simulate only the mean flow and model the effects of all turbulent scales. These are generally unacceptable for urban CT modeling because they are unable to capture unsteady plume dynamics. Large Eddy Simulation (LES) constitutes an effective intermediate approach between DNS and the RANS methods. LES is capable of simulating flow features that cannot be handled with RANS, such as significant flow unsteadiness, and provides higher accuracy than the industrial methods at reasonable cost.

The main assumptions of LES are: (i) that transport is largely governed by large-scale unsteady convective features that can be resolved, (ii) that the less-demanding accounting of the small-scale flow features can be undertaken by using suitable sub-grid scale (SGS) models. Because the larger scale unsteady features of the flow are expected to govern the unsteady plume dynamics in urban geometries, the LES approximation has the potential to capture many key features which the RANS methods and the various Gaussian plume methodologies cannot.

2 Monotonically Integrated LES

Traditional LES approaches seek sufficiently high-order discretization and grid resolution to ensure that effects due to numerics are sufficiently small, so that crucial LES turbulence ingredients (filtering and SGS modeling) can be resolved. In the absence of an accepted universal theory of turbulence, the development and improvement of SGS models are unavoidably pragmatic and based on the rational use of empirical information. Classical approaches [9] have included many proposals ranging from inherently-limited eddy-viscosity formulations to more sophisticated mixed models combining dissipative eddy-viscosity models with the more accurate but less stable Scale-Similarity Model. The main drawback of mixed models relates to their computational complexity and cost for the practical flows of interest at moderate-to-high Re. The shortcomings of LES methods have led many researchers to abandon the classical LES formulations and shift focus directly to the SGS modeling implicitly provided by nonlinear (monotone) convection algorithms (see, e.g., [10], for a recent survey). The idea that a suitable SGS reconstruction might be implicitly provided by discretization in a particular class of numerical schemes [11] lead to proposing the Monotonically Integrated LES (MILES) approach [12, 13]. Later theoretical studies show clearly that certain nonlinear (flux-limiting) algorithms with dissipative leading order terms have appropriate built-in (i.e. "implicit") Sub-Grid Scale (SGS) models [14–16]. Our formal analysis and numerous tests have demonstrated that the MILES implicit tensorial SGS model is appropriate for both free shear flows and wall bounded flows. These are the conditions of most importance for CBR transport in cities.

As discussed further below, the MILES concept can be effectively used as a solid basis for CFD-based contaminant transport simulation in urban-scale scenarios, where conventional LES methods are far too expensive and RANS methods are inadequate.

3 MILES for Urban Scale Simulations

The FAST3D-CT three-dimensional flow simulation model [1, 17, 18] is based on the scalable, low dissipation Flux-Corrected Transport (FCT) convection algorithm [19, 20]. FCT is a high-order, monotone, positivity-preserving method for solving generalized continuity equations with source terms. The required monotonicity is achieved by introducing a diffusive flux and later correcting the calculated results with an antidiffusive flux modified by a flux limiter. The specific version of the convection algorithm implemented in FAST3D-CT is documented in [21].

Additional physical processes to be modeled include providing a consistent stratified urban boundary layer, realistic wind fluctuations and solar heating including shadows from buildings and trees. We must also model aerodynamic drag and heat losses due to the presence of trees, surface absorption variations and turbulent heat transport. Additional features include multi-group droplet and particle distributions with turbulent transport to surfaces as well as gravitational settling, solar chemical degradation, evaporation of airborne droplets, relofting of particles on the ground and ground evaporation of liquids. Incorporating specific models for these processes in the simulation codes is a challenge but can be accomplished with reasonable sophistication. The primary difficulty is the effective calibration and validation of all these physical models since much of the input needed from field measurements or experiments on these processes is typically insufficient or even nonexistent. Furthermore, even though the individual models can all be validated to some extent, the larger problem of validating the overall code has to be tackled as well. Some of the principally fluid dynamics related issues are elaborated further below.

3.1 Urban Flow Modeling Issues

3.1.1 Atmospheric Boundary Layer Specification

We have to deal with a finite domain and so precise planetary boundary layer characterization upstream of this domain greatly affects the boundary-condition prescription required in the simulations. The weather, time-of-day, cloud cover and humidity all determine if the boundary layer is thermally stable or unstable and thus determine the level and structure of velocity fluctuations. Moreover, the fluctuating winds, present in the real world but usually not known quantitatively, are known to be important because of sensitivity studies.

In FAST3D-CT the time average of the urban boundary layer is specified analytically with parameters chosen to represent the overall thickness and inflection points characteristic of the topography and buildings upstream of the computational domain. These parameters can be determined self-consistently by computations over a wider domain, since the gross features of the urban boundary layer seem to establish themselves in a kilometer or so, but this increases the cost of simulations considerably.

A non-periodic deterministic realization of the wind fluctuations is currently being superimposed on the average velocity profiles. This realization is specified as a suitable nonlinear superposition of modes with several wavelengths and amplitudes. Significant research issues remain unresolved in this area, both observationally and computationally. Deterministic [22] and other [23] approaches to formulating turbulent inflow boundary conditions are currently being investigated in this context. The strength of the wind fluctuations, along with solar heating as described just below, are shown to be major determinants of how quickly the contaminant density flushes from the domain in time and this in turn is extremely important in emergency applications as it determines overall dosage.

Fig. 1 Effect of fluctuations on trapping of contaminants

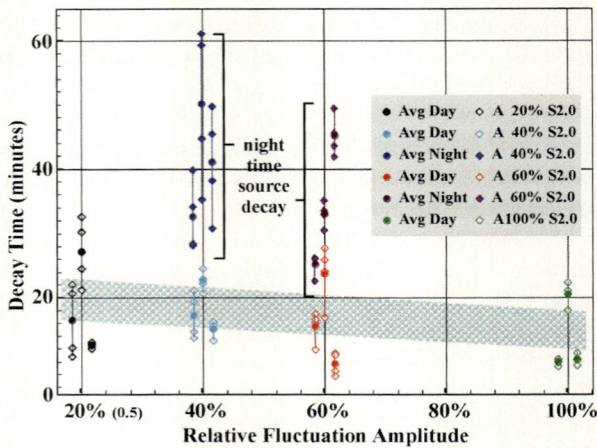

3.1.2 Solar Heating Effects

An accurate ray-tracing algorithm that properly respects the building and tree geometry computes solar heating in FAST3D-CT. The trees and buildings cast shadows depending on the instantaneous angle to the sun. Reducing the solar constant slightly can represent atmospheric absorption above the domain of the simulation and the model will even permit emulating a time-varying cloud cover. The geometry database has a land-use variable defining the ground composition. Our simulations to date identify only two conditions, ground and water, though the model can deal with the differences between grass, dirt, concrete and blacktop given detailed enough land-use data. The simulated interaction of these various effects in actual urban scenarios has been extensively illustrated in [1].

Figure 1 shows that the rate that a contaminant is flushed out of a city by the winds can vary by a factor of four or more due to solar heating variations from day to night and due to variations in the relative strength of the wind gusts. The horizontal axis of the figure indicates the relative strength of the gusting fluctuations at the boundaries, from about 20% on the left to about 100% on the right. For each of six different "environmental" conditions, twelve ground-level sources were released, four independent realizations at each of three source locations around the urban geometry. These source locations and the scale lengths of all the wind fluctuations were held fixed for the six different runs. The value of the exponential decay time in minutes is plotted for each source and realization as a diamond-shaped symbol. The figure shows that the decay time is two or three times longer for release at night compared to the day for otherwise identical conditions. The dark blue diamonds (decay times) should be compared with the light blue and the purple diamonds compared with the red. One can also see that the decay times get systematically shorter as the wind fluctuation amplitude is increased from left to right. This is emphasized by the light blue shaded bar through the center of the four daytime data sets.

3.1.3 Tree Effects

Although we can resolve individual trees if they are large enough, their effects (i.e., aerodynamic drag, introduction of turbulent velocity fluctuations, and heat losses) are represented through modified forest canopy models [24] including effects due to the presence of foliage. For example, an effective drag-force source term for the momentum equations can be written as $F = -C_d\, a(z)|\mathbf{v}|\mathbf{v}$, where $C_d = 0.15$ is an isotropic drag coefficient, $a(z)$ is a seasonally-adjusted leaf area density, z is the vertical coordinate, and $\mathbf{v}$ is the local velocity. The foliage density is represented in a fractal-like way so that fluctuations will appear even in initially laminar flows through geometrically regular stands of trees.

3.1.4 Geometry Specification

An efficient and readily accessible data stream is available to specify the building geometry data to FAST3D-CT. High-resolution (1 m or smaller) vector geometry data in the ESRI ARCVIEW data format is commercially available for most major cities. From these data, building heights are determined on a regular mesh of horizontal locations with relatively high resolution (e.g., 1 m). Similar tables for terrain, vegetation, and other land use variables can be extracted. These tables are interrogated during the mesh generation to determine which cells in the computational domain are filled with building, vegetation, or terrain. This masking process is a very efficient way to convert a simple geometric representation of an urban area to a computational grid.

This grid masking approach is used to indicate which computational cells are excluded from the calculation as well as to determine where suitable wall boundary conditions are to be applied. However, the grid masking approach is too coarse to represent rolling terrain, for which a shaved cell approach is applied. The terrain surface is represented by varying the location of the lower interface of the bottom cell. Even though this results in a terrain surface that is ultimately discontinuous, the jump between adjacent cells is very small. Operational results show that this approach works reasonably well and allow gradual changes in terrain height.

A more accurate representation of the geometry is possible with the VCE approach [25] in which all cell volume and interface areas are allowed to vary. This level of detail now begins to approach that of conventional aerodynamics CFD but it is not seen that this is necessary.

3.1.5 Wall Boundary Conditions

Appropriate wall boundary conditions must be provided so that the airflow goes around the buildings. It is not possible with the available resolution to correctly model the boundary layer on the surface. Therefore, rough-wall boundary layer models [26] are used for the surface stress, i.e., $\tau = \rho C_D (U_{//})^2$, and for the heat transfer from the wall, $H_o = \rho C_p C_H U_{\parallel}(\Theta - \Theta_o)$, where ρ is the mass density, C_D and C_H are coefficients characterizing the roughness and thermal properties of the walls or ground surface, $U_{\parallel}$ is the tangential velocity at the near-wall (first grid point adjacent to the wall), C_p is the specific heat at constant pressure, and Θ and Θ_o are the potential temperature at the wall, and near-wall, respectively.

4 The MILES Implicit SGS Model

Historically, flux-limiting (flux-correcting) methods have been of particular interest in the MILES context. A flux-limiter $0 \leq \Gamma \leq 1$ combines a high-order convective flux-function $\mathbf{v}_f^H$ that is well behaved in smooth flow regions, with a low-order dispersion-free flux-function $\mathbf{v}_f^L$ that is well behaved near sharp gradients. Thus the total flux-function with the limiter Γ becomes $\mathbf{v}_f = \mathbf{v}_f^H - (1 - \Gamma)[\mathbf{v}_f^H - \mathbf{v}_f^L]$. Properties of the implicit SGS model in MILES are related to the choice of Γ, $\mathbf{v}_f^L$, and $\mathbf{v}_f^H$, as well as to other specific features of the algorithm [15, 16]. This is quite similar to choosing/adjusting an (explicit) SGS model in the context of conventional LES.

Because of its inherently less-diffusive nature, prescribing Γ based on local monotonicity constraints is a more attractive choice in developing MILES [14–16]. This is supported by our comparative channel flow studies [16] of the global performance of MILES as a function of flux limiter. For example, the van-Leer TVD limiter (e.g., [27]) was found to be too diffusive as compared to FCT [19] and GAMMA [28] limiters which produce velocity profiles that agree well with the reference DNS data.

4.1 Street Crossings

Another approach to controlling unwanted numerical diffusion is through the appropriate choice of low and high order transport algorithms. In our simulations of urban areas, the typical grid resolution is of the order of 5 to 10 meters. While this resolution is adequate to represent the larger features of the city, many of the smaller features are resolved with only one to two cells. This is true of smaller streets found in cities, which are about 10 to 20 m wide. Alleyways are even smaller. These smaller streets may be represented by only one or two cells in our computation, putting a tremendous demand on the numerical convection not to diffuse and retard the flow down these narrow streets.

By using the rough-wall boundary conditions discussed above instead of no-slip boundary conditions, the flow can proceed unhampered down a single street even for streets that are only one cell wide. However, if there is another street intersecting the first, it was found that the flow essentially stagnates at this intersection. The problem only occurs when dealing with streets which are 1–2 cells wide and not with wider

Fig. 2 Advected quantity as function of grid index for both the conventional and modified low-order schemes

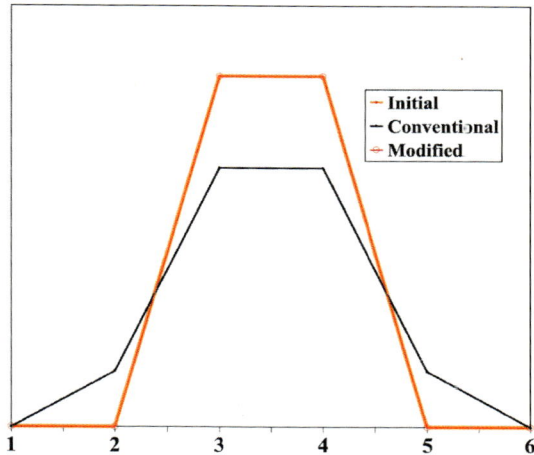

streets. After careful inspection, it was determined that the problem arose due to the form of the diffusion term in the low-order solution in the standard FCT algorithm, LCPFCT [21], used in the FAST3D-CT code.

The traditional low-order component of FCT introduces numerical diffusion even when the velocity goes to zero (as in the cross street) [21]. In normal situations, the flux limiter is able to locate an adjacent cell that has not been disturbed by the diffusion in the low-order method and is able to restore the solution to its original undiffused value. However, when the streets are 1–2 cells wide, the region of high velocity is diffused by low-order transport and there are no cells remaining at the higher velocity (Fig. 2). Thus the flux limiter cannot restore the solution in these cells to the original high value.

A solution to this problem lay in changing the form of the diffusion in the low-order method. In LCPFCT, the algorithmic diffusion coefficient for the low-order scheme is given by $\nu = \frac{1}{6} + \frac{1}{3}\varepsilon^2$, where the Courant number $\varepsilon = |U|\Delta t/\Delta x$. Note that ν does not go to zero even when U goes to zero (as in the cross street). The simplest less-diffusive low-order algorithm which ensures monotonicity is the upwind method previously used in the formal MILES analysis (e.g., [16]) for which the diffusion coefficient is given by $\nu_{upwind} = \frac{1}{2}|\varepsilon|$, which has the desired form for ν. When the diffusion coefficient in the low-order component of FCT is replaced by ν_{upwind}, the flow no longer stagnates at the intersection of streets (Fig. 2). This variation of the low-order method is only used for the momentum equations. It is not required for the density equation, since the density is almost constant everywhere. With this modification of the low-order method, the global properties of the transport algorithm were altered sufficiently to address this problem peculiar of under-resolved flows in urban areas.

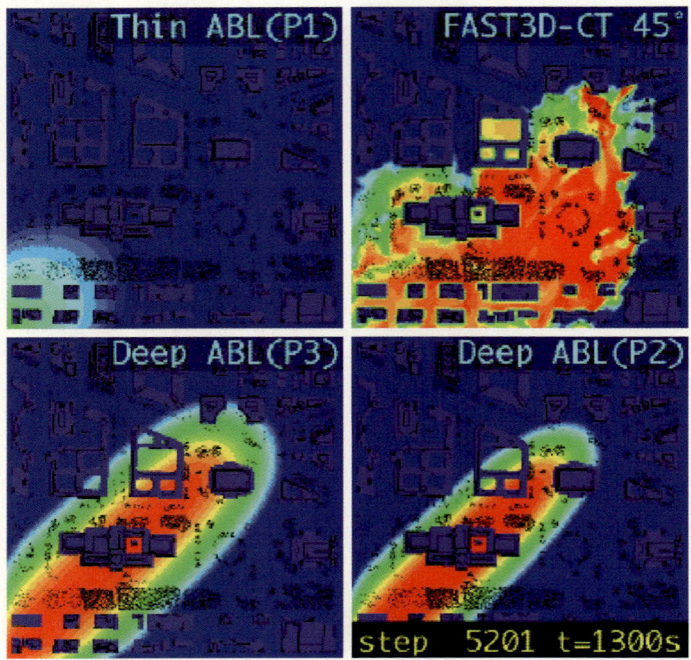

Fig. 3 Comparison of Gaussian Plume and FAST3D-CT simulations

5 Practical Examples

5.1 *Gaussian Lagrangian vs. Unsteady 3D Solutions*

Gaussian atmospheric transport and dispersion schemes are characterized by some initial direct spreading of the contaminant upwind by the diffusion, regardless of wind speed. The characteristic differences between the three Gaussian similarity solutions in Fig. 3 are similar to the differences between different Gaussian plume/puff models. None of these approximate, idealized solutions has the correct shape, trapping behavior, or plume width when compared to the FAST3D-CT simulation shown in the upper-right panel of the Fig. 3. The contaminant gets trapped in the recirculation zones behind buildings and continues to spread laterally long after simpler models say the cloud has moved on.

More detailed comparisons using actual "common use" puff/plume models (e.g., [29]) show a range of results depending on how much of the 3D urban boundary layer information from the detailed simulation is incorporated in the Gaussian model. Though building-generated aerodynamic asymmetries cannot be replicated, crosswind spreading and downwind drift can be approximately matched given enough free parameters. However, because the detailed simulations show that the plume expands like an angular sector away from the source, Gaussian models show

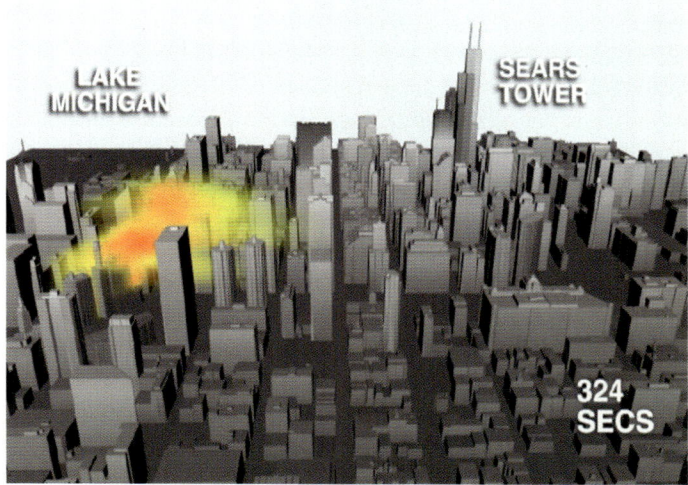

Fig. 4 View of contaminant release looking east toward downtown Chicago

too rapid a lateral spreading in the vicinity of the source to provide a plume that is approximately the correct width downwind.

5.2 Unsteady 3D Solution—Chicago

The city of Chicago is typical of a large, densely populated metropolitan area in the United States. The streets in the downtown area are laid out in a grid-like fashion, and are relatively narrow. The buildings are very tall but with small footprints. For example, the Sears tower is now the tallest building in the U.S.

Figures 4–6 show different views of a contaminant cloud from a FAST3D-CT simulation of downtown Chicago using a $360 \times 360 \times 55$ grid (6 m resolution). A 3 m/s wind off the lake from the east blows contaminant across a portion of the detailed urban geometry data set required for accurate flow simulations. One feature that is very apparent from these figures is that the contaminant is lofted rapidly above the tops of the majority of the buildings. This vertical spreading of the contaminant is solely due to the geometrical effect of the buildings. This behavior has also been observed in other simulations in which the buildings are not as tall.

Placement of the contaminant source can have a very nonlinear effect on the dispersion characteristics. Figures 7 and 8 show results of identical simulations with the exception of the contaminant release locations, which are shown by the red markers. The blue markers show the release location in the other simulation. Although the release locations differed by less than 0.5 km the dispersion characteristics are markedly different. The narrower dispersion pattern in Fig. 7 is likely caused by

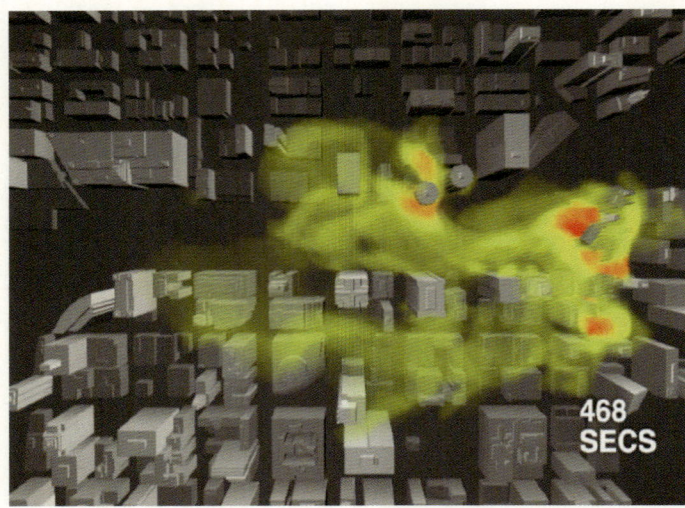

Fig. 5 Overhead view of contaminant concentrations over Chicago River

Fig. 6 View of contaminant from the Sears tower

a channeling effect of the Chicago River where velocities are higher. The wider dispersion pattern in Fig. 8 is likely due to a combination of flow deflection and recirculation of the flow from the building geometry. This behavior may also be dependent on release time. Work is continuing to determine the function dependence on location and release time. However, it is clear that spatially averaged parameteri-

Large Scale Urban Simulations with FCT 103

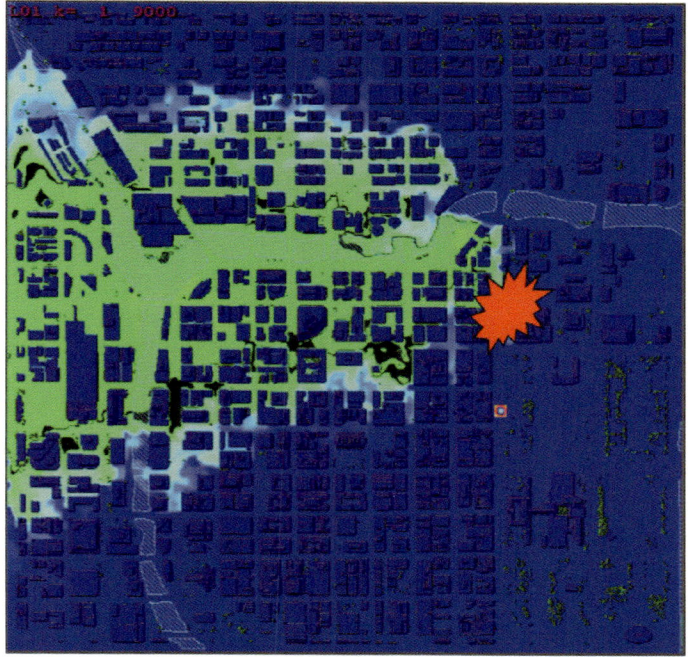

Fig. 7 Contaminant dispersion at ground level for release close to Chicago River

zations of urban surface characteristics will be unable to account for these nonlinear effects.

Additional simulations for Chicago were used to examine the effect of the modified low-order component of FCT as described in Sect. 4.1. Figure 9 shows the contaminant at ground level 9 minutes after release using the standard FCT algorithm LCPFCT [21]. The channeling effect of the Chicago River is quite dominant, though some lateral spreading occurs as well. The calculations were then repeated with the modified low-order method. These results are shown in Fig. 10. It is immediately apparent that the lateral spreading is much larger in this second case and that the cloud also propagates more rapidly downstream. These effects can be attributed to the lowered numerical diffusion in the cross-stream direction and the consequent lowering of numerical diffusion overall. Figure 11 shows the velocity (averaged horizontally over the computational domain) and RMS fluctuation profiles for both LCPFCT and the modified low-order scheme. The modified method has higher values for both velocity and RMS fluctuation. This is consistent with the observation of increased downstream and cross-stream propagation of the contaminant.

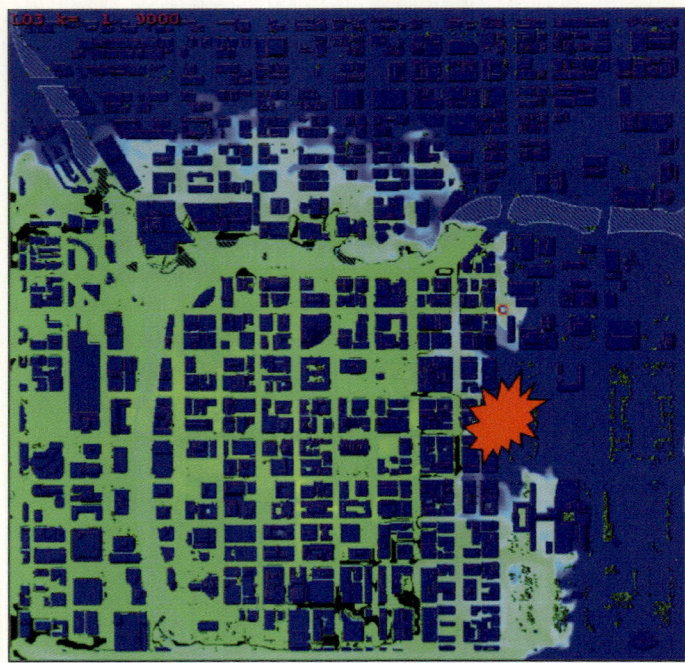

Fig. 8 Contaminant dispersion at ground level for release further from Chicago River

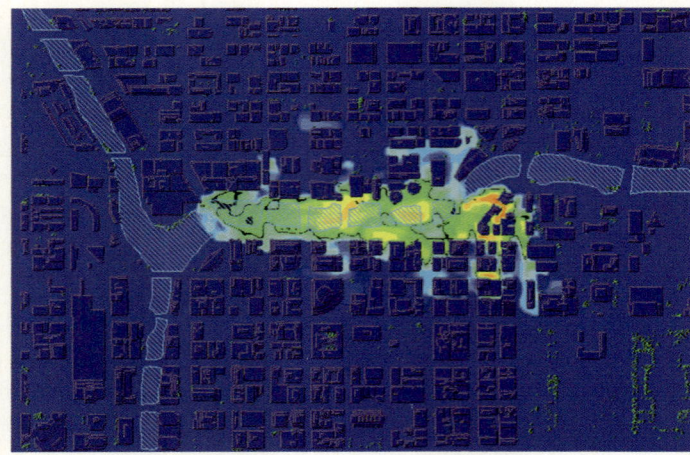

Fig. 9 Contaminant dispersion using standard low-order method

Fig. 10 Contaminant dispersion using modified low-order method

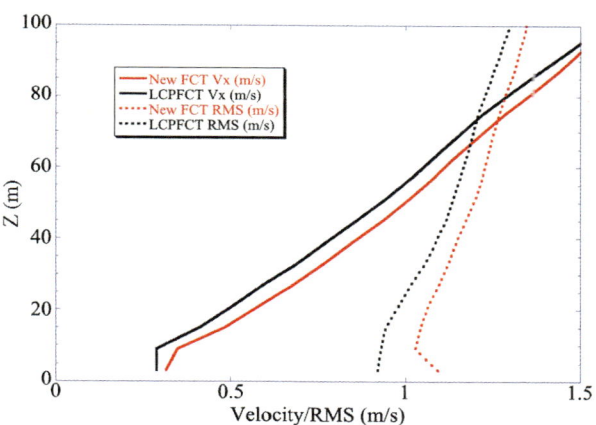

Fig. 11 Comparison of the standard and modified low-order method: velocity and RMS fluctuation profiles

5.3 Unsteady 3D Solution—Baghdad

Baghdad is rather typical of capital cities—it has large, spread-out governmental buildings, parks and monuments. There is no large dense urban core with tall buildings but there are several greater than 20 story buildings that are spread out. Residential areas are mostly suburban with some high-rise housing. This city structure is quite different from Chicago with its skyscrapers. A limited amount of high-resolution building data was available from a government-related source; however this data only included the largest buildings and covered a fraction of the area of the city. Large portions, especially residential areas, were not covered. Also, land-use data (trees, water, etc.) were not available in high-resolution form.

The missing data was constructed manually, primarily from commercially available satellite photographs of the city. These photographs had sufficient resolution to discern trees, water, and even types of housing. "Synthetic" buildings were generated to represent areas not covered by the available high-resolution data. Typical building heights and shapes found in suburbs were assigned at random to suburban regions. One of the difficulties not typically found in CFD calculations which proved to be a challenge was to ensure proper geo-referencing of the data, i.e., ensure everything lined up. This is especially severe when working from photographs that do not have a uniform resolution, and may sometimes not have the proper desired orientation.

5.3.1 In-Situ Validation

One of the obvious difficulties that arise for simulations of urban areas is that of validation of results. Experimental data is rarely available, and what little that is available is extremely limited in scope and coverage. The type and extent of data that is available restrict the quality of the validation effort. For Baghdad, no specific field measurements are available. However, just prior to the start of the war in Iraq, large trenches filled with oil were set ablaze in hope that the smoke would obscure targets. The smoke from one such fire provided an opportunity to at least visually "validate" our plume calculations. Figure 12 is a satellite photograph (courtesy DigitalGlobe) of the smoke from a trench fire near the monument to the Unknown Soldier. Figures 13 and 14 show the results from our simulations of the event. Color contours of the tracer gas are shown. The weather conditions for that day were given as "light wind from northwest." The simulations were carried with nominal wind speed of 3 m/s at 340°. An important unknown that had to be estimated is the level of fluctuation in the wind. The simulation depicted in Fig. 13 used a low level of fluctuation, which is consistent with the light steady winds typically found in March in the area. In order to investigate the importance of wind fluctuations, a higher level of fluctuations was simulated as shown in Fig. 14, which had fluctuations four times as high in amplitude as the baseline case (Fig. 13). As expected, the plume does spread slightly further. However, for low wind fluctuations, the spreading is largely controlled by the geometry of the city—an effect that becomes more dominant in dense urban areas. These calculations show that while a good knowledge of the weather is required for accurate predictions, in order to predict a worst-case scenario it is possible to select the appropriate parameters without perfect knowledge of all input conditions.

5.4 Detailed Validation Study—Hamburg, Germany

In a systematic study FAST3D-CT results of turbulent flow in the inner city of Hamburg, Germany, are being compared to reference measurements from a boundary-layer wind tunnel experiment. The urban structure is characteristic for northern and central European cities with complex crossings and courtyards. The focus of the validation exercise is the comparison of time-series information and the characterization of turbulent flow structures within and above the urban canopy.

Large Scale Urban Simulations with FCT

Fig. 12 Smoke plume from oil fire in Baghdad

5.4.1 Experimental and Numerical Methods

Laboratory measurements in specialized boundary-layer wind tunnels can provide an ideal validation data basis supplementary to information from field sites. Well definable and controllable boundary conditions together with the potential to repeat experimental runs under the same constraints result in high statistical confidence levels of the measured quantities. The reference measurements were performed in the boundary-layer wind tunnel facility at the University of Hamburg. The wind-tunnel model comprises the city center of Hamburg together with industrial harbor sites that are separated from the downtown area by the river Elbe. In total, the model domain encompasses an area of 3.7 km × 1.4 km in full-scale dimension. The physical model was built on a scale of 1:350, including terrain and a 3.5 m high water front. Effects of urban greenery are not accounted for. Figure 15 shows a photograph

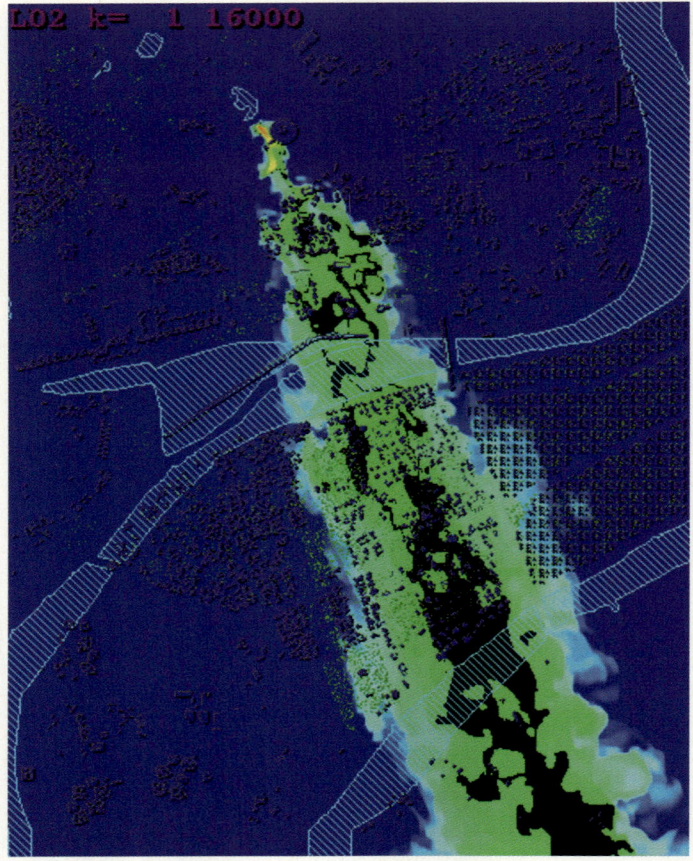

Fig. 13 Simulation of smoke plume. Low wind fluctuations

of the wind-tunnel model. The flow is approaching from the southwest (235°), mirroring a quite frequent meteorological condition for that area. The inflow boundary layer profiles were physically modeled to feature urban (i.e. very rough) turbulence characteristics (wind profile exponent $\alpha \approx 0.29$; roughness length $z_0 \approx 1.5$ m) under neutral atmospheric stratification. All flow measurements were conducted using non-intrusive 2D laser Doppler velocimetry.

The 3D FAST3D-CT simulation for Hamburg was performed on a 4 km × 4 km region of the inner city with 2.5 m grid resolution. The calculation was run on 64 CPUs of a SGI Altix computer and took more than three weeks to generate over 4 hours of real time data. The average wind direction is 235° rotated clockwise from due south. The wind speed was approximately 7 m/s at a height of 190 m. To match the FAST3D-CT conditions with the wind-tunnel experiments as closely as possible, all temperature related effects such as buoyancy and surface heating as

Large Scale Urban Simulations with FCT

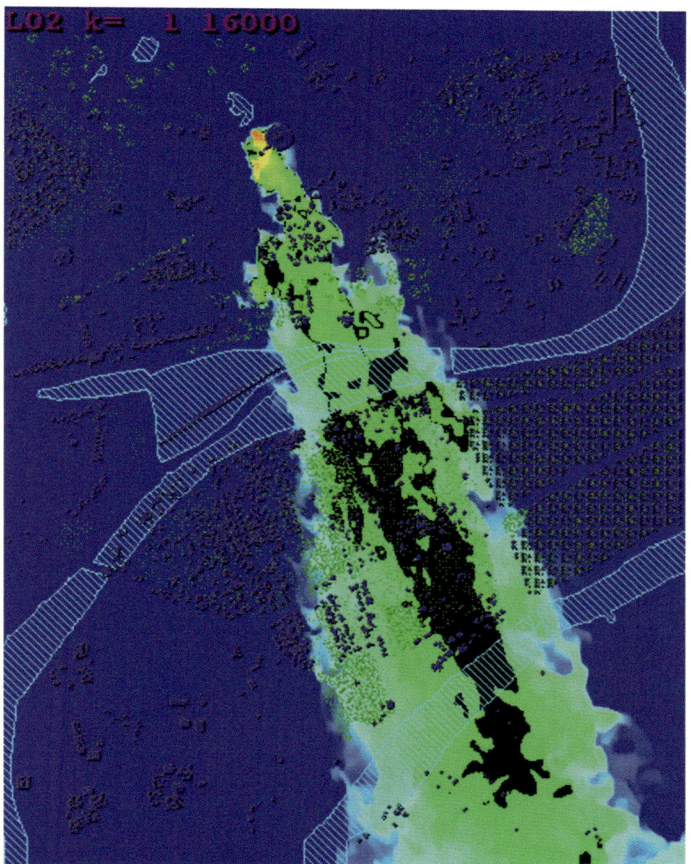

Fig. 14 Simulation of smoke plume. High wind fluctuations

well as drag effects of trees have been turned off. Time-dependent wind data were collected every 0.5 seconds at various heights up to 130 m.

For the validation exercise, 22 measurement locations within the model domain were chosen, including narrow street canyons, complex intersections, and measurement points close to the ground. This selection was made to represent areas of the city that are characteristic of urban flow situations and also pose challenges to numerical models. Velocity measurements were made in the numerical calculations to match the specified locations in the wind-tunnel experiment as closely as possible. The nearest neighbor extraction was chosen in order to avoid contamination of the results by interpolating data in order to have an exact spatial match. This procedure in some cases led to slight offsets of the x, y, and z positions of the comparison points that were in the range of a few centimeters up to a maximum of 1.75 m. Experimental and numerical data were normalized by referencing all velocities and their derivatives to a reference wind speed at a fixed location. This monitoring point

Fig. 15 Urban model of the inner city of Hamburg mounted in the boundary-layer wind-tunnel. View is from the inflow direction of 235°. Courtesy of the Environmental Wind Tunnel Laboratory at the University of Hamburg

was defined at a height of 49 m above the river Elbe at approximately 1 km upstream from the city center.

5.5 Mean Flow Comparison

The validation started from a comparison of mean flow characteristics in terms of time-averaged velocities. Figure 16 shows comparisons of vertical profiles of the streamwise velocity component from wind-tunnel measurements and FAST3D-CT simulations. In these and the following figures measurement locations are indicated by red dots on the city maps. The profile positions differ in the arrangement of the surrounding buildings. Figure 16(a) shows velocity profiles above the river Elbe (the (x, y)-location is identical with the reference point). Being situated well upstream of the densely built-up city center, the good agreement between experimental and numerical profiles mirrors a good match of the mean inflow conditions. A good agreement is also found for positions at which the flow is strongly influenced by the building structure. Figure 16(b) shows a profile measured in a very narrow street canyon. In Fig. 16(c), the measurement position is located in an open plaza exhibiting a strong recirculation regime that is captured quite well by the code. Measurements shown in Figs. 16(d)–(f) were conducted at intersections that trigger complex flow behavior. At elevations below the mean building height (approx. $H_{\text{mean}} \approx 35$ m by averaging over the city center) there is a slight trend towards an underprediction of velocities, whereas higher wind speeds than in the reference are observed at heights larger than 2.5 H_{mean}. The close proximity of building walls and the wall model used in the simulation might explain the slight offsets found within the street

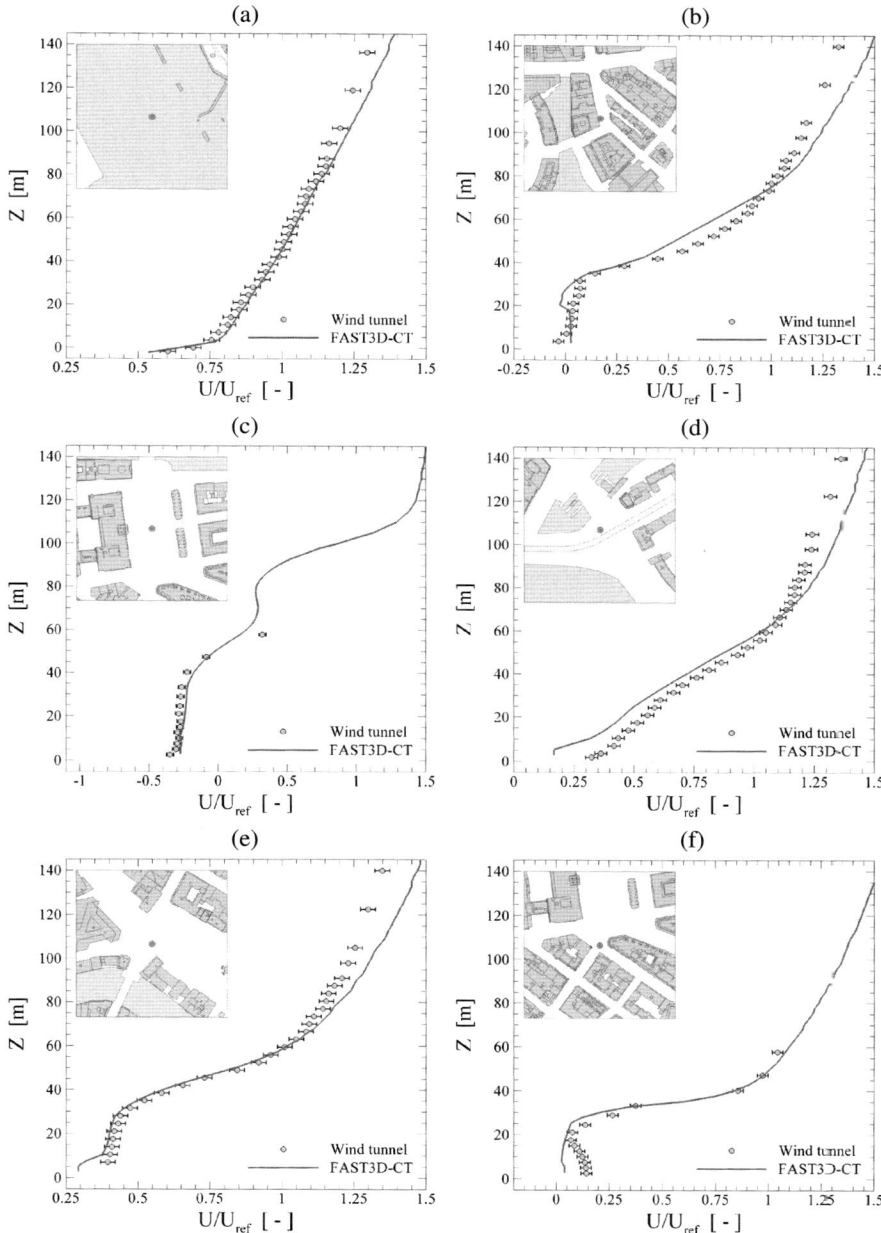

Fig. 16 Comparison of mean streamwise velocity profiles from wind-tunnel measurements (*circles*) and numerical simulations with FAST3D-CT (*lines*) at different locations within the city (**a**)–(**f**). Scatter bars attached to the experimental values represent the reproducibility of the data based on repetition measurements. The incoming flow is from left to right

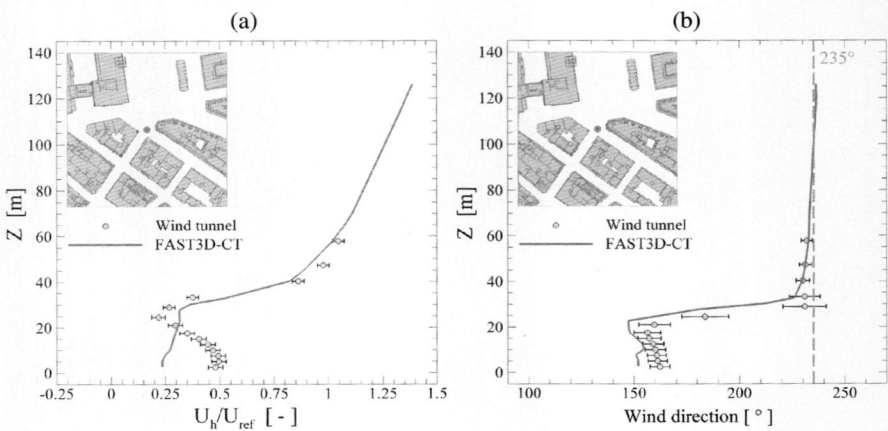

Fig. 17 Mean horizontal wind speed (**a**) and wind direction profiles (**b**) from wind-tunnel measurements (*circles*) and FAST3D-CT calculations (*lines*)

canyon. The stronger acceleration well above the canopy might reflect an excess of TKE in the numerical inflow prescription or the specific implementation of the upper numerical boundary.

5.6 Time Series Analysis

Next, experimental and numerical time series were analyzed in terms of frequency distributions and turbulent energy spectra. It has to be noted that both signals differ in their length and their time resolution under full-scale conditions. While the 170 s measurement time in the wind tunnel results in a full-scale duration of 16.5 h, the duration of the numerical time series is 4.5 h. Especially at low elevations within street canyons the full-scale temporal resolution of 2 Hz of the FAST3D-CT signals is better than the scaled wind-tunnel data rate that is strongly affected by the local flow seeding conditions.

First, the frequency distributions of instantaneous horizontal wind speeds and wind directions were evaluated. The mean horizontal wind speeds U_h and wind directions are compared in terms of vertical profiles shown in Figs. 17(a) and 17(b), respectively. At each of the profile heights, the fluctuations about these means were investigated. Figure 18 shows wind-rose diagrams of horizontal wind speeds and directions that were observed (Fig. 18(a)) and simulated (Fig. 18(b)) at four different heights within and above the street canyon. The wind-rose bars display the fractional frequency at which certain wind speeds (color-coded) were observed from the respective class of wind directions. At first view, the graphs show that the model predicts the deflection of wind directions inside the canopy quite well, together with the adjustment to the wind direction of the inflow at rooftop level and well above at 57.75 m (i.e. 1.65 H_{mean}). The spread about the central direction is largest at rooftop

Large Scale Urban Simulations with FCT 113

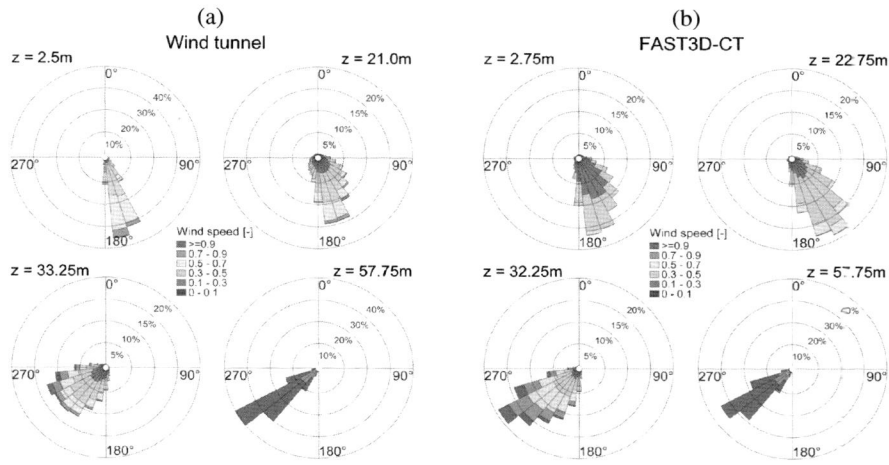

Fig. 18 Wind-rose diagrams showing frequency distributions of horizontal wind speeds and wind directions for wind-tunnel measurements (**a**) and FAST3D-CT simulations (**b**) at four different heights within and above a street canyon

height and smallest at the highest elevation in both the experiment and the simulation. However, discrepancies in velocity magnitudes are observed inside the canopy, especially for the lowermost point at 2.5 m (experimental) and 2.75 m (simulation), respectively. As discussed earlier in connection with the mean flow validation, the lower magnitudes are most likely due to the influence of wall boundary conditions prescribed at the ground and at upright building surfaces. Despite these differences the analysis indicates that the LES code is able to reproduce the directional fluctuation levels caused by unsteady flow effects quite reliably.

Auto-spectral energy densities of the turbulent streamwise velocity component are studied in order to analyze the spectral content associated with different eddy structures in the flow. The spectra were obtained using an FFT algorithm. Figure 19 shows scaled frequency spectra obtained from numerical and experimental velocities at various locations at heights of 17.5 m ($\approx 0.5\ H_{mean}$) in Figs. 19(a) and 19(b) and 45.5 m ($\approx 1.3\ H_{mean}$) in Figs. 19(c) and 19(d). A very good agreement of the production and energy-containing range of the spectra is found at all positions. The energetic peaks associated with integral length scale eddies coincide very well for the measurements shown in Figs. 19(b) and 19(d), whereas at the other positions the peaks are shifted for more than a decade towards higher frequencies. In order to investigate this, further analyses might concentrate on comparisons of integral length scales that can be determined from autocorrelation time scales invoking Taylor's hypothesis.

Common to all of the numerical spectra is their fast roll-off in the high frequency range that marks the onset of the influence of the nonlinear flux-limiting (MILES) and numerical dissipation. At most of the investigated locations this influence becomes noticeable approximately one decade after the spectral peak was reached

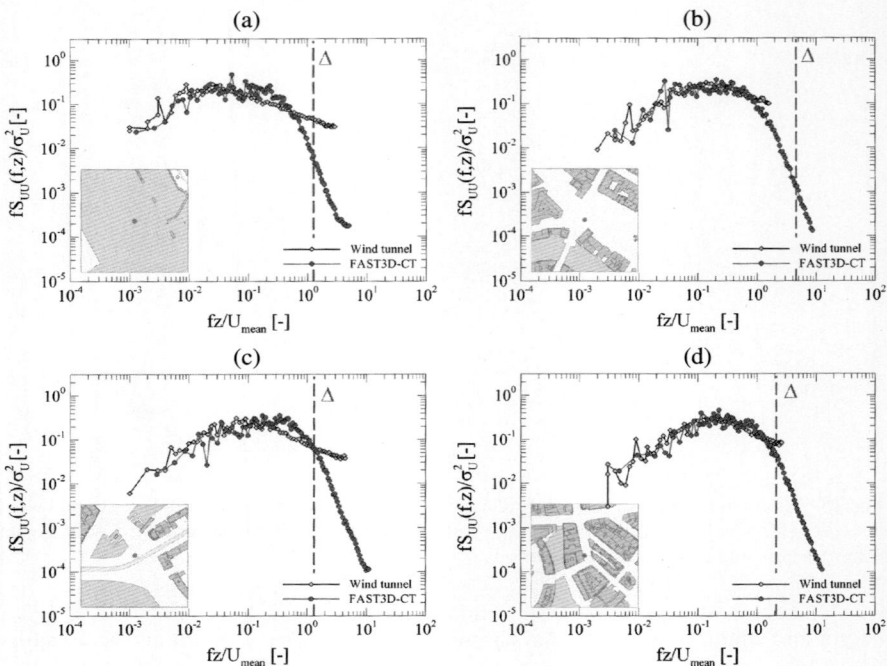

Fig. 19 Auto-spectral energy densities of the fluctuating streamwise velocity component from wind-tunnel measurements and simulations with FAST3D-CT at various locations within the city at heights of (**a**)–(**b**) 17.5 m ($\approx$0.5 H_{mean}) and (**c**)–(**d**) 45.5 m ($\approx$1.3 H_{mean}). The *dashed lines* separate the low frequency parts of the spectra that can be directly resolved by the numerical model given the grid resolution of $\Delta = 2.5$ m and the respective mean wind speeds from the subgrid-scales affected by numerical diffusion

resulting in a shortened extent of the inertial range. These urban flows are characterized by local production of turbulence at scales very close to the grid cutoff.

In consideration of the fact that FAST3D-CT was particularly designed to simulate dispersion processes in urban areas, the very good match of the energy-containing ranges associated with eddies that play a dominant role for scalar transport confirms the model's fitness for that purpose. However, it should be studied whether an extension of the inertial range is possible in order to add to the physical fidelity of the LES, even though this is not expected to contribute appreciably to the dispersion of contaminants.

6 Concluding Remarks

Physically realistic urban simulations are now possible but still require some compromises due to time, computer, and manpower resource limitations. The necessary trade-offs result in sometimes using simpler models, numerical algorithms, or geometry representations than we would wish. We know that the quality of the spatially and time-varying boundary conditions imposed, that is, the fluctuating winds, require improvement. However, detailed time-dependent wind field observations at key locations can be processed suitably to provide initial and boundary conditions and, at the least, can be used for global validation (e.g., [23]).

We believe that the building and large-scale fluid dynamics effects that can be presently captured govern the turbulent dispersion, and expect that the computed predictions will get better in time because the MILES methodology is convergent. However, there is considerable room to improve both the numerical implementation and the understanding of the backscatter that is included implicitly by through the MILES methodology.

Inherent uncertainties in simulation inputs and model parameters beyond the environmental variability also lead to errors that need to be further quantified by comparison with high quality reference data. Judicious choice of test problems for calibration of models and numerical algorithms are essential and sensitivity analyses help to determine the most important processes requiring improvement. In spite of inherent uncertainties and model trade-offs it is possible to achieve a significant degree predictability. For example, today, the biggest errors in comparing to field trials are associated with determining the actual wind profile and direction during the trial.

The FAST3D-CT simulation model can be used to simulate sensor and system response to postulated threats, to evaluate and optimize new systems, and to conduct sensitivity studies for relevant processes and parameters. Moreover, the simulations can constitute a virtual test range for micro- and nano-scale atmospheric fluid dynamics and aerosol physics, to interpret and support field experiments, and to evaluate, calibrate, and support simpler models.

Figure 5 illustrates the critical dilemma in the CT context: unsteady 3D urban-scenario flow simulations are currently feasible—but they are still expensive and require a degree of expertise to perform. First responders and emergency managers on site for contaminant release threats cannot afford to wait while actual simulations and data post-processing are being carried out. A concept addressing this problem [1, 30] carries out 3D unsteady simulations in advance and pre-computes compressed databases for specific urban areas based on suitable (e.g., historical, seasonally adjusted) assumed weather, wind conditions, and distributed test-sources. The relevant information is summarized as *Dispersion nomograph*^{TM} data so that it can be readily used through portable devices, in conjunction with urban sensors providing current observational information regarding local contaminant concentrations, wind speed, direction, and relative strength of wind fluctuations.

Acknowledgements The authors wish to thank Bob Doyle and the many members of NRL's LCP&FD for helpful technical discussions and scientific contributions to this effort. Further thanks are expressed to the members of the Environmental Wind Tunnel Laboratory at the University of Hamburg that contributed to the work within the Hamburg Pilot Project. Aspects of the work presented here were supported by ONR through NRL, the DoD High Performance Computing Modernization Office, DARPA, MDA and the German Federal Office of Civil Protection and Disaster Assistance (BBK) as well as the City of Hamburg, Germany.

References

1. Boris, J.P.: The threat of chemical and biological terrorism: preparing a response. Comput. Sci. Eng., **4**, 22–32 (2002)
2. Hazard prediction and assessment capability. http://www.dtra.mil/td/acecenter/td_hpac_fact.html
3. Bauer, T., Wolski, M.: Software User's Manual for the Chemical/Biological Agent Vapor, Liquid, and Solid Tracking (VLSTRACK) Computer Model, Version 3.1. NSWCDD/TR-01/83 (April 2001)
4. ALOHA Users Manual: (1999). Available for download at: http://www.epa.gov/ceppo/cameo/pubs/aloha.pdf. Additional information: http://response.restoration.noaa.gov/cameo/aloha.html
5. Leone, J.M., Nasstrom, J.S., Maddix, D.M., Larsen, D.J., Sugiyama, G.: LODI User's Guide, Version 1.0. Lawrence Livermore National Laboratory, Livermore (2001)
6. Aliabadi, S., Watts, M.: Contaminant propagation in battlespace environments and urban areas. AHPCRC Bull. **12**(4) (2002). http://www.ahpcrc.org/publications/archives/v12n4/Story3/
7. Chan, S.: FEM3C—an Improved Three-Dimensional Heavy-Gas Dispersion Model: User's Manual. UCRL-MA-116567, Lawrence Livermore National Laboratory, Livermore (1994) Rev. 1
8. Camelli, F., Löhner, R.: Assessing maximum possible damage for release events. In: Seventh Annual George Mason University Transport and Dispersion Modeling Workshop, June 2003
9. Sagaut, P.: Large Eddy Simulation for Incompressible Flows, 3rd edn. Springer, New York (2006)
10. Grinstein, F.F., Margolin, L.G., Rider, W.J. (eds.): Implicit Large Eddy Simulation: Computing Turbulent Flow Dynamics, 2nd Printing. Cambridge University Press, New York (2010)
11. Boris, J.P.: On large eddy simulation using subgrid turbulence models. In: Lumley, J.L. (ed.) Whither Turbulence? Turbulence at the Crossroads, p. 344. Springer, New York (1989)
12. Boris, J.P., Grinstein, F.F., Oran, E.S., Kolbe, R.J.: New insights into large eddy simulation. Fluid Dyn. Res. **10**, 199–228 (1992)
13. Oran, E.S., Boris, J.P.: Computing turbulent shear flows—a convenient conspiracy. Comput. Phys. **7**(5), 523–533 (1993)
14. Fureby, C., Grinstein, F.F.: Monotonically integrated large eddy simulation of free shear flows. AIAA J. **37**, 544–556 (1999)
15. Fureby, C., Grinstein, F.F.: Large eddy simulation of high Reynolds-number free & wall-bounded flows. J. Comput. Phys. **181**, 68–97 (2002)
16. Grinstein, F.F., Fureby, C.: Recent progress on MILES for high Reynolds-number flows. J. Fluids Eng. **124**, 848–861 (2002)
17. Cybyk, B.Z., Boris, J.P., Young, T.R., Lind, C.A., Landsberg, A.M.: A detailed contaminant transport model for facility hazard assessment in urban areas. AIAA Paper 99-3441 (1999)
18. Cybyk, B.Z., Boris, J.P., Young, T.R., Emery, M.H., Cheatham, S.A.: Simulation of fluid dynamics around complex urban geometries. AIAA Paper 2001-0803 (2001)
19. Boris, J.P., Book, D.L.: Flux-corrected transport I, SHASTA, a fluid transport algorithm that works. J. Comput. Phys. **11**, 8–69 (1973)
20. Boris, J.P., Book, D.L.: Solution of the continuity equation by the method of flux-corrected transport. Methods Comput. Phys. **16**, 85–129 (1976)
21. Boris, J.P., Landsberg, A.M., Oran, E.S., Gardner, J.H.: LCPFCT—a flux-corrected transport algorithm for solving generalized continuity equations. U.S. Naval Research Laboratory Memorandum Report NRL/MR/6410-93-7192 (1993)
22. Mayor, S.D., Spalart, P.R., Tripoli, G.J.: Application of a perturbation recycling method in the large-eddy simulation of a mesoscale convective internal boundary layer. J. Atmos. Sci. **59**, 2385–2395 (2002)
23. Bonnet, J.P., Coiffet, C., Delville, J., Druault, Ph., Lamballais, E., Largeau, J.F., Lardeau, S., Perret, L.: The generation of realistic 3D unsteady inlet conditions for LES. AIAA 2003-0065 (2002)

24. Dwyer, M.J., Patton, E.G., Shaw, R.H.: Turbulent kinetic energy budgets from a large-eddy simulation of flow above and within a forest canopy. Bound.-Layer Meteorol. **84**, 23–43 (1997)
25. Landsberg, A.M., Young, T.R., Boris, J.P.: An efficient parallel method for solving flows in complex three-dimensional geometries. AIAA Paper 94-0413 (1994)
26. Pal Arya, S.: Introduction to Micrometeorology. Academic Press, San Diego (1998)
27. Hirsch, C.: Numerical Computation of Internal and External Flows. Wiley, New York (1999)
28. Jasak, H., Weller, H.G., Gosman, A.D.: High resolution NVD differencing scheme for arbitrarily unstructured meshes. Int. J. Numer. Methods Fluids **31**, 431–449 (1999)
29. Pullen, J., Boris, J.P., Young, T.R., Patnaik, G., Iselin, J.P.: Comparing studies of plume morphology using a puff model and an urban high-resolution model. In: Seventh Annual George Mason University Transport and Dispersion Modeling Workshop, June 2003
30. Boris, J.P., Obenschain, K., Patnaik, G., Young, T.R.: CT-ANALYST^{TM}, fast and accurate CBR emergency assessment. In: Proceedings of the 2nd International Conference on Battle Management, Williamsburg, VA, November 4–8, 2002

40 Years of FCT: Status and Directions

Rainald Löhner and Joseph D. Baum

Abstract A somewhat historical perspective of the use of FCT for fluid dynamics is given. The particular emphasis is on large-scale blast problems. A comparison with other high-resolution CFD solvers is included to highlight the differences between them, as well as the relative cost. Results from test runs, as well as several relevant production runs are shown. Outstanding issues that deserve further investigation are identified.

1 Introduction

By the early 70's, computers had become sufficiently fast to allow the simulation of unsteady compressible flows with 'complex physics' (nonlinear source terms, arbitrary equations of state, large density/pressure/temperature ranges, etc.). A particularly disturbing observation made time and again was that low-order schemes, although overly dissipative, in many cases gave better results than high-order schemes. High-order schemes tended to have over/undershoots or ripples in regions of high gradients, which in turn could lead to completely unphysical results (for example, premature ignition for combustion calculations). Ironically, while a tremendous amount of effort was still being devoted to perfecting the 'ultimate linear scheme', Godunov [15] by the end of the 50's had already proven that any such effort would be futile. No linear scheme of order higher than one could give monotonicity preserving results. The resolution of this quandary came with the birth of nonlinear schemes.

R. Löhner (✉)
Center for Computational Fluid Dynamics, MS 6A2, George Mason University, Fairfax, VA 22030-4444, USA
e-mail: rlohner@gmu.edu

J.D. Baum
Advanced Technology Group, SAIC, McLean, VA 22102, USA
e-mail: joseph.d.baum@saic.com

Any high-order scheme used to advance the solution either in time or between iterations towards steady-state may be written as

$$u^{n+1} = u^n + \Delta u = u^n + \Delta u^l + \left(\Delta u^h - \Delta u^l\right) = u^l + \left(\Delta u^h - \Delta u^l\right). \qquad (1)$$

Here Δu^h and Δu^l denote the increments obtained by the high- and low-order scheme respectively, and u^l is the monotone solution at time $t = t^{n+1}$ of the low-order scheme. The idea behind any nonlinear scheme such as FCT is to limit the second term on the right-hand side of (1)

$$u^{n+1} = u^l + \lim\left(\Delta u^h - \Delta u^l\right), \qquad (2)$$

in such a way that no new over/undershoots are created. Note that even though the original PDE as well as the low- and high-order schemes are linear, the resulting overall scheme is nonlinear, as it depends on the local behavior of u.

While it became clear that the TVD concept was only enforceable in 1-D (see [16]: 'except in certain trivial cases, any method that is TVD in two space dimensions is at most first-order accurate'; 1-D TVD is being used unchanged to this day on an edge/face basis in 3-D production codes, and TVD has been supplanted by LED as an aim), the FCT concept realized a huge step forward with Zalesak's generalization [54] to schemes of arbitrary order and dimensions. Here was a way to apply FCT to schemes of any order, dimension, and, perhaps most importantly from an application point of view, grid topology. Zalesak's ideas were ported to Finite Element grids by Parrott and Christie [39] for scalar equations. Building on this work, Löhner, Morgan, Peraire and Vahdati [28, 29] added synchronization of limiters for systems of equations, found a steepener for contact discontinuities and added a Lapidus-type artificial viscosity [27] that removed the terracing problem for expansion fans. Further progress in the effective use of vector-machines by renumbering and grouping of elements, fast adaptive refinement with useful error indicators for transient problems [30], the switch from element- to edge-based data structures [35], extension to arbitrary Lagrangian-Eulerian (ALE) frames for problems with moving bodies [5] and mesh embedding [33] led to codes with the inherent geometrical flexibility of unstructured grids that were cost-competitive with structured grid codes. The remarkable combination of faster techniques and more powerful computers led to ever increasing problem size and application scope, as can be seen from Table 1. It is interesting to note that throughout a 25 year period the basic FEM-FCT algorithm for the simulation of transient shock problems has remained essentially unaltered.

The increase in compute power also shifted the CFD bottleneck in industry from schemes (80's: CPU time as the competitive differentiator) to grid generation (90's: surface definition to mesh as the competitive differentiator) to process integration (00's: CAD to accurate solution as the competitive differentiator). By the beginning of the 21st century, all CFD codes used in industry were based on some form of unstructured grid. At the same time FCT was placed on a sound theoretical basis by Kuzmin et al. [21–24], who also extended FCT to fully implicit time-marching and iterative limiting (for semi-implicit implementations of FCT and applications, see [40, 41, 47]). A recent successful extension into aeroacoustics based on large-eddy simulations [26] shows that this class of schemes remains very competitive.

Table 1 Increase of problem size

Size	Year	Problem	Machine
$> 10^2$	1983	Airfoil	ICL
$> 10^3$	1985	Forebody	CDC-205
$> 10^4$	1986	Train	Cray-XMP
$> 10^5$	1989	Train	Cray-2
$> 10^6$	1991	T-62	Cray-2
$> 10^7$	1994	WTC	Cray-M90
$> 10^8$	1998	Village	SGI-O2000

2 Basic Principles of FCT

Consider the system of conservation laws

$$\mathbf{u}_{,t} + \mathbf{F}^i_{,i} = \mathbf{S}, \tag{3}$$

where $\mathbf{u}, \mathbf{F}, \mathbf{S}$ denote the unknowns, fluxes and source-terms. Any finite volume or finite element discretization will yield a discrete system of the form:

$$\mathbf{M}^{ij} \hat{u}^j_{,t} = R^i_s + C^{ij} \mathscr{F}_{ij}. \tag{4}$$

Here, $\mathbf{M}, \hat{u}^j, R_s, C^{ij}, F^{ij}$ denote the mass-matrix, vector of unknowns, right-hand side due to sources, edge-coefficients for fluxes and edge-fluxes respectively. Let us consider first the traditional TVD approach. For the standard Galerkin approximation we have

$$\mathscr{F}_{ij} = \mathbf{f}_i + \mathbf{f}_j, \tag{5}$$

i.e. an equal weighting of fluxes at the end-point of an edge. This (high-order) combination of fluxes is known to lead to an unstable discretization, and must be augmented by stabilizing terms to achieve a stable, low-order scheme. In what follows, we enumerate the most commonly used options in order to compare them to FCT. We start with those schemes that limit before evaluating fluxes in order to contrast them to FCT, where the limiting is performed after evaluating fluxes.

2.1 Limiting Before Flux Evaluation

If we assume that the flow variables are constant in the vicinity of the edge endpoints i, j, a discontinuity will occur at the edge midpoint. The evolution in time of this local flowfield was first obtained analytically by Riemann [44], and consists of a shock, a contact discontinuity and an expansion wave. More importantly, the flux at the discontinuity remains constant in time. One can therefore replace the average flux of the Galerkin approximation by this so-called Riemann flux. This stable scheme, which uses the flux obtained from an exact Riemann solver, was first proposed by Godunov [15]. The flux is given by

$$\mathscr{F}_{ij} = 2\mathbf{f}(\mathbf{u}_{ij}^R),\tag{6}$$

where $\mathbf{u}_{ij}^R$ is the local exact solution of the Riemann problem to the Euler equations, expressed as

$$\mathbf{u}_{lr}^R = Rie(\mathbf{u}_l, \mathbf{u}_r)\tag{7}$$

where

$$\mathbf{u}_r = \mathbf{u}_i, \quad \mathbf{u}_l = \mathbf{u}_j.\tag{8}$$

This scheme represents what one may call the 'ultimate first order scheme'. All waves are taken into account, and the basic underlying physics are well reproduced. In order to achieve a higher order scheme, the amount of inherent dissipation must be reduced. This implies reducing the magnitude of the difference $\mathbf{u}_i - \mathbf{u}_j$ by 'guessing' a smaller difference of the unknowns at the location where the Riemann flux is evaluated (i.e. the middle of the edge). The assumption is made that the function behaves smoothly in the vicinity of the edge. This allows the construction or 'reconstruction' of alternate values for the unknowns at the middle of the edge. The additional information required to achieve a scheme of higher order via these improved values at the middle of the edge can be obtained in a variety of ways:

- Through continuation and interpolation from neighboring elements [7];
- Via extension along the most aligned edge [51]; or
- By evaluation of gradients [34, 35, 52].

The last option is the one most commonly used, but carries a considerable computational overhead: 15 gradients for the unknowns in 3-D can account for a large percentage of CPU time.

The inescapable fact stated in Godunov's theorem that no linear scheme of order higher than one is free of oscillations implies that with these higher order extensions, some form of limiting will be required. For a review of these, see [49]. It is important to note that this form of limiting is done *before flux evaluation*, and that, strictly speaking, it should be performed with characteristic variables. A typical Godunov-based scheme therefore has four main cost components:

- Solution of the exact Riemann problem;
- Gradient-based reconstruction of higher order approximations to the left and right states;
- Forward/backward transformation from conservative to characteristic variables; and
- Limiting.

In the sequel, we will enumerate possible simplifications to each of these cost components, thereby deriving a whole spectrum of commonly used schemes.

The solution of the (nonlinear) Riemann problem requires an iterative procedure which is expensive. Therefore, a considerable amount of effort has been devoted to obtain faster 'approximate Riemann solvers' that still retain as much of the physics as the basic Riemann problem [38, 45, 50]. A widely used solver of this class is the one derived by Roe [45], which may be written as:

$$\mathcal{F}_{ij} = \mathbf{f}_i + \mathbf{f}_j - |\mathbf{A}^{ij}|(\mathbf{u}_i - \mathbf{u}_j) \qquad (9)$$

where $|\mathbf{A}^{ij}|$ denotes the standard Roe matrix evaluated in the direction d^{ij}. Note that, as before, reducing the magnitude of the difference $\mathbf{u}_i - \mathbf{u}_j$ via reconstruction and limiting leads to schemes of higher order.

A further possible simplification can be made by replacing the Roe matrix by its spectral radius. This leads to a numerical flux function of the form

$$\mathcal{F}_{ij} = \mathbf{f}_i + \mathbf{f}_j - |\lambda^{ij}|(\mathbf{u}_i - \mathbf{u}_j), \qquad (10)$$

where

$$|\lambda^{ij}| = |v_{ij}^k \cdot S_k^{ij}| + c^{ij}, \qquad (11)$$

and v_{ij}^k and c^{ij} denote edge values, computed as nodal averages, of the fluid velocity and speed of sound respectively, and S_k^{ij} is the unit normal vector associated with the edge (i.e. in 3-D the unit normal of the finite volume surface associated with the edge). This can be considered as a centered difference scheme plus a second order dissipation operator, leading to a first order, monotone scheme. As before, a higher order scheme can be obtained by a better approximation to the 'right' and 'left' states of the 'Riemann problem'. Given that for smooth problems through the use of limiters the second order dissipation $|\mathbf{u}_i - \mathbf{u}_j|$ reverts to fourth order dissipation [19, 32], and that limiting requires a considerable number of operations, the next possible simplification is to replace the limiting procedure by a pressure sensor function. A scheme of this type may be written as

$$\mathcal{F}_{ij} = \mathbf{f}_i + \mathbf{f}_j - |\lambda_{ij}|\left[\mathbf{u}_i - \mathbf{u}_j + \frac{\beta}{2}\mathbf{l}_{ji} \cdot (\nabla \mathbf{u}_i + \nabla \mathbf{u}_j)\right], \qquad (12)$$

where $0 < \beta < 1$ denotes a pressure sensor function of the form [42]

$$\beta = 1 - \frac{p_i - p_j + 0.5\mathbf{l}_{ji} \cdot (\nabla p_i + \nabla p_j)}{|p_i - p_j| + |0.5\mathbf{l}_{ji} \cdot (\nabla p_i + \nabla p_j)|} \qquad (13)$$

and $\mathbf{l}_{ji} = \mathbf{x}_j - \mathbf{x}_i$. For $\beta = 0, 1$, second and fourth order damping operators are obtained respectively. Several forms are possible for the sensor function β [37]. Although this discretization of the Euler fluxes looks like a blend of second and fourth order dissipation, it has no adjustable parameters. The scalar dissipation operator presented above still requires the evaluation of gradients. This can be quite costly for Euler simulations: for a typical multistage scheme, more than 40% of the CPU-time is spent in gradient-operations, even if a new dissipation operator is only required at every other stage. The reason lies in the very large number of gradients required: 15 for the unknowns in 3-D, and an additional 3 for the pressure. An alternative would be to simplify the combination of second- and fourth order damping operators by writing out explicitly these operators:

$$d_2 = \lambda_{ij}(1-\beta)[\mathbf{u}_i - \mathbf{u}_j], \qquad d_4 = \lambda_{ij}\beta\left[\mathbf{u}_i - \mathbf{u}_j + \frac{\mathbf{l}_{ji}}{2}\cdot(\nabla\mathbf{u}_i + \nabla\mathbf{u}_j)\right]. \tag{14}$$

Performing a Taylor series expansion in the direction of the edge, we have

$$\mathbf{u}_i - \mathbf{u}_j + \frac{\mathbf{l}_{ji}}{2}\cdot(\nabla\mathbf{u}_i + \nabla\mathbf{u}_j) \approx \frac{\mathbf{l}_{ji}^2}{4}\left[\left.\frac{\partial^2\mathbf{u}}{\partial l^2}\right|_j - \left.\frac{\partial^2\mathbf{u}}{\partial l^2}\right|_i\right]. \tag{15}$$

This suggests the following simplification, which neglects the off-diagonal terms of the tensor of second derivatives:

$$\frac{\mathbf{l}_{ji}^2}{4}\left[\left.\frac{\partial^2\mathbf{u}}{\partial l^2}\right|_j - \left.\frac{\partial^2\mathbf{u}}{\partial l^2}\right|_i\right] \approx \frac{\mathbf{l}_{ji}^2}{4}[\nabla^2\mathbf{u}_j - \nabla^2\mathbf{u}_i], \tag{16}$$

and leads to the familiar blend of second and fourth order damping operators [20, 36]

$$\mathscr{F}_{ij} = \mathbf{f}_i + \mathbf{f}_j - |\lambda_{ij}|(1-\beta)[\mathbf{u}_i - \mathbf{u}_j] - |\lambda_{ij}|\beta\frac{\mathbf{l}_{ji}^2}{4}[\nabla^2\mathbf{u}_j - \nabla^2\mathbf{u}_i]. \tag{17}$$

2.1.1 Lax-Wendroff/Taylor-Galerkin

The essential feature of Lax-Wendroff/ Taylor-Galerkin schemes is the combination of time and space discretizations, leading to second order accuracy in both time and space. An edge-based two-step Taylor-Galerkin scheme can readily be obtained by setting the numerical flux to

$$\mathscr{F}_{ij} = 2\mathbf{f}(\mathbf{u}_{ij}^{n+\frac{1}{2}}), \tag{18}$$

where

$$\mathbf{u}_{ij}^{n+\frac{1}{2}} = \frac{1}{2}(\mathbf{u}_i + \mathbf{u}_j) - \frac{\Delta t}{2}\left.\frac{\partial \mathbf{f}^k}{\partial x_k}\right|_{ij}, \tag{19}$$

and $\left.\frac{\partial \mathbf{f}^k}{\partial x_k}\right|_{ij}$ is computed on each edge and given by either

$$\left.\frac{\partial \mathbf{f}^k}{\partial x_k}\right|_{ij} \approx \frac{\mathbf{l}_{ij}}{\mathbf{l}_{ij}^2}\cdot(\mathbf{F}^i - \mathbf{F}^j) \quad \text{or} \quad \left.\frac{\partial \mathbf{f}^k}{\partial x_k}\right|_{ij} \approx \frac{\mathbf{D}_{ij}}{\mathbf{D}_{ij}^2}\cdot(\mathbf{F}^i - \mathbf{F}^j), \tag{20}$$

where $\mathbf{D}_{ij}$ denotes the edge-coefficients for the advective terms obtained from the Galerkin approximation. The major advantage of this scheme lies in its speed, since there is no requirement of gradient computations, as well as limiting procedures for smooth flows. An explicit numerical dissipation (e.g. in the form of a Lapidus viscosity [27]) is needed to model flows with discontinuities. Taylor-Galerkin schemes by themselves are of little practical use for problems with strong shocks or other discontinuities. However, they provide high order schemes with the best cost/performance ratio for the flux-corrected transport schemes presented below.

2.2 Limiting After Flux Evaluation

Limiting after flux evaluation is the key idea inherent to all FCT schemes. If we focus on high order schemes of the Lax-Wendroff/Taylor-Galerkin family, the high order increment may be written as

$$\mathbf{M}_l \Delta \mathbf{u}^h = \mathbf{r} + (\mathbf{M}_l - \mathbf{M}_c) \Delta \mathbf{u}^h. \tag{21}$$

Here $\mathbf{M}_l$ denotes the diagonal, lumped mass-matrix, and $\mathbf{M}_c$ the consistent finite element mass-matrix. The low order scheme is simply given by

$$\mathbf{M}_l \Delta \mathbf{u}^l = \mathbf{r} + c_d (\mathbf{M}_c - \mathbf{M}_l) \mathbf{u}^n, \tag{22}$$

i.e. lumped mass-matrix plus sufficient diffusion to keep the solution monotonic. Subtracting these two equations yields the antidiffusive edge contributions

$$\left(\Delta \mathbf{u}^h - \Delta \mathbf{u}^l \right) = \mathbf{M}_l^{-1} (\mathbf{M}_l - \mathbf{M}_c) \left(c_d \mathbf{u}^n + \Delta \mathbf{u}^h \right). \tag{23}$$

Note that no physical fluxes appear in the antidiffusive edge contributions. This may also be interpreted as: advance the physical fluxes with extra diffusion, thus assuring transport, conservation, etc. Thereafter, perform the antidiffusive step to enhance the solution as much as possible without violating monotonicity principles. The simplicity of the antidiffusive edge contributions for this class of scheme makes it both fast and very general, and has been one of the main reasons why this scheme has served the CFD community for more than 25 years without major alterations.

Let us treat in more detail limiting after flux evaluation. If we consider an isolated point surrounded by elements, the task of the limiting procedure is to insure that the increments or decrements due to the antidiffusive contributions do not exceed a prescribed tolerance (see Fig. 1).

In the most general case, the contributions to a point will be a mix of positive and negative contributions. Given that the antidiffusive element/edge contributions (*AEC*'s) will be limited, i.e. multiplied by a number $0 \leq C_{el} \leq 1$, it may happen that after limiting all positive or negative contributions vanish. The largest increment (decrement) will occur when only the positive (negative) contributions are considered. For this reason, we must consider what happens if only positive or only negative contributions are added to a point. The comparison of the allowable increments

Fig. 1 Limiting procedure

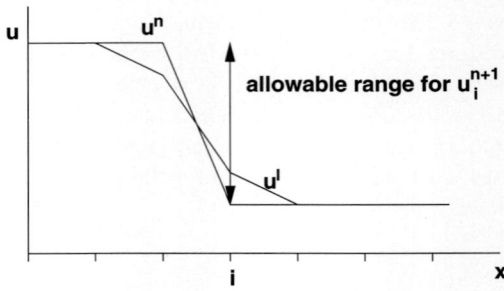

and decrements with these all-positive and all-negative contributions then yields the maximum allowable percentage of the *AEC*'s that may be added or subtracted to a point. On the other hand, an element/edge may contribute to a number of nodes, and in order to maintain strict conservation, the limiting must be performed for all the element/edge node contributions in the same way. Therefore, a comparison for all the nodes of an element/edge is performed, and the smallest of the evaluated percentages that applies is retained. Defining the following quantities:

$P_i^{\pm}$: the sum of all positive (negative) element/edge contributions to node i

$$P_i^{\pm} = \sum_{el} \begin{Bmatrix} \max \\ \min \end{Bmatrix} (0, AEC_{el}); \qquad (24)$$

$Q_i^{\pm}$: the maximum (minimum) increment node i is allowed to achieve

$$Q_i^{\pm} = u_i^{\max/\min} - u_i^l; \qquad (25)$$

the ratio of positive and negative contributions that ensure monotonicity is given by

$$R_i^{\pm} := \begin{cases} \min(1, Q_i^{\pm}/P_i^{\pm}), & P_i^+ > 0 > P_i^-, \\ 0, & \text{otherwise.} \end{cases} \qquad (26)$$

For the elements/edges, the final value taken is the most conservative:

$$C_{el} = \min(element/edge\ nodes) \begin{cases} R_i^+ & \text{if } EC > 0, \\ R_i^- & \text{if } EC < 0. \end{cases} \qquad (27)$$

The allowed value $u_i^{\max/\min}$ is taken between each point and its nearest neighbors. For element-based schemes, it may be obtained in three steps as follows:

(a) Maximum (minimum) nodal unknowns of u^n and u^l:

$$u_i^* = \begin{Bmatrix} \max \\ \min \end{Bmatrix} (u_i^l, u_i^n); \qquad (28)$$

Table 2 Ingredients of CFD solvers

Solver	Riemann	Gradient	Char. Transf.	Limiting
Classic Godunov	Yes	Yes	Yes	Yes
Consvar Godunov	Yes	Yes	No	Yes
Consvar Roe	Approx	Yes	No	Yes
Scal. Dissip.	No	Yes	No	Yes
Scal. Edge 2/4	No	Yes	No	No
Scal. Lapl 2/4	No	No	No	No
Taylor-Galerkin	No	No	No	No
TG-FCT	No	No	No	Yes

(b) Maximum (minimum) nodal value of element/edge:

$$u_e^* = \begin{Bmatrix} \max \\ \min \end{Bmatrix} (u_A^*, u_B^*, \ldots, u_C^*); \tag{29}$$

(c) Maximum (minimum) unknowns of all elements/edges surrounding node i:

$$u_i^{\substack{\max \\ \min}} = \begin{Bmatrix} \max \\ \min \end{Bmatrix} (u_1^*, u_2^*, \ldots, u_m^*). \tag{30}$$

A number of variations are possible for $u_i^{\substack{\max \\ \min}}$. For example, the so-called 'clipping limiter' is obtained by setting $u_i^* = \{{\max \atop \min}\} u_i^l$ in (a), i.e., by not looking back to the solution at the previous timestep or iteration, but simply comparing nearest neighbor values at the new timestep or iteration for the low-order scheme. As remarked before, the limiting is based solely on the unknowns u, not on a ratio of differences as in most TVD schemes. Table 2 summarizes the main ingredients of high-resolution schemes, indirectly comparing the cost of most current flow solvers.

2.3 Iterative Limiting

Given that the antidiffusive element contributions surrounding a point can have different signs, and that a 'most conservative' compromise has to be reached for positive and negative contributions, the remaining (i.e. not yet added) antidiffusive element contributions may still be added to the new solution without violating monotonicity constraints. This may be achieved by the following iterative limiting procedure:

- For iterations $j = 1, k$:
  - Perform limiting procedure:

$$u^{n+1} = u^l + \sum_{el} C_{el} \cdot AEC;$$

- Update remaining antidiffusive element contributions:

$$AEC \leftarrow (1 - C_{el})AEC;$$

- Re-define the low-order solution at t^{n+1}:

$$u^l = u^{n+1}.$$

Experience indicates that for explicit schemes, the improvements obtained by this iterative limiting procedure are modest. However, for implicit schemes, the gains are considerable and well worth the extra computational effort [21–25].

2.4 Test Cases

We include two simple test cases that demonstrate the performance of the edge-based FEM-FCT Euler solver. In both cases, FCT was run with limiter synchronization on the density and energy.

(a) *Shock Tube*: This is a classic example, which was used repeatedly to compare different Euler solvers [48]. Initially, a membrane separates two fluid states given by $\rho_1 = 1.0$, $\mathbf{v}_1 = 0.0$, $p_1 = 1.0$ and $\rho_2 = 0.1$, $\mathbf{v}_2 = 0.0$, $p_2 = 0.1$. The membrane ruptures, giving rise to a shock, a contact discontinuity and a rarefaction wave. Figure 2(a) shows the surface mesh and the surface contours of the density. A line-cut through the 3-D mesh is compared to the exact solution in Fig. 2(b). Note that the number of points appearing here corresponds to faces of tetrahedra being cut, i.e., is 2–3 times the number of actual points. One can see that the shock and contact discontinuities are captured over 2 or 4 elements respectively.

(b) *Shock Diffraction Over a Wall*: The second example shown is typical of some of the large-scale blast simulations carried out with FCT schemes over the last decade [1–4, 31, 46], and is taken from [43]. The outline of the domain is shown in Fig. 3(a). Surface pressures for an axisymmetric and 3-D run, together with comparison to photographs from experiments is given in Figs. 3(b), (c). The comparison to experimental results at different stations is shown in Fig. 3(d). As one can see, FCT schemes yield excellent results for this class of problems.

For more verification runs, see [3, 12, 28, 29, 31].

3 Some Landmark Runs

In this section we list a few memorable runs that were conducted with FEM-FCT over the years.

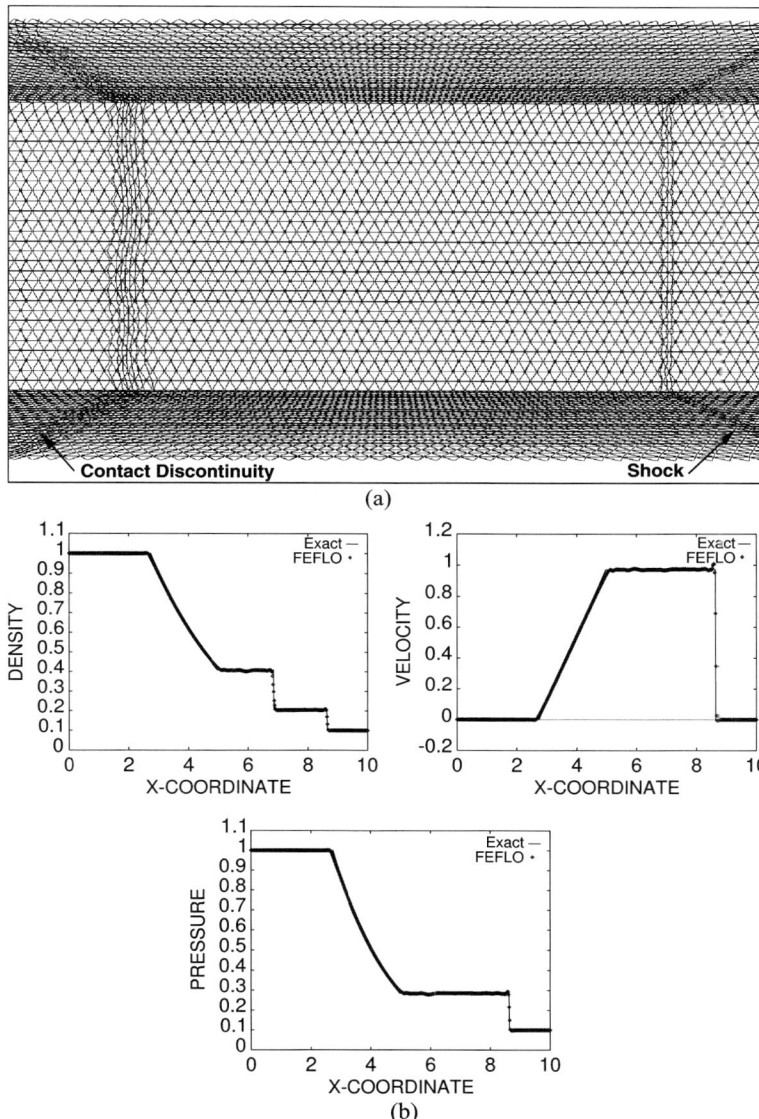

Fig. 2 Shock tube: (**a**) surface mesh and density; (**b**) comparison to exact values

(a) *T-62*: This simulation considers the impact of a strong shockwave on a main battlefield tank. The particular tank happens to be a T-62, of which there seem to be an abundance in the Middle East. The aim of the run was to gauge the possible effects of strong shockwaves on such tanks. This run was the first to exceed 2 Mtets, and was conducted on a CRAY-2. It was also the first run of this

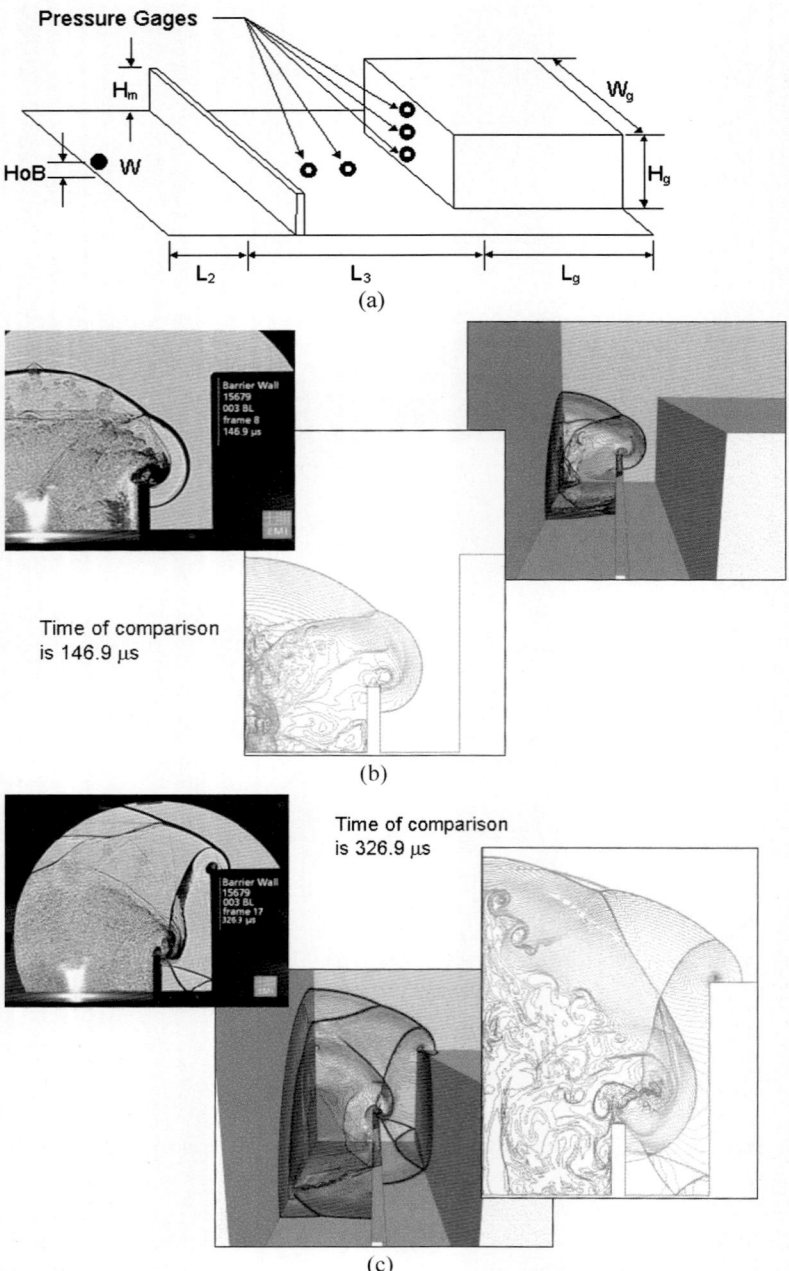

Fig. 3 (a) Shock diffraction over a wall. (b) Results at time $t = 146$ μsec. (c) Results at time $t = 327$ μsec. (d) Comparison to experimental values

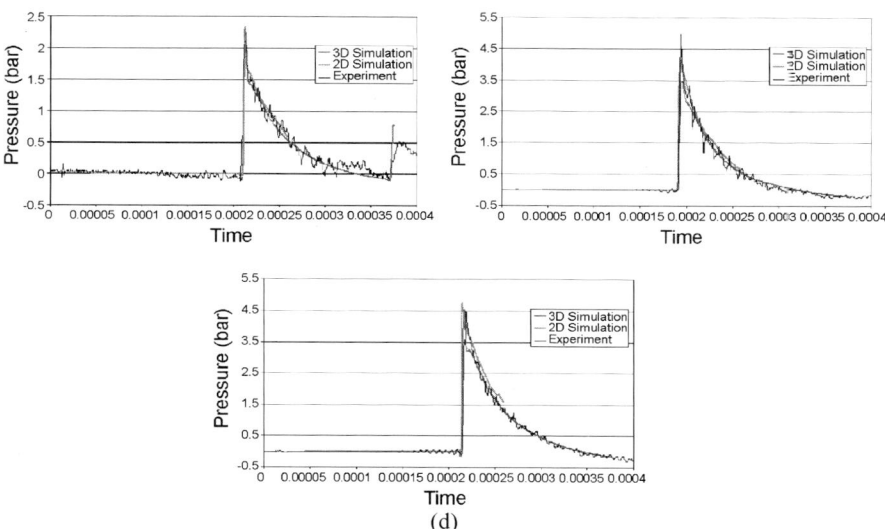

Fig. 3 (Continued)

magnitude to make extensive use of adaptive refinement. Even with an initial mesh grading that exhibited smaller elements close to the vehicle, the mesh was adaptively refined (and coarsened) to 2 levels every 5 timesteps. Figure 4 shows the surface discretization and pressure at different times. The passage of the shockwave over the vehicle is readily visible. For more information, see [1].

(b) *World Trade Center*: The World Trade Center has always been a symbol of global capitalism, and has therefore been a target for terrorist attacks almost since its inception. In 1993 a powerful car bomb was detonated in one of the lower parking levels, causing loss of life and extensive damage. In order to aid in the forensic studies that followed, a simulation was conducted. Data was assembled from a variety of sources: drawings, blueprints, videos, CAD-data. The distribution of cars (more than 400 of them) was set from a statistical distribution of 10 typical models. This run was the first to exceed 20 Mtets, and was conducted on a CRAY-2M. Figure 5 clearly shows the shock wave at approximately $T = 96$ msec. For more information, see [4].

(c) *Truck*: This simulation considers the interaction of a strong shockwave with a typical command and control center truck. The aim of the run was to demonstrate the feasibility of conducting fully coupled fluid-structure interaction calculations for this class of problems. The mesh was allowed to move close to the vehicle, and the FEM-FCT algorithm was cast in an Arbitrary Lagrangian-Eulerian (ALE) frame of reference. Approximately 10 global remeshings and countless local remeshings were required to accommodate the severe deformation of the structure. The response may be seen in Fig. 6. For more information, see [6].

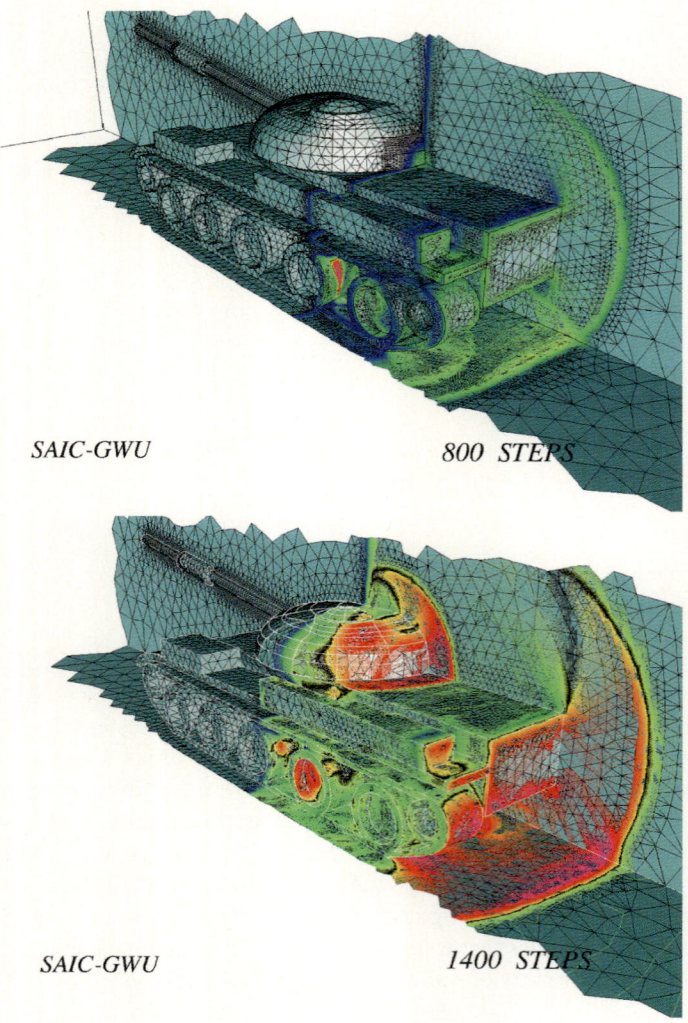

Fig. 4 Shock interaction with T-62

(d) *Blast in City*: This simulation considers a strong explosion in a typical city setting. The particular location chosen corresponds to Nairobi, where a powerful bomb was detonated close to the American embassy in 1998. This run was the first to reach 500 Mtets in a fully adaptive, transient setting, and was carried out on a multiprocessor SGI O2000 machine. The propagation of the blast wave can be discerned from Fig. 7(a). The adaptive refinement of the mesh can be seen in Fig. 7(b).

(d) *Generic Weapon Fragmentation*: This simulation considers the detonation and fragmentation of a generic weapon. It is performed using a JWL model for the

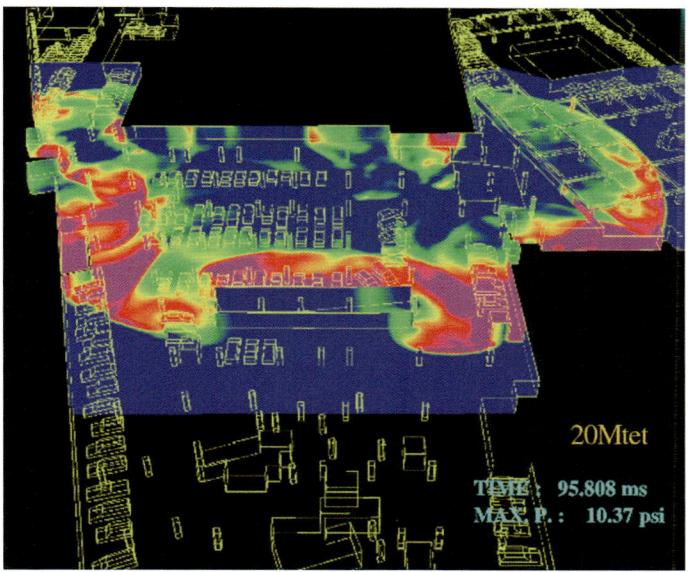

Fig. 5 Shock propagation in World Trade Center

fluid and FEM-FCT as the basic flow solver, and a large-deformation structural dynamics code for the casing. The flow mesh is not moving, i.e. the structure inside the flowfield is treated using the embedded adaptive approach [33]. The CSD domain was modeled with approximately 66 Khex elements corresponding to 1,555 fragments whose mass distribution matches statistically the mass distribution encountered in experiments. The structural elements were assumed to fail once the average strain in an element exceeded 60%.

The CFD mesh was refined to 3 levels in the vicinity of the solid surface. Additionally, the mesh was refined based on the modified interpolation error indicator proposed in [30], using the density as indicator variable. Adaptive refinement was invoked every 5 timesteps during the coupled CFD/CSD run. The CFD mesh started with 39 Mtet, and ended with 72 Mtet. Figures 8(a), (b) show the structure as well as the pressure contours in a cut plane at two times during the run. The detonation wave is clearly visible, as well as the thinning of the structural walls and the subsequent fragmentation.

4 Outstanding Issues

In this section, we list some of the outstanding issues in CFD solvers based on FCT.

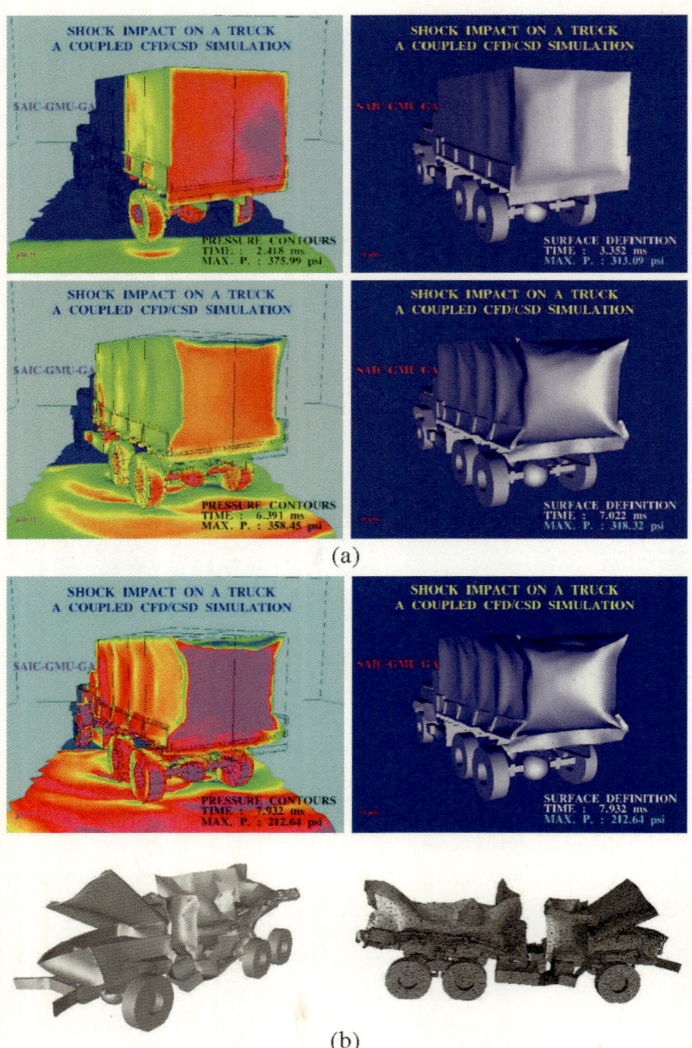

Fig. 6 Shock interaction with a truck

4.1 Steepening

The antidiffusive step in FCT is designed to steepen the low-order solution obtained at the new timestep or iteration. In some cases, particularly for high-order schemes of order greater than two, the antidiffusive step can flatten the profile of the solution even further, or lead to an increase of wiggles and noise. This is the case even though the solution remains within its allowed limits. A situation where this is the case can be seen from Fig. 9.

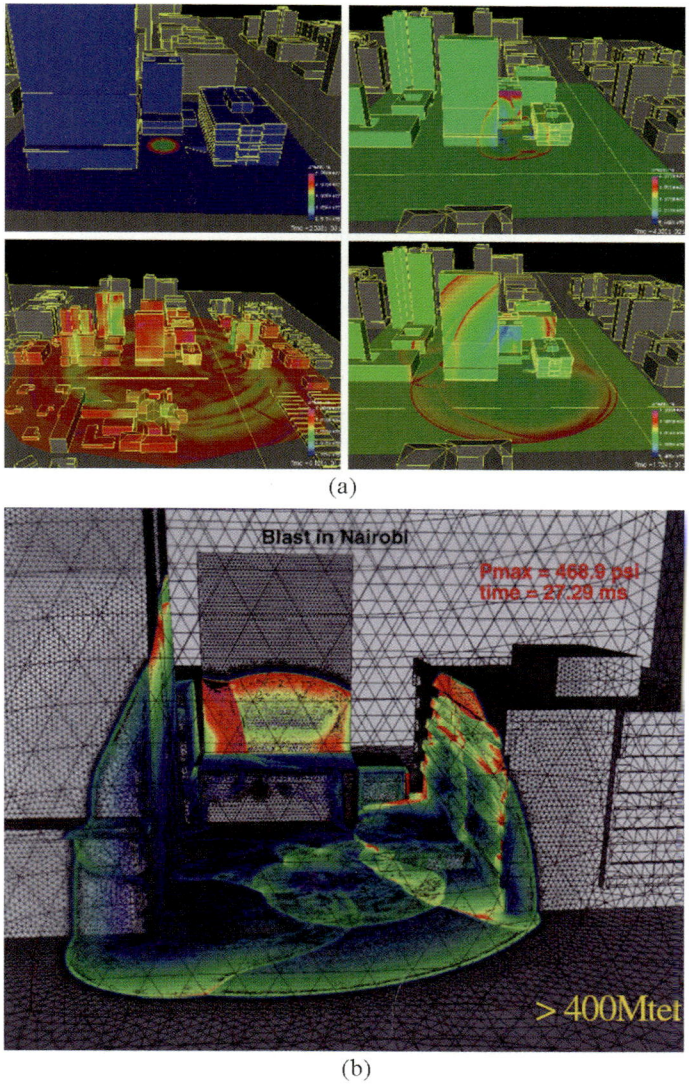

Fig. 7 (**a**) Blast in city. (**b**) Detail showing adaptive refinement

This type of behavior can be avoided if the antidiffusive flux is either set to zero or reversed. A simple way to decide when to reverse the antidiffusive fluxes is to compute the scalar product of the low-order solution at the new timestep or iteration and the antidiffusive element/edge contributions:

$$\nabla u^l \cdot AEC < 0 \quad \Longrightarrow \quad AEC = -\alpha \cdot AEC, \tag{31}$$

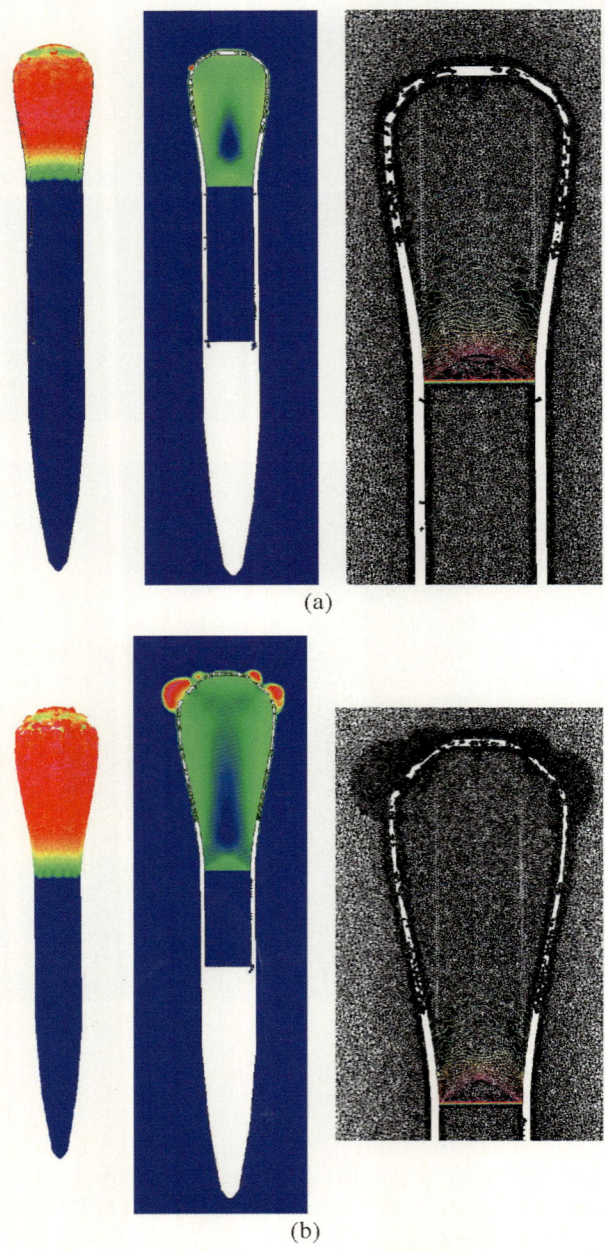

Fig. 8 CSD/flow velocity/mesh at: (**a**) 68 ms; (**b**) 102 ms

Fig. 9 Steepener for FCT

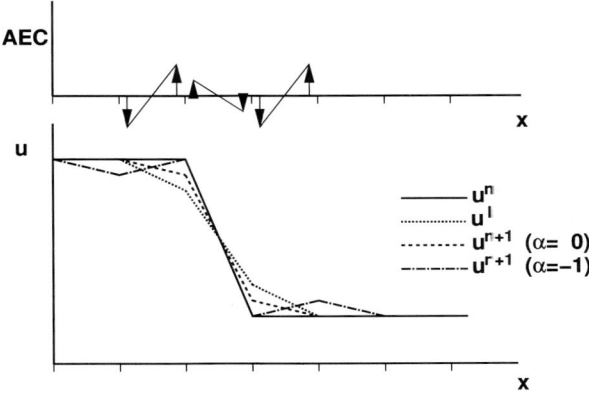

with $0 < \alpha < 1$. Values of α greater than unity lead to steepening. This can be beneficial in some cases, but is highly dangerous, as it can lead to unphysical solutions (e.g. spurious contact discontinuities) in complex applications. Although this type of steepening works well in practice, a more formal analysis/treatment would be beneficial.

4.2 Limiting for Systems of Equations

The results available in the literature [8–10, 28, 29, 39, 54, 55] indicate that with FCT results of excellent quality can be obtained for a single PDE, e.g. the scalar advection equation. However, in the attempt to extend the limiting process to systems of PDEs no immediately obvious or natural limiting procedure becomes apparent. Obviously, for 1-D problems one could advect each simple wave system separately, and then assemble the solution at the new time step. However, for multidimensional problems such a splitting is not possible, as the acoustic waves are circular in nature. Finite-Difference FCT codes used for production runs [13, 14] have so far limited each equation separately, invoking operator-splitting arguments. This approach does not always give very good results, as may be seen from [48] comparison of schemes for the Riemann problem, and has been a point of continuing criticism by those who prefer to use the more costly Riemann-solver-based, essentially one-dimensional TVD-schemes [11, 18, 38, 45, 49, 50, 53]. An attractive alternative is to introduce 'system character' for the limiter by combining the limiters for all equations of the system. Many variations are possible and can be implemented, giving different performance for different problems. Some of the possibilities are listed here, with comments where empirical experience is available.

(a) *Independent treatment of each equation as in operator-split FCT*: This is the least diffusive method, tending to produce an excessive amount of ripples in the non-conserved quantities (and ultimately also in the conserved quantities).

(b) *Use of the same limiter* (C_{el}) *for all equations*: This produces much better results, seemingly because the phase errors for all equations are 'synchronized'. This was also observed by Harten and Zwaas [17] and Zhmakin and Fursenko [56] for a class of schemes very similar to FCT. We mention the following possibilities:

- Use of a certain variable as 'indicator variable' (e.g. density, pressure, entropy).
- Use of the minimum of the limiters obtained for the density and the energy ($C_{el} = \min(C_{el}(\rho), C_{el}(\rho e))$): this produces acceptable results, although some undershoots for very strong shocks are present. This option is currently the preferred choice for strongly unsteady flows characterized by propagating and/or interacting shock waves.
- Use of the minimum of the limiters obtained for the density and the pressure ($C_{el} = \min(C_{el}(\rho), C_{el}(p))$): this again produces acceptable results, particularly for steady-state problems.

4.2.1 Limiting Any Set of Quantities

A general algorithm to limit any set of quantities may be formulated as follows:

- Define a new set of (non-conservative) variables $\mathbf{u}'$;
- Transform: $\mathbf{u}, \mathbf{u}^l \rightarrow \mathbf{u}', \mathbf{u}'^l$ and see how much $\mathbf{u}'$ can change at each point $\Rightarrow$ $\Delta\mathbf{u}'|_{\min}^{\max}$;
- Define a mean value of $\mathbf{u}^l$ in the elements and evaluate:

$$\Delta\mathbf{u}' = \mathbf{A}(\overline{\mathbf{u}}^l) \cdot \Delta\mathbf{u}, \qquad (32)$$

- Limit the transformed increments $\Delta\mathbf{u}' \Rightarrow C'_{el} \Rightarrow \Delta\mathbf{u}'' = \lim(\Delta\mathbf{u}')$;
- Transform back the variables and add:

$$\Delta\mathbf{u}^* = \mathbf{A}^{-1}(\overline{\mathbf{u}}^l) \cdot \Delta\mathbf{u}'' = \mathbf{A}^{-1}(\overline{\mathbf{u}}^l) \cdot C'_{el} \cdot \mathbf{A}(\overline{\mathbf{u}}^l) \cdot \Delta\mathbf{u}. \qquad (33)$$

4.3 Terracing

Terracing typically occurs when smooth profiles are transported over many gridpoints. Without incurring a loss of monotonicity, the profiles become ragged, exhibiting terraces (hence the name). In most cases, terracing occurs due to antidiffusive fluxes that are too strong, as in the case discussed above for steepeners. A usual way to suppress steepening is by adding a small amount of background dissipation, typically in the form of a fourth-order damping or via Lapidus smoothing [27]. While these empirical fixes work well, it would be highly desirable to have a formal understanding of this phenomenon.

4.4 Low-Order Schemes

An important observation which is very often overlooked in steady flow problems (e.g. classical aerodynamics) is that the creation of new extrema during a run can only occur at the rate of the low-order scheme. Therefore, one should strive for a low-order scheme with the lowest possible diffusion that still satisfies the monotonicity constraints of the physics. The effects of overdiffusion and/or bad limiting can often be seen in blast simulations, where shock rise times are too long and peak pressures too low. Consider the Taylor-Galerkin based FCT shown before. The antidiffusive edge contributions are given by:

$$(\Delta \mathbf{u}^h - \Delta \mathbf{u}^l) = \mathbf{M}_l^{-1}(\mathbf{M}_l - \mathbf{M}_c)(c_d \mathbf{u}^n + \Delta \mathbf{u}^h). \tag{34}$$

The usual procedure is to use a constant factor c_d throughout the mesh. Consider now a transient problem, integrated with an explicit scheme on a mesh with a large variation of mesh size and maximum eigenvalues of the flow (velocity, speed of sound). In regions where the local allowable timestep Δt_l is much larger than the minimum timestep over the mesh Δt chosen to advance the solution, the factor c_d could be lowered according to

$$c_d^* = \frac{\Delta t}{\Delta t_l} c_d \tag{35}$$

without violating monotonicity. Not reducing the diffusion coefficient by this ratio of timesteps can lead to an overdiffused solution with larger shock rise times and lower shock peaks. The example shown in Fig. 10 is a typical case. It considers a blast in an urban canyon. The geometry and blast location is shown in Fig. 10(a). For symmetry reasons, only half of the domain is required for the calculation. The solution, shown at different times in Fig. 10(b), was initialized from a detailed 1-D run. Figure 10(c) shows the comparison of pressure and impulse for the station marked in Fig. 10(a). The notation is as follows: FCTUSUAL is the usual edge-based FEM-FCT with Taylor-Galerkin and a constant c_d; FCTDTRAT is the same, except that c_d is multiplied by the ratio of allowable timesteps as shown above; FCTDTRI2 is FCTDTRAT with a second iterative antidiffusion pass; and FCTCARI2 is the same as FCTDTRI2, except that now the grid has a Cartesian point distribution (to see if this had an effect). One can see that, as expected, the peak pressures are higher when the dissipation of the low-order is diminished. We remark that these differences would not be seen in the shock-tube cases commonly used to compare schemes, as the solution remains constant after the shock has passed.

5 Conclusions and Outlook

FCT algorithms have proven to be an invaluable ingredient of many production codes over the last four decades. They have been used in a large variety of fields, such as fluid dynamics, plasma physics, petroleum engineering and electromagnetics. The present paper has given a somewhat historical perspective for fluid dynam-

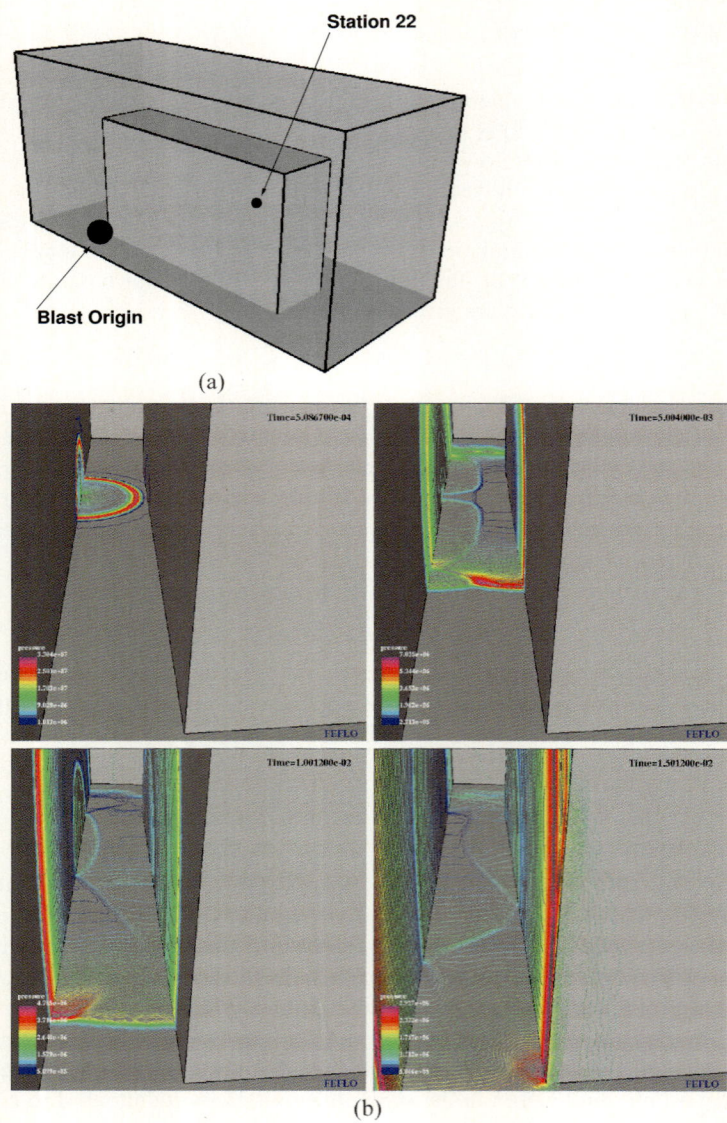

Fig. 10 (**a**) Blast in urban canyon. (**b**) Blast in urban canyon: surface pressures. (**c**) Blast in urban canyon: comparison of pressures and impulses

ics, with particular emphasis on large-scale blast problems. A comparison with other high-resolution CFD solvers has been included to highlight the differences between them, as well as the relative cost. Results from test runs, as well as several relevant production runs have been shown. Finally, several outstanding issues that deserve further investigation have been identified.

Fig. 10 (Continued)

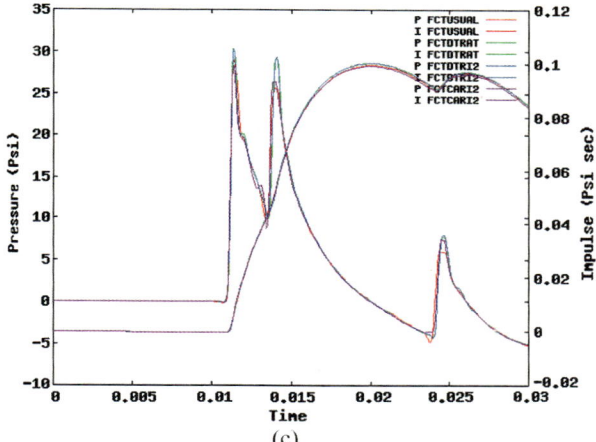

(c)

While many other CFD techniques have appeared in the last two decades, FCT, due to its very favorable cost/accuracy ratio, has been able to survive almost unchanged. The increase in compute power may, perhaps, shift the emphasis to more costly, refined schemes. However, for the truly large-scale problems at present there exist few alternatives to FCT.

Acknowledgements It is our great pleasure to acknowledge the input and stimulus provided by the many colleagues with whom we had the opportunity to work over the years. In particular the teams at GMU, Swansea (Wales, UK), NRL-LCP&FD, SAIC (ATG) and NASA (LARC, GSFC). We would also like to take the opportunity to thank Cray Research, IBM and SGI for many free hours on their machines over the years. The work compiled here would not have been possible without the steady support received from such organizations as the Defense Nuclear Agency, the Defense Threat Reduction Agency and the Air Force Office of Scientific Research. This support is gratefully acknowledged.

References

1. Baum, J.D., Löhner, R.: Numerical simulation of shock interaction with a modern main battlefield tank. AIAA-91-1666 (1991)
2. Baum, J.D., Luo, H., Löhner, R.: Numerical simulation of a blast inside a Boeing 747. AIAA-93-3091 (1993)
3. Baum, J.D., Löhner, R.: Numerical simulation of shock-box interaction using an adaptive finite element scheme. AIAA J. **32**(4), 682–692 (1994)
4. Baum, J.D., Luo, H., Löhner, R.: Numerical simulation of blast in the World Trade Center. AIAA-95-0085 (1995)
5. Baum, J.D., Luo, H., Löhner, R.: Validation of a New ALE, Adaptive unstructured moving body methodology for multi-store ejection simulations. AIAA-95-1792 (1995)
6. Baum, J.D., Luo, H., Löhner, R., Yang, C., Pelessone, D., Charman, C.: A coupled fluid/structure modeling of shock interaction with a truck. AIAA-96-0795 (1996)
7. Billey, V., Périaux, J., Perrier, P., Stoufflet, B.: 2-D and 3-D Euler computations with finite element methods in aerodynamic. In: International Conference on Hypersonic Problems, Saint-Etienne, Jan. 13–17 (1986)

8. Book, D.L., Boris, J.P., Hain, K.: Flux-corrected transport. II. Generalizations of the method. J. Comput. Phys. **18**, 248–283 (1975)
9. Boris, J.P., Book, D.L.: Flux-corrected transport. I. SHASTA, a transport algorithm that works. J. Comput. Phys. **11**, 38–69 (1973)
10. Boris, J.P., Book, D.L.: Flux-corrected transport. III. Minimal-error FCT algorithms. J. Comput. Phys. **20**, 397–431 (1976)
11. Colella, P.: Multidimensional upwind methods for hyperbolic conservation laws. Preprint LBL-17023 (1983)
12. de Fainchtein, R., Zalesak, S.T., Löhner, R., Spicer, D.S.: Finite element simulation of a turbulent MHD system: comparison to a pseudo-spectral simulation. Comput. Phys. Commun. **86**, 25–39 (1995)
13. Fry, M.A., Book, D.L.: Adaptation of flux-corrected transport codes for modelling dusty flows. In: Archer, R.D., Milton, B.E. (eds.) Proc. 14th Int.Symp. on Shock Tubes and Waves, New South Wales University Press, Sydney (1983)
14. Fyfe, D.E., Gardner, J.H., Picone, M., Fry, M.A.: In: Fast Three-Dimensional Flux-Corrected Transport Code for Highly Resolved Compressible Flow Calculations. Springer Lecture Notes in Physics, vol. 218, pp. 230–234. Springer, Berlin (1985)
15. Godunov, S.K.: Finite difference method for numerical computation of discontinuous solutions of the equations of fluid dynamics. Mat. Sb. **47**, 271–306 (1959)
16. Goodman, J.B., LeVeque, R.J.: On the accuracy of stable schemes for 2D scalar conservation laws. Math. Comput. **45**, 15–21 (1985)
17. Harten, A., Zwaas, G.: Self-adjusting hybrid schemes for shock computations. J. Comput. Phys. **6**, 568–583 (1972)
18. Harten, A.: High resolution schemes for hyperbolic conservation laws. J. Comput. Phys. **49**, 357–393 (1983)
19. Hirsch, C.: Numerical Computation of Internal and External Flow. Wiley, New York (1991)
20. Jameson, A., Schmidt, W., Turkel, E.: Numerical solution of the Euler equations by finite volume methods using Runge-Kutta time-stepping schemes. AIAA-81-1259 (1981)
21. Kuzmin, D.: Positive finite element schemes based on the flux-corrected transport procedure. In: Computational Fluid and Solid Mechanics, pp. 887–888. Amsterdam, Elsevier (2001)
22. Kuzmin, D., Turek, S.: Explicit and implicit high-resolution finite element schemes based on the flux-corrected-transport algorithm. In: Brezzi, F., et al. (eds.) Proc. 4th European Conf. Num. Math. and Advanced Appl, pp. 133–143. Springer, Berlin (2002)
23. Kuzmin, D., Turek, S.: Flux correction tools for finite elements. J. Comput. Phys. **175**, 525–558 (2002)
24. Kuzmin, D., Möller, M., Turek, S.: Multidimensional FEM-FCT schemes for arbitrary time-stepping. Int. J. Numer. Methods Fluids **42**, 265–295 (2003)
25. Kuzmin, D., Löhner, R., Turek, S. (eds.): Flux-Corrected Transport. Springer, Berlin (2005)
26. Liu, J., Kailasanath, K., Ramamurti, R., Munday, D., Gutmark, E., Löhner, R.: Large-eddy simulations of a supersonic jet and its near-field acoustic properties. AIAA J. **47**(8), 1849–1864 (2009)
27. Löhner, R., Morgan, K., Peraire, J.: A simple extension to multidimensional problems of the artificial viscosity due to lapidus. Commun. Appl. Numer. Methods **1**, 141–147 (1985)
28. Löhner, R., Morgan, K., Peraire, J., Vahdati, M.: Finite element flux-corrected transport (FEM-FCT) for the Euler and Navier-Stokes equations. Int. J. Numer. Methods Fluids **7**, 1093–1109 (1987)
29. Löhner, R., Morgan, K., Vahdati, M., Boris, J.P., Book, D.L.: FEM-FCT: combining unstructured grids with high resolution. Commun. Appl. Numer. Methods **4**, 717–730 (1988)
30. Löhner, R., Baum, J.D.: Adaptive h-refinement on 3-D unstructured grids for transient problems. Int. J. Numer. Methods Fluids **14**, 1407–1419 (1992)
31. Löhner, R., Yang, Chi, Baum, J.D., Luo, H., Pelessone, D., Charman, C.: The numerical simulation of strongly unsteady flows with hundreds of moving bodies. Int. J. Numer. Methods Fluids **31**, 113–120 (1999)
32. Löhner, R.: Applied CFD Techniques. Wiley, New York (2001)

33. Löhner, R., Baum, J.D., Mestreau, E.L., Sharov, D., Charman, Ch., Pelessone, D.: Adaptive embedded unstructured grid methods. AIAA-03-1116 (2003)
34. Luo, H., Baum, J.D., Löhner, R., Cabello, J.: Adaptive edge-based finite element schemes for the Euler and Navier-Stokes equations. AIAA-93-0336 (1993)
35. Luo, H., Baum, J.D., Löhner, R.: Edge-based finite element scheme for the Euler equations. AIAA J. **32**(6), 1183–1190 (1994)
36. Mavriplis, D.: Three-dimensional unstructured multigrid for the Euler equations. AIAA-91-1549-CP (1991)
37. Mestreau, E., Löhner, R., Aita, S.: TGV tunnel-entry simulations using a finite element code with automatic remeshing. AIAA-93-0890 (1993)
38. Osher, S., Solomon, F.: Upwind difference schemes for hyperbolic systems of conservation laws. Math. Comput. **38**, 339–374 (1982)
39. Parrott, A.K., Christie, M.A.: FCT applied to the 2-D finite element solution of tracer transport by single phase flow in a porous medium. In: Morton, K.W., Baines, M.J. (eds.) Proc. ICFD-Conf. Num. Meth. in Fluid Dyn. Academic Press, Reading (1986)
40. Patnaik, G., Guirguis, R.H., Boris, J.P., Oran, E.S.: A barely implicit correction for flux-corrected transport. J. Comput Phys. (1988)
41. Patnaik, G., Kailasanath, K., Laskey, K.J., Oran, E.S.: Detailed numerical simulations of cellular flames. In: Proc. 22nd Symposium (Int.) on Combustion, The Combustion Institute, Pittsburgh (1989)
42. Peraire, J., Peiro, J., Morgan, K.: A three-dimensional finite element multigrid solver for the Euler equations. AIAA-92-0449 (1992)
43. Rice, D.L., Giltrud, M.E., Baum, J.D., Luo, H., Mestreau, E.: Experimental and numerical investigation of shock diffraction about blast walls. In: Proc. 16th Int. Symp. Military Aspects of Blast and Shocks, Keble College, Oxford, UK, September 10–15 (2000)
44. Riemann, G.F.B.: Über die Fortpflanzung ebener Luftwellen von Endlicher Schwingungweite. Abhandlungen der Königlichen Gesellschaft der Wissenschaften zu Göttingen **8** (1860)
45. Roe, P.L.: Approximate Riemann solvers, parameter vectors and difference schemes. J. Comput. Phys. **43**, 357–372 (1981)
46. Sivier, S., Loth, E., Baum, J.D., Löhner, R.: Vorticity produced by shock wave diffraction. Shock Waves **2**, 31–41 (1992)
47. Scannapieco, A.J., Ossakow, S.L.: Nonlinear equatorial spread F. Geophys. Res. Lett. **3**, 451–454 (1976)
48. Sod, G.: A survey of several finite difference methods for systems of nonlinear hyperbolic conservation laws. J. Comput. Phys. **27**, 1–31 (1978)
49. Sweby, P.K.: High resolution schemes using flux limiters for hyperbolic conservation laws. SIAM J. Numer. Anal. **21**, 995–1011 (1984)
50. van Leer, B.: Towards the ultimate conservative scheme. II. Monotonicity and conservation combined in a second order scheme. J. Comput. Phys. **14**, 361–370 (1974)
51. Weatherill, N.P., Hassan, O., Marchant, M.J., Marcum, D.L.: Adaptive inviscid flow solutions for aerospace geometries on efficiently generated unstructured tetrahedral meshes. AIAA-93-3390 (1993)
52. Whitaker, D.L., Grossman, B., Löhner, R.: Two-dimensional Euler computations on a triangular mesh using an upwind, finite-volume scheme. AIAA-89-0365 (1989)
53. Woodward, P., Colella, P.: The numerical simulation of two-dimensional fluid flow with strong shocks. J. Comput. Phys. **54**, 115–173 (1984)
54. Zalesak, S.T.: Fully multidimensional flux-corrected transport algorithm for fluids. J. Comput. Phys. **31**, 335–362 (1979)
55. Zalesak, S.T., Löhner, R.: Minimizing numerical dissipation in modern shock capturing schemes. In: Chung, T.J., Karr, G. (eds.) Proc. 7th Int. Conf. Finite Elements in Flow Problems, Huntsville, AL, pp. 1205–1211 (1989)
56. Zhmakin, A.I., Fursenko, A.A.: A class of monotonic shock-capturing difference schemes. NRL Memo. Rep. 4567 (1981)

Algebraic Flux Correction I

Scalar Conservation Laws

Dmitri Kuzmin

Abstract This chapter is concerned with the design of high-resolution finite element schemes satisfying the discrete maximum principle. The presented *algebraic flux correction* paradigm is a generalization of the flux-corrected transport (FCT) methodology. Given the standard Galerkin discretization of a scalar transport equation, we decompose the antidiffusive part of the discrete operator into numerical fluxes and limit these fluxes in a conservative way. The purpose of this manipulation is to make the antidiffusive term local extremum diminishing. The available limiting techniques include a family of implicit FCT schemes and a new linearity-preserving limiter which provides a unified treatment of stationary and time-dependent problems. The use of Anderson acceleration makes it possible to design a simple and efficient quasi-Newton solver for the constrained Galerkin scheme. We also present a linearized FCT method for computations with small time steps. The numerical behavior of the proposed algorithms is illustrated by a grid convergence study for convection-dominated transport problems and anisotropic diffusion equations.

1 Introduction

A major bottleneck in finite element simulation of transport phenomena is the inability of the standard Galerkin discretization to satisfy the relevant maximum principles and/or maintain positivity on general meshes. This deficiency manifests itself in spurious undershoots and overshoots that pop up in regions of insufficient mesh resolution. Discontinuous weak solutions to hyperbolic conservation laws are particularly difficult to compute using continuous finite elements. The Galerkin "best approximations" to elliptic and parabolic transport equations may also exhibit nonphysical artifacts in proximity to unresolved small-scale features [48, 50]. An effective remedy to this problem must be found when it comes to the development of general-purpose finite element codes for Computational Fluid Dynamics.

D. Kuzmin (✉)
Applied Mathematics III, University Erlangen-Nuremberg, Cauerstr. 11, 91058, Erlangen, Germany
e-mail: kuzmin@am.uni-erlangen.de

Many modern high-resolution schemes for the equations of fluid mechanics are based on the flux-corrected transport (FCT) algorithm [6, 78] or use total variation diminishing (TVD) limiters [26, 73] to enforce the discrete maximum principle. The basic idea boils down to using a high-order scheme in smooth regions and a nonoscillatory low-order scheme elsewhere. The implementation of FCT and TVD in explicit finite element codes dates back to the late 1980s [3, 54, 55, 65, 68, 69]. The author and his coworkers developed the first implicit FCT schemes for continuous (linear and multilinear) finite elements [35, 39, 45, 47]. In the first edition of this book, we introduced the *algebraic flux correction* paradigm [43], a general framework for the design of multidimensional flux limiters. This approach leads to many useful generalizations of FCT and TVD-like methods [36, 41, 42].

The recent comparative study by John and Schmeyer [32] indicates that FEM-FCT is superior to mainstream stabilization techniques when it comes to solving unsteady convection problems with linear finite elements. However, flux correction of FCT type is inappropriate for steady-state computations since the results depend on the pseudo-time step, and severe convergence problems may occur. Flux limiters of TVD type [36, 37, 43, 46, 57] are free of these drawbacks but require mass lumping. As an alternative to FCT and TVD, we developed a linearity-preserving flux limiter that can handle stationary and time-dependent problems equally well [42]. In this algorithm, the same strategy is used to constrain the convective term, anisotropic diffusion, and the consistent mass matrix. Furthermore, linearity preservation implies consistency and second-order accuracy for smooth data [10, 59].

The cost of algebraic flux correction depends on the number of iterations required to obtain a converged solution. In our experience, this cost can be significantly reduced using a linearization of the antidiffusive term [39] or convergence acceleration techniques for iterative solvers. In particular, we recommend *Anderson mixing* [1, 17, 18, 77] (also known as *Anderson acceleration*) which combines a number of iterates in a GMRES-like fashion. As shown by Eyert [17], the accelerated solver belongs to the Broyden family of Jacobian-free quasi-Newton methods.

This chapter summarizes our work on algebraic flux correction schemes inspired by Zalesak's FCT algorithm [78]. Due to many recent developments, the presentation of this material differs considerably from the first edition of the book. In Sect. 2, we briefly review the continuous maximum principles for linear convection-diffusion equations. The discrete maximum principles and sufficient conditions of positivity preservation are formulated in Sect. 3. In Sects. 4 and 5, we analyze the standard Galerkin discretization and explain the philosophy behind algebraic flux correction. The generalized FCT algorithm and the linearity-preserving flux limiter are presented in Sects. 6 and 7, respectively. In Sect. 8, we address the design and acceleration of iterative solvers for the nonlinear system. A grid convergence study for 2D test problems is presented in Sect. 9. Finally, we summarize the results and outline some promising directions for further research.

Table 1 Taxonomy of scalar transport equations

Parabolic type	$\frac{\partial u}{\partial t} + \nabla \cdot (\mathbf{v}u - \mathsf{D}\nabla u) = 0$	$\frac{\partial u}{\partial t} - \nabla \cdot (\mathsf{D}\nabla u) = 0$
Elliptic type	$\nabla \cdot (\mathbf{v}u - \mathsf{D}\nabla u) = 0$	$-\nabla \cdot (\mathsf{D}\nabla u) = 0$
Hyperbolic type	$\frac{\partial u}{\partial t} + \nabla \cdot (\mathbf{v}u) = 0$	$\nabla \cdot (\mathbf{v}u) = 0$

2 Analysis of the Continuous Problem

The model problem that will serve as a vehicle for the presentation of our high-resolution finite element schemes is the linear convection-diffusion equation

$$\frac{\partial u}{\partial t} + \nabla \cdot (\mathbf{v}u - \mathsf{D}\nabla u) = 0 \quad \text{in } \Omega \tag{1}$$

which describes the transport of a conserved scalar quantity $u(\mathbf{x}, t)$ in a bounded domain $\Omega \subset \mathbb{R}^d$, $d \in \{1, 2, 3\}$. The velocity $\mathbf{v}$ and diffusion tensor D are given.

If all terms are present, equation (1) is parabolic Also of interest are steady-state solutions ($\frac{\partial u}{\partial t} = 0$) as well as the limiting cases of pure convection ($\mathsf{D} = 0$) and pure diffusion ($\mathbf{v} = 0$). The PDE type for each model is listed in Table 1.

In the case of unsteady transport, we prescribe an initial condition of the form

$$u(\mathbf{x}, 0) = u_0(\mathbf{x}), \quad \forall \mathbf{x} \in \Omega. \tag{2}$$

The Dirichlet-Neumann boundary conditions for our model problem are given by

$$u = u_D \quad \text{on } \Gamma_D, \tag{3}$$
$$\mathbf{n} \cdot \nabla u = 0 \quad \text{on } \Gamma_N, \tag{4}$$

where $\mathbf{n}$ is the unit outward normal to the boundary $\Gamma = \partial \Omega$. In the presence of diffusion, we have $\Gamma_D \cup \Gamma_N = \Gamma$. In the hyperbolic case, we have $\Gamma_N = \emptyset$ and

$$\Gamma_D = \{\mathbf{x} \in \Gamma \mid \mathbf{v} \cdot \mathbf{n} < 0\}.$$

Definition 1 Let Σ be the set of points where initial/boundary conditions are prescribed, i.e., $\Sigma := \Gamma_D \cup \Gamma_N$ in the steady case and $\Sigma := \{(\mathbf{x}, t) \mid \mathbf{x} \in \Gamma_D \cup \Gamma_N \vee t = 0\}$ in the unsteady case. The (continuous) *maximum principle* holds if

$$\min_{\Sigma} u \leq u \leq \max_{\Sigma} u. \tag{5}$$

If the initial and boundary conditions are nonnegative, then the maximum principle implies that $u \geq 0$. This yields another useful a priori estimate of u in terms of $u|_\Sigma$.

Definition 2 The solution of a scalar transport equation is *positivity-preserving* if

$$\min_{\Sigma} u \geq 0 \quad \implies \quad u \geq 0. \tag{6}$$

At the continuous level, the solution of a scalar transport equation without sources or sinks is always positivity-preserving but a proof of the maximum principle is available only for the case of an incompressible velocity field ($\nabla \cdot \mathbf{v} = 0$).

Theorem 1 *The following a priori estimates hold for all PDEs listed in Table* 1:

(i) $\nabla \cdot \mathbf{v} = 0 \Longrightarrow \min_\Sigma u \leq u \leq \max_\Sigma u$ *(maximum principle)*;
(ii) $u|_\Sigma \geq 0 \Longrightarrow u \geq 0$ *(positivity preservation)*.

A formal proof of this theorem for each PDE type, its generalization to equations with source terms, and some useful corollaries can be found, e.g., in [41].

The maximum principle and positivity preservation are important for several reasons. On the one hand, the a priori bounds may represent certain physical constraints. For example, concentrations of chemical species are known to lie between 0 and 1. On the other hand, some useful information about the solutions of differential solutions becomes available, although these solutions are generally unknown. Upper/lower bounds, uniqueness proofs, and comparison principles can be obtained using elementary calculus. Last but not least, discrete maximum principles play an important role in the development of numerical methods for transport equations.

3 Analysis of the Discrete Problem

Of course, a good numerical scheme must respect the known properties of exact solutions. In this section, we review algebraic constraints which imply a discrete maximum principle and/or ensure positivity preservation. In the next sections, we will use these sufficient conditions to constrain the Galerkin discretization of the unsteady convection-diffusion equation (1). We tacitly assume that one or two terms in this equation may be missing, so that it represents all models listed in Table 1.

3.1 Semi-discrete Problem

Any space discretization of (1) produces a system of differential algebraic equations

$$M \frac{du}{dt} = Qu, \tag{7}$$

where $u(t)$ is the vector of time-dependent nodal values, $M = \{m_{ij}\}$ is the so-called *mass matrix*, and $Q = \{q_{ij}\}$ is the discrete transport operator. The properties of M and Q depend on the computational mesh and on the discretization method.

As in the continuous case, the unknown solution values are known to be bounded under certain assumptions. The semi-discrete equation for u_i reads

$$\sum_j \left(m_{ij} \frac{du_j}{dt} \right) = \sum_j q_{ij} u_j. \tag{8}$$

To prevent spurious undershoots/overshoots, we impose the following constraints which imply a semi-discrete maximum principle and positivity preservation.

Theorem 2 *Consider a semi-discrete scheme of the form* (8). *Suppose that*

$$m_{ii} > 0, \quad m_{ij} = 0, \quad q_{ij} \geq 0, \quad \forall j \neq i. \tag{9}$$

Then the following a priori estimates hold for the solution value u_i:

(i) $\sum_j q_{ij} = 0$ and $u_i \geq u_j$, $\forall j \neq i$ $\implies$ $\dfrac{du_i}{dt} \leq 0$ (*semi-DMP*);

(ii) $u_j(0) \geq 0$, $\forall j$ $\implies$ $u_i(t) \geq 0$, $\forall t > 0$ (*positivity preservation*).

Proof A comparison of Theorems 1 and 2 reveals that the zero row sum property is a discrete version of the incompressibility constraint. If $\sum_j q_{ij} = 0$, then

$$\frac{du_i}{dt} = \frac{1}{m_{ii}} \sum_{j \neq i} q_{ij}(u_j - u_i). \tag{10}$$

Suppose that $u_i = \max_j u_j$. By assumption, we have $m_{ii} > 0$ and $q_{ij}(u_j - u_i) \leq 0$, $\forall j \neq i$. Thus $\frac{du_i}{dt} \leq 0$, i.e., a maximum cannot increase. This proves (i).

To prove (ii), suppose that $u_i(t) = 0$ and $u_j(t) \geq 0$ for all $j \neq i$. It follows that

$$\frac{du_i}{dt} = \frac{1}{m_{ii}} \sum_{j \neq i} q_{ij} u_j, \tag{11}$$

where $q_{ij} u_j \geq 0$, $\forall j \neq i$. Thus, the solution value u_i cannot become negative. □

Definition 3 *A space discretization of the form* (10) *with* $m_{ii} > 0$ *and* $q_{ij} \geq 0$ *for all* i *and* $j \neq i$ *is called* local extremum diminishing (LED).

The LED criterion was introduced by Jameson [29, 30] in the context of finite volume methods for unstructured grids. It is consistent with the FCT philosophy [6]: *no new extrema can form and existing extrema cannot grow*. The word *local* refers to the fact that the coefficient matrices are sparse, so only the nearest neighbors of node i make a nonzero contribution to (8) and define the bounds for u_i.

It is easy to prove that a LED scheme is total variation diminishing (TVD) in 1D. By the Godunov theorem [22], a linear positivity-preserving/LED discretization of a hyperbolic transport equation can be at most first-order accurate. The order barrier for a linear LED approximation of the diffusive term is 2 (see [28], pp. 118–120). Hence, the conditions of Theorem 2 are very restrictive in the linear case but they turn out to be a handy tool for the design of *nonlinear* high-resolution schemes.

3.2 Fully Discrete Problem

The fully discrete counterpart of problem (7) is a sparse linear system of the form

$$\begin{pmatrix} A_\Omega & A_\Gamma \\ 0 & I \end{pmatrix} \begin{pmatrix} u_\Omega \\ u_\Gamma \end{pmatrix} = \begin{pmatrix} B_\Omega & B_\Gamma \\ 0 & I \end{pmatrix} \begin{pmatrix} g_\Omega \\ g_\Gamma \end{pmatrix}, \tag{12}$$

where I is the identity matrix and $u_\Gamma = g_\Gamma$ is the vector of Dirichlet boundary values.

For a two-level time-stepping scheme, $u_\Omega = u_\Omega^{n+1}$ is the vector of unknowns and $g_\Omega = u_\Omega^n$ is the vector of solution values from the last time step. For stationary problems $B_\Omega = 0$ and $B_\Gamma = 0$. The general form of the i-th equation reads

$$a_{ii} u_i = b_{ii} g_i + \sum_{j \in S_i} (b_{ij} g_j - a_{ij} u_j), \tag{13}$$

where $S_i := \{j \neq i \mid a_{ij} \neq 0 \vee b_{ij} \neq 0\}$ is the set of nearest neighbors of node i.

Definition 4 The solution to (13) satisfies the *local* discrete maximum principle if

$$u_i^{\min} \leq u_i \leq u_i^{\max}, \tag{14}$$

where

$$u_i^{\max} := \max\left\{ \max_{j \in S_i \cup \{i\}} g_j, \max_{k \in S_i} u_k \right\}, \tag{15}$$

$$u_i^{\min} := \min\left\{ \min_{j \in S_i \cup \{i\}} g_j, \min_{k \in S_i} u_k \right\} \tag{16}$$

are the largest and smallest solution values that appear in the right-hand side of (13).

The so-defined local DMP implies that u_i should not decrease as result of increasing any other nodal value that contributes to the discretized equation for node i [66]. Conversely, u_i should not increase if another nodal value is decreased, all other things being fixed. If the given solution values are all nonnegative, then so is u_i.

Definition 5 The solution to (13) is said to be *locally* positivity-preserving if

$$u_i^{\min} \geq 0 \quad \Longrightarrow \quad u_i \geq 0. \tag{17}$$

The following theorem presents sufficient conditions of local positivity preservation and an additional constraint which guarantees the validity of (14).

Theorem 3 *Suppose that the coefficients of the discrete problem* (13) *satisfy*

$$a_{ii} > 0, \quad b_{ii} \geq 0, \quad a_{ij} \leq 0, \quad b_{ij} \geq 0, \quad \forall j \in S_i. \tag{18}$$

Then the following a priori estimates hold for the solution value u_i:

(i) $\sum_j a_{ij} = \sum_j b_{ij} \implies u_i^{\min} \leq u_i \leq u_i^{\max}$ *(local DMP)*;

(ii) $u_i^{\min} \geq 0 \implies u_i \geq 0$ *(positivity preservation)*.

Proof To prove (i), we define $w_j := u_j - u_i^{\max}$ and $v_j := g_j - u_i^{\max}$ such that

$$w_j \leq 0, \quad \forall j \in S_i, \quad v_j \leq 0, \quad \forall j \in S_i \cup \{i\}. \tag{19}$$

Using the row sum condition $\sum_j a_{ij} = \sum_j b_{ij}$, we can express (13) as follows:

$$a_{ii} w_i = b_{ii} v_i + \sum_{j \in S_i} (b_{ij} v_j - a_{ij} w_j). \tag{20}$$

By (18) and (19), the right-hand side of (20) is nonpositive. Since $a_{ii} > 0$, we have $w_i \leq 0$ or, equivalently, $u_i \leq u_i^{\max}$. The proof for $u_i \geq u_i^{\min}$ is similar.

To prove (ii), suppose that $u_i^{\min} \geq 0$, i.e., $u_j \geq 0$, $\forall j \in S_i$ and $g_j \geq 0$, $\forall j \in S_i \cup \{i\}$. By (18), the right-hand side of (13) is nonnegative, so $a_{ii} > 0 \Rightarrow u_i \geq 0$. □

The theorem implies that u_i is bounded by the solution values in a neighborhood of node i. If the local DMP holds for all nodes, then global maxima and minima must occur on the Dirichlet boundary or at the previous time level. Likewise, local positivity preservation for all nodes implies global positivity preservation.

Definition 6 The solution to (12) satisfies the *global* discrete maximum principle if

$$\min g \leq u \leq \max g, \tag{21}$$

where u denotes the vector of unknowns and g is the vector of given solution values.

Definition 7 The solution to (12) is said to be *globally* positivity-preserving if

$$g \geq 0 \quad \Longrightarrow \quad u \geq 0. \tag{22}$$

A typical proof of (21) and (22) is based on the theory of monotone matrices [76].

Definition 8 A regular matrix A is called *monotone* if $A^{-1} \geq 0$ or, equivalently, if

$$u \geq 0 \quad \Longrightarrow \quad Au \geq 0.$$

Definition 9 A monotone matrix A with $a_{ij} \leq 0$, $\forall j \neq i$ is called an *M-matrix*.

Theorem 4 *Consider a fully discrete scheme of the form $Au = Bg$. Suppose that the coefficients of $A = \{a_{ij}\}$ and $B = \{b_{ij}\}$ satisfy conditions (18) for all i.*

If A is strictly or irreducibly diagonally dominant, then A is an M-matrix and

(i) *the global DMP holds if $\sum_j a_{ij} = \sum_j b_{ij}$, $\forall i$;*
(ii) *the scheme is globally positivity-preserving.*

Proof We refer to Varga [76] for a proof of the M-matrix property. To prove the global DMP, define the vectors $w := u - \max g$ and $v := g - \max g$. Invoking (20), we obtain a linear system of the form $Aw = Bv$, where A is monotone, $B \geq 0$, and $v \leq 0$. Hence $w = A^{-1} Bv \leq 0$, which implies $u \leq \max g$. Similarly, the solution to $Au = Bg$ proves positivity-preserving since $u = A^{-1} Bg \geq 0$ whenever $g \geq 0$. □

4 Galerkin Discretization

Some finite element approximations are known to satisfy the conditions of Theorems 3 and 4 unconditionally or under mild restrictions on the geometric properties of the mesh (no obtuse angles, no thin elements) [12, 19, 33, 34]. However, these sufficient conditions become too restrictive in the case of high-order finite elements, convection-dominated transport equations, and anisotropic diffusion problems. The usual remedy is to add a certain amount of artificial diffusion in order to compensate the contribution of matrix entries that have a wrong sign. Many shock capturing techniques, including algebraic flux correction, are based on this approach.

We will explain the principles of algebraic flux correction in the finite element context. To begin with, let us discretize the generic transport equation (1) using the (continuous) Galerkin approximation which delivers optimal accuracy in smooth regions but tends to produce spurious undershoots and overshoots elsewhere.

The variational form of our Dirichlet-Neumann boundary value problem reads

$$\int_\Omega w \left(\frac{\partial u}{\partial t} + \nabla \cdot (\mathbf{v}u) \right) d\mathbf{x} + \int_\Omega \nabla w \cdot (D\nabla u) \, d\mathbf{x} = 0 \qquad (23)$$

for all admissible test functions w vanishing on the Dirichlet boundary Γ_D. We assume sufficient regularity without giving a formal definition of Sobolev spaces.

Let $\{\varphi_j\}$ be a finite set of piecewise-linear or multilinear basis functions. The numerical solution $u_h \approx u$ is defined as a linear combination thereof

$$u_h = \sum_j u_j \varphi_j. \qquad (24)$$

The unknown degrees of freedom are the coefficients u_j which represent the (possibly time-dependent) values of u_h at the vertices of the mesh.

Instead of differentiating the convective flux, we replace it with the interpolant

$$(\mathbf{v}u)_h = \sum_j (\mathbf{v}_j u_j) \varphi_j, \qquad (25)$$

where $\mathbf{v}_j$ denotes the velocity at node j. This approach is known as the *group finite element* formulation [20, 21]. The divergence of (25) is given by

$$\nabla \cdot (\mathbf{v}u)_h = \sum_j u_j (\mathbf{v}_j \cdot \nabla \varphi_j). \qquad (26)$$

The contribution of the diffusive flux is evaluated using the consistent gradient

$$\nabla u_h = \sum_j u_j \nabla \varphi_j. \qquad (27)$$

To obtain a semi-discrete equation for the solution value u_i, substitute approximations (24), (26), and (27) into (23) with the test function $w_h := \varphi_i$. This gives

$$\sum_j \left(\int_\Omega \varphi_i \varphi_j \, d\mathbf{x} \right) \frac{du_j}{dt} = -\sum_j \mathbf{v}_j \cdot \left(\int_\Omega \varphi_i \nabla \varphi_j \, d\mathbf{x} \right) u_j$$

$$-\sum_j \left(\int_\Omega \nabla \varphi_i \cdot (D \nabla \varphi_j) \, dx \right) u_j. \tag{28}$$

The resultant semi-discrete problem can be written in the generic matrix form

$$M_C \frac{du}{dt} = (K - L) u, \tag{29}$$

where $M_C = \{m_{ij}\}$ denotes the consistent mass matrix, $K = \{k_{ij}\}$ is the convective part of the discrete transport operator, and $L = \{l_{ij}\}$ is the contribution of the diffusive term. By (28) the coefficients of the three matrices are given by

$$m_{ij} = \int_\Omega \varphi_i \varphi_j \, dx, \qquad l_{ij} = \int_\Omega \nabla \varphi_i \cdot (D \nabla \varphi_j) \, dx, \tag{30}$$

$$k_{ij} = -\mathbf{v}_j \cdot \mathbf{c}_{ij}, \qquad \mathbf{c}_{ij} = \int_\Omega \varphi_i \nabla \varphi_j \, dx. \tag{31}$$

In the case of an unsteady velocity field, the convective part K must be updated at each time step. If the mesh is fixed, then the coefficients $\mathbf{c}_{ij}$ of the discrete gradient operator do not change and need to be evaluated just once. Hence, the group finite element formulation makes it possible to update K in a very efficient way.

Let $0 = t_0 < t^1 < t^2 < \cdots < t^M = T$ be a sequence of discrete time levels for the time integration of (29). For simplicity, we assume that the time step $\Delta t := t^{n+1} - t^n$ is constant so that $t^n = n \Delta t$. By the Fundamental Theorem of Calculus

$$M_C (u^{n+1} - u^n) = \int_{t^n}^{t^{n+1}} (K - L) u \, dt.$$

The integral is approximated using a suitable quadrature rule. In particular, we will consider the fully discrete problem for the standard θ-scheme

$$[M_C - \theta \Delta t (K - L)] u^{n+1} = [M_C + (1 - \theta) \Delta t (K - L)] u^n, \tag{32}$$

where $\theta \in [0, 1]$ is the degree of implicitness. The forward Euler ($\theta = 0$) version is unstable for convection-dominated transport problems and gives rise to severe time step restrictions in the case of dominating diffusion. For this reason, we restrict ourselves to the unconditionally stable Crank-Nicolson ($\theta = \frac{1}{2}$) and backward Euler ($\theta = 1$) time stepping. If a fully explicit treatment is desired, we recommend the family of strong stability-preserving Runge-Kutta methods [23, 24] which guarantee the local and global DMP if the underlying space discretization is LED and the time steps are sufficiently small. Other explicit schemes can generate spurious oscillations even if the space discretization satisfies the conditions of Theorem 2.

5 Algebraic Flux Correction

The fully discrete scheme (32) is a linear system of the form $Au^{n+1} = Bu^n$. The diagonal entries of the matrices A and B are positive, at least for sufficiently small time steps Δt. However, a violation of the DMP conditions (18) may be caused by

- positive off-diagonal entries of the consistent mass matrix M_C;
- negative off-diagonal entries of the discrete convection operator K;
- positive off-diagonal entries of the discrete diffusion operator L.

In the process of algebraic flux correction, we constrain the contribution of these entries trying to stay as close as possible to the original Galerkin discretization.

The "good" part of the Galerkin scheme (29) is an ODE system of the form

$$M_L \frac{du}{dt} = (\tilde{K} - \tilde{L})u, \tag{33}$$

where M_L and $\tilde{K} - \tilde{L}$ satisfy the conditions of Theorem 2. We define these matrices in Sect. 5.1. The "bad" *antidiffusive* part of (29) is given by

$$f(u) = (M_L - M_C)\frac{du}{dt} + (K - \tilde{K})u - (L - \tilde{L})u. \tag{34}$$

To prevent a possible violation of the (semi-)discrete maximum principle, we decompose the antidiffusive term $f(u)$ into numerical fluxes and limit the magnitude of these fluxes in regions where they threaten to create an undershoot or overshoot. To this end, each flux is multiplied by a solution-dependent correction factor. In contrast to mainstream stabilization techniques for finite elements, there are no free parameters. The constrained Galerkin scheme is guaranteed to be positivity-preserving and satisfy the DMP if it holds for the solution of the continuous problem.

In this section, we review the design philosophy behind algebraic flux correction. Some generalizations [39, 45, 47] of the multidimensional FCT algorithm [78] and a new linearity-preserving flux limiter [42, 48] are presented in the next section.

5.1 Artificial Diffusion Operators

The derivation of (33) begins with *row-sum mass lumping*. In explicit finite element codes, the mass matrix M_C is frequently replaced with the diagonal approximation

$$M_L := \text{diag}\{m_i\}, \quad m_i = \sum_j m_{ij}. \tag{35}$$

For linear finite elements, this conservative modification is equivalent to inexact evaluation of M_C with a low-order Newton-Cotes quadrature rule [25].

The negative off-diagonal entries of the (nonsymmetric) convection operator K are eliminated by adding a suitably designed artificial diffusion operator D.

Definition 10 A symmetric matrix $D = \{d_{ij}\}$ is called a *discrete diffusion operator* if D has zero row and column sums [45]. That is,

$$d_{ij} = d_{ji}, \quad \sum_j d_{ij} = \sum_i d_{ij} = 0. \tag{36}$$

To make sure that $\tilde{K} := K + D$ has no negative off-diagonal entries, we define

$$d_{ij} := \max\{-k_{ij}, 0, -k_{ji}\}, \quad \forall j \neq i. \tag{37}$$

Remark 1 Artificial diffusion coefficients that enforce positivity in this way were used to construct low-order schemes for FCT as early as in the mid-1970s [8].

Definition (37) implies $d_{ij} = d_{ji}$. To comply with the zero row sum condition, let

$$d_{ii} := -\sum_{j \neq i} d_{ij}. \tag{38}$$

By symmetry, the column sums are also equal to zero, so D satisfies conditions (36).

In practice, there is no need to assemble the global matrix D. Instead, artificial diffusion can be built into K in a loop over the edges of its sparsity graph. By definition, each edge is a pair of nodes $\{i, j\}$ that corresponds to a pair of nonzero off-diagonal coefficients k_{ij} and k_{ji}. The required update is as follows:

$$\begin{aligned} k_{ii} &:= k_{ii} - d_{ij}, & k_{ij} &:= k_{ij} + d_{ij}, \\ k_{ji} &:= k_{ji} + d_{ij}, & k_{jj} &:= k_{jj} - d_{ij}. \end{aligned} \tag{39}$$

Without loss of generality, the edges of the sparsity graph are oriented so that

$$k_{ij} \leq k_{ji}. \tag{40}$$

This orientation convention implies that node i is located 'upwind' and corresponds to the row number of the negative off-diagonal entry to be eliminated.

Physical diffusion can be taken into account before or after the assembly of D. If some off-diagonal entries of L are strictly positive, we split it into the antidiffusive part $L^+ = \{l_{ij}^+\}$ and the remainder $\tilde{L} := L - L^+$. The entries of L^+ are given by

$$l_{ii}^+ := -\sum_{j \neq i} l_{ij}^+, \qquad l_{ij}^+ := \max\{0, l_{ij}\}, \quad \forall j \neq i. \tag{41}$$

The conservative elimination of l_{ij}^+ can also be performed edge-by-edge

$$\begin{aligned} l_{ii} &:= l_{ii} + l_{ij}^+, & l_{ij} &:= l_{ij} - l_{ij}^+, \\ l_{ji} &:= l_{ji} - l_{ij}^+, & l_{jj} &:= l_{jj} + l_{ij}^+. \end{aligned} \tag{42}$$

If all off-diagonal entries of L are nonpositive, then $L^+ = 0$ and $\tilde{L} = L$. However, the standard Galerkin approximation may fail to satisfy the DMP conditions if the mesh and/or the diffusion tensor are highly anisotropic [50]. In this case, $\tilde{L}$ is a monotone but possibly inconsistent approximation to L. The lack of consistency must be compensated in the course of flux correction (see Sect. 7).

Example 1 To clarify the implications of (39), consider the 1D convection equation

$$\frac{\partial u}{\partial t} + v \frac{\partial u}{\partial x} = 0 \quad \text{in } \Omega = (0, 1), \tag{43}$$

where v is a positive constant. The inflow boundary condition is given by

$$u(0) = g.$$

On a uniform mesh of linear finite elements, the standard Galerkin method yields

$$K = \frac{1}{2} \begin{pmatrix} \cdots & & & & \\ & v & 0 & -v & \\ & & v & 0 & -v \\ & & & v & 0 & -v \\ & & & & & \cdots \end{pmatrix}.$$

For any interior node, $m_i = \Delta x$, where Δx is the constant mesh size. Hence, the lumped-mass version of (29) is equivalent to the central difference scheme

$$\frac{du_i}{dt} + v \frac{u_{i+1} - u_{i-1}}{2\Delta x} = 0.$$

Since $k_{ij} = -\frac{v}{2}$ for $j = i+1$, the artificial diffusion coefficient (37) is $d_{ij} = \frac{v}{2}$ and

$$\tilde{K} = \begin{pmatrix} \cdots & & & & \\ & v & -v & 0 & \\ & & v & -v & 0 \\ & & & v & -v & 0 \\ & & & & & \cdots \end{pmatrix},$$

which corresponds to the first-order accurate upwind difference approximation

$$\frac{du_i}{dt} + v \frac{u_i - u_{i-1}}{\Delta x} = 0.$$

Thus, the elimination of negative off-diagonal entries from a skew-symmetric operator K can be interpreted as *discrete upwinding* [43]. For any pair of nodes i and $j = i+1$ numbered in accordance with (40), the grid point x_i lies upstream of x_j.

After the discretization in time by the standard θ-scheme, the upwind difference method proves positivity-preserving under the CFL-like condition [45]

$$v \frac{\Delta t}{\Delta x} \leq \frac{1}{1-\theta}, \quad 0 \leq \theta < 1. \tag{44}$$

According to this formula, there is no time step restriction for the backward Euler method ($\theta = 1$) which corresponds to first-order 'upwinding' in time.

5.2 Conservative Flux Decomposition

The replacement of the high-order Galerkin scheme (29) by the perturbed system (33) ensures positivity preservation but creates a lot of numerical diffusion. The next ingredient of an algebraic flux correction scheme is a decomposition of (34) into a sum of numerical fluxes. These antidiffusive fluxes enable us to remove artificial diffusion in regions where the Galerkin solution is sufficiently smooth.

The antidiffusive term (34) represents the difference between the residuals of systems (29) and (33). By definition of the matrices M_L, $\tilde{K}$, and $\tilde{L}$, we have

$$f(u) = (M_L - M_C)\frac{du}{dt} - Du - L^+ u. \tag{45}$$

By construction, the matrices $M_C - M_L$, D, and L^+ are discrete diffusion operators in the sense of Definition 10. Using the zero row sum property, we obtain

$$(M_C u - M_L u)_i = \sum_j m_{ij} u_j - u_i \sum_j m_{ij} = \sum_{j \neq i} m_{ij}(u_j - u_i), \tag{46}$$

$$(Du)_i = \sum_j d_{ij} u_j = \sum_{j \neq i} d_{ij} u_j + d_{ii} u_i = \sum_{j \neq i} d_{ij}(u_j - u_i), \tag{47}$$

$$(L^+ u)_i = \sum_j l_{ij}^+ u_j = \sum_{j \neq i} l_{ij}^+ u_j + l_{ii}^+ u_i = \sum_{j \neq i} l_{ij}^+ (u_j - u_i). \tag{48}$$

The right-hand sides of (47)–(48) resemble that of a LED scheme. By symmetry, the components of the sums over $j \neq i$ can be interpreted as numerical fluxes that describe a conservative mass exchange between a pair of nodes. Let

$$f_{ij} = \left(m_{ij}\frac{d}{dt} + d_{ij} + l_{ij}^+\right)(u_i - u_j), \quad \forall j \neq i \tag{49}$$

denote the *raw antidiffusive flux* from node j into node i. In the fully discrete version, the time derivative is replaced with a finite difference.

The net antidiffusion received by node i admits the following decomposition

$$f_i = \sum_{j \neq i} f_{ij}, \quad f_{ji} = -f_{ij}. \tag{50}$$

Since $f_{ij} + f_{ji} = 0$ by definition, the antidiffusive term does not change the total mass of the discrete solution. The mass added to node i is subtracted from its neighbors.

5.3 Limited Antidiffusive Correction

Some of the raw antidiffusive fluxes f_{ij} are harmless but others may create an undershoot or overshoot. The contribution of these "bad" fluxes must be limited so as to keep the antidiffusive term local extremum diminishing for a given solution.

The flux-corrected counterpart of (29) is a semi-discrete problem of the form

$$M_L \frac{du}{dt} = (\tilde{K} - \tilde{L})u + \bar{f}(u), \tag{51}$$

where the (nonlinear) term $\bar{f}(u)$ stands for the sum of limited antidiffusive fluxes

$$\bar{f}_i = \sum_{j \neq i} \bar{f}_{ij}, \quad \bar{f}_{ji} = -\bar{f}_{ij}. \tag{52}$$

A well-designed flux limiter produces $\bar{f}_{ij} = f_{ij}$ in smooth regions and $\bar{f}_{ij} = 0$ in interior or boundary layers. The unconstrained Galerkin scheme (29) and its nonoscillatory part (33) correspond to $\bar{f} = f$ and $\bar{f} = 0$, respectively.

In general, the best definition of $\bar{f}_{ij}$ satisfying the LED constraint is given by the solution of a constrained optimization problem [5]. A nonoptimal but cost-effective alternative is the multiplication by a solution-dependent correction factor

$$\bar{f}_{ij} := \alpha_{ij} f_{ij}, \quad 0 \leq \alpha_{ij} \leq 1. \tag{53}$$

This kind of flux correction traces its origins to the FCT algorithm and forms the basis for the construction of our algebraic flux correction schemes.

The following criterion guarantees that the antidiffusive term (52) is LED

$$\sum_{j \neq i} q_{ij} \min\{0, u_j - u_i\} \leq \sum_{j \neq i} \alpha_{ij} f_{ij} \leq \sum_{j \neq i} q_{ij} \max\{0, u_j - u_i\} \tag{54}$$

for a given set of bounded nonnegative coefficients q_{ij}. The upper and lower bounds may consist of a single term associated with a local maximum or minimum

$$u_i^{\max} := \max\left\{u_i, \max_{j \in S_i} u_j\right\}, \tag{55}$$

$$u_i^{\min} := \min\left\{u_i, \min_{j \in S_i} u_j\right\}. \tag{56}$$

Introducing $q_i := \sum_{j \neq i} q_{ij}$, we can replace (54) with the weakened LED constraint

$$q_i \left(u_i^{\min} - u_i\right) \leq \sum_{j \neq i} \alpha_{ij} f_{ij} \leq q_i \left(u_i^{\max} - u_i\right). \tag{57}$$

If u_i is a local maximum, then (54) and (57) imply the cancellation of all positive fluxes. Similarly, all negative fluxes are canceled if u_i is a local maximum. Hence, the sum of $\bar{f}_{ij} := \alpha_{ij} f_{ij}$ cannot create an undershoot or overshoot at node i.

The above criteria provide a general framework for the design of algebraic flux correction schemes that differ in the definition of the LED bounds for the sum of limited antidiffusive fluxes. The best choice of q_{ij} and q_i is dictated by accuracy and efficiency considerations. Obviously, increasing the value of these parameters makes the bounds less restrictive. However, this may cause divergence of iterative solvers for the resultant nonlinear system. For accuracy reasons, it is essential to guarantee that $\alpha_{ij} = 1$ is acceptable whenever the solution varies linearly in a neighborhood of node i. This design principle is called *linearity preservation* [5, 10, 59].

We use a generalization of Zalesak's FCT algorithm [78] to calculate α_{ij} satisfying (57). The same limiting strategy is used to enforce the LED bounds defined by (54) in algebraic flux correction schemes of TVD type [37, 41, 46]. In the following sections, we address the design of multidimensional flux limiters and the iterative solution of nonlinear systems produced by the constrained Galerkin schemes.

6 Generalized FCT Algorithms

FCT was the first nonlinear high-resolution scheme to be equipped with a flux limiter. The classical FCT algorithms of Boris, Book, and Hain [6, 8, 9] belong to the class of *diffusion-antidiffusion* (DAD) methods [13] that involve two steps:

1. Advance the solution in time with an explicit low-order scheme containing enough numerical diffusion to suppress undershoots and overshoots.
2. Correct the solution using antidiffusive fluxes limited in such a way that no new maxima or minima can form and existing extrema cannot grow.

The numerical diffusion of the low-order method makes it possible to maintain positivity and improves the phase accuracy of an explicit approximation to the convective term. The limited antidiffusive correction reduces the amplitude errors in a LED manner. In contrast to TVD methods [26, 73], the upper and lower bounds for the FCT limiter are defined in terms of the low-order predictor and designed to accept as much antidiffusion as possible without violating the positivity constraint.

Zalesak's fully multidimensional FCT algorithm [78] is based on blending explicit high- and low-order approximations so as to constrain the maximum and minimum increments to each nodal value. The work of Zalesak has formed the basis for the development of all algebraic flux correction schemes to be presented in this chapter. The combination of FCT with finite elements and unstructured meshes dates back to the explicit algorithms of Parrott and Christie [65] and Löhner et al. [54, 55]. A number of implicit FEM-FCT schemes were published by the author and his coworkers [35, 39, 45, 47]. The rationale for the use of an implicit time discretization stems from the fact that the CFL stability condition becomes too restrictive in the case of nonuniform velocity fields and locally refined meshes.

In this section, we begin with a presentation of predictor-corrector FCT algorithms in which the antidiffusive fluxes are linearized about a provisional low-order solution. The linearized FCT scheme [39] is recommended for evolutionary problems that call for the use of small time steps. We also present the nonlinear version of this scheme which requires iterative flux correction. In particular, we describe an algorithm for 'recycling' the rejected antidiffusion step-by-step [47]. Finally, we summarize the pros and cons of the FCT approach to algebraic flux correction.

6.1 Linearized FCT Scheme

After the discretization in time by the two-level θ-scheme, the constrained Galerkin discretization (51) produces a nonlinear algebraic system of the form

$$Au^{n+1} = Bu^n + \bar{f}, \tag{58}$$

where $\bar{f} = \bar{f}(u^{n+1}, u^n)$ denotes the limited antidiffusive term. The matrices

$$A = \frac{1}{\Delta t} M_L - \theta(\tilde{K} - \tilde{L}) \tag{59}$$

and

$$B = \frac{1}{\Delta t}M_L + (1-\theta)(\tilde{K} - \tilde{L}) \tag{60}$$

represent the nonoscillatory low-order part of the original Galerkin scheme. If the governing equation is nonlinear or the velocity field is time-dependent, then the coefficients of A and B may change as the solution evolves.

If the time step Δt is relatively small, it is worthwhile to linearize (58) using a predictor-corrector strategy. At the first step of the linearized FCT algorithm [39], we disregard the antidiffusive term $\bar{f}$ and solve the linear system

$$Au^L = Bu^n. \tag{61}$$

By construction, the off-diagonal entries of $\tilde{K}$ and $\tilde{L}$ are nonnegative and nonpositive, respectively. The diagonal coefficients of these matrices have the opposite sign (except in the case of a strongly compressible velocity field). By Theorem 4, our low-order scheme (61) is positivity-preserving under the CFL-like condition

$$\Delta t \leq \frac{1}{1-\theta} \frac{m_i}{\tilde{l}_{ii} - \tilde{k}_{ii}}, \quad \forall i. \tag{62}$$

Furthermore, the discrete maximum principle holds if A and B have equal row sums.

Remark 2 Van Slingerland [74, 75] proposed a variable-order θ-scheme in which (62) is used to determine the optimal degree of implicitness $\theta_{ij} \in [0, 1]$ individually for each pair of nodes. This approach requires a conservative flux decomposition not only for the antidiffusive term but also for the low-order operator.

Remark 3 The two-level θ-scheme can be replaced with any other time integration scheme, e.g., a strong stability-preserving (TVD) Runge-Kutta method [23, 24]. Clearly, the time step restriction will depend on the time integration method.

The low-order predictor u^L is used to evaluate the raw antidiffusive fluxes

$$f_{ij} = m_{ij}(\dot{u}_i^L - \dot{u}_j^L) + d_{ij}(u_i^L - u_j^L), \quad j \neq i \tag{63}$$

where $\dot{u}^L$ is an approximation to the vector of time derivatives at the time level t^{n+1}.

For example, the semi-discrete low-order scheme (33) with $u := u^L$ yields

$$\dot{u}^L = M_L^{-1}\big[(\tilde{K} - \tilde{L})u^L\big]. \tag{64}$$

The so-defined approximation is smooth but diffusive. Another option is

$$\dot{u}^L = M_C^{-1}\big[(K - L)u^L\big]. \tag{65}$$

This formula follows from (29). The well-conditioned mass matrix M_C can be 'inverted' with 3–5 cycles of the preconditioned Richardson iteration [16, 39].

The raw antidiffusive fluxes f_{ij} are passed to the multidimensional FCT limiter (see Sect. 6.4) which returns a set of correction factors α_{ij}. This gives

$$\bar{f}_i = \sum_{j \neq i} \alpha_{ij} f_{ij}, \quad 0 \leq \alpha_{ij} \leq 1. \tag{66}$$

After flux limiting, the final solution u^{n+1} is obtained with the explicit correction

$$M_L u^{n+1} = M_L u^L + \Delta t \bar{f}. \qquad (67)$$

Remark 4 Due to the linearization about u^L, the unconstrained ($\alpha_{ij} := 1$) version of the above algorithm is no longer equivalent to the original Galerkin scheme.

6.2 Nonlinear FCT Scheme

Linearization errors are avoided if the nonlinear system (58) is solved in an iterative fashion. This approach leads to a FEM-FCT algorithm in which the antidiffusive fluxes f_{ij} and the corresponding correction factors α_{ij} are updated step-by-step until the residuals or relative changes become smaller than a prescribed tolerance.

Let $\{u^{(m)}\}$ be a sequence of successive approximations to the flux-corrected Galerkin solution u^{n+1}. A reasonable initial guess is $u^{(0)} = u^n$ or $u^{(0)} = 2u^n - u^{n-1}$. These settings correspond to the constant and linear extrapolation in time, respectively. Given the current iterate $u^{(m)}$ and the vector of approximate time derivatives

$$\dot{u}^{(m)} := \frac{u^{(m)} - u^n}{\Delta t}, \qquad (68)$$

we recalculate the implicit part of the raw antidiffusive fluxes given by

$$f_{ij}^{(m)} = m_{ij}\left(\dot{u}_i^{(m)} - \dot{u}_j^{(m)}\right) + \theta\left(d_{ij} + l_{ij}^+\right)\left(u_i^{(m)} - u_j^{(m)}\right)$$
$$+ (1-\theta)\left(d_{ij} + l_{ij}^+\right)\left(u_i^n - u_j^n\right), \quad j \neq i. \qquad (69)$$

Then we apply the FCT limiter (see Sect. 6.4) and solve the linear system

$$A u^{(m+1)} = B u^n + \bar{f}^{(m)}. \qquad (70)$$

Each solution update of the form (70) can be split into three steps [35, 39]

1. Compute an explicit low-order approximation to $u^{n+1-\theta}$ by solving

$$M_L \tilde{u}^{(0)} = B u^n. \qquad (71)$$

2. Apply limited antidiffusive fluxes to the intermediate solution $\tilde{u}$

$$M_L \tilde{u}^{(m+1)} = M_L \tilde{u}^{(0)} + \Delta t \, \bar{f}^{(m)}. \qquad (72)$$

3. Solve the linear system for the new approximation to u^{n+1}

$$A u^{(m+1)} = M_L \tilde{u}^{(m+1)}. \qquad (73)$$

Note that the auxiliary solution $\tilde{u}^{(0)}$ is independent of the iteration number m, so it needs to be determined just once per time step (for $m = 0$). For its computation to be positivity-preserving, the time step Δt must satisfy (62). The flux limiting procedure presented in Sect. 6.4 guarantees that $\tilde{u}^{(0)} \geq 0 \implies \tilde{u}^{(m+1)} \geq 0$. The last solution update is positivity-preserving by the M-matrix property of A. Thus

$$u^{(0)} \geq 0 \implies \tilde{u}^{(0)} \geq 0 \implies \tilde{u}^{(m+1)} \geq 0 \implies u^{(m+1)} \geq 0 \qquad (74)$$

provided that the CFL-like condition (62) holds for the given Δt and $\theta \in (0, 1]$.

6.3 Iterative FCT Scheme

The implicit FCT scheme (71)–(73) 'forgets' the history of previous flux correction steps when it comes to the assembly of $\bar{f}^{(m)}$. Hence, it tends to reject more antidiffusion than necessary to enforce the positivity constraint. As shown by Schär and Smolarkiewicz [63], an iterative 'recycling' of the rejected antidiffusive fluxes may significantly improve the accuracy of an FCT algorithm in some cases.

An iterative limiting strategy for maximizing the amount of accepted antidiffusion in implicit FCT schemes was developed in [47]. Replacing (72) with

$$M_L \tilde{u}^{(m+1)} = M_L \tilde{u}^{(m)} + \Delta t \bar{f}^{(m)}, \tag{75}$$

we perform flux limiting in terms of $\tilde{u}^{(m)}$ rather than $\tilde{u}^{(0)}$. The sum of all previous corrections is built into $\tilde{u}^{(m)}$, so only the remainder of $f_{ij}^{(m)}$ needs to be limited

$$f_{ij}^{(m)} := f_{ij}^{(m)} - \sum_{k=0}^{m-1} \alpha_{ij}^{(k)} f_{ij}^{(k)} \tag{76}$$

for all $m > 0$. This simplifies the job of the flux limiter and enables it to accept more antidiffusion. For a detailed description of the algorithm, we refer to [47].

Iterative FCT is more accurate than (71)–(73) but converges very slowly. For this reason, we do not recommend its use unless it is justified by unusually stringent accuracy requirements. For many problems of practical interest, the predictor-corrector approach presented in Sect. 6.1 offers the best cost/accuracy ratio.

6.4 Zalesak's FCT Limiter

In this section, we present Zalesak's limiter [78] that we use to calculate the correction factors α_{ij} for all FCT schemes. Consider a solution update of the form

$$m_i u_i = m_i \tilde{u}_i + \Delta t \sum_{j \neq i} \alpha_{ij} f_{ij}, \tag{77}$$

where $\tilde{u}$ is a nonoscillatory intermediate solution. Let $u_i^{\max}$ and $u_i^{\min}$ denote the local extrema of $\tilde{u}$. The objective is to find the best value of α_{ij} such that

$$u_i^{\min} \leq u_i \leq u_i^{\max}. \tag{78}$$

This condition implies that (77) satisfies the local discrete maximum principle.

6.4.1 Prelimiting Step

The process of flux correction begins with the optional elimination of fluxes that have the same sign as $\tilde{u}_j - \tilde{u}_i$. Such fluxes flatten the solution profile instead of steepening it. As a consequence, the flux-corrected solution may exhibit spurious ripples within the bounds allowed by the limiter [14]. In the original Boris-Book limiter [6], a wrong sign is reversed, and the magnitude of the antidiffusive flux is limited in the usual way. This fix works well for discontinuities but may distort a smooth profile. A safer remedy is to cancel the "diffusive" fluxes by setting

$$f_{ij} := 0, \quad \text{if } f_{ij}(\tilde{u}_j - \tilde{u}_i) > 0. \tag{79}$$

This optional adjustment is called *prelimiting* because it must be performed before the computation of the correction factors α_{ij} and flux limiting [14, 78].

Zalesak [78] argued that the effect of (79) is marginal and cosmetic in nature since the vast majority of antidiffusive fluxes have the right sign. This remark might have led many readers to disregard equations (14) and (14′) in [78]. Two decades later, the need for prelimiting of the form (79) was emphasized by DeVore [14] who explained its ramifications and demonstrated that it may lead to a marked improvement of accuracy. In our experience, prelimiting is particularly useful in the context of finite element approximations because the contribution of the consistent mass matrix may change the sign of the raw antidiffusive flux and render it diffusive. The cancellation of such outliers is essential for keeping the solution free of ripples.

6.4.2 Limiting Strategy

In accordance with the LED criterion (57), the choice of the correction factors α_{ij} should ensure that positive antidiffusive fluxes cannot create an overshoot, while negative ones cannot create an undershoot. Assuming the worst-case scenario, we enforce condition (78) using Zalesak's multidimensional FCT algorithm [78]:

1. Compute the sums of positive/negative antidiffusive fluxes into node i

$$P_i^+ = \sum_{j \neq i} \max\{0, f_{ij}\}, \qquad P_i^- = \sum_{j \neq i} \min\{0, f_{ij}\}. \tag{80}$$

2. Determine the distance to a local maximum/minimum and the bounds

$$Q_i^+ = \frac{m_i}{\Delta t}(u_i^{\max} - \tilde{u}_i), \qquad Q_i^- = \frac{m_i}{\Delta t}(u_i^{\min} - \tilde{u}_i). \tag{81}$$

3. Evaluate the nodal correction factors for the net increment to node i

$$R_i^+ = \min\left\{1, \frac{Q_i^+}{P_i^+}\right\}, \qquad R_i^- = \min\left\{1, \frac{Q_i^-}{P_i^-}\right\}. \tag{82}$$

4. Check the sign of the raw antidiffusive flux f_{ij} and multiply it by

$$\alpha_{ij} = \begin{cases} \min\{R_i^+, R_j^-\}, & \text{if } f_{ij} > 0, \\ \min\{R_i^-, R_j^+\}, & \text{if } f_{ij} < 0. \end{cases} \tag{83}$$

Remark 5 It is worthwhile to set $R_i^\pm := 1$ if a Dirichlet boundary condition is imposed at node i and, therefore, the value of u_i does not depend on α_{ij}.

The above definition of α_{ij} guarantees that sum of limited antidiffusive fluxes satisfies (57) with $q_i = \frac{m_i}{\Delta t}$. The LED property (78) follows from the estimate

$$u_i^{\min} = \tilde{u}_i + \frac{\Delta t}{m_i} Q_i^- \leq u_i \leq \tilde{u}_i + \frac{\Delta t}{m_i} Q_i^+ = u_i^{\max}.$$

The presence of the time step Δt in the denominator of $Q_i^\pm$ is a blessing or a curse, depending on the purpose of simulation. On the one hand, the LED constraints become less restrictive and, consequently, a larger portion of the raw antidiffusive flux f_{ij} is retained as the time step is refined. This makes FCT the method of choice for transient computations. On the other hand, the use of large Δt results in a loss of accuracy, and severe convergence problems may occur in the steady state limit.

6.4.3 Edge-Based Implementation

The practical implementation of Zalesak's FCT limiter depends on the employed data structures, storage techniques, and software development concepts. In the pseudo-code labeled Algorithm 1, we take advantage of the fact that $f_{ji} = -f_{ij}$ and $\alpha_{ji} = \alpha_{ij}$. The flux sums $P_i^\pm$ and the corresponding bounds $Q_i^\pm$ are assembled in a loop over all neighbors $j \in S_i$ such that $j > i$. The values of $P_j^\mp$ and $Q_j^\mp$ are updated in the same j-loop. The nodal correction factors $R_i^\pm$ are evaluated in the next i-loop. When it comes to the assembly of the antidiffusive term, flux limiting is performed in another loop over $j > i$. The flux $\bar{f}_{ij} := \alpha_{ij} f_{ij}$ is added to $\bar{f}_i$ and subtracted from $\bar{f}_j$. This implementation of FCT calls for the use of edge-based data structures [4, 67, 70] which operate with pairs of nodes, just like finite volume schemes.

The advantages of edge-based finite element solvers include algorithmic simplicity, low memory requirements, and a major reduction in indirect addressing [52, 53]. Moreover, edge-based data structures are well-suited for large-scale parallel computing [11, 51, 58]. Last but not least, the equivalence between linear finite elements and vertex-centered finite volumes can be exploited to develop a unified framework for edge-based flux/slope limiting on unstructured meshes [57, 71].

Of course, algebraic flux correction schemes can also be implemented in an existing finite element code based on traditional element-by-element matrix assembly. In our code, we use edge-based data structures for limiting purposes only.

6.4.4 Clipping and Terracing

A well-known problem associated with flux correction of FCT type is *clipping* [7, 78]. Since the sum of limited antidiffusive fluxes is forced to be local extremum diminishing, existing peaks lose a little bit of amplitude during each time step. To alleviate peak clipping, Zalesak [78] defined $u_i^{\max}$ and $u_i^{\min}$ as the local extrema of u^n or $\tilde{u}$. This adjustment is consistent with the local discrete maximum principle

Algorithm 1 Edge-based implementation of FCT

$P^\pm := 0, \ Q^\pm := 0, \ \bar{f} := 0$

For all i **do**

    For all $j \in S_i, \ j > i$ **do**

$$P_i^\pm := P_i^\pm + {\max \atop \min}\{0, f_{ij}\}$$

$$P_j^\pm := P_j^\pm + {\max \atop \min}\{0, -f_{ij}\}$$

$$Q_i^\pm := {\max \atop \min}\{Q_i^\pm, \tfrac{m_i}{\Delta t}(u_j - u_i)\}$$

$$Q_j^\pm := {\max \atop \min}\{Q_j^\pm, \tfrac{m_j}{\Delta t}(u_i - u_j)\}$$

For all i **do**

$$R_i^\pm := \min\{1, \tfrac{Q_i^\pm}{P_i^\pm}\}$$

For all i **do**

    For all $j \in S_i, \ j > i$ **do**

$$\alpha_{ij} := \min\{R_i^\pm, R_j^\mp\}$$

$$\bar{f}_{ij} := \alpha_{ij} f_{ij}$$

$$\bar{f}_i := \bar{f}_i + \bar{f}_{ij}$$

$$\bar{f}_j := \bar{f}_j - \bar{f}_{ij}$$

(Definition 4). However, it may produce an overshoot or undershoot if the transport equation contains source terms that change the definition of the local DMP.

In our experience, a nonclipping flux limiter can be designed using information about the higher-order derivatives. In [40] we developed such a limiter for quadratic finite elements within the framework of a discontinuous Galerkin (DG) method. In the case of linear or multilinear finite element FCT schemes, the second derivatives are not available, which makes it more difficult to distinguish between spurious spikes ('wiggles') and smooth peaks. The use of Hessian recovery techniques produces a smooth approximation which is no longer a reliable shock detector.

Another infamous byproduct of FCT manifests itself in distortions of a smooth profile. This phenomenon is known as *terracing* and represents 'an integrated, nonlinear effect of residual phase errors' [64] or, loosely speaking, 'the ghosts of departed ripples' [7]. A particularly severe form of terracing is caused by the linear instability of the high-order scheme. For this reason, we do not recommend the use of the forward Euler time-stepping ($\theta = 0$) even though the flux-corrected scheme proves positivity-preserving under the CFL-like time step restriction (62).

Terracing can also be caused by the lack of information about the solution behavior in the exterior of Ω. Figure 1 displays a zoom of the FCT solution to the 1D convection equation (43) with $v = 1$ an $u^0 = x$ in $\Omega = (0, 1)$. The standard

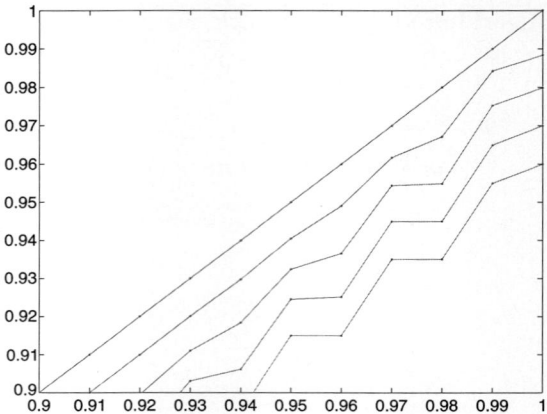

Fig. 1 Terracing in the neighborhood of a hyperbolic outlet

Galerkin method would produce excellent results for a linear profile but the FCT version gives rise to terracing in a neighborhood of the (artificial) open boundary $x = 1$. This happens because the solution value at the last node is treated as a peak, although it is not a peak if we make Ω a little longer [60]. This example indicates that the FCT limiter is not linearity-preserving, which makes it particularly prone to terracing.

6.5 Evaluation of FCT

In our experience, FCT produces excellent results for strongly time-dependent problems. The use of small time steps increases the amount of accepted antidiffusion and justifies the linearization of the antidiffusive flux which leads to a simple and efficient predictor-corrector algorithm. The cost of an implicit FCT scheme depends on the choice of iterative methods, parameter settings, and stopping criteria. If the time step is very small, then a good initial guess is available and the sparse linear system can be solved with 1–2 iterations of the Jacobi or Gauß-Seidel method. Thus, the cost *per time step* approaches that of an explicit finite difference or finite volume scheme. As the time step increases, so does the number of iterations, and advanced linear algebra tools (smoothers, preconditioners, convergence acceleration techniques) may need to be employed. Moreover, the use of large time steps degrades the accuracy of an FCT algorithm since $Q_i^\pm \to 0$ as $\Delta t \to \infty$. Other potential drawbacks include clipping, terracing, and the ad hoc nature of the prelimiting procedure.

We conclude that algebraic flux correction of FCT type is the method of choice for evolutionary problems. No other shock capturing technique performs better when it comes to solving an unsteady hyperbolic equation with linear finite elements [32]. For steady-state computations, we recommend the linearity-preserving limiter presented in the next section. It is similar to FCT in many ways but its derivation is based on variational gradient recovery, and all high-order operators (M_C, K, and L) are constrained independently using the same general-purpose limiting strategy.

7 Linearity-Preserving Limiters

As an alternative to FCT, we developed several multidimensional flux limiters which are independent of the time step and produce a TVD scheme in the 1D case [36, 37, 43]. As this methodology evolved and matured, we realized that the definition of the upper and lower bounds for a generalized TVD scheme must guarantee *linearity preservation* on arbitrary meshes. In other words, the constrained approximation must reduce to the underlying Galerkin scheme if the solution is a linear function. This property implies consistency and second-order accuracy for smooth data [10, 59]. In the context of algebraic flux correction, it can be enforced using variational gradient recovery to obtain the LED bounds for the edge-based slope limiter [48].

Another open problem in the design of TVD-like schemes for finite elements was the treatment of the consistent mass matrix which is essential for maintaining the high accuracy of the Galerkin scheme for time-dependent problems. Our multidimensional limiters of TVD type were designed to constrain the entries of the discrete convection operator, and our first attempts to limit the consistent mass matrix independently were rather unsuccessful. This has led us to marry FCT and 'TVD' within the framework of a general-purpose flux limiter [36]. Unfortunately, the resulting scheme inherited not only the advantages but also some drawbacks of the two limiting techniques (dependence on the time step, lack of linearity preservation, artificial coupling between the antidiffusive fluxes associated with different discrete operators). Moreover, the increased complexity of the algorithm has made it too expensive for practical purposes. For some time, we continued using the more efficient special-purpose limiting techniques: FCT for time-dependent problems and lumped-mass 'TVD' for steady-state computations. In this section, we introduce an algebraic flux correction scheme that can handle both situations equally well.

The algorithm to be presented is a fully multidimensional counterpart of the edge-based slope limiter we developed in [48] for anisotropic diffusion problems. In what follows, we extend it to steady and unsteady convective transport. The contribution of the consistent mass matrix is taken into account by applying the limiter to the vector of discretized time derivatives. Furthermore, we constrain the sum of raw antidiffusive fluxes instead of individual fluxes or slopes. This revision results in a marked gain of accuracy as compared to edge-by-edge slope limiting.

Another major improvement is a new iterative solver for the nonlinear algebraic system. We present a nonlinear SSOR scheme which updates the nodal values of the numerical solution and the limited antidiffusive fluxes in a single loop over the nodes of the computational mesh. To speed up convergence, we use *Anderson acceleration* [1, 77], also known as *Anderson mixing* [17, 18]. The efficiency of this approach is confirmed by our numerical study for an anisotropic diffusion equation. On fine meshes, the number of SSOR iterations is reduced by a factor of 60 and more.

7.1 Flux Splitting

So far we have limited all components of the raw antidiffusive flux f_{ij} using a common correction factor $\alpha_{ij} \in [0, 1]$. Let us now replace definition (53) with

$$\bar{f}_{ij} := \alpha_{ij}^M f_{ij}^M + \alpha_{ij}^K f_{ij}^K + \alpha_{ij}^L f_{ij}^L, \tag{84}$$

where α_{ij}^M, α_{ij}^K, and α_{ij}^L denote the individual correction factors for the fluxes

$$f_{ij}^M = m_{ij}(\dot{u}_i - \dot{u}_j), \tag{85}$$

$$f_{ij}^K = d_{ij}(u_i - u_j), \tag{86}$$

$$f_{ij}^L = l_{ij}^+(u_i - u_j). \tag{87}$$

The limited antidiffusive term (52) proves local extremum diminishing if

$$q_i^M \left(\dot{u}_i^{\min} - \dot{u}_i \right) \leq \sum_{j \neq i} \alpha_{ij}^M f_{ij}^M \leq q_i^M \left(\dot{u}_i^{\max} - \dot{u}_i \right), \tag{88}$$

$$q_i^K \left(u_i^{\min} - u_i \right) \leq \sum_{j \neq i} \alpha_{ij}^K f_{ij}^K \leq q_i^K \left(u_i^{\max} - u_i \right), \tag{89}$$

$$q_i^L \left(u_i^{\min} - u_i \right) \leq \sum_{j \neq i} \alpha_{ij}^L f_{ij}^L \leq q_i^L \left(u_i^{\max} - u_i \right) \tag{90}$$

for some positive constants q_i^M, q_i^K, and q_i^L independent of u. In this section, we use criterion (88)–(90) to determine the values of α_{ij}^M, α_{ij}^K, and α_{ij}^L.

Without loss of generality, we consider $f_{ij} := f_{ij}^K$ and present the limiting strategy that delivers α_{ij} satisfying (57) for a given $q_i > 0$. The fluxes f_{ij}^L and f_{ij}^M are limited in the same way but the bounds for f_{ij}^M are defined in terms of $\dot{u}$ rather than u.

7.2 Gradient-Based Slope Limiting

To get started, we present the symmetric linearity-preserving (LP) slope limiter we developed in [48] in the context of steady anisotropic diffusion. This algorithm belongs to the family of edge-based stencil reconstruction methods that constrain the jumps of the gradient along the line connecting two nodes [29, 51, 57, 67].

A raw antidiffusive flux of the form $f_{ij} = d_{ij}(u_i - u_j)$ requires limiting if the difference between u_i and u_j is "too large." Introducing the limited slope

$$\bar{s}_{ij} := \alpha_{ij}(u_i - u_j), \quad 0 \leq \alpha_{ij} \leq 1, \tag{91}$$

we define

$$\bar{f}_{ij} := d_{ij} \bar{s}_{ij} = \alpha_{ij} f_{ij}. \tag{92}$$

Thus, the multiplication of f_{ij} by α_{ij} is equivalent to replacing $u_i - u_j$ with $\bar{s}_{ij}$. An algorithm that produces $\bar{s}_{ij}$ rather than α_{ij} is called a *slope limiter*.

The definition of $\bar{s}_{ij}$ must guarantee that $\bar{f}_{ij}$ is LED. This will be the case if

$$s_{ij}^{\min} \leq \bar{s}_{ij} \leq s_{ij}^{\max}, \tag{93}$$

where the upper and lower bounds are given by

$$s_{ij}^{\max} = \gamma_{ij}(u_i^{\max} - u_i), \tag{94}$$

$$s_{ij}^{\min} = \gamma_{ij}(u_i^{\min} - u_i) \tag{95}$$

for some bounded $\gamma_{ij} \geq 0$. That is, each slope is constrained in the same manner as the sum of antidiffusive fluxes in (57). This approach is used in many edge-based extensions of 1D high-resolution schemes to unstructured meshes [29, 51, 57].

To construct the LED bounds $s_{ij}^{\max}$ and $s_{ij}^{\min}$, consider the linear approximation

$$u_i - u_j \approx s_{ij} := (\nabla u)_i \cdot (\mathbf{x}_i - \mathbf{x}_j). \tag{96}$$

The value of $(\nabla u)_i$ is obtained using numerical differentiation. A variety of gradient reconstruction techniques based on averaging or superconvergent patch recovery are available for this purpose. In our method, we use the lumped-mass L^2 projection

$$(\nabla u)_i = \frac{1}{m_i} \sum_k \mathbf{c}_{ik} u_k, \tag{97}$$

where m_i is a diagonal entry of the lumped mass matrix M_L, and $\mathbf{c}_{ik}$ is a vector-valued coefficient of the discrete gradient operator $\mathbf{C}$ given by (31). Since the gradients of Lagrange basis functions sum to zero, we have

$$\mathbf{c}_{ii} = -\sum_{k \neq i} \mathbf{c}_{ik}.$$

Thus

$$(\nabla u)_i = \frac{1}{m_i} \sum_{k \neq i} \mathbf{c}_{ik}(u_k - u_i). \tag{98}$$

We will use this representation to derive the LED bounds for the extrapolated slope s_{ij}, and then we will use these bounds to define γ_{ij} in (94) and (95).

Plugging (98) into the definition of s_{ij}, we obtain the following estimates

$$s_{ij} \leq \frac{1}{m_i} \sum_{k \neq i} |\mathbf{c}_{ik} \cdot (\mathbf{x}_i - \mathbf{x}_j)| (u_i^{\max} - u_i), \tag{99}$$

$$s_{ij} \geq \frac{1}{m_i} \sum_{k \neq i} |\mathbf{c}_{ik} \cdot (\mathbf{x}_i - \mathbf{x}_j)| (u_i^{\min} - u_i). \tag{100}$$

To make the bounds for s_{ij} less restrictive, we multiply them by 2 and define

$$\gamma_{ij} := \frac{2}{m_i} \sum_{k \neq i} |\mathbf{c}_{ik} \cdot (\mathbf{x}_i - \mathbf{x}_j)|. \tag{101}$$

This nonnegative coefficient is used to determine the bounds (94) and (95) for the slope limiter. The localized LED constraint (93) can be enforced by setting

$$\bar{s}_{ij} = \begin{cases} \min\{s_{ij}^{\max}, u_i - u_j\}, & \text{if } u_i > u_j, \\ \max\{s_{ij}^{\min}, u_i - u_j\}, & \text{if } u_i < u_j. \end{cases} \quad (102)$$

The one-sided limiting strategy is sufficient if the slope $\bar{s}_{ji} := -\bar{s}_{ij}$ cannot violate the LED principle for node j. In particular, this is the case if j is a node on the Dirichlet boundary or a downwind neighbor of node i (see Sect. 7.3). In all other cases, the slope limiter must enforce not only (93) but also $s_{ji}^{\min} \leq \bar{s}_{ji} \leq s_{ji}^{\max}$.

The following definition of $\bar{s}_{ij}$ guarantees the LED property for both nodes [48]

$$\bar{s}_{ij} = \begin{cases} \min\{s_{ij}^{\max}, u_i - u_j, -s_{ji}^{\min}\}, & \text{if } u_i > u_j, \\ \max\{s_{ij}^{\min}, u_i - u_j, -s_{ji}^{\max}\}, & \text{if } u_i < u_j. \end{cases} \quad (103)$$

This symmetric limiting strategy corresponds to a double application of the one-sided slope limiter. In the following Theorem, we prove linearity preservation.

Theorem 5 *If u_h is linear, then the lumped-mass L^2 projection (97) is exact and*

$$s_{ij} = u_i - u_j = \bar{s}_{ij}.$$

Proof If u_h is a linear, then its gradient is constant and $u_i - u_j = \nabla u_h \cdot (\mathbf{x}_i - \mathbf{x}_j)$. It follows that $s_{ij} = u_i - u_j$ if $(\nabla u)_i = \nabla u_h$. According to (97), we have

$$(\nabla u)_i = \frac{1}{m_i} \int_\Omega \varphi_i \nabla u_h \, d\mathbf{x} = \nabla u_h \left(\frac{1}{m_i} \int_\Omega \varphi_i \, d\mathbf{x} \right) = \nabla u_h \quad (104)$$

since the diagonal entry of the lumped mass matrix is given by

$$m_i = \sum_j m_{ij} = \int_\Omega \varphi_i \left(\sum_j \varphi_j \right) d\mathbf{x} = \int_\Omega \varphi_i \, d\mathbf{x}.$$

Thus, the L^2 projection is exact and $s_{ij} = u_i - u_j$. By definition of γ_{ij}, the slope $\bar{s}_{ij} = s_{ij}$ satisfies the imposed constraints, whence no limiting is performed. □

Linearity preservation implies that $\bar{f}_{ij} \to f_{ij}$ as $h \to 0$. Therefore, the constrained Galerkin scheme is consistent even if the low-order scheme is inconsistent.

Example 2 To illustrate the relationship of the linearity-preserving slope limiter to classical TVD schemes [26, 73], consider a 1D mesh with uniform spacing Δx. In this case, the coefficients of (97) are given by $m_i = \Delta x$ and $c_{i\pm 1/2} = \pm 1/2$.

The resulting formula for u'_i is equivalent to the second-order central difference

$$u'_i = \frac{1}{2} \left(\frac{u_i - u_{i-1}}{\Delta x} + \frac{u_{i+1} - u_i}{\Delta x} \right) = \frac{u_{i+1} - u_{i-1}}{2\Delta x}.$$

For any interior node, the local maxima and minima of the grid function are

$$u_i^{\max} = \max\{u_{i-1}, u_i, u_{i+1}\}, \qquad u_i^{\min} = \min\{u_{i-1}, u_i, u_{i+1}\}.$$

Furthermore, $\gamma_{ij} = 2$ for $j = i + 1$ since estimate (99)–(100) corresponds to

$$u_i^{\min} - u_i \leq \Delta x u_i' \leq u_i^{\max} - u_i.$$

The one-sided slope limiter (102) can be written as a single-line formula

$$\bar{s}_{ij} = \text{minmod}\{2(u_{i-1} - u_i), u_i - u_{i+1}\},$$

and the corresponding formula for the symmetric slope limiter (103) reads

$$\bar{s}_{ij} = \text{minmod}\{2(u_{i-1} - u_i), u_i - u_{i+1}, 2(u_{i+1} - u_{i+2})\}.$$

The *minmod* limiter function returns the argument with the smallest magnitude if all arguments have the same sign and zero otherwise. That is,

$$\text{minmod}\{a, b, \ldots\} = \begin{cases} \min\{a, b, \ldots\}, & \text{if } a > 0, b > 0, \ldots, \\ \max\{a, b, \ldots\}, & \text{if } a < 0, b < 0, \ldots, \\ 0, & \text{otherwise.} \end{cases}$$

It follows that the proposed slope limiter is activated only if two consecutive gradients have opposite signs or their magnitudes differ by a factor of 2 and more.

7.3 Symmetric Flux Limiter

In contrast to the fully multidimensional FCT method, the linearity-preserving (LP) slope limiter presented in Sect. 7.2 constrains the antidiffusive flux f_{ij} independently of all other fluxes into node i. This is convenient but the results are quite sensitive to the orientation of mesh edges. In this section, we convert the edge-based slope limiter into an FCT-like limiter for the sum of antidiffusive fluxes. The LED constraint (57) can be enforced using the following generalization of (80)–(83).

1. Compute the sums of positive/negative antidiffusive fluxes to be limited

$$P_i^+ = \sum_{j \neq i} \max\{0, f_{ij}\}, \qquad P_i^- = \sum_{j \neq i} \min\{0, f_{ij}\}. \tag{105}$$

2. Define local extremum diminishing upper/lower bounds of the form

$$Q_i^+ = q_i(u_i^{\max} - u_i), \qquad Q_i^- = q_i(u_i^{\min} - u_i). \tag{106}$$

3. Compute the nodal correction factors for positive/negative fluxes

$$R_i^+ = \min\left\{1, \frac{Q_i^+}{P_i^+}\right\}, \qquad R_i^- = \min\left\{1, \frac{Q_i^-}{P_i^-}\right\}. \tag{107}$$

4. Limit the fluxes f_{ij} and f_{ji} using the common correction factor

$$\alpha_{ij} = \begin{cases} \min\{R_i^+, R_j^-\}, & \text{if } f_{ij} > 0, \\ \min\{R_i^-, R_j^+\}, & \text{if } f_{ij} < 0. \end{cases} \tag{108}$$

As in the case of FCT, this definition of the correction factor α_{ij} implies that

$$Q_i^- \leq R_i^- P_i^- \leq \sum_{j \neq i} \alpha_{ij} f_{ij} \leq R_i^+ P_i^+ \leq Q_i^+. \tag{109}$$

To maintain linearity preservation, we define $Q_i^\pm$ as the sum of the LED bounds we imposed on individual slopes/fluxes in Sect. 7.2. That is, we set

$$q_i := \sum_{j \neq i} \gamma_{ij} d_{ij}. \tag{110}$$

In contrast to FCT, the resulting formula for $Q_i^\pm$ is independent of the time step.

7.4 One-Sided Flux Limiter

Algorithm (105)–(108) is ideally suited for constraining a symmetric operator like L or M_C. In the latter case, the antidiffusive fluxes (85) and the bounds $Q_i^\pm$ be defined in terms of $\dot{u}$ rather than u. At the fully discrete level, the time derivative is replaced with the finite difference approximation $\dot{u} \approx (u^{n+1} - u^n)/\Delta t$. Note that the same correction factor α_{ij}^M is applied to the explicit and implicit part of f_{ij}^M.

In principle, the discrete convection operator K can also be constrained using (105)–(108). However, it turns out that the LED constraint for node j is satisfied automatically if $k_{ji} > 0$. To take advantage of this fact, we limit the convective part in an upwind-biased fashion [43, 46]. In accordance with our upwind-downwind edge orientation convention (40), we assume that $k_{ij} \leq k_{ji}$. As long as

$$\bar{k}_{ji} := k_{ji} + (1 - \alpha_{ij}) d_{ij}$$

is nonnegative for all $\alpha_{ij} \in [0, 1]$, it is enough to make sure that (57) holds for node i.

In the one-sided version of (105)–(108), we begin with the prelimiting step

$$f_{ij} := \left(d_{ij} + \max\{0, k_{ji}\}\right)(u_i - u_j) \tag{111}$$

which is required to enforce the LED constraint for node j in the unlikely case of $k_{ji} < 0$. After this prelimiting, the correction factors α_{ij} are calculated as follows:

1. Compute the sums of positive/negative antidiffusive fluxes to be limited

$$P_i^+ = \sum_{k_{ij} \leq k_{ji}} \max\{0, f_{ij}\}, \qquad P_i^- = \sum_{k_{ij} \leq k_{ji}} \min\{0, f_{ij}\}. \tag{112}$$

2. Compute q_i and the local extremum diminishing upper/lower bounds

$$Q_i^+ = q_i(u_i^{\max} - u_i), \qquad Q_i^- = q_i(u_i^{\min} - u_i). \tag{113}$$

3. Compute the nodal correction factors for positive/negative fluxes

$$R_i^+ = \min\left\{1, \frac{Q_i^+}{P_i^+}\right\}, \qquad R_i^- = \min\left\{1, \frac{Q_i^-}{P_i^-}\right\}. \tag{114}$$

4. Multiply f_{ij} and f_{ji} by the nodal correction factor for the *upwind* node i

$$k_{ij} \leq k_{ji} \quad \Longrightarrow \quad \alpha_{ij} = \begin{cases} R_i^+, & \text{if } f_{ij} \geq 0, \\ R_i^-, & \text{if } f_{ij} < 0, \end{cases} \quad \alpha_{ji} := \alpha_{ij}. \tag{115}$$

We have used this one-sided limiting strategy to design algebraic flux correction schemes based on a generalization of upwind TVD limiters [36, 37, 43, 46].

8 Solution of Nonlinear Systems

After the discretization in time, the flux-corrected discrete problem can be written in the form (58). Since the antidiffusive term depends on the unknown solution, the nonlinear discrete problem must be solved in an iterative way. In contrast to the nonlinear FCT algorithm presented in Sect. 6.2, only the fully converged solution is guaranteed to be nonoscillatory. Therefore, it is essential to make sure that iterations converge. Moreover, convergence must be fast enough to keep the cost of algebraic flux correction reasonable. Thus, the robustness and efficiency of the iterative solver for the nonlinear system are just as important as the flux limiting procedure.

8.1 Defect Correction Scheme

As in the case of FCT, the structure of the nonlinear system (58) suggests the use of a fixed-point iteration with a lagged evaluation of the antidiffusive term

$$Au^{(m+1)} = Bu^n + \bar{f}^{(m)}. \tag{116}$$

A more general class of defect correction schemes can be formally written as

$$u^{(m+1)} = u^{(m)} + \omega \tilde{A}^{-1} r^{(m)}, \tag{117}$$

where $\tilde{A}$ is an approximation to the Jacobian of the nonlinear system, $\omega \in [0, 1]$ is a relaxation parameter, and $r^{(m)}$ is the residual vector given by

$$r^{(m)} = Bu^n - Au^{(m)} + \bar{f}^{(m)}. \tag{118}$$

In practice, the matrix $\tilde{A}$ is 'inverted' by solving a linear system (see Algorithm 2). The iteration process is typically terminated when certain norms of $u^{(m+1)} - u^{(m)}$ and/or $r^{(m+1)}$ become smaller than a prescribed tolerance. More elaborate stopping criteria based on the finite element theory can be found in [2]. Clearly, the rates of convergence and the overall efficiency of the above defect correction scheme are strongly influenced by the choice of the 'preconditioner' $\tilde{A}$.

The default setting is $\omega := 1$ and $\tilde{A} := A$, which corresponds to (116). By construction, the low-order operator A is an M-matrix. This property results in fast convergence of inner iterations. If the time step Δt is very small, the solution can be updated in a fully explicit fashion using the diagonal preconditioner $\tilde{A} := \text{diag}(A)$. As few as 1–3 outer iterations may suffice if good initial guess is available. Thus,

---
Algorithm 2 Defect correction scheme

Set $u^{(0)} := u^n$

For all $m = 0, 1, \ldots$ **do**

    Solve the linear system $\tilde{A} \Delta u^{(m+1)} = r^{(m)}$

    Update the solution $u^{(m+1)} := u^{(m)} + \omega \Delta u^{(m+1)}$

    Exit if the stopping criteria are satisfied

Set $u^{n+1} := u^{(m+1)}$

---

the cost per time step might be comparable to that of an explicit algorithm. On the other hand, such a solver may fail to converge if the time step is too large.

Some advanced preconditioning and underrelaxation techniques are discussed in [41, 48, 61]. In quasi-Newton methods, $\tilde{A}$ must be a good approximation to the Jacobian of (58). Due to the complex structure and nondifferentiability of the limited antidiffusive term, the assembly of such preconditioners is very complicated and expensive. Thus, Jacobian-free solvers are to be preferred. In particular, the convergence acceleration method described in Sect. 8.3 leads to a Newton-like scheme in which the memory effect is exploited to avoid numerical differentiation.

8.2 Nonlinear SSOR Scheme

A major drawback of fixed-point methods like (116) is the fully explicit treatment of the antidiffusive term. An attempt to build implicit antidiffusion into the preconditioner $\tilde{A}$ aggravates convergence problems if all correction factors are taken from the previous outer iteration. This has led us to update the solution values, the antidiffusive fluxes, and the correction factors simultaneously in a loop over nodes. The resulting algorithm can be classified as a nonlinear Gauß-Seidel / SSOR method.

The i-th equation of the flux-corrected Galerkin scheme (58) can be written as

$$\sum_j a_{ij} u_j = b_i + \bar{f}_i, \qquad (119)$$

where the antidiffusive term $\bar{f}_i$ depends on $u = u^{n+1}$, whereas $b_i = \sum_j b_{ij} u_j^n$ is known.

The calculation of $u_i^{(m+1)} \approx u_i^{n+1}$ begins with the assembly of $\bar{f}_i$. In the forward sweep, the new values of u_j are already available for all $j < i$. Thus

$$u_j = \begin{cases} u_j^{(m+1)}, & \text{if } j < i, \\ u_j^{(m)}, & \text{if } j \geq i. \end{cases} \qquad (120)$$

In the backward sweep, the solution values are updated in the reverse order, so the i-th step begins with $u_j = u_j^{(m+1)}$ for $j > i$ and $u_j = u_j^{(m)}$ otherwise.

Given the array of current solution values u_i, we recalculate the raw antidiffusive fluxes f_{ij}, apply the flux limiter, and add the result to $\bar{f}_i$ (see Algorithm 3).

Algorithm 3 Assembly of $\bar{f}_i$ (symmetric version)

For all i **do**

$$P_i^\pm := 0, \quad Q_i^\pm := 0, \quad \bar{f}_i := 0$$

  For all $j \in S_i$ **do**

$$P_i^\pm := P_i^\pm + {\max \atop \min}\{0, f_{ij}\}$$

$$Q_i^\pm := {\max \atop \min}\{Q_i^\pm, \tfrac{m_i}{\Delta t}(u_j - u_i)\}$$

$$R_i^\pm := \min\{1, Q_i^\pm / P_i^\pm\}$$

  For all $j \in S_i$ **do**

$$\alpha_{ij} := \min\{R_i^\pm, R_j^\mp\}$$

$$\bar{f}_i := \bar{f}_i + \alpha_{ij} f_{ij}$$

Since the value of α_{ij} depends not only on $R_i^\pm$ but also on $R_j^\mp$, we store the updated nodal correction factors, so that they are readily available when it comes to calculating α_{ij}. Due to the lag in evaluation of $\bar{f}_{ij}$ and $\bar{f}_{ji}$, intermediate approximations may be nonconservative but $\bar{f}_{ji} = -\bar{f}_{ij}$ when the algorithm converges.

Given the updated value of $\bar{f}_i$, the old solution value u_i is overwritten by

$$u_i := u_i + \frac{1}{\tilde{a}_{ii}}\left(b_i - \sum_j a_{ij}u_j + \bar{f}_i\right), \tag{121}$$

where $\tilde{a}_{ii} \geq a_{ii}$. Setting $\tilde{a}_{ii} := a_{ii}$, one obtains the symmetric Gauß-Seidel (SGS) method which may fail to converge if the implicit part of $\bar{f}_i$ is too large compared to $\sum_j a_{ij}u_j$. A possible remedy is implicit underrelaxation of the form

$$\tilde{a}_{ii} := \frac{a_{ii}}{\omega}, \quad 0 < \omega \leq 1.$$

Equivalently, the SSOR scaling factor $\tilde{a}_{ii}$ can be defined by adding a nonnegative number to the diagonal entry. In our numerical experiments, we used

$$\tilde{a}_{ii} := a_{ii} + \theta \sum_{j \neq i}(d_{ij} + l_{ij}^+).$$

The flow chart of the nonlinear SSOR method for solving (58) is labeled Algorithm 4. The forward sweep can be written as $(\tilde{D} + \tilde{L})\Delta u^* := r$, where r is the residual, $\tilde{D} = \text{diag}\{\tilde{a}_{ii}\}$ is a diagonal matrix of scaling factors, and $\tilde{L}$ is the strict lower triangular part of A plus limited antidiffusion. Likewise, the backward sweep can be written as $(\tilde{D} + \tilde{U})\Delta u := \Delta u^*$, where $\tilde{U}$ is a strict upper triangular matrix. Thus, Algorithm 4 can written in the form (117) with $\omega = 1$ and

$$\tilde{A} = (\tilde{D} + \tilde{L})\tilde{D}^{-1}(\tilde{D} + \tilde{U}).$$

Algorithm 4 Nonlinear SSOR iteration

For all $i = 1, \ldots, N-1, N$ **do**   *(forward sweep)*

For all $i = N, N-1, \ldots, 1$ **do**   *(backward sweep)*

   Update the antidiffusive term $\bar{f}_i$ using Algorithm 3

   Calculate the new solution value u_i using (121)

Luo et al. [56] used this sort of defect correction as a preconditioner for a linear GMRES solver. A nonlinear version of this solution strategy is recovered when the method presented in Sect. 8.3 is employed to accelerate Algorithm 4.

In the iterative solver for steady transport equations, we set $\theta := 1$ and $b_i := 0$. Furthermore, the contribution of the mass matrix is removed, which corresponds to using an infinitely large pseudo-time step Δt. It is also possible to march the solution to the steady state using $\theta := 1$ and local time stepping. In either case, a usable initial guess can be obtained by solving the linear system with $\bar{f}_i = 0$ or $\bar{f}_i = f_i$.

8.3 Anderson Acceleration

Since the cost of recalculating the correction factors for the flux limiter is rather high, slow convergence of an iterative method can make algebraic flux correction very expensive. The fixed-point defect correction scheme (117) and the nonlinear SSOR iteration (121) generate a sequence of successive approximations but only the last iterate $u^{(m)}$ is used when it comes to the computation of $u^{(m+1)}$. It turns out that including information from a number of previous iterates may dramatically improve the convergence behavior. This idea is exploited in many vector extrapolation techniques for vector sequences (see, e.g., [31, 72]). In this work, we employ the convergence acceleration technique known as *Anderson mixing* [1, 17, 18, 77]. As shown in [17], this approach is equivalent to the Broyden scheme for the inverse Jacobian but is easier to implement and explain. On linear problems, the accelerated fixed point iteration is related to the preconditioned GMRES method [77].

Following Walker and Ni [77], we formulate Anderson acceleration as shown in Algorithm 5. In practice, it is worthwhile to calculate the weights by solving an equivalent unconstrained least squares problem [77]. Furthermore, Anderson acceleration may need to be restarted if the vectors $\Delta u^{(m)}$ become (almost) linearly dependent, or if the norm of $\Delta u^{(m)}$ is much greater than that of $\Delta u^{(m-1)}$. We refer to [17, 18, 62, 77] for a discussion of various improvements and practical implementation details.

Algorithm 5 Anderson acceleration

For all $m = 0, 1, \ldots$ **do**

    Compute $\tilde{u}^{(m)} := g(u^{(m)})$ with (117) or (121)

    Store $\tilde{u}^{(m)}$ and $\Delta u^{(m)} := \tilde{u}^{(m)} - u^{(m)}$

    Given $k \leq m$ iterates, determine the weights
$$\omega^{(m)} = \left(\omega_1^{(m)}, \ldots, \omega_k^{(m)}\right)^T$$
    by solving the constrained least-squares problem
$$\min_{\omega^{(m)}} \left\| \sum_{i=1}^{k} \omega_i^{(m)} \Delta u^{(m-k+i)} \right\|_2 \quad \text{s.t.} \quad \sum_{i=1}^{k} \omega_i^{(m)} = 1$$
    Set $u^{(m+1)} := \sum_{i=1}^{k} \omega_i^{(m)} \tilde{u}^{(m-k+i)}$
---

9 Numerical Examples

A properly designed high-resolution scheme should be (i) at least second-order accurate for smooth data and (ii) capable of resolving small-scale features without excessive smearing or steepening. To evaluate the accuracy and efficiency of our linearity-preserving limiting techniques, we apply them to three representative benchmark problems which have already been studied using other algebraic flux correction schemes [39, 41, 43, 48] as well as variational shock capturing [32], monotone finite volume schemes [49, 50], and slope limiters for discontinuous Galerkin methods [40]. Thus, a quantitative comparison of the results is possible.

Given a reference solution u and a numerical approximation u_h, we define

$$E_1(h) = \sum_i m_i |u(\mathbf{x}_i) - u_i| \approx \|u - u_h\|_1, \tag{122}$$

$$E_2(h) = \sqrt{\sum_i m_i |u(\mathbf{x}_i) - u_i|^2} \approx \|u - u_h\|_2, \tag{123}$$

where $m_i = \int_\Omega \varphi_i \, d\mathbf{x}$ stands for a diagonal coefficient of the lumped mass matrix M_L.

The objective of the below numerical study is to investigate the dependence of the errors E_1 and E_2 on the mesh size h and on the choice of the limiting strategy. In particular, we will use the numerical solutions computed on the two finest meshes to estimate the expected order of accuracy by the formula [49]

$$p = \log_2\left(\frac{E_1(2h)}{E_1(h)}\right). \tag{124}$$

In the last two examples, we compare the convergence behavior of the global defect correction scheme to that of nonlinear SSOR with Anderson acceleration.

9.1 Solid Body Rotation

The solid body rotation test [49, 78] is often used to evaluate numerical advection schemes. The problem to be solved is the continuity equation

$$\frac{\partial u}{\partial t} + \nabla \cdot (\mathbf{v}u) = 0 \quad \text{in } \Omega = (0,1) \times (0,1). \tag{125}$$

The velocity $\mathbf{v}$ describes a counterclockwise rotation about the center of Ω

$$\mathbf{v}(x,y) = (0.5 - y, x - 0.5). \tag{126}$$

After each full revolution, the exact solution u coincides with the given initial data u_0. Hence, the challenge of this test is to preserve the shape of u_0.

Following LeVeque [49], we simulate solid body rotation of the profile displayed in Fig. 2. The geometry of each body is described by a given function $G(x,y)$ defined on a circle of radius $r_0 = 0.15$ centered at some point $(x_0, y_0) \in \Omega$. Let

$$r(x,y) = \frac{1}{r_0}\sqrt{(x-x_0)^2 + (y-y_0)^2}$$

be the normalized distance from the point (x_0, y_0). Then $r(x,y) \leq 1$ inside the circle.

The slotted cylinder is centered at the point $(x_0, y_0) = (0.5, 0.75)$ and

$$G(x,y) = \begin{cases} 1, & \text{if } |x-x_0| \geq 0.025 \text{ or } y \geq 0.85, \\ 0, & \text{otherwise.} \end{cases}$$

The sharp cone is centered at $(x_0, y_0) = (0.5, 0.25)$, and its shape is given by

$$G(x,y) = 1 - r(x,y).$$

The smooth hump is centered at $(x_0, y_0) = (0.25, 0.5)$, and the shape function is

$$G(x,y) = \frac{1 + \cos(\pi r(x,y))}{4}.$$

In the rest of the domain, the solution to (125) is initialized by zero, and homogeneous Dirichlet boundary conditions are prescribed at the inlets.

The snapshots presented in Figs. 2–5 show the shape of the solution at the final time $T = 2\pi$, which corresponds to one full rotation. All computations were performed on a uniform mesh of 128×128 bilinear elements using the Crank-Nicolson time-stepping with the time step $\Delta t = 10^{-3}$. The results obtained with $\alpha_{ij} := 1$ and $\alpha_{ij} := 0$ are displayed in Figs. 3 and 4, respectively. As expected, the unconstrained Galerkin solution exhibits spurious oscillations, while its low-order counterpart is too diffusive. The solution shown in Fig. 5 was computed using linearized FCT (see Sect. 6.1) with $\dot{u}^L$ given by (65). A detailed numerical study of FCT schemes (explicit vs. implicit, linearized vs. nonlinear) can be found in [39].

In the captions to Figs. 6–9, the abbreviations LPSL and LPFL refer to the limiting techniques described in Sects. 7.2 and 7.3, respectively (LP := *Linearity Preserving*, SL:= *Slope Limiting*, FL := *Flux Limiting*). The results shown in Figs. 6 and 7 indicate that LPFL is more accurate than LPSL and almost as accurate as FCT. This

Fig. 2 Solid body rotation: initial data / exact solution at $t = 2\pi$

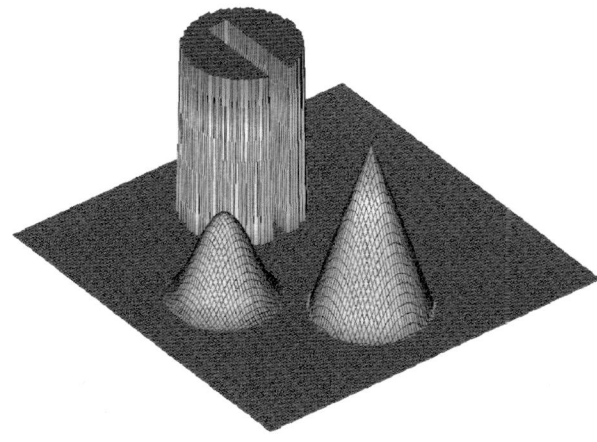

Fig. 3 Solid body rotation: Galerkin solution at $t = 2\pi$

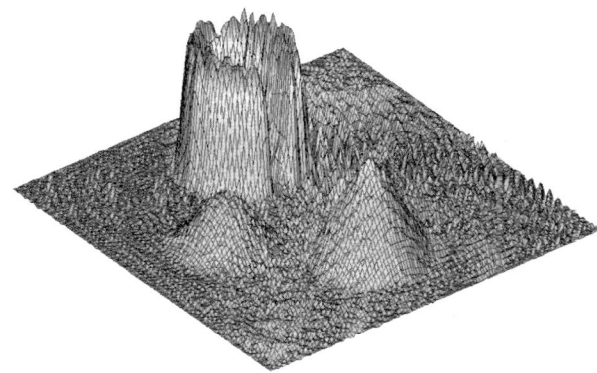

Fig. 4 Solid body rotation: low-order solution at $t = 2\pi$

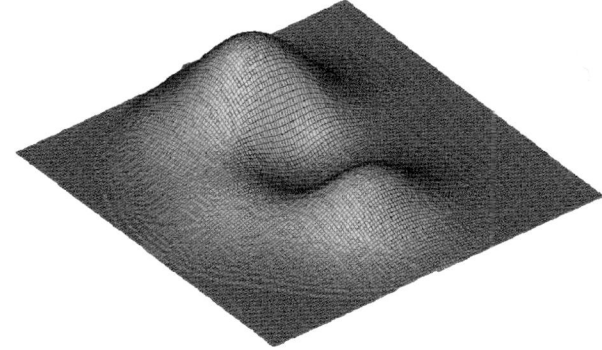

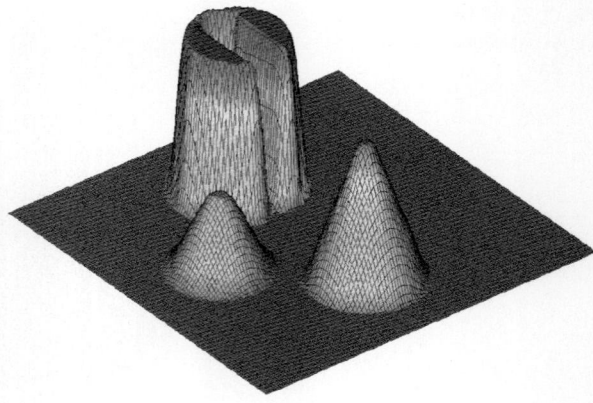

Fig. 5 Solid body rotation: FEM-FCT solution at $t = 2\pi$

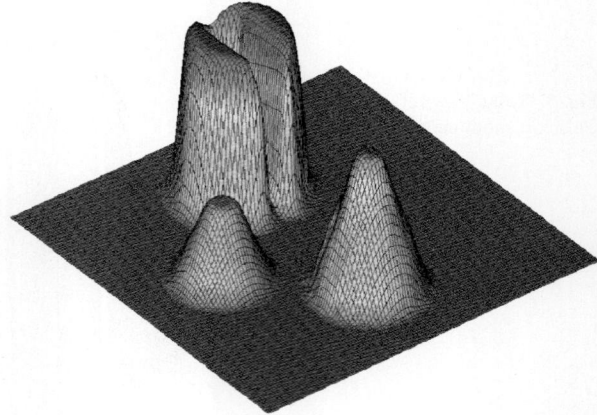

Fig. 6 Solid body rotation: consistent-mass LPSL solution at $t = 2\pi$

is good news since the solid body rotation test belongs to the class of problems that FCT can handle much better than other shock-capturing methods [32].

In contrast to flux limiters of TVD type [36, 37], LPSL and LPFL are applicable to the antidiffusive part of the consistent mass matrix which makes it possible to attain fourth-order accuracy with linear finite elements (see [15], p. 96). To demonstrate the importance of this result, we present the numerical solutions obtained with the lumped mass matrix ($\alpha_{ij}^M := 0$) in Figs. 8 and 9. The diagram in Fig. 10 depicts the E_1 convergence history for the consistent and lumped-mass versions of LPSL and LPFL. The numerical values of E_1 and E_2 are listed in Tables 2 and 3. The local Courant number $\nu = |\mathbf{v}|\frac{\Delta t}{h}$ equals zero at the center of the square domain and attains its largest value $\nu_{\max} = \frac{1}{\sqrt{2}}\frac{\Delta t}{h}$ at the corners. In the process of mesh refinement, the time step was adjusted to maintain the fixed ratio $\frac{\Delta t}{h} = 0.128$.

The expected order of accuracy p is estimated using (124) with $h = 1/256$. The rates of convergence for the LP algorithms used in Figs. 6–9 are given by $p = 0.96, 0.90, 0.77$, and 0.77, respectively. The consistent-mass LPFL produces

Fig. 7 Solid body rotation: consistent-mass LPFL solution at $t = 2\pi$

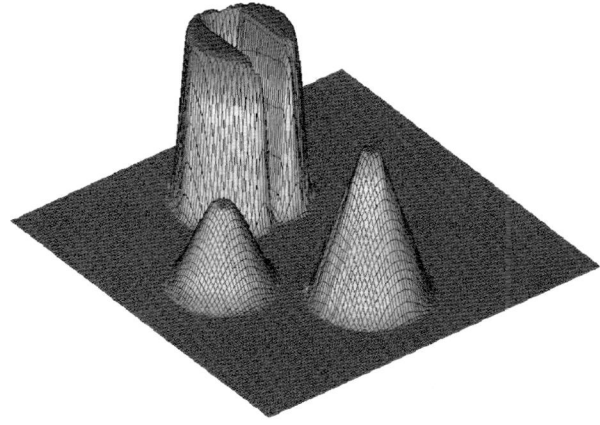

Fig. 8 Solid body rotation: lumped-mass LPSL solution at $t = 2\pi$

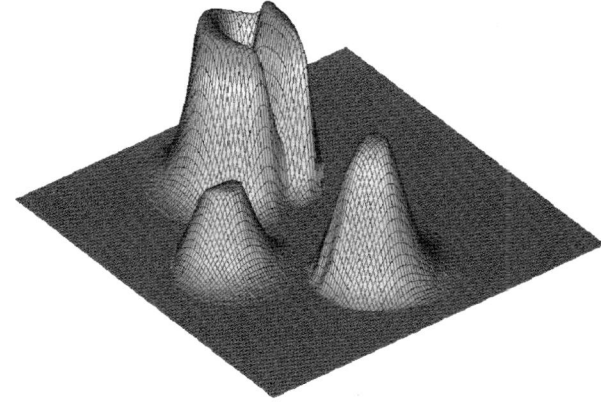

Fig. 9 Solid body rotation: lumped-mass LPFL solution at $t = 2\pi$

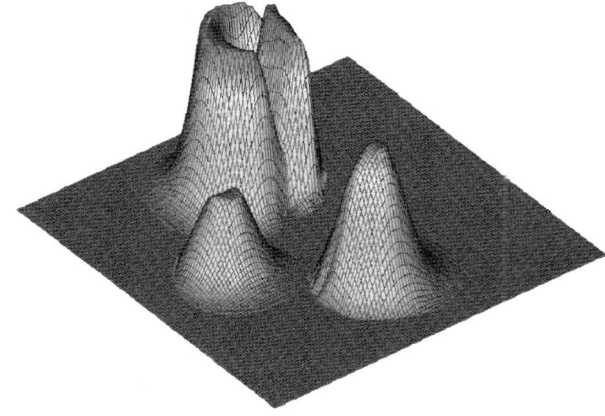

Fig. 10 Solid body rotation, convergence history for LP limiters

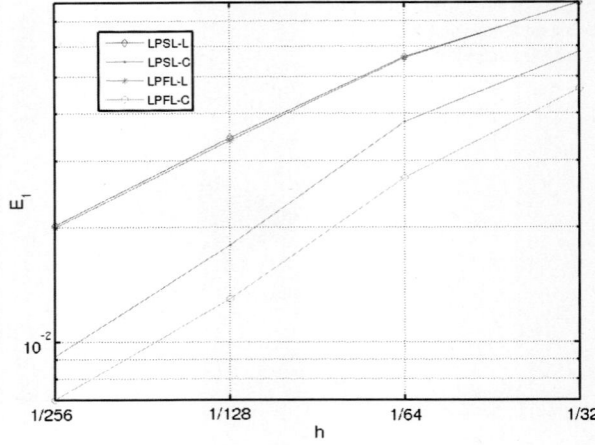

Table 2 Solid body rotation: LPSL grid convergence

h	LPSL, lumped mass		LPSL, consistent mass	
	E_1	E_2	E_1	E_2
1/32	0.783E−01	0.163E+00	0.582E−01	0.135E+00
1/64	0.564e−01	0.144e+00	0.380E−01	0.111E+00
1/128	0.346e−01	0.109e+00	0.180E−01	0.704E−01
1/256	0.203e−01	0.803e−01	0.919E−02	0.509E−01

Table 3 Solid body rotation: LPFL grid convergence

h	LPSL, lumped mass		LPSL, consistent mass	
	E_1	E_2	E_1	E_2
1/32	0.785E−01	0.165E+00	0.465E−01	0.125E+00
1/64	0.560E−01	0.147E+00	0.271E−01	0.907E−01
1/128	0.340E−01	0.110E+00	0.130E−01	0.612E−01
1/256	0.200E−01	0.806E−01	0.705E−02	0.459E−01

smaller errors than LPSL. However, there is hardly any difference if mass lumping is performed. In this case, both algorithms converge at the rate $p = 0.77$, which is a typical value for a TVD scheme that delivers $p = 2$ for smooth data. The use of the consistent mass matrix results in a significant gain of accuracy and faster grid convergence. This justifies the additional effort invested in the computation of α_{ij}^M.

Table 4 Circular convection: LPSL grid convergence

h	Smooth data		Discontinuous data	
	E_1	E_2	E_1	E_2
1/32	0.318E−01	0.551E−01	0.821E−01	0.152E+00
1/64	0.104E−01	0.204E−01	0.449E−01	0.108E+00
1/128	0.251E−02	0.595E−02	0.259E−01	0.860E−01
1/256	0.537E−03	0.160E−02	0.138E−01	0.601E−01

Table 5 Circular convection: LPFL grid convergence

h	Smooth data		Discontinuous data	
	E_1	E_2	E_1	E_2
1/32	0.146E−01	0.266E−01	0.540−01	0.131E+00
1/64	0.377E−02	0.801E−02	0.295E−01	0.893E−01
1/128	0.944E−03	0.230E−02	0.185E−01	0.757E−01
1/256	0.218E−03	0.632E−03	0.104E−01	0.519E−01

9.2 Circular Convection

The second test problem is taken from [27]. Consider the hyperbolic PDE

$$\nabla \cdot (\mathbf{v}u) = 0 \quad \text{in } \Omega = (-1, 1) \times (0, 1) \tag{127}$$

which describes steady circular convection if the velocity field is defined as

$$\mathbf{v}(x, y) = (y, -x).$$

The exact solution and inflow boundary conditions for this test are given by

$$u(x, y) = \begin{cases} G(r), & \text{if } 0.35 \leq r = \sqrt{x^2 + y^2} \leq 0.65, \\ 0, & \text{otherwise}, \end{cases}$$

where $G(r)$ is a given function that defines the shape of the solution profile.

To evaluate the performance of LPSL and LPFL for smooth data and discontinuous solutions, we consider the following shape functions

$$G_1(r) = \cos^2\left(5\pi \frac{2r-1}{3}\right), \qquad G_2(r) \equiv 1.$$

As before, computations are performed on a uniform mesh of bilinear finite elements which is successively refined to perform a grid convergence study.

The exact solution to the circular convection problem is constant along the streamlines of the stationary velocity field. Figure 11 displays the results for $G = G_1$ and $G = G_2$ computed using the LPFL algorithm with $h = 1/64$. The convergence history for LPSL and LPFL is presented in Tables 4 and 5, respectively. In the case of the smooth profile G_1, the E_1 errors for LPSL are approximately twice as large as those for LPFL. The expected orders of accuracy are 2.22 and 2.11, respectively.

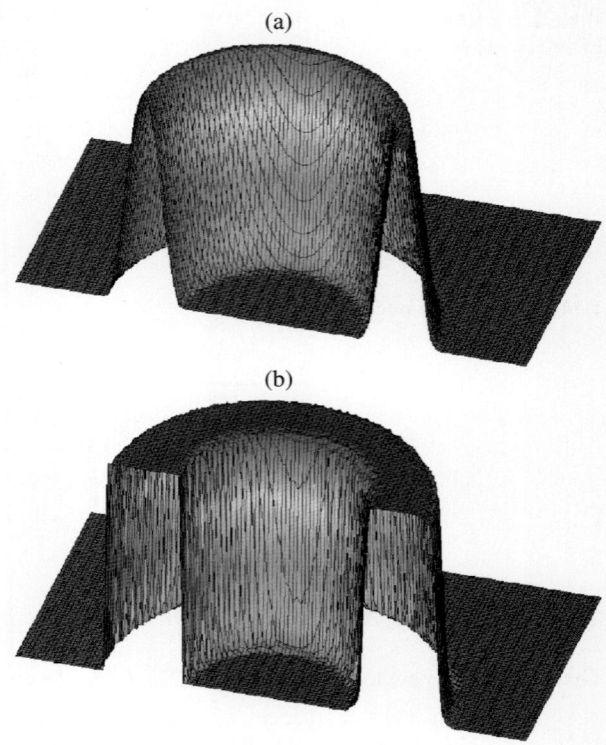

Fig. 11 Circular convection: LPFL results for (**a**) smooth and (**b**) discontinuous data

In the case of the discontinuous profile G_2, the convergence rates drop to 0.91 for LPSL and 0.83 for LPFL. The absolute values of the E_1 errors differ by a factor of 1.5. We conclude that the revised limiting strategy leads to a marked improvement not only for transient convection problems but also in steady-state computations.

The iterative solver was configured to run until the absolute norm of the residual becomes smaller than 10^{-6}. This stopping criterion is more stringent than necessary to obtain an accurate solution. However, it is important to make sure that the residuals go to zero. The methods under investigation are the global defect correction scheme (with $\tilde{a}_{ii} = 2a_{ii}$ and $\tilde{a}_{ij} = a_{ij}$ for $j \neq i$) and the nonlinear SSOR method (with $\tilde{a}_{ii} = \sum_j |a_{ij}|$). The same subroutine was used to evaluate the residuals for both schemes. To prevent division by zero, the LP limiter was implemented using $R_i^{\pm} = \min\{1, \frac{Q_i^{\pm} \pm \epsilon}{P_i^{\pm} \pm \epsilon}\}$, where ϵ is a multiple of the machine precision.

In the circular convection test with the discontinuous profile, the residuals begin to oscillate, and convergence stalls if no Anderson acceleration is performed. The number of nonlinear iterations for the accelerated schemes is presented in Table 6, where AA(k) stands for Anderson acceleration applied to k iterates. The defect correction scheme with $k = 5$ fails to converge in most cases. The total number of iterations for $k = 10$ is twice as large as that for nonlinear SSOR. Moreover, the cost of a defect correction cycle is higher than that of an SSOR iteration.

Table 6 Circular convection: number of nonlinear iterations

h	Defect correction		Nonlinear SSOR	
	$AA(5)$	$AA(10)$	$AA(5)$	$AA(10)$
1/32	576	574	274	287
1/64	—	975	487	501
1/128	—	1866	908	940
1/256	—	—	—	1893

9.3 Anisotropic Diffusion

In the last example, we consider a steady anisotropic diffusion equation

$$-\nabla \cdot (\mathcal{D}\nabla u) = 0 \quad \text{in } \Omega, \tag{128}$$

where $\Omega = (0,1)^2 \setminus [4/9, 5/9]^2$ is a square domain with a hole in the middle.

The outer and inner boundary of Ω are denoted by Γ_0 and Γ_1, respectively (see Fig. 12(a)). The following Dirichlet boundary conditions are prescribed

$$u(x,y) = \begin{cases} -1, & \text{if } (x,y) \in \Gamma_0, \\ 1, & \text{if } (x,y) \in \Gamma_1. \end{cases} \tag{129}$$

The diffusion tensor $\mathcal{D}$ is a symmetric positive definite matrix defined as

$$\mathcal{D} = R(-\theta)\begin{pmatrix} k_1 & 0 \\ 0 & k_2 \end{pmatrix}\mathcal{R}(\theta), \tag{130}$$

where k_1 and k_2 are the positive eigenvalues and $\mathcal{R}(\theta)$ is a rotation matrix

$$\mathcal{R}(\theta) = \begin{pmatrix} \cos\theta & \sin\theta \\ -\sin\theta & \cos\theta \end{pmatrix}. \tag{131}$$

The eigenvalues of $\mathcal{D}$ represent the diffusion coefficients associated with the axes of the Cartesian coordinate system rotated by the angle θ. Let

$$k_1 = 100, \quad k_2 = 1, \quad \theta = -\frac{\pi}{6}.$$

By the continuous maximum principle, the exact solution to the above Dirichlet problem is bounded by the prescribed boundary data $u|_\Gamma = \pm 1$. However, the diffusion tensor (130) is highly anisotropic, which may result in a violation of the DMP even if a regular mesh of acute/non-narrow type is employed.

The above benchmark problem was introduced by Lipnikov et al. [50]. The results obtained with LPSL can be found in [48]. In this section, we discretize the anisotropic diffusion equation (128) using LPFL and linear finite elements on uniform triangular meshes. Since no exact solution is available, the reference solution depicted in Fig. 12(b) is calculated with the standard Galerkin method on a very fine mesh ($h = 1/1152$). This solution is bounded by the prescribed Dirichlet boundary values, as required by the maximum principle. The unconstrained Galerkin solutions computed on coarser meshes exhibit spurious undershoots shown as the

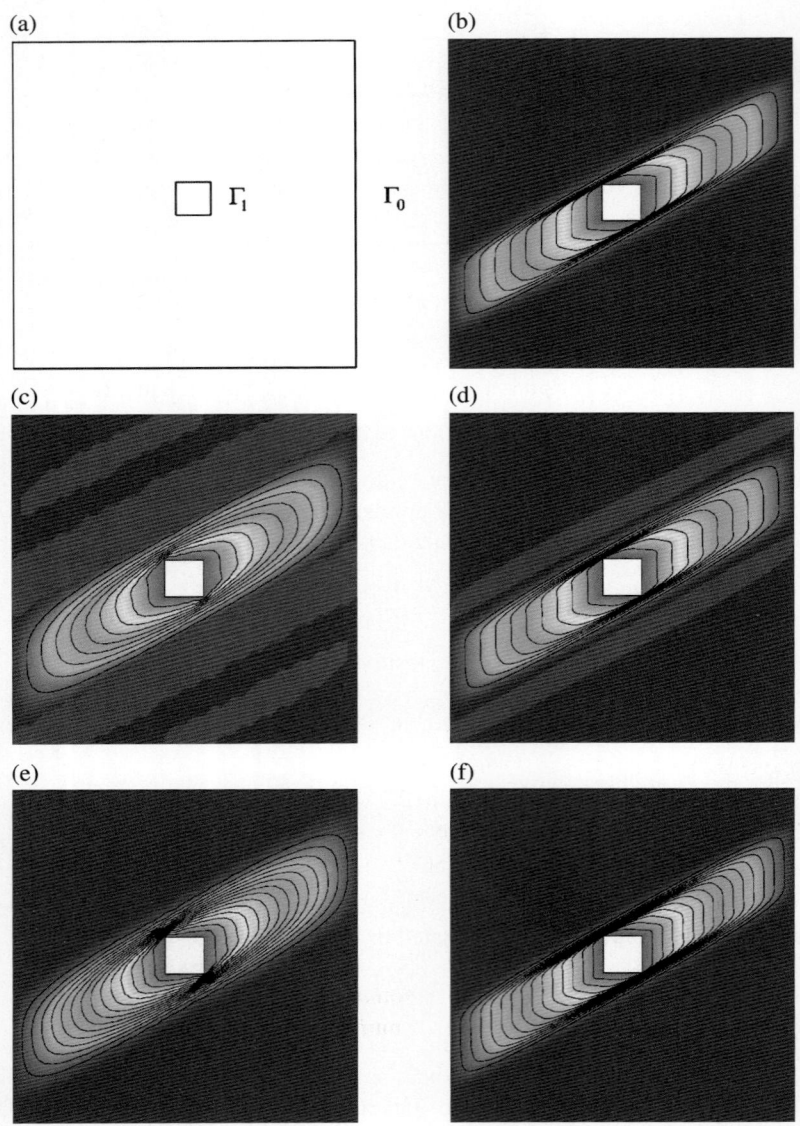

Fig. 12 Anisotropic diffusion: (**a**) domain geometry, (**b**) reference solution, (**c**) Galerkin, $h = 1/36$, (**d**) Galerkin, $h = 1/288$, (**e**) LPFL, $h = 1/36$, (**f**) LPFL, $h = 1/288$

dark blue regions in Figs. 12(c) and 12(d). Algebraic flux correction based on the LPFL algorithm makes it possible to enforce the DMP constraint without excessive smearing. The solutions for $h = 1/36$ and $h = 1/288$ are presented in Figs. 12(e) and 12(f).

Table 7 Anisotropic diffusion: Galerkin grid convergence

h	E_1	E_2	p	$u_{\min}$	$u_{\max}$
1/18	0.826E−01	0.194E+00		−1.06565	1.00000
1/36	0.514E−01	0.136E+00	0.68	−1.05527	1.00000
1/72	0.298E−01	0.904E−01	0.79	−1.03944	1.00000
1/144	0.155E−01	0.544E−01	0.94	−1.01818	1.00000
1/288	0.684E−02	0.278E−01	1.18	−1.00133	1.00000
1/576	0.225E−02	0.103E−01	1.60	−1.00000	1.00000

Table 8 Anisotropic diffusion: LPFL grid convergence

h	E_1	E_2	p	NNL-A	NNL
1/18	0.741E−01	0.181E+00		70	258
1/36	0.441E−01	0.128E+00	0.75	293	1,136
1/72	0.257E−01	0.874E−01	0.78	448	4,904
1/144	0.143E−01	0.547E−01	0.85	951	20,375
1/288	0.712E−02	0.292E−01	1.01	1,094	51,73
1/576	0.245E−02	0.111E−01	1.54	1,976	120,213

The results of the grid convergence study are summarized in Tables 7 and 8. On coarse meshes, the LPFL algorithm produces smaller errors than the underlying Galerkin scheme. As the mesh is refined, the undershoots produced by the latter method become smaller and eventually disappear. In the fourth column, we list the rate of convergence (124) for each pair of meshes. Note that the value of p increases monotonically as the mesh size h goes to zero.

The nonlinearity of the algebraic system associated with the flux-corrected Galerkin discretization of the anisotropic diffusion equation is more severe than in the case of pure convection. This phenomenon was first discovered in [48]. The last two columns in Table 8 list the total number of nonlinear SSOR iterations required to make the maximum norm of the residual smaller than $\epsilon = 10^{-6}$. It is worth mentioning that the values of E_1 and E_2 converged at early stages of the iteration process. Hence, a better choice of stopping criteria would make the iterative solver more efficient [2]. The numbers in the column labeled NNL-A were obtained with Anderson acceleration, as described in Sect. 8.3. If it is switched off, a dramatic increase in the number of nonlinear iterations NNL is observed (see the last column in Table 8). The accelerated version is 60 times faster on the finest mesh.

In the current implementation of Anderson acceleration, we always mix $k = 5$ iterates and calculate the corresponding weights using the LAPACK subroutine DGELS to solve the (unconstrained) least squares problem. The improvements proposed in [17, 18, 62, 77] are likely to result in a further gain of efficiency.

10 Summary and Outlook

The algebraic flux correction paradigm presented in this chapter provides a set of general rules, concepts, and tools for enforcing the discrete maximum principle and positivity preservation in the context of low-order finite element approximations on arbitrary meshes. The presented methodology is based on a generalization of FCT. In particular, we addressed the design of implicit FCT schemes, developed a linearity-preserving slope limiter and converted it into a fully multidimensional format. In contrast to FCT, the new approach to flux correction is well-suited not only for time-dependent problems but also for steady transport equations.

The use of flux limiting gives rise to a nonlinear system which must be linearized or solved in an iterative way. The former approach has led us to an efficient predictor-corrector algorithm for computations with small time steps. In the case of stationary transport equations or large time steps, the linearization of antidiffusive fluxes about a low-order predictor would degrade the accuracy of the algebraic flux correction scheme and inhibit convergence. Hence, there is no way to replace the iterative solution of a nonlinear system with a single postprocessing step. Our results for the anisotropic diffusion equation indicate that Anderson acceleration is a very useful tool for the design of efficient quasi-Newton iterative solvers. The nonlinear SSOR method presented in this paper can also be used as a smoother within the framework of a full multigrid/full approximation scheme (FMG-FAS).

The generality of algebraic flux correction makes it very powerful. The same limiter routine can be employed to enforce positivity constraints in 2D and 3D, on structured and unstructured meshes. The origin of discrete operators makes no difference as far as the M-matrix property is concerned. However, the flux limiter must be designed to keep the perturbation of the discrete problem as small as possible. The demand for high resolution is particularly difficult to meet in the case of higher-order finite elements because the fluxes may depend on solution values at more than two nodes, and even the construction of an optimal low-order scheme becomes a nontrivial task [38]. This has led us to believe that higher-order Galerkin schemes must be constrained within the framework of hp-adaptivity. In regions where the derivatives of order $p \geq 1$ are smooth, no limiting is required. Otherwise, the polynomial degree p must be reduced until a smooth derivative is found [40] or a (multi-) linear approximation ($p = 1$) is recovered in a given element. In the latter case, flux limiting can be performed using the methodology presented in this chapter.

The unavoidable loss of accuracy around internal and boundary layers can be compensated using h-adaptation, i.e., local mesh refinement. The Galerkin orthogonality error produced by the flux limiter is computable and easy to localize. Thus, it provides valuable feedback for goal-oriented mesh adaptation [44].

In the next two chapters, we extend algebraic flux correction to systems of conservation laws including the compressible Euler and incompressible Navier-Stokes equations. The topics to be addressed include the construction of artificial viscosity operators, flux limiting in terms of nonconservative variables, synchronization of the correction factors, and failsafe control of the solution behavior. We also discuss the treatment of source/sink terms in the context of the k–ε turbulence model.

References

1. Anderson, D.G.: Iterative procedures for nonlinear integral equations. J. Assoc Comput. Mach. **12**, 547–560 (1965)
2. Arioli, M., Loghin, D., Wathen, A.J.: Stopping criteria for iterations in finite element methods. Numer. Math. **99**, 381–410 (2006)
3. Arminjon, P., Dervieux, A.: Construction of TVD-like artificial viscosities on 2-dimensional arbitrary FEM grids. INRIA Research Report 1111 (1989)
4. Baum, J.D., Löhner, R.: Numerical simulation of pilot/seat ejection from an F-16. AIAA Paper, 93-0783 (1993)
5. Bochev, P., Ridzal, D., Scovazzi, G., Shashkov, M.: Constrained-optimization based data transfer: a new perspective on flux correction. Chap. 10 in this volume. doi:10 1007/978-94-007-4038-9_10
6. Boris, J.P., Book, D.L.: Flux-Corrected Transport: I. SHASTA, a fluid transport algorithm that works. J. Comput. Phys. **11**, 38–69 (1973)
7. Book, D.L.: The conception, gestation, birth, and infancy of FCT. Chap. 1 in this volume. doi:10.1007/978-94-007-4038-9_1
8. Book, D.L., Boris, J.P., Hain, K.: Flux-corrected transport: II. Generalizations of the method. J. Comput. Phys. **18**, 248–283 (1975)
9. Boris, J.P., Book, D.L.: Flux-Corrected Transport: III. Minimal-error FCT algorithms. J. Comput. Phys. **20**, 397–431 (1976)
10. Carette, J.-C., Deconinck, H., Paillère, H., Roe, P.L.: Multidimensional upwinding: its relation to finite elements. Int. J. Numer. Methods Fluids **20**, 935–955 (1995)
11. Catabriga, L., Coutinho, A.L.G.A.: Implicit SUPG solution of Euler equations using edge-based data structures. Comput. Methods Appl. Mech. Eng. **191**, 3477–3490 (2002)
12. Ciarlet, P.G., Raviart, P.-A.: Maximum principle and convergence for the finite element method. Comput. Methods Appl. Mech. Eng. **2**, 17–31 (1973)
13. Dietachmayer, G.S.: A comparison and evaluation of some positive definite advection schemes. In: Noyle, J., May, R. (eds.) Computational Techniques and Applications, pp. 217–232. Elsevier, Amsterdam (1986)
14. DeVore, C.R.: An improved limiter for multidimensional flux-corrected transport. NASA Technical Report AD-A360122 (1998)
15. Donea, J., Huerta, A.: Finite Element Methods for Flow Problems. Wiley, Chichester (2003)
16. Donea, J., Giuliani, S., Laval, H., Quartapelle, L.: Time-accurate solution of advection-diffusion equations by finite elements. Comput. Methods Appl. Mech. Eng. **193**, 123–145 (1984)
17. Eyert, V.: A comparative study on methods for convergence acceleration of iterative vector sequences. J. Comput. Phys. **124**, 271–285 (1996)
18. Fang, H., Saad, Y.: Two classes of multisecant methods for nonlinear acceleration. Numer. Linear Algebra Appl. **16**, 197–221 (2009)
19. Faragó, I., Horváth, R., Korotov, S.: Discrete maximum principle for linear parabolic problems solved on hybrid meshes. Appl. Numer. Math. **53**, 249–264 (2005)
20. Fletcher, C.A.J.: The group finite element formulation. Comput. Methods Appl. Mech. Eng. **37**, 225–243 (1983)
21. Fletcher, C.A.J.: A comparison of finite element and finite difference solutions of the one- and two-dimensional Burgers' equations. J. Comput. Phys. **51**, 159–188 (1983)
22. Godunov, S.K.: Finite difference method for numerical computation of discontinuous solutions of the equations of fluid dynamics. Mat. Sb. **47**, 271–306 (1959)
23. Gottlieb, S., Shu, C.W.: Total variation diminishing Runge-Kutta schemes. Math. Comput. **67**, 73–85 (1998)
24. Gottlieb, S., Shu, C.-W., Tadmor, E.: Strong stability-preserving high-order time discretization methods. SIAM Rev. **43**, 89–112 (2001)
25. Hansbo, P.: Aspects of conservation in finite element flow computations. Comput. Methods Appl. Mech. Eng. **117**, 423–437 (1994)

26. Harten, A.: High resolution schemes for hyperbolic conservation laws. J. Comput. Phys. **49**, 357–393 (1983)
27. Hubbard, M.E.: Non-oscillatory third order fluctuation splitting schemes for steady scalar conservation laws. J. Comput. Phys. **222**, 740–768 (2007)
28. Hundsdorfer, W., Verwer, J.G.: Numerical Solution of Time-Dependent Advection-Diffusion-Reaction Equations. Springer, Berlin (2003)
29. Jameson, A.: Computational algorithms for aerodynamic analysis and design. Appl. Numer. Math. **13**, 383–422 (1993)
30. Jameson, A.: Analysis and design of numerical schemes for gas dynamics 1. Artificial diffusion, upwind biasing, limiters and their effect on accuracy and multigrid convergence. Int. J. Comput. Fluid Dyn. **4**, 171–218 (1995)
31. Jemcov, A., Maruszewski, J.P.: Algorithm stabilization and acceleration in computational fluid dynamics: exploiting recursive properties of fixed point algorithms. In: Amano, R.S., Sundén, B. (eds.) Computational Fluid Dynamics and Heat Transfer. WIT Press, Southampton (2010)
32. John, V., Schmeyer, E.: On finite element methods for 3D time-dependent convection-diffusion-reaction equations with small diffusion. Comput. Methods Appl. Mech. Eng. **198**, 475–494 (2008)
33. Karátson, J., Korotov, S.: Discrete maximum principles for finite element solutions of nonlinear elliptic problems with mixed boundary conditions. Numer. Math. **99**, 669–698 (2005)
34. Karátson, J., Korotov, S., Křížek, M.: On discrete maximum principles for nonlinear elliptic problems. Math. Comput. Simul. **76**, 99–108 (2007)
35. Kuzmin, D.: Positive finite element schemes based on the flux-corrected transport procedure. In: Bathe, K.J. (ed.) Computational Fluid and Solid Mechanics, pp. 887–888. Elsevier, Amsterdam (2001)
36. Kuzmin, D.: On the design of general-purpose flux limiters for implicit FEM with a consistent mass matrix. I. Scalar convection. J. Comput. Phys. **219**, 513–531 (2006)
37. Kuzmin, D.: Algebraic flux correction for finite element discretizations of coupled systems. In: Oñate, E., Papadrakakis, M., Schrefler, B. (eds.) Computational Methods for Coupled Problems in Science and Engineering II, CIMNE, Barcelona, pp. 653–656 (2007)
38. Kuzmin, D.: On the design of algebraic flux correction schemes for quadratic finite elements. J. Comput. Appl. Math. **218**(1), 79–87 (2008)
39. Kuzmin, D.: Explicit and implicit FEM-FCT algorithms with flux linearization. J. Comput. Phys. **228**, 2517–2534 (2009)
40. Kuzmin, D.: A vertex-based hierarchical slope limiter for p-adaptive discontinuous Galerkin methods. J. Comput. Appl. Math. **233**, 3077–3085 (2010)
41. Kuzmin, D.: A Guide to Numerical Methods for Transport Equations. University Erlangen-Nuremberg, Erlangen (2010). http://www.mathematik.uni-dortmund.de/~kuzmin/Transport.pdf
42. Kuzmin, D.: Linearity-preserving flux correction and convergence acceleration for constrained Galerkin schemes. J. Comput. Appl. Math. (2012, to appear)
43. Kuzmin, D., Möller, M.: Algebraic flux correction I. Scalar conservation laws. In: Kuzmin, D., et al. (eds.) Flux-Corrected Transport: Principles, Algorithms, and Applications, pp. 155–206. Springer, Berlin (2005)
44. Kuzmin, D., Möller, M.: Goal-oriented mesh adaptation for flux-limited approximations to steady hyperbolic problems. J. Comput. Appl. Math. **233**, 3113–3120 (2010)
45. Kuzmin, D., Turek, S.: Flux correction tools for finite elements. J. Comput. Phys. **175**, 525–558 (2002)
46. Kuzmin, D., Turek, S.: High-resolution FEM-TVD schemes based on a fully multidimensional flux limiter. J. Comput. Phys. **198**, 131–158 (2004)
47. Kuzmin, D., Möller, M., Turek, S.: High-resolution FEM-FCT schemes for multidimensional conservation laws. Comput. Methods Appl. Mech. Eng. **193**, 4915–4946 (2004)
48. Kuzmin, D., Shashkov, M.J., Svyatskiy, D.: A constrained finite element method satisfying the discrete maximum principle for anisotropic diffusion problems. J. Comput. Phys. **228**, 3448–3463 (2009)

49. LeVeque, R.J.: High-resolution conservative algorithms for advection in incompressible flow. SIAM J. Numer. Anal. **33**, 627–665 (1996)
50. Lipnikov, K., Shashkov, M., Svyatskiy, D., Vassilevski, Yu.: Monotone finite volume schemes for diffusion equations on unstructured triangular and shape-regular polygonal meshes. J. Comput. Phys. **227**, 492–512 (2007)
51. Löhner, R.: Applied CFD Techniques: An Introduction Based on Finite Element Methods, 2nd edn. Wiley, Chichester (2008)
52. Löhner, R.: Edges, stars, superedges and chains. Comput. Methods Appl. Mech. Eng. **111**, 255–263 (1994)
53. Löhner, R., Galle, M.: Minimization of indirect addressing for edge-based field solvers. Commun. Numer. Methods Eng. **18**(5), 335–343 (2002)
54. Löhner, R., Morgan, K., Peraire, J., Vahdati, M.: Finite element flux-corrected transport (FEM-FCT) for the Euler and Navier-Stokes equations. Int. J. Numer. Methods Fluids **7**, 1093–1109 (1987)
55. Löhner, R., Morgan, K., Vahdati, M., Boris, J.P., Book, D.L.: FEM-FCT: combining unstructured grids with high resolution. Commun. Appl. Numer. Methods **4**, 717–729 (1988)
56. Luo, H., Baum, J.D., Löhner, R., Fast, A.: Matrix-free implicit method for compressible flows on unstructured grids. J. Comput. Phys. **146**, 664–690 (1998)
57. Lyra, P.R.M.: Unstructured grid adaptive algorithms for fluid dynamics and heat conduction. PhD thesis, University of Wales, Swansea (1994)
58. Lyra, P.R.M., Willmersdorf, R.B., Martins, M.A.D., Coutinho, A.L.G.A.: Parallel implementation of edge-based finite element schemes for compressible flow on unstructured grids. In: Proceedings of the 3rd International Meeting on Vector and Parallel Processing, Faculdade de Engenharia da Universidade do Porto, Porto, Portugal, 21–23 Juni (1998)
59. Mer, K.: Variational analysis of a mixed element/volume scheme with fourth-order viscosity on general triangulations. Comput. Methods Appl. Mech. Eng. **153**, 45–62 (1998)
60. Möller, M.: Hochauflösende FEM-FCT-Verfahren zur Diskretisierung von konvektionsdominanten Transportproblemen mit Anwendung auf die kompressiblen Eulergleichungen. Diploma thesis, University of Dortmund (2003)
61. Möller, M.: Efficient solution techniques for implicit finite element schemes with flux limiters. Int. J. Numer. Methods Fluids **55**, 611–635 (2007)
62. Ni, P.: Anderson acceleration of fixed-point iteration with applications to electronic structure computations. PhD thesis, Worcester Polytechnic Institute (2009)
63. Schär, C., Smolarkiewicz, P.K.: A synchronous and iterative flux-correction formalism for coupled transport equations. J. Comput. Phys. **128**, 101–120 (1996)
64. Oran, E.S., Boris, J.P.: Numerical Simulation of Reactive Flow, 2nd edn. Cambridge University Press, Cambridge (2001)
65. Parrott, A.K., Christie, M.A.: FCT applied to the 2-D finite element solution of tracer transport by single phase flow in a porous medium. In: Numerical Methods for Fluid Dynamics, pp. 609–619. Oxford University Press, London (1986)
66. Patankar, S.V.: Numerical Heat Transfer and Fluid Flow. McGraw-Hill, New York (1980)
67. Peraire, J., Vahdati, M., Peiro, J., Morgan, K.: The construction and behavior of some unstructured grid algorithms for compressible flows. In: Numerical Methods for Fluid Dynamics, IV, pp. 221–239. Oxford University Press, Oxford (1993)
68. Selmin, V.: Finite element solution of hyperbolic equations. I. One-dimensional case. INRIA Research Report 655 (1987)
69. Selmin, V.: Finite element solution of hyperbolic equations. II. Two-dimensional case. INRIA Research Report 708 (1987)
70. Selmin, V.: The node-centred finite volume approach: bridge between finite differences and finite elements. Comput. Methods Appl. Mech. Eng. **102**, 107–138 (1993)
71. Selmin, V., Formaggia, L.: Unified construction of finite element and finite volume discretizations for compressible flows. Int. J. Numer. Methods Eng. **39**, 1–32 (1996)
72. Smith, D.A., Ford, W.F., Sidi, A.: Extrapolation methods for vector sequences. SIAM Rev. **29**, 199–233 (1987)

73. Sweby, P.K.: High resolution schemes using flux limiters for hyperbolic conservation laws. SIAM J. Numer. Anal. **21**, 995–1011 (1984)
74. van Slingerland, P.: An accurate and robust finite volume method for the advection diffusion equation. M.Sc. thesis, Delft University of Technology (June 2007)
75. van Slingerland, P., Borsboom, M., Vuik, C.: A local theta scheme for advection problems with strongly varying meshes and velocity profiles. Report 08-17, Department of Applied Mathematical Analysis, Delft University of Technology (June 2008)
76. Varga, R.S.: Matrix Iterative Analysis. Prentice-Hall, Englewood Cliffs (1962)
77. Walker, H.W., Ni, P.: Anderson acceleration for fixed-point iterations. SIAM J. Numer. Anal. **49**, 1715–1735 (2011)
78. Zalesak, S.T.: Fully multidimensional flux-corrected transport algorithms for fluids. J. Comput. Phys. **31**, 335–362 (1979)

Algebraic Flux Correction II

Compressible Flow Problems

Dmitri Kuzmin, Matthias Möller, and Marcel Gurris

Abstract Flux limiting for hyperbolic systems requires a careful generalization of the design principles and algorithms introduced in the context of scalar conservation laws. In this chapter, we develop FCT-like algebraic flux correction schemes for the Euler equations of gas dynamics. In particular, we discuss the construction of artificial viscosity operators, the choice of variables to be limited, and the transformation of antidiffusive fluxes. An a posteriori control mechanism is implemented to make the limiter failsafe. The numerical treatment of initial and boundary conditions is discussed in some detail. The initialization is performed using an FCT-constrained L^2 projection. The characteristic boundary conditions are imposed in a weak sense, and an approximate Riemann solver is used to evaluate the fluxes on the boundary. We also present an unconditionally stable semi-implicit time-stepping scheme and an iterative solver for the fully discrete problem. The results of a numerical study indicate that the nonlinearity and non-differentiability of the flux limiter do not inhibit steady state convergence even in the case of strongly varying Mach numbers. Moreover, the convergence rates improve as the pseudo-time step is increased.

1 Introduction

The first successful finite element schemes for compressible flow problems were developed by the Swansea and INRIA groups in the 1980s. The most prominent

---

D. Kuzmin (✉)
Applied Mathematics III, University Erlangen-Nuremberg, Cauerstr. 11, 91058 Erlangen, Germany
e-mail: kuzmin@am.uni-erlangen.de

M. Möller · M. Gurris
Institute of Applied Mathematics (LS III), TU Dortmund, Vogelpothsweg 87, 44227 Dortmund, Germany

M. Möller
e-mail: matthias.moeller@math.tu-dortmund.de

M. Gurris
e-mail: marcel.gurris@math.tu-dortmund.de

representative of these schemes is the two-step Taylor-Galerkin method [1, 43] and its combination with FCT [42, 56, 57]. The early 1990s have witnessed the advent of edge-based data structures [6, 44, 53, 58] that offer a number of significant advantages compared to the traditional element-based implementation. In the case of P_1 finite elements, the edge-based formulation is equivalent to a vertex-centered finite volume scheme [58, 59]. This equivalence makes it possible to implement approximate Riemann solvers and slope limiters in the context of finite element discretizations on simplex meshes [45–48, 51, 53]. However, the resulting schemes require mass lumping and are sensitive to the orientation of mesh edges.

All classical high-resolution FEM are explicit and, therefore, subject to time step restrictions. Implicit schemes have the potential of being unconditionally stable but rely on the quality of the iterative solver for the nonlinear system. In particular, a careful linearization/preconditioning of the discrete Jacobian operator is essential. A semi-implicit solution strategy [9, 14, 66] and weak imposition of characteristic boundary conditions [18] lead to an algorithm that converges to steady state solutions at arbitrarily large CFL numbers [18, 19]. This is a remarkable result since the use of nondifferentiable limiters is commonly believed to inhibit convergence.

The development of flux-corrected transport schemes for systems of equations is more difficult than in the scalar case. A limiter designed to control the local maxima and minima of the conservative variables does not guarantee that the pressure or internal energy will stay nonnegative. Likewise, the velocity is not directly constrained and may exhibit spurious fluctuations. Since the rate of transport depends on the oscillatory velocity and pressure fields, undershoots and overshoots eventually carry over to the conservative variables. As a typical consequence, the speed of sound becomes negative, indicating that the simulation is going to crash.

In this chapter, we review some recent advances in the design of implicit algebraic flux correction schemes for the Euler equations [18, 19, 32–34, 49]. After the presentation of the standard Galerkin scheme, we discuss various forms of artificial dissipation and the above difficulties associated with flux limiting for systems of equations. In particular, we present a synchronized FCT limiter that features a node-based transformation to primitive variables and a failsafe control mechanism inspired by the recent work of Zalesak [73]. Also, we address the treatment of nonlinearities and the implementation of initial/boundary conditions. A numerical study is performed for a number of steady and unsteady inviscid flow problems in 2D.

2 The Euler Equations

The Euler equations of gas dynamics represent a system of conservation laws for the mass, momentum, and energy of an inviscid compressible fluid

$$\frac{\partial \rho}{\partial t} + \nabla \cdot (\rho \mathbf{v}) = 0, \qquad (1)$$

$$\frac{\partial (\rho \mathbf{v})}{\partial t} + \nabla \cdot (\rho \mathbf{v} \otimes \mathbf{v} + p \mathscr{I}) = 0, \qquad (2)$$

$$\frac{\partial(\rho E)}{\partial t} + \nabla \cdot (\rho E \mathbf{v} + p\mathbf{v}) = 0, \qquad (3)$$

where ρ is the density, $\mathbf{v}$ is the velocity, p is the pressure, E is the total energy, and $\mathscr{I}$ is the identity tensor. The system is closed with the equation of state

$$p = (\gamma - 1)\left(\rho E - \frac{\rho|\mathbf{v}|^2}{2}\right) \qquad (4)$$

for an ideal polytropic gas with the heat capacity ratio γ. The default is $\gamma = 1.4$ (air).

The nonlinear system (1)–(3) can be written in the generic divergence form

$$\frac{\partial U}{\partial t} + \nabla \cdot \mathbf{F} = 0, \qquad (5)$$

where

$$U = \begin{bmatrix} \rho \\ \rho \mathbf{v} \\ \rho E \end{bmatrix}, \qquad \mathbf{F} = \begin{bmatrix} \rho \mathbf{v} \\ \rho \mathbf{v} \otimes \mathbf{v} + p\mathscr{I} \\ \rho E \mathbf{v} + p\mathbf{v} \end{bmatrix} \qquad (6)$$

are the vectors of conservative variables and fluxes. It can be shown that [69]

$$\mathbf{F} = AU, \qquad (7)$$

where $\mathbf{A} = \frac{\partial \mathbf{F}}{\partial U}$ is the Jacobian tensor associated with the quasi-linear form of (5)

$$\frac{\partial U}{\partial t} + \mathbf{A} \cdot \nabla U = 0. \qquad (8)$$

Due to the hyperbolicity of the Euler equations, any directional Jacobian matrix $\mathbf{e} \cdot \mathbf{A}$ is diagonalizable and admits the factorization [24, 37, 69]

$$\mathbf{e} \cdot \mathbf{A} = R\Lambda R^{-1}, \qquad (9)$$

where $\Lambda(\mathbf{e})$ is the diagonal matrix of eigenvalues and $R(\mathbf{e})$ is the matrix of right eigenvectors. In the 3D case, the eigenvalues of the 5×5 matrix $\mathbf{e} \cdot \mathbf{A}$ are given by

$$\lambda_1 = \mathbf{e} \cdot \mathbf{v} - c, \qquad (10)$$

$$\lambda_2 = \lambda_3 = \lambda_4 = \mathbf{e} \cdot \mathbf{v}, \qquad (11)$$

$$\lambda_5 = \mathbf{e} \cdot \mathbf{v} + c, \qquad (12)$$

where $c = \sqrt{\gamma p/\rho}$ is the speed of sound. Thus, the solution to a Riemann problem is a superposition of three waves traveling at speed $\mathbf{e} \cdot \mathbf{v}$ and two waves propagating at speeds $\pm c$ relative to the gas. Closed-form expressions for the eigenvectors associated with each characteristic speed can be found, e.g., in [55].

Let $\Omega \subset \mathbb{R}^n$, $n \in \{1, 2, 3\}$ be a bounded domain. The solution to the unsteady Euler equations is initialized by a given distribution of all variables

$$U(\mathbf{x}, t) = U_0(\mathbf{x}) \quad \text{in } \Omega. \tag{13}$$

Given a vector of "free stream" solution values U_∞, characteristic boundary conditions of Dirichlet or Neumann type can be defined in terms of the solution to the Riemann problem with the interior state U and exterior state U_∞, see Sect. 11.

In general, we impose Dirichlet boundary conditions on the boundary part Γ_D

$$U = G(U, U_\infty) \quad \text{on } \Gamma_D \tag{14}$$

and Neumann (normal flux) boundary conditions on the boundary part Γ_N

$$\mathbf{n} \cdot \mathbf{F} = F_n(U, U_\infty) \quad \text{on } \Gamma_N, \tag{15}$$

where $\mathbf{n}$ is the unit outward normal. Note that the solution to the Riemann problem depends not only on the prescribed boundary data but also on the unknown solution.

3 Group FEM Approximation

To begin with, we discretize the Euler equations using linear or multilinear finite elements. After integration by parts, the variational formulation of (5) becomes

$$\int_\Omega \left(w \frac{\partial U}{\partial t} - \nabla w \cdot \mathbf{F} \right) d\mathbf{x} + \int_\Gamma w F_n \, ds = 0, \quad \forall w. \tag{16}$$

Since the test function w vanishes on Γ_D, the surface integral reduces to that over Γ_N.

Within the framework of Fletcher's [16] group finite element formulation, the approximate solution $U_h \approx U$ and the numerical flux function $\mathbf{F}_h \approx \mathbf{F}$ are interpolated using the same set of piecewise-polynomial basis functions $\{\varphi_i\}$. That is,

$$U_h(\mathbf{x}, t) = \sum_j \mathrm{U}_j(t) \varphi_j(\mathbf{x}), \tag{17}$$

$$\mathbf{F}_h(\mathbf{x}, t) = \sum_j \mathbf{F}_j(t) \varphi_j(\mathbf{x}). \tag{18}$$

Inserting these approximations into the Galerkin weak form (16), one obtains a system of semi-discretized equations for the time-dependent nodal values

$$\sum_j \left(\int_\Omega \varphi_i \varphi_j \, d\mathbf{x} \right) \frac{d\mathrm{U}_j}{dt} = \sum_j \left(\int_\Omega \nabla \varphi_i \varphi_j \, d\mathbf{x} \right) \cdot \mathbf{F}_j - \int_\Gamma \varphi_i F_n \, ds. \tag{19}$$

By the homogeneity property (7) of the Euler fluxes, we have

$$\mathbf{F}_j = \mathbf{A}_j \mathbf{U}_j.$$

Thus, the matrix form of the semi-discrete problem can be written as follows:

$$M_C \frac{d\mathbf{U}}{dt} = K\mathbf{U} + \mathbf{S}(\mathbf{U}). \tag{20}$$

The $(n+2) \times (n+2)$ blocks of the consistent mass matrix $M_C = \{\mathbf{M}_{ij}\}$ are defined by $\mathbf{M}_{ij} = m_{ij}\mathbf{I}$, where $\mathbf{I}$ stands for the identity matrix and

$$m_{ij} = \int_\Omega \varphi_i \varphi_j \, d\mathbf{x}. \tag{21}$$

Furthermore, the vector of boundary loads associated with node i is given by

$$\mathbf{S}_i = -\int_\Gamma \varphi_i \mathbf{F}_n \, ds, \tag{22}$$

and the formula for entries of the discrete Jacobian operator $K = \{\mathbf{K}_{ij}\}$ reads

$$\mathbf{K}_{ij} = \mathbf{c}_{ji} \cdot \mathbf{A}_j, \quad \mathbf{c}_{ij} = \int_\Omega \varphi_i \nabla \varphi_j \, d\mathbf{x}. \tag{23}$$

Since $\sum_j \varphi_j \equiv 1$, the matrix of discrete derivatives $\mathbf{C} := \{\mathbf{c}_{ij}\}$ has zero row sums

$$\sum_j \mathbf{c}_{ij} = \mathbf{0}. \tag{24}$$

Furthermore, integration by parts reveals that the coefficients $\mathbf{c}_{ij}$ and $\mathbf{c}_{ji}$ satisfy

$$\mathbf{c}_{ji} = -\mathbf{c}_{ij} + \int_\Gamma \varphi_i \varphi_j \mathbf{n} \, ds. \tag{25}$$

The boundary term is symmetric and corresponds to an entry of the mass matrix for the surface triangulation of Γ. In the case of (multi-)linear finite elements, the basis function φ_i vanishes on Γ, unless $\mathbf{x}_i$ is a boundary node. It follows that

$$\mathbf{c}_{ji} = -\mathbf{c}_{ij}, \quad \mathbf{c}_{ii} = \mathbf{0}, \quad \mathbf{S}_i = \mathbf{0} \tag{26}$$

in the interior of Ω. The above properties of the discrete gradient operator $\mathbf{C}$ play an important role in the derivation of edge-based data structures [27, 40, 59].

4 Edge-Based Representation

Properties (24) and (26) make it possible to express the components of $K\mathrm{U}$ in terms of edge contributions. The following representation is valid inside Ω

$$(K\mathrm{U})_i = \sum_{j\neq i} \mathbf{e}_{ij} \cdot (\mathbf{F}_j - \mathbf{F}_i), \quad \mathbf{e}_{ij} = \frac{\mathbf{c}_{ji} - \mathbf{c}_{ij}}{2}. \tag{27}$$

The numerical fluxes for an edge-based implementation are defined by [34, 59]

$$(K\mathrm{U})_i = -\sum_{j\neq i} G_{ij}, \quad G_{ij} = \mathbf{c}_{ij} \cdot \mathbf{F}_i - \mathbf{c}_{ji} \mathbf{F}_j. \tag{28}$$

For the derivation of the above flux decomposition for $K\mathrm{U}$, we refer to [27, 34, 59]. As shown by Roe [54], the flux difference can be linearized as follows

$$\mathbf{F}_j - \mathbf{F}_i = \mathbf{A}_{ij}(\mathrm{U}_j - \mathrm{U}_i). \tag{29}$$

The edge Jacobian matrix $\mathbf{A}_{ij} := \mathbf{A}(\rho_{ij}, \mathbf{v}_{ij}, H_{ij})$ is associated with a special set of density-averaged variables known as the *Roe mean values*

$$\rho_{ij} = \sqrt{\rho_i \rho_j}, \tag{30}$$

$$\mathbf{v}_{ij} = \frac{\sqrt{\rho_i}\mathbf{v}_i + \sqrt{\rho_j}\mathbf{v}_j}{\sqrt{\rho_i} + \sqrt{\rho_j}}, \tag{31}$$

$$H_{ij} = \frac{\sqrt{\rho_i}H_i + \sqrt{\rho_j}H_j}{\sqrt{\rho_i} + \sqrt{\rho_j}}, \tag{32}$$

where $H = E + \frac{p}{\rho}$ denotes the stagnation enthalpy. The speed of sound is given by

$$c_{ij} = \sqrt{(\gamma - 1)\left(H_{ij} - \frac{|\mathbf{v}_{ij}|^2}{2}\right)}. \tag{33}$$

By virtue of (27) and (29), the following relationship holds for internal nodes

$$\mathrm{K}_{ii} = -\sum_{j\neq i} \mathrm{K}_{ij}, \quad \mathrm{K}_{ij} = \mathbf{e}_{ij} \cdot \mathbf{A}_{ij}, \quad j \neq i. \tag{34}$$

This representation of K_{ij} turns out to be very useful when it comes to the design of artificial viscosity operators for algebraic flux correction schemes (see the next section). However, the assembly of K should be performed using definition (23).

By the hyperbolicity of the Euler equations, the directional Roe matrix $\mathbf{e}_{ij} \cdot \mathbf{A}_{ij}$ is diagonalizable with real eigenvalues. Invoking (9), we obtain the factorization

$$\mathbf{e}_{ij} \cdot \mathbf{A}_{ij} = |\mathbf{e}_{ij}|\, \mathrm{R}_{ij}\, \Lambda_{ij}\, \mathrm{R}_{ij}^{-1}. \tag{35}$$

According to (10)–(12) the entries of the eigenvalue matrix Λ_{ij} are given by

$$\lambda_1 = v_{ij} - c_{ij}, \tag{36}$$

$$\lambda_2 = \lambda_3 = \lambda_4 = v_{ij}, \tag{37}$$

$$\lambda_5 = v_{ij} + c_{ij}. \tag{38}$$

Here c_{ij} is the speed of sound (33) for Roe's approximate Riemann solver, while

$$v_{ij} = \frac{\mathbf{e}_{ij} \cdot \mathbf{v}_{ij}}{|\mathbf{e}_{ij}|}$$

is the density-averaged velocity along the (virtual) edge connecting nodes i and j.

5 Artificial Viscosity Operators

In the chapter on algebraic flux correction for scalar conservation laws [31], we constructed a nonoscillatory low-order scheme using row-sum mass lumping

$$M_L := \text{diag}\{m_i I\}, \quad m_i = \sum_j m_{ij} \tag{39}$$

and conservative postprocessing of the Galerkin operator $K = \{\mathrm{K}_{ij}\}$. For systems of conservation laws, each block K_{ij} is an $(n+2) \times (n+2)$ matrix. The blocks of the artificial diffusion operator $D := \{\mathrm{D}_{ij}\}$ are matrices of the same size. As in the scalar case, the discrete Jacobian operator is modified edge-by-edge thus:

$$\begin{aligned}
\mathrm{K}_{ii} &:= \mathrm{K}_{ii} - \mathrm{D}_{ij}, & \mathrm{K}_{ij} &:= \mathrm{K}_{ij} + \mathrm{D}_{ij}, \\
\mathrm{K}_{ji} &:= \mathrm{K}_{ji} + \mathrm{D}_{ij}, & \mathrm{K}_{jj} &:= \mathrm{K}_{jj} - \mathrm{D}_{ij}.
\end{aligned} \tag{40}$$

Replacing K with $L := K + D$, one obtains the low-order approximation to (20)

$$M_L \frac{d\mathrm{U}}{dt} = L\mathrm{U} + \mathrm{S}(\mathrm{U}). \tag{41}$$

If all off-diagonal matrix blocks L_{ij} are positive semi-definite, then such a low-order scheme proves local extremum diminishing (LED) with respect to local *characteristic variables* [34]. This condition is a generalization of the LED criterion for scalar transport equations. In the case of a hyperbolic system it is less restrictive than the requirement that all off-diagonal entries of L be nonnegative.

According to (34) and (35), the negative eigenvalues of K_{ij} and K_{ji} can be eliminated by adding tensorial artificial dissipation of the form [34]

$$\mathrm{D}_{ij} = |\mathbf{e}_{ij} \cdot \mathbf{A}_{ij}| := |\mathbf{e}_{ij}| R_{ij} |\Lambda_{ij}| R_{ij}^{-1}, \tag{42}$$

where $|\Lambda_{ij}|$ is a diagonal matrix containing the absolute values of the eigenvalues.

Flux limiting in terms of characteristic variables requires that the diffusive and antidiffusive fluxes be defined separately for each component of $\mathbf{e}_{ij} = (e_{ij}^1, \ldots, e_{ij}^n)$ and $\mathbf{A}_{ij} = (A_{ij}^1, \ldots, A_{ij}^n)$. Thus, the above definition of D_{ij} should be replaced with

$$D_{ij} = |e_{ij}^1 A_{ij}^1| + \cdots + |e_{ij}^n A_{ij}^n|. \tag{43}$$

In the 1D case, the low-order scheme with artificial viscosity of the form (42) or (43) reduces to Roe's approximate Riemann solver (see Appendix).

The cost of evaluating the Roe matrix $\mathbf{A}_{ij}$ is rather high. An inexpensive alternative is the computation of D_{ij} using the Jacobian at the arithmetic mean state

$$\mathbf{A}_{ij} := \mathbf{A}\left(\frac{\mathbf{U}_j + \mathbf{U}_i}{2}\right). \tag{44}$$

Banks et al. [5] present a numerical study of methods that use this linearization. In particular, the expected order of accuracy is verified numerically. Importantly, the replacement of the Roe mean values with the arithmetic mean does not make the scheme nonconservative if this approximation is used in the definition of D_{ij} only.

In particularly sensitive applications, the minimal artificial viscosity based on the characteristic decomposition of $\mathbf{A}_{ij}$ may fail to suppress spurious oscillations. This is unacceptable if the flux limiter relies on the assumption that the local extrema of the low-order solution constitute physically legitimate upper and lower bounds.

A possible remedy is the use of Rusanov-like scalar dissipation proportional to the fastest characteristic speed [5, 73]. The straightforward definition is

$$D_{ij} = d_{ij}\mathbf{I}, \quad d_{ij} = |\mathbf{e}_{ij}|\max_i |\lambda_i|, \tag{45}$$

where $\max_i |\lambda_i| = |\mathbf{e}_{ij}|(|v_{ij}| + c_{ij})$ is the spectral radius of the Roe matrix. In our experience, a more robust and efficient low-order scheme is obtained with [33]

$$D_{ij} = \max\{d_{ij}, d_{ji}\}\mathbf{I}, \quad d_{ij} = |\mathbf{e}_{ij} \cdot \mathbf{v}_j| + |\mathbf{e}_{ij}|c_j, \tag{46}$$

where $c_i = \sqrt{\gamma p_i/\rho_i}$ is the speed of sound at node i. In the context of implicit schemes, scalar dissipation may be used for preconditioning purposes even if tensorial artificial viscosity of the form (42) or (43) is favored for accuracy reasons.

6 Algebraic Flux Correction

The semi-discrete Galerkin scheme (20) admits a conservative splitting into the nonoscillatory low-order part (41) and an antidiffusive correction:

$$M_C \frac{d\mathbf{U}}{dt} = K\mathbf{U} + \mathbf{S}(\mathbf{U}) \quad \Longleftrightarrow \quad M_L \frac{d\mathbf{U}}{dt} = L\mathbf{U} + \mathbf{S}(\mathbf{U}) + \mathbf{F}(\mathbf{U}), \tag{47}$$

where $F(U)$ is the vector of raw antidiffusive fluxes. By definition of M_L and D

$$F_i = \sum_{j \neq i} F_{ij}, \qquad F_{ij} = m_{ij}\left(\frac{dU_i}{dt} - \frac{dU_j}{dt}\right) + D_{ij}(U_i - U_j). \qquad (48)$$

In the process of flux correction, F_i is replaced with its limited counterpart

$$\bar{F}_i = \sum_{j \neq i} \bar{F}_{ij}, \qquad \bar{F}_{ij} := \alpha_{ij} F_{ij}, \quad 0 \leq \alpha_{ij} \leq 1. \qquad (49)$$

In Sect. 9, we discuss various generalizations of scalar limiting techniques to systems. All of them produce a constrained semi-discrete problem of the form

$$M_L \frac{dU}{dt} = R(U), \qquad (50)$$

where $R(U) = LU + S(U) + \bar{F}(U)$ incorporates the nonlinear antidiffusive correction.

Let U^n denote the vector of solution values at the time level $t^n = n\Delta t$, where Δt is a constant time step. Integration in time by the two-level θ-scheme yields

$$M_L \frac{U^{n+1} - U^n}{\Delta t} = \theta R(U^{n+1}) + (1-\theta) R(U^n), \qquad (51)$$

where $\theta \in (0, 1]$ is the implicitness parameter. In the fully discrete form of (48), the time derivative $\frac{dU_i}{dt}$ is replaced with $\frac{U_i^{n+1} - U_i^n}{\Delta t}$ and $D_{ij}(U_i - U_j)$ becomes

$$\theta D_{ij}^{n+1}(U_i^{n+1} - U_j^{n+1}) + (1-\theta) D_{ij}^n (U_i^n - U_j^n).$$

The structure of the constrained flux $\bar{F}_{ij}$ depends on the adopted limiting strategy.

7 Solution of Nonlinear Systems

Following a common practice [9, 14, 66], we linearize the contribution of $R(U^{n+1})$ to the right-hand side of (51) about U^n using the Taylor series expansion

$$R(U^{n+1}) \approx R(U^n) + \left(\frac{\partial R}{\partial U}\right)^n (U^{n+1} - U^n). \qquad (52)$$

Plugging this approximation into (51), one obtains the linear algebraic system

$$\left[\frac{M_L}{\Delta t} - \theta \left(\frac{\partial R}{\partial U}\right)^n\right](U^{n+1} - U^n) = R(U^n). \qquad (53)$$

If the steady-state solution is of interest, we use the backward Euler method ($\theta = 1$) and gradually increase the pseudo-time step Δt. When the solution begins to approach the steady state ($R(U) = 0$), the removal of the mass matrix can greatly speed up the convergence process since (53) reduces to Newton's method

$$-\left(\frac{\partial R}{\partial U}\right)^n (U^{n+1} - U^n) = R(U^n) \qquad (54)$$

in the limit of infinitely large (pseudo)-time steps. On the other hand, removing the mass matrix too soon may have an adverse effect on the convergence rates [61].

Trépanier et al. [66] found it useful to freeze the Jacobian after the residuals reach a prescribed tolerance. This can significantly reduce the cost of matrix assembly.

Neglecting the nonlinearity of $L = K + D$, we approximate the Jacobian by [18]

$$\frac{\partial R}{\partial U} \approx K + D + \frac{\partial S}{\partial U} + \frac{\partial \bar{F}}{\partial U}. \qquad (55)$$

If the blocks of the Galerkin transport operator K are defined by (23), the use of $K(U^n)$ instead of $K(U^{n+1})$ boils down to replacing the flux $\mathbf{F}_j = \mathbf{A}_j^{n+1} U_j^{n+1}$ with the flux $\mathbf{F}_j = \mathbf{A}_j^n U_j^{n+1}$. Thus, the above linearization about U^n is conservative.

Since the vector of boundary fluxes $S(U)$ depends on the solution of a Riemann problem, its differentiation is a rather laborious process. For details, we refer to Gurris [18] who derived a formula for $\frac{\partial S}{\partial U}$ using a repeated application of the chain rule. His numerical study indicates that the implicit treatment of the weakly imposed boundary conditions makes it possible to achieve unconditional stability.

The use of a non-differentiable flux limiter rules out the derivation of closed-form expressions for $\frac{\partial \bar{F}}{\partial U}$. In principle, the antidiffusive term can be differentiated numerically using finite differencing [49, 50]. However, the significant overhead cost and the sensitivity to the choice of the free parameter restrict the practical utility of this approach. Moreover, the resultant matrix is not as sparse as the low-order Jacobian since the use of limiters widens the computational stencils. For this reason, we currently favor a semi-explicit treatment of limited antidiffusion.

Instead of linearizing the nondifferentiable antidiffusive term about U^n, one can update it in an iterative fashion. Given an approximate solution $U^{(m)} \approx U^{n+1}$ to (53), a new approximation $U^{(m+1)}$ is obtained by solving the linear system

$$J(U^{(m)})(U^{(m+1)} - U^n) = R(U^n) + \theta(\bar{F}(U^{(m)}) - \bar{F}(U^n)), \qquad (56)$$

$$J(U) = \frac{M_L}{\Delta t} - \theta\left(L(U) + \frac{\partial S}{\partial U}\right). \qquad (57)$$

Due to the semi-explicit treatment of $\bar{F}(U^{n+1})$, the so-defined defect correction scheme may converge rather slowly. However, it can be converted into a quasi-Newton method using the Anderson convergence acceleration technique [26].

The repeated evaluation of the antidiffusive term can be avoided using a linearization about the solution of the low-order system. This predictor-corrector strategy is appropriate if the transient flow behavior dictates the use of small time steps. In this case, the following algorithm [28, 33] is a cost-effective alternative to (53)

1. Calculate the end-of-step solution $\mathrm{U}^L \approx \mathrm{U}^{n+1}$ to the low-order system

$$J(\mathrm{U}^n)(\mathrm{U}^L - \mathrm{U}^n) = L(\mathrm{U}^n)\mathrm{U}^n + \mathrm{S}(\mathrm{U}^n). \tag{58}$$

2. Calculate the vector of raw antidiffusive fluxes F_{ij} linearized about U^L

$$\mathrm{F}_{ij} = m_{ij}(\dot{\mathrm{U}}_i^L - \dot{\mathrm{U}}_j^L) + \mathrm{D}_{ij}(\mathrm{U}_i^L - \mathrm{U}_j^L), \tag{59}$$

where $\dot{\mathrm{U}}_i^L$ is a low-order approximation to the time-derivative at node i

$$\dot{\mathrm{U}}^L = M_L^{-1}[L(\mathrm{U}^L)\mathrm{U}^L + \mathrm{S}(\mathrm{U}^L)]. \tag{60}$$

3. Apply the flux limiter and calculate the final solution U^{n+1}

$$\mathrm{U}_i^{n+1} = \mathrm{U}_i^L + \frac{1}{m_i}\sum_{j\neq i}\bar{\mathrm{F}}_{ij}. \tag{61}$$

8 Solution of Linear Systems

In the 3D case, there are 5 unknowns (density, 3 momentum components, and energy) per mesh node. Hence, each linear system to be solved can be written as

$$\begin{bmatrix} J_{11} & J_{12} & J_{13} & J_{14} & J_{15} \\ J_{21} & J_{22} & J_{23} & J_{24} & J_{25} \\ J_{31} & J_{32} & J_{33} & J_{34} & J_{35} \\ J_{41} & J_{42} & J_{43} & J_{44} & J_{45} \\ J_{51} & J_{52} & J_{53} & J_{54} & J_{55} \end{bmatrix} \begin{bmatrix} \Delta u_1 \\ \Delta u_2 \\ \Delta u_3 \\ \Delta u_4 \\ \Delta u_5 \end{bmatrix} = \begin{bmatrix} r_1 \\ r_2 \\ r_3 \\ r_4 \\ r_5 \end{bmatrix}. \tag{62}$$

Simultaneous update of all variables is costly in terms of CPU time and memory requirements. The coupled system can be split into smaller subproblems using an iterative method of block-Jacobi or block-Gauss-Seidel type. In the former case, the new value of Δu_k is calculated using Δu_l from the last outer iteration:

$$J_{kk}\Delta u_k^{(m+1)} = r_k - \sum_{k\neq l} J_{kl}\Delta u_l^{(m)}, \qquad \Delta u^{(0)} := 0, \tag{63}$$

where m is the iteration counter and k is the subproblem index. Replacing $\Delta u_l^{(m)}$ with $\Delta u_l^{(m+1)}$ for $l < k$, one obtains the block-Gauss-Seidel method

$$J_{kk}\Delta u_k^{(m+1)} = r_k - \sum_{l>k} J_{kl}\Delta u_l^{(m)} - \sum_{l<k} J_{kl}\Delta u_l^{(m+1)}. \tag{64}$$

This segregated solution strategy is easy to implement but may require many iterations per time step. A more robust iterative solver for (62) can be designed using a Krylov-subspace or multigrid method equipped with a smoother/preconditioner that involves solution of small coupled problems on elements/patches. In the next chapter, we will use such a method to solve the discrete saddle point problem for the finite element discretization of the incompressible Navier-Stokes equations.

9 Flux Limiting for Systems

The design of flux limiters for hyperbolic systems is more involved than that for scalar conservation laws. If the density, momentum, and energy increments are limited separately, undershoots/overshoots are likely to arise in all quantities of interest. The following remedies to this problem have been proposed [41, 42, 71–73]

- synchronization of the correction factors for selected control variables;
- transformations to nonconservative (primitive, characteristic) variables;
- a posteriori control and postprocessing of the flux-corrected solution.

In the synchronized version of the FCT limiter [41, 42], all components of the raw antidiffusive flux F_{ij} are multiplied by the same correction factor α_{ij}. No synchronization of α_{ij} is required if a transformation to the local characteristic variables is performed. However, this sort of flux correction is computationally expensive and requires dimensional splitting for the diffusive and antidiffusive fluxes.

Limiters that constrain the primitive (density, velocity, pressure) or characteristic variables are typically quite reliable but the involved linearizations may also cause them to fail, no matter how carefully they are designed. While it is impossible to rule out the formation of spurious maxima and minima a priori, they can be easily detected and removed at a postprocessing step. This approach was introduced by Zalesak [73] who used it to maintain the nonnegativity of pressures and internal energies in a characteristic FCT method for the compressible Euler equations.

9.1 Transformation of Variables

We begin with the presentation of a symmetric limiter for a general set of dependent quantities. In classical high-resolution schemes for the Euler equations, the required transformations between the conservative and nonconservative variables are usually performed edge-by-edge [40, 71–73]. The solution-dependent transformation matrix $T_{ij} = T_{ji}$ is evaluated using a suitably defined average of U_i and U_j.

A very general limiting strategy for systems was proposed by Löhner [40]. Given a tentative solution U and the corresponding vector of raw antidiffusive fluxes

$$F_{ij} = \left[f_{ij}^{\rho}, \mathbf{f}_{ij}^{\rho v}, f_{ij}^{\rho E} \right]^T, \tag{65}$$

the following algorithm can be used to calculate the synchronized correction factors α_{ij} for a given set of possibly nonconservative control variables:

1. Initialize the three auxiliary arrays for the generalized Zalesak limiter

$$P_i^\pm := 0, \qquad Q_i^\pm := 0, \qquad R_i^\pm := 1. \tag{66}$$

2. For each pair of neighbor nodes, perform the local change of variables

$$\hat{F}_{ij} := T_{ij} F_{ij}, \qquad \Delta w_{ij} := T_{ij}(u_j - u_i), \tag{67}$$

$$\hat{F}_{ji} := -\hat{F}_{ij}, \qquad \Delta w_{ji} := -\Delta w_{ij}. \tag{68}$$

3. Update the sums of positive/negative components to be limited

$$P_{i,k}^\pm := P_{i,k}^\pm + {\max \atop \min}\{0, \hat{f}_{ij}^k\}, \qquad P_{j,k}^\pm := P_{j,k}^\pm + {\max \atop \min}\{0, \hat{f}_{ji}^k\}. \tag{69}$$

4. Update the upper/lower bounds for the sum of limited increments

$$Q_{i,k}^\pm := {\max \atop \min}\{Q_{i,k}^\pm, \Delta w_{ij}^k\}, \qquad Q_{j,k}^\pm := {\max \atop \min}\{Q_{j,k}^\pm, \Delta w_{ji}^k\}. \tag{70}$$

5. Calculate the nodal correction factors for positive/negative edge contributions

$$R_{i,k}^\pm = \min\left\{1, \frac{\gamma_i Q_{i,k}^\pm}{P_{i,k}^\pm}\right\}, \tag{71}$$

where $\gamma_i > 0$ is a positive scaling factor ($\gamma_i = m_i/\Delta t$ for generalized FCT).

6. Determine the edge correction factors for the given quantity of interest

$$\alpha_{ij}^k = \min\{R_{ij}^k, R_{ji}^k\}, \qquad R_{ij}^k = \begin{cases} R_{i,k}^+, & \text{if } \hat{f}_{ij}^k \geq 0, \\ R_{i,k}^-, & \text{if } \hat{f}_{ij}^k < 0. \end{cases} \tag{72}$$

7. Multiply all components of F_{ij} and F_{ji} by the synchronized correction factor

$$\alpha_{ij} = \min_k \alpha_{ij}^k. \tag{73}$$

Instead of calculating α_{ij}^k independently and taking the minimum, one can redefine α_{ij}^k as the correction factor for the raw antidiffusive flux [33]

$$F_{ij}^k := \alpha_{ij}^{k-1} F_{ij}^{k-1}. \tag{74}$$

This sequential limiting procedure amounts to the multiplication of $F_{ij}^0 := F_{ij}$ by

$$\alpha_{ij} = \alpha_{ij}^k \cdot \alpha_{ij}^{k-1} \cdots \alpha_{ij}^1. \tag{75}$$

In contrast to (73), the result depends on the order in which the correction factors α_{ij}^k are calculated. However, the raw antidiffusive fluxes (74) already include the net effect of previous corrections, which makes the limiter less diffusive.

In our experience, averaging across shocks and contact discontinuities may give rise to unbounded solutions in some particularly sensitive problems. This has led us to prefer a node-based approach to the transformation of variables for the synchronized flux limiter [33]. In the revised version, we replace (67) and (68) with

$$\hat{F}_{ij} := T_i F_{ij}, \qquad \Delta W_{ij} := T_j U_j - T_i U_i, \tag{76}$$

$$\hat{F}_{ji} := -T_j F_{ij}, \qquad \Delta W_{ji} := -\Delta W_{ij}. \tag{77}$$

Since the transformation matrices T_i and T_j are generally different, the transformed antidiffusive fluxes are no longer skew-symmetric, i.e., $\hat{F}_{ji} \neq -\hat{F}_{ij}$. However, the flux-limited scheme remains conservative since the synchronized correction factor α_{ij} is applied to the vector of original fluxes (65). It is neither necessary nor desirable to require that the increments to nonconservative variables be skew-symmetric.

The node-based approach makes the limiter more robust. First, the transformation matrix T_i is the same for all antidiffusive fluxes into node i. Second, the upper and lower bounds are defined using the correct nodal values of the nonconservative variables. Moreover, the revised algorithm requires less arithmetic operations.

9.2 Limiting Primitive Variables

In this section, we describe the synchronized FCT limiter with node-based transformations to the primitive variables. The flux-corrected solution is given by

$$m_i U_i = m_i U_i^L + \sum_{j \neq i} \alpha_{ij} F_{ij}, \tag{78}$$

where U^L denotes the low-order predictor. To calculate α_{ij}, we define [33]

$$\mathbf{v}_i = \frac{(\rho \mathbf{v})_i}{\rho_i}, \qquad p_i = (\gamma - 1)\left[(\rho E)_i - \frac{|(\rho \mathbf{v})_i|^2}{2\rho_i}\right], \tag{79}$$

$$\mathbf{f}_{ij}^v = \frac{\mathbf{f}_{ij}^{\rho v} - \mathbf{v}_i f_{ij}^\rho}{\rho_i}, \qquad f_{ij}^p = (\gamma - 1)\left[f_{ij}^{\rho E} + \frac{|\mathbf{v}_i|^2}{2} f_{ij}^\rho - \mathbf{v}_i \cdot \mathbf{f}_{ij}^{\rho v}\right]. \tag{80}$$

Let u_i^L be the low-order approximation to ρ, v, or p. The raw antidiffusive 'flux' from node j into node i is denoted by f_{ij}^u. In accordance with the FCT philosophy, the choice of the correction factor α_{ij}^u must ensure that the limited antidiffusive correction does not increase the local maxima and minima of u^L. The node-based approach to computation of α_{ij}^u involves the following algorithmic steps [33]:

1. Compute the sums of positive/negative antidiffusive increments to node i

$$P_i^+ = \sum_{j \neq i} \max\{0, f_{ij}^u\}, \qquad P_i^- = \sum_{j \neq i} \min\{0, f_{ij}^u\}. \tag{81}$$

2. Compute the distance to a local maximum/minimum of the low-order solution

$$Q_i^+ = u_i^{\max} - u_i^L, \qquad Q_i^- = u_i^{\min} - u_i^L. \tag{82}$$

3. Compute the nodal correction factors for the net increment to node i

$$R_i^\pm := \min\left\{1, \frac{m_i Q_i^\pm}{\Delta t P_i^\pm}\right\}. \tag{83}$$

4. Define $\alpha_{ij}^u = \alpha_{ji}^u$ so as to satisfy the LED constraints for nodes i and j

$$\alpha_{ij}^u = \min\{R_{ij}, R_{ji}\}, \qquad R_{ij} = \begin{cases} R_i^+, & \text{if } f_{ij}^u \geq 0, \\ R_i^-, & \text{if } f_{ij}^u < 0. \end{cases} \tag{84}$$

If all primitive variables are selected for limiting, the synchronized correction factor α_{ij} for the explicit solution update (78) can be defined as [32, 41, 42]

$$\alpha_{ij} = \min\{\alpha_{ij}^\rho, \alpha_{ij}^v, \alpha_{ij}^p\} \tag{85}$$

or

$$\alpha_{ij} = \alpha_{ij}^\rho \alpha_{ij}^v \alpha_{ij}^p. \tag{86}$$

In the multidimensional case, small velocity fluctuations in the crosswind direction may result in the cancellation of the entire flux. To avoid this, we set $\alpha_{ij}^1 := 1$ or define α_{ij}^v as the correction factor for the streamline velocity [33].

Since the change of variables in (79) and (80) involves a linearization about u_i^L, there is no guarantee that the flux-corrected solution given by (61) will stay within the original bounds, especially in the presence of large jumps. Therefore, our FCT limiting strategy includes a postprocessing step in which all undershoots and overshoots are detected and removed. The first 'failsafe' flux limiter of this kind was proposed by Zalesak (see [73], pp. 36 and 56). His recipe is very simple: "if, after flux limiting, either the density or the pressure in a cell is negative, all the fluxes into that cell are set to their low order values, and the grid point values are recalculated." It is tacitly assumed that the low-order solution is free of nonphysical values.

A similar approach can be used to enforce *local* FCT constraints in a failsafe manner [33]. The flux-corrected value u_i of the control variable u is acceptable if

$$u_i^{\min} \leq u_i \leq u_i^{\max}. \tag{87}$$

If any quantity of interest (density, velocity, pressure) has an undershoot/overshoot at node i, then a fixed percentage of the added antidiffusive fluxes $\alpha_{ij} F_{ij}$ and $\alpha_{ji} F_{ji}$ is removed until the offense is eliminated [33]. The number of correction cycles N depends on the effort invested in the calculation of α_{ij}. If the synchronized FCT limiter is applied to all primitive variables, then undershoots and overshoots are an exception, so that $N = 1$ is optimal. On the other hand, 3–5 cycles may be appropriate in the case $\alpha_{ij} = \alpha_{ij}^\rho$ or $\alpha_{ij} = \alpha_{ij}^p$. The choice of N affects only the amount of rejected antidiffusion. The bounds of the low-order solution are guaranteed to be preserved even for $\alpha_{ij} \equiv 1$. Hence, the failsafe corrector can not only reinforce but also replace the synchronized FCT limiter, as demonstrated by the numerical study in [33].

9.3 Limiting Characteristic Variables

The idea of flux limiting in terms of local characteristic variables dates back to the work of Yee et al. [71, 72] on total variation diminishing (TVD) schemes for the Euler equations. The traditional approach to implementation of such high-resolution schemes in edge-based finite element codes is based on the reconstruction of local 1D stencils [2, 10, 40, 45, 56, 57]. The development of a genuinely multidimensional characteristic limiter is complicated by the fact that the eigenvalues and eigenvectors of the Jacobian matrices $\mathbf{e}_{ij} \cdot \mathbf{A}_{ij}$ depend on the orientation of $\mathbf{e}_{ij}$, whereas all components of the sums $P_i^\pm$ must correspond to the same set of local characteristic variables. For this reason, we use artificial viscosity of the form (43) and limit the antidiffusive fluxes associated with each coordinate direction independently.

In contrast to the synchronized FCT algorithm for primitive variables, it is worthwhile to use different correction factors for different waves. In this case, an edge-based transformation of variables is required to keep the scheme conservative.

The multiplication by the matrix of left eigenvectors $\mathrm{L}_{ij} = \mathrm{R}_{ij}^{-1}$ of a directional Jacobian A_{ij}^d, $1 \leq d \leq n$ transforms $\mathrm{U}_j - \mathrm{U}_i$ into the characteristic difference

$$\Delta \mathrm{w}_{ij} = \mathrm{R}_{ij}^{-1}(\mathrm{U}_j - \mathrm{U}_i).$$

Since the local characteristic variables are essentially decoupled, the components of $\Delta \mathrm{w}_{ij}$ can be limited separately. If a one-sided limiting strategy is adopted, the sign of the eigenvalue λ_k determines the upwind direction for the k-th wave. Let

$$I = \begin{cases} i, & \text{if } \lambda_k \geq 0, \\ j, & \text{if } \lambda_k < 0. \end{cases} \tag{88}$$

In the process of flux limiting, a nodal correction factor $R_{I,k}^\pm$ is applied to Δw_{ij}^k

$$\widehat{\Delta w_{ij}^k} = \begin{cases} R_{I,k}^+ \Delta w_{ij}^k, & \text{if } \Delta w_{ij}^k \leq 0, \\ R_{I,k}^- \Delta w_{ij}^k, & \text{if } \Delta w_{ij}^k > 0. \end{cases} \tag{89}$$

The multiplication by the matrix of right eigenvectors transforms the remaining artificial viscosity (if any) to the conservative variables. The flux to be added is

$$\Phi\left(e_{ij}^d \mathrm{A}_{ij}^d, \mathrm{U}_j - \mathrm{U}_i\right) := \left|e_{ij}^d\right| \mathrm{R}_{ij} |\Lambda_{ij}| (\Delta \mathrm{w}_{ij} - \widehat{\Delta \mathrm{w}_{ij}}). \tag{90}$$

Clearly, the use of dimensional splitting makes this sort of algebraic flux correction more expensive than the synchronized FCT algorithm. However, flux limiting in terms of local characteristic variables is very reliable and produces accurate results. We refer to Zalesak [73] for a presentation of characteristic FCT limiters.

10 Constrained Initialization

The initialization of data is an important ingredient of numerical algorithms for systems of conservation laws. If the initial data are prescribed analytically, it is essential to guarantee that the numerical solution has the right total mass, momentum, and energy when the simulation begins. The pointwise definition of nodal values

$$U_i^0 = U_0(\mathbf{x}_i) \tag{91}$$

is generally nonconservative. This may result in significant errors if the computational mesh is too coarse in regions where U_0 is discontinuous. On the other hand, conservative high-order projections tend to produce undershoots and overshoots.

The first use of FCT in the context of constrained data projection (initialization, interpolation, remapping) dates back to the work of Smolarkiewicz and Grell [62] who introduced a class of nonconservative monotone interpolation schemes. Conservative FCT interpolations were employed by Váchal and Liska [67] and Liska et al. [39]. Farrell et al. [12] introduced a bounded L^2 projection operator for globally conservative interpolation between unstructured meshes. In the monograph by Löhner ([40], pp. 257–260), the FCT limiter is applied to the difference between the consistent and lumped-mass L^2 projections. The latter serves as the low-order method that satisfies the maximum principle for linear finite elements [12].

A general approach to synchronized FCT projections for systems of conserved variables was presented in [33]. Let U denote the initial data or a numerical solution from an arbitrary finite element space. The standard L^2 projection is defined by

$$\int_\Omega w_h U_h^H \, d\mathbf{x} = \int_\Omega w_h U \, d\mathbf{x}, \quad \forall w_h. \tag{92}$$

The nodal values of the high-order approximation U_h^H satisfy the linear system

$$M_C \mathrm{U}^H = \mathrm{R}, \tag{93}$$

where $M_C = \{m_{ij}1\}$ is the consistent mass matrix and R is the load vector

$$\mathrm{R}_i = \int_\Omega \varphi_i U \, d\mathbf{x}. \tag{94}$$

If the functions φ_i and U are defined on different meshes, numerical integration can be performed using a *supermesh* that represents the union of the two meshes [12].

The lumped-mass approximation to (93) is a linear system with a diagonal matrix

$$M_L \mathrm{U}^L = \mathrm{R}. \tag{95}$$

The so-defined low-order solution U_h^L has the same 'mass' as U_h^H but is free of undershoots and overshoots, at least in the case of linear finite elements [12].

The difference between u_i^H and u_i^L admits the conservative flux decomposition

$$\mathrm{u}_i^H = \mathrm{u}_i^L + \sum_{j \neq i} F_{ij}, \quad F_{ij} = m_{ij}\left(\mathrm{u}_i^H - \mathrm{u}_j^H\right). \tag{96}$$

The process of flux limiting involves the same algorithmic steps as the FEM-FCT scheme for the Euler equations. The use of failsafe postprocessing is optional.

11 Boundary Conditions

The implementation of boundary conditions for the Euler equations is an issue of utmost importance. The solution to a hyperbolic system is a superposition of several waves traveling in certain directions at finite speeds. Hence, the proper choice of boundary conditions depends on the wave propagation pattern [17, 24, 60, 69]. In this section, we review the underlying theory and discuss the numerical treatment of characteristic boundary conditions in an implicit finite element formulation.

11.1 Physical Boundary Conditions

The number of physical boundary conditions (PBC) to be imposed is determined using a transformation to the local characteristic variables associated with the unit outward normal $\mathbf{n}$. The result is a set of five decoupled convection equations

$$\frac{\partial w_k}{\partial t} + \lambda_k \frac{\partial w_k}{\partial n} = 0, \quad k = 1, \ldots, 5, \tag{97}$$

where w_k are the so-called *Riemann invariants* and λ_k are the eigenvalues of the directional Jacobian $\mathbf{n} \cdot \mathbf{A}$. The matrix-vector form of system (97) reads

$$\frac{\partial W}{\partial t} + \Lambda \frac{\partial W}{\partial n} = 0. \tag{98}$$

The matrix $\Lambda = \mathrm{diag}\{\lambda_1, \ldots, \lambda_5\}$ and vector $W = [w_1, \ldots, w_5]^T$ are given by [69]

$$\Lambda = \mathrm{diag}\{v_n - c, v_n, v_n, v_n, v_n + c\} \tag{99}$$

and

$$W = \left[v_n - \frac{2c}{\gamma - 1}, s, v_\xi, v_\eta, v_n + \frac{2c}{\gamma - 1}\right]^T. \tag{100}$$

Here $v_n = \mathbf{n} \cdot \mathbf{v}$ is the normal velocity, v_ξ and v_η are the two components of the tangential velocity $\boldsymbol{\tau} \cdot \mathbf{v}$, c is the speed of sound, and $s = c_v \log(\frac{p}{\rho^\gamma})$ is the entropy.

Since the evolution of the Riemann invariants is governed by pure convection equations, a boundary condition is required for each incoming wave. Hence, the number of PBC equals the number of negative eigenvalues N_λ. By virtue of (99), the sign of λ_k depends on v_n, as well as on the local Mach number

$$M = \frac{|v_n|}{c}.$$

The following types of boundaries may need to be considered when it comes to formulating a well-posed boundary-value problem for the Euler equations:

- Supersonic inlet: $v_n < 0$, $M > 1$. All eigenvalues are negative, so $N_\lambda = 5$.
- Supersonic outlet: $v_n > 0$, $M > 1$. All eigenvalues are positive, so $N_\lambda = 0$.
- Subsonic inlet: $v_n < 0$, $M < 1$. Only $\lambda_5 = v_n + c$ is nonnegative, so $N_\lambda = 4$.
- Subsonic outlet: $v_n > 0$, $M < 1$. Only $\lambda_1 = v_n - c$ is negative, so $N_\lambda = 1$.
- Solid wall boundary: $v_n = M = 0$. Only $\lambda_1 = -c$ is negative, so $N_\lambda = 1$.

In many cases, the N_λ boundary conditions are given in terms of the conservative or primitive variables. It is also possible to prescribe the total enthalpy, entropy, temperature, or inclination angle. These data define the "free stream" state U_∞ for the computation of the Dirichlet/Neumann boundary conditions (14) and (15).

11.2 Numerical Boundary Conditions

The need for numerical boundary conditions (NBC) arises whenever $0 < N_\lambda < 5$ so that the boundary values and normal fluxes cannot be determined using the prescribed PBC alone. The missing information is obtained by solving a Riemann problem. The internal state U is defined as the numerical solution to the Euler equations at the given point. The external state U_∞ can be obtained as follows [14, 60, 69]:

1. Convert the given numerical solution U to the Riemann invariants W.
2. Set $W_\infty := W$ and overwrite the incoming Riemann invariants by PBC.
3. Given the modified vector W_∞, calculate the free stream values U_∞.

In contrast to cell-centered finite volume methods, there is no need for extrapolation because the values of U_h are readily available at each boundary point.

The right-hand side $G(U, U_\infty)$ of the Dirichlet boundary condition (14) is defined as the exact or approximate solution to the boundary Riemann problem associated with the states U and U_∞. Likewise, the normal flux $F_n(U, U_\infty)$ for the Neumann boundary condition (14) can be calculated using Toro's [64] exact Riemann solver or Roe's approximate Riemann solver [54]. The latter approach yields

$$F_n(U, U_\infty) = \mathbf{n} \cdot \frac{\mathbf{F}(U) + \mathbf{F}(U_\infty)}{2} - \frac{1}{2}|\mathbf{n} \cdot \mathbf{A}(U, U_\infty)|(U_\infty - U), \quad (101)$$

where $\mathbf{A}(U, U_\infty)$ is the Roe matrix for the states U and U_∞. This approach to weak imposition of characteristic boundary conditions is closely related to their numerical treatment in finite volume and discontinuous Galerkin methods [9, 66].

11.3 Practical Implementation

In a practical implementation, it is worthwhile to initialize W_∞ by the vector of free stream values and overwrite the Riemann invariants associated with nonnegative eigenvalues by the corresponding components of W. Such an algorithm is well-suited for boundaries of any type since it determines the direction of wave propagation and the upstream values of the characteristic variables automatically.

The transformation of the internal state U to the vector of Riemann invariants W is performed using definition (100). The inverse transformation is given by [60, 69]

$$\rho = \left[\frac{c^2}{\gamma}\exp\left(-\frac{w_2}{c_v}\right)\right]^{\frac{1}{\gamma-1}}, \tag{102}$$

$$\rho \mathbf{v} = \rho(v_n \mathbf{n} + v_\xi \boldsymbol{\tau}_\xi + v_\eta \boldsymbol{\tau}_\eta), \tag{103}$$

$$\rho E = \frac{p}{\gamma-1} + \frac{\rho}{2}(v_n^2 + v_\xi^2 + v_\eta^2), \tag{104}$$

where $\boldsymbol{\tau}_\xi$ and $\boldsymbol{\tau}_\eta$ are two unit vectors spanning the tangential plane, and

$$v_n = \frac{w_5 - w_1}{2}, \qquad v_\xi = w_3, \qquad v_\eta = w_4,$$

$$c = \frac{\gamma - 1}{4}(w_5 - w_1), \qquad p = \frac{\rho c^2}{\gamma}.$$

If the physical boundary conditions are given in terms of primitive variables or other quantities, a conversion to the Riemann invariants is required. The practical implementation of such boundary conditions depends on the type of the boundary.

11.3.1 Open Boundary Conditions

At a supersonic inlet, the free stream values of the conservative variables U_∞ can be prescribed without transforming to the Riemann invariants. At a supersonic outlet, the exterior state is given by $U_\infty = U$ so that the Roe flux (101) reduces to

$$F_n(U, U) = \mathbf{n} \cdot \mathbf{F}(U).$$

At a subsonic inlet, it is common to prescribe the density ρ_{in}, pressure p_{in}, and tangential velocity $\boldsymbol{\tau} \cdot \mathbf{v}_{in}$. In this case, the Riemann invariants w_3 and w_4 are given, whereas $w_2 = c_v \log(\frac{p_{in}}{\rho_{in}^\gamma})$ is computable. The last incoming Riemann invariant is

$$w_1 = w_5 - \frac{4}{\gamma - 1}\sqrt{\frac{\gamma p_{in}}{\rho_{in}}}. \tag{105}$$

In the case of a subsonic outlet with a prescribed exit pressure p_{out}, we have [60]

$$w_1 = w_5 - \frac{4}{\gamma - 1}\sqrt{\frac{\gamma p_{out}}{\rho_{out}}}, \qquad (106)$$

where ρ_{out} depends on the calculated interior density ρ and pressure p as follows:

$$\rho_{out} = \rho \left(\frac{p_{out}}{p}\right)^{\frac{1}{\gamma}}.$$

The outgoing Riemann invariant w_5 is evaluated using the trace of the finite element solution. The open boundary conditions (105) and (106) are generally regarded as more physical than a prescribed upstream value of the Riemann invariant w_1.

11.3.2 Wall Boundary Conditions

At a solid surface, there is no convective flux across the boundary. Hence, the normal velocity v_n must vanish. The so-defined *no-penetration/free slip* condition

$$\mathbf{n} \cdot \mathbf{v} = 0 \qquad (107)$$

constrains a linear combination of the three velocity components. The numerical implementation of this condition in an implicit scheme presents a considerable difficulty if the boundary is not aligned with the axes of the coordinate system.

In finite element methods for incompressible flow problems, the free slip condition (107) is usually imposed in the strong sense using element-by-element transformations to a local reference frame spanned by the normal and tangential vectors [7, 11]. The same effect can be achieved using an iterative projection of the velocity vector on the tangential plane [32]. However, the semi-explicit treatment of the wall boundary condition slows down the iterative solver and may result in a lack of robustness. Therefore, a fully implicit treatment is to be preferred.

In the weak form of the free slip condition, the free stream values for the computation of $F_n(U, U_\infty)$ are calculated using the mirror (reflection) condition

$$\mathbf{n} \cdot (\mathbf{v}_\infty + \mathbf{v}) = 0.$$

The density, tangential velocity, and total energy remain unchanged. Thus

$$U_\infty = \begin{bmatrix} \rho \\ \rho \mathbf{v}_\infty \\ \rho E \end{bmatrix}, \quad \mathbf{v}_\infty = \mathbf{v} - 2\mathbf{n}(\mathbf{v} \cdot \mathbf{n}).$$

Another popular weak form of the zero flux boundary condition is given by

$$F_n = \begin{bmatrix} 0 \\ \mathbf{n}p \\ 0 \end{bmatrix}. \qquad (108)$$

This version does not involve the solution of a Riemann problem and has been used in FEM codes with considerable success [4, 9, 59]. However, the Roe flux (101) constitutes a more physical wall boundary condition than (108). In any case, the weak imposition of the free slip condition may give rise to a nonzero normal velocity on the wall. This problem can be fixed by adding a penalty term [18].

11.3.3 Calculation of Surface Integrals

The imposition of natural boundary conditions requires the evaluation of the numerical flux $F_n(U, U_\infty)$ at each quadrature point $\hat{\mathbf{x}}_i$. The exterior state U_∞ is associated with a ghost node $\hat{\mathbf{x}}_{i,\infty}$ located on the other side of the boundary. The ghost nodes provide the free stream values of the Riemann invariants and play the same role as *image cells* in cell-centered finite volume schemes for the Euler equations [66].

If a curved boundary is approximated using isoparametric (linear or bilinear) finite elements, then the normal vector **n** is generally discontinuous at the vertices and edges of the surface triangulation. The boundary integrals can be assembled element-by-element using the unique normal to the boundary of each cell [18]. However, the value of $F_n(U, U_\infty)$ at $\hat{\mathbf{x}}_i$ should be obtained by interpolating the (unique) nodal values to ensure consistency with the group FEM approximation (18). Otherwise, numerical side effects may arise in the boundary layer and pollute the solution in the interior of the computational domain. As a remedy, a unique normal direction can be determined using a suitable averaging procedure [11, 49] or an analytical description of the curved boundary. For a detailed discussion of solid wall boundary conditions in curved geometries, we refer to Krivodonova and Berger [25].

12 Numerical Examples

The results presented in this section illustrate some properties of our algebraic flux correction schemes for the Euler equations. We consider a suite of 2D benchmark problems covering a relatively wide range of Mach numbers and boundary conditions. The objective of the below numerical study is to investigate the dependence of the error on the mesh size h and on the choice of the limiting strategy.

The accuracy of a numerical solution $u_h \approx u$ is measured in the global norms

$$E_1(u, h) = \sum_i m_i |u(\mathbf{x}_i) - u_i| \approx \|u - u_h\|_1, \tag{109}$$

$$E_2(u, h) = \sqrt{\sum_i m_i |u(\mathbf{x}_i) - u_i|^2} \approx \|u - u_h\|_2. \tag{110}$$

The rate of grid convergence is illustrated by the expected order of accuracy

$$p = \log_2\left(\frac{E_i(u, 2h)}{E_i(u, h)}\right), \quad i = 1, 2. \tag{111}$$

To begin with, we will evaluate the performance of the linearized FCT algorithm for unsteady compressible flow problems [28]. Next, we will investigate the convergence behavior of a characteristic TVD-like limiter for steady-state computations [29]. In this work, stationary solutions are obtained using pseudo-time-stepping. For additional numerical examples, the interested reader is referred to [18, 33].

12.1 Shock Tube Problem

Sod's shock tube problem [63] is a standard benchmark for the unsteady Euler equations. The domain $\Omega = (0, 1)$ is initially separated by a membrane into two sections. When the membrane is removed, the gas begins to flow into the region of lower pressure. The initial condition for the nonlinear Riemann problem is given by

$$\begin{bmatrix} \rho_L \\ \mathbf{v}_L \\ p_L \end{bmatrix} = \begin{bmatrix} 1.0 \\ 0.0 \\ 1.0 \end{bmatrix}, \quad \begin{bmatrix} \rho_R \\ \mathbf{v}_R \\ p_R \end{bmatrix} = \begin{bmatrix} 0.125 \\ 0.0 \\ 0.1 \end{bmatrix}, \quad (112)$$

where the subscripts refer to the subdomains $\Omega_L = (0, 0.5)$ and $\Omega_R = (0.5, 1)$. The reflective wall boundary conditions are prescribed at the endpoints of Ω.

The removal of the membrane at $t = 0$ releases a shock wave that propagates to the right with velocity satisfying the Rankine-Hugoniot conditions. All of the primitive variables are discontinuous across the shock which is followed by a contact discontinuity. The latter represents a moving interface between the regions of different densities but constant velocity and pressure. A rarefaction wave propagates in the opposite direction providing a smooth transition to the original values of the state variables in the left part of the domain. Hence, the flow pattern in the shock tube is characterized by three waves traveling at different speeds [35].

The dashed lines in Fig. 1 show the exact solution to the Riemann problem (112) at the final time $T = 0.231$. This solution was calculated using the exact Riemann solver HE-E1RPEXACT [65]. The numerical solution U_h^0 was initialized by means of the FCT-constrained data projection (96) and advanced in time by the semi-implicit Crank-Nicolson scheme with the time step $\Delta t = h/10$. For each algorithm under consideration, a grid convergence study was performed on a sequence of uniform grids with mesh spacing $h = 1/N$ for $N = 100, 200, 400, 800, 1600, 3200$.

All numerical solutions shown in Figs. 1(a)–(h) were calculated on a uniform mesh of 100 linear finite elements. The results produced by the low-order schemes ($\alpha_{ij} = 0$) are nonosillatory but the excess numerical diffusion gives rise to strong smearing of the moving fronts. In this example, Roe's approximate Riemann solver (Fig. 1(a)) performs slightly better than the Rusanov scalar dissipation (Fig. 1(b)).

The linearized FCT algorithm (58)–(61) produced the snapshots displayed in Figs. 1(c)–(f). In this study, the correction factors α_{ij} were calculated via sequential limiting of the control variables listed in the parentheses. Similar results were obtained with synchronization of the form (73). The computation of the low-order predictor using Roe's formula (42) was found to generate undershoots/overshoots

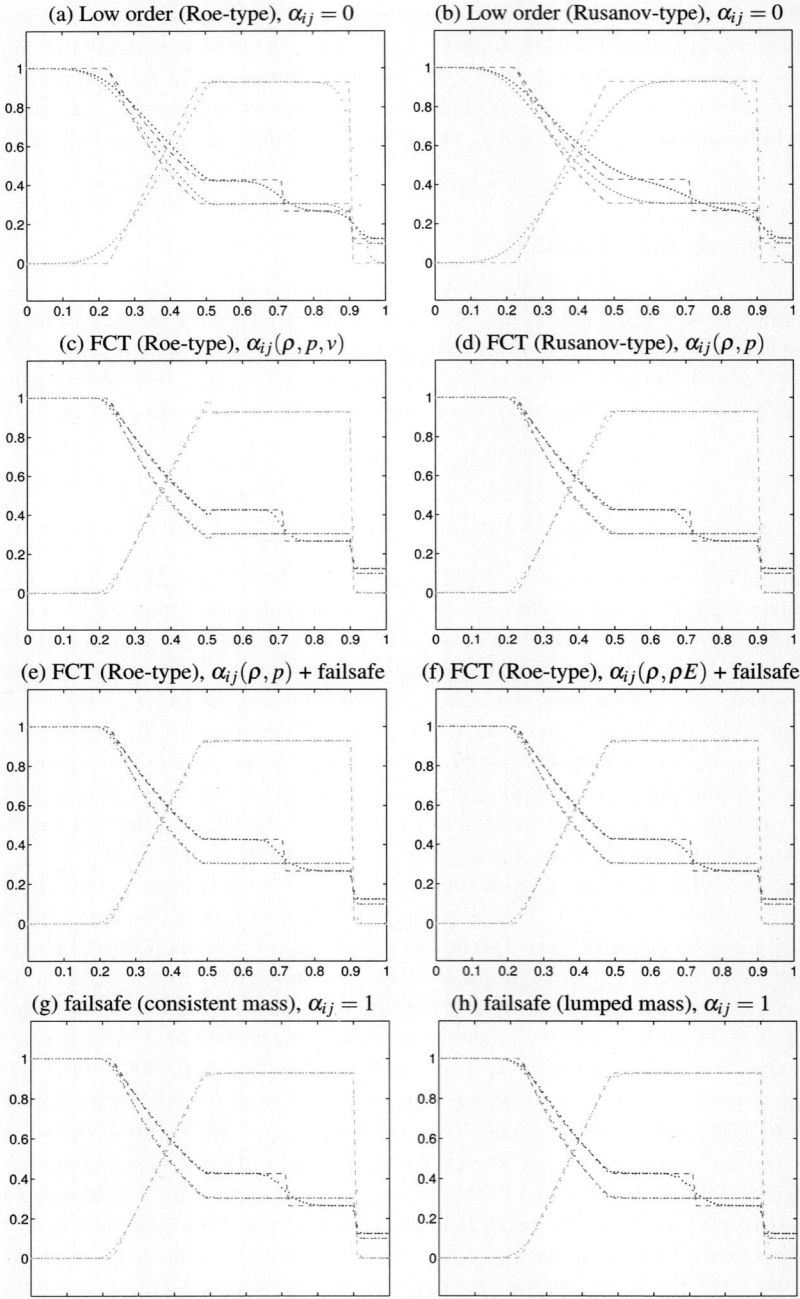

Fig. 1 Shock tube problem: $h = 10^{-2}$, $\Delta t = 10^{-3}$. Snapshots of the density (*blue*), velocity (*green*), and pressure (*red*) distribution at the final time $T = 0.231$ (Color figure online)

Table 1 Shock tube problem: grid convergence of the low-order schemes ($\alpha_{ij} = 0$)

h	Roe's scheme			Rusanov's scheme		
	$E_1(\rho, h)$	$E_1(u, h)$	$E_1(p, h)$	$E_1(\rho, h)$	$E_1(u, h)$	$E_1(p, h)$
1/100	2.1788e−02	4.2199e−02	1.9789e−02	2.8687e−02	5.4016e−02	2.6282e−02
1/200	1.3969e−02	2.4904e−02	1.1995e−02	1.9468e−02	3.2518e−02	1.6138e−02
1/400	8.8233e−03	1.3737e−02	7.0753e−03	1.2659e−02	1.8557e−02	9.6411e−03
1/800	5.5562e−03	7.5261e−03	4.1293e−03	8.1083e−03	1.0478e−02	5.6589e−03
1/1600	3.4900e−03	4.0920e−03	2.3835e−03	5.1423e−03	5.8427e−03	3.2668e−03
1/3200	2.2003e−03	2.2017e−03	1.3609e−03	3.2579e−03	3.2178e−03	1.8593e−03
	$p = 0.67$	$p = 0.89$	$p = 0.81$	$p = 0.66$	$p = 0.86$	$p = 0.81$

that carry over to the FCT solution even if the limiter is applied to *all* primitive variables (Fig. 1(c)). Synchronized limiting of all conservative variables was the only FCT method to produce satisfactory results (not shown here, see Table 2) with the Roe-type low-order scheme. In the case of the Rusanov scalar dissipation, nonoscillatory solutions were obtained with the density-pressure FCT limiter (Fig. 1(d)).

The failsafe control of density and pressure (see Sect. 9.2) makes the solutions less sensitive to the choice of control variables for the base limiter. The nonoscillatory results shown in Figs. 1(e)–(f) were obtained using the density-pressure postprocessing for the Roe-FCT scheme with $\alpha_{ij}(\rho, p)$ and $\alpha_{ij}(\rho, \rho E)$, respectively. The results in Figs. 1(g)–(h) indicate that it is even possible to deactivate the main limiter, i.e., set $\alpha_{ij} := 1$ and remove (a fraction of) the antidiffusive flux in regions where the local FCT constraints (87) are violated. However, this practice is not generally recommended since it might trigger aggressive limiting at the postprocessing step.

In contrast to high-resolution schemes of TVD type, the raw antidiffusive flux (59) includes a contribution of the consistent mass matrix. The lumped-mass version ($\dot{U}^L := 0$) of the FCT algorithm produces the solution shown in Fig. 1(h). The superior phase accuracy of the consistent-mass Galerkin discretization justifies the additional effort invested in the computation of the approximate time derivative (60).

The error norms for the density, pressure, and velocity fields calculated with the above algorithms are listed in Tables 1, 2, 3 and 4. The expected order of accuracy p was estimated by formula (111) using the solutions computed on the two finest meshes. As expected, the largest errors are observed for the low-order approximations. The accuracy of Roe's approximate Riemann solver is marginally better than that of the Rusanov scheme. The rate of grid convergence for the density approaches $p = 2/3$, which is in good agreement with the results presented in [5]. Tables 2, 3 and 4 confirm that the linearized FCT algorithm converges much faster than the underlying low-order scheme. The expected order of accuracy attains values in the range 0.9–1.1.

The presented grid convergence study sheds some light on various aspects of flux limiting for the unsteady Euler equations. As a rule of thumb, constraining the density and pressure or total energy is a good choice in the context of synchronous

Table 2 Shock tube problem: grid convergence of FCT without failsafe correction

h	Roe-type predictor, $\alpha_{ij}(\rho,\rho E,\rho v)$			Rusanov-type predictor, $\alpha_{ij}(\rho,p)$		
	$E_1(\rho,h)$	$E_1(u,h)$	$E_1(p,h)$	$E_1(\rho,h)$	$E_1(u,h)$	$E_1(p,h)$
1/100	7.2976e−03	1.1244e−02	5.6132e−03	9.2527e−03	1.0041e−02	4.6990e−03
1/200	3.8044e−03	6.8249e−03	2.9742e−03	5.1909e−03	6.2159e−03	2.5124e−03
1/400	1.9693e−03	3.3300e−03	1.4743e−03	2.8313e−03	3.0024e−03	1.2358e−03
1/800	1.0334e−03	1.5903e−03	7.2550e−04	1.4237e−03	1.4209e−03	6.0422e−04
1/1600	5.3461e−04	7.3201e−04	3.5412e−04	7.0374e−04	6.4491e−04	2.9243e−04
1/3200	2.8770e−04	3.3918e−04	1.7761e−04	3.5707e−04	2.9345e−04	1.4587e−04
	$p=0.89$	$p=1.11$	$p=1.00$	$p=0.98$	$p=1.13$	$p=1.00$

Table 3 Shock tube problem: grid convergence of FCT with failsafe correction

h	Roe-type predictor, $\alpha_{ij}(\rho,p)$			Roe-type predictor, $\alpha_{ij}(\rho,\rho E)$		
	$E_1(\rho,h)$	$E_1(u,h)$	$E_1(p,h)$	$E_1(\rho,h)$	$E_1(u,h)$	$E_1(p,h)$
1/100	8.4389e−03	9.0475e−03	4.2509e−03	7.9186e−03	8.8199e−03	4.2180e−03
1/200	5.0820e−03	5.7128e−03	2.2982e−03	4.7613e−03	5.6378e−03	2.2879e−03
1/400	3.0545e−03	2.7739e−03	1.1361e−03	2.6349e−03	2.7383e−03	1.1309e−03
1/800	1.8737e−03	1.3183e−03	5.5785e−04	1.4212e−03	1.2924e−03	5.5345e−04
1/1600	1.1502e−03	5.9718e−04	2.7068e−04	7.0388e−04	5.8257e−04	2.6794e−04
1/3200	6.6805e−04	2.7102e−04	1.3611e−04	3.5656e−04	2.6259e−04	1.3413e−04
	$p=0.78$	$p=1.14$	$p=0.99$	$p=0.98$	$p=1.15$	$p=1.00$

Table 4 Shock tube problem: grid convergence of Roe-type failsafe FCT for $\alpha_{ij}=1$

h	Consistent mass matrix			Lumped mass matrix		
	$E_1(\rho,h)$	$E_1(u,h)$	$E_1(p,h)$	$E_1(\rho,h)$	$E_1(u,h)$	$E_1(p,h)$
1/100	8.4725e−03	9.2123e−03	4.3338e−03	8.9680e−03	1.0579e−02	5.0899e−03
1/200	5.1763e−03	5.8569e−03	2.3466e−03	5.4849e−03	6.3070e−03	2.6797e−03
1/400	3.0879e−03	2.8643e−03	1.1668e−03	3.2170e−03	3.0904e−03	1.3348e−03
1/800	1.9700e−03	1.3755e−03	5.7960e−04	1.9142e−03	1.4843e−03	6.5940e−04
1/1600	1.2025e−03	6.3730e−04	2.8494e−04	1.1215e−03	6.9581e−04	3.2554e−04
1/3200	7.5109e−04	3.0126e−04	1.4721e−04	6.4676e−04	3.3016e−04	1.6544e−04
	$p=0.68$	$p=1.08$	$p=0.95$	$p=0.79$	$p=1.08$	$p=0.98$

FCT. The failsafe feature improves the robustness of the algorithm but may increase the amount of numerical diffusion. To achieve optimal phase accuracy for time-dependent problems, the raw antidiffusive flux must include the contribution of the

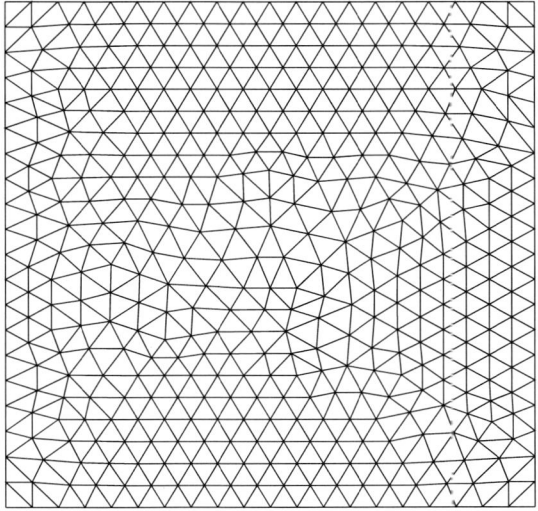

Fig. 2 Radially symmetric Riemann problem: coarse mesh, 824 triangles, 453 vertices

consistent mass matrix. Of course, the overall performance of the algorithm also depends on the accuracy of the time-stepping scheme and on the time step size.

12.2 Radially Symmetric Riemann Problem

The second transient benchmark [36] is a radially symmetric 2D counterpart of the shock tube problem. Before an impulsive start, an imaginary membrane separates the square domain $\Omega = (-0.5, 0.5) \times (-0.5, 0.5)$ into the inner circle

$$\Omega_L = \{(x, y) \in \Omega \mid \sqrt{x^2 + y^2} < 0.13\}$$

and the complement $\Omega_R = \Omega \setminus \Omega_L$. Reflective boundary conditions are prescribed on the boundary of Ω. The gas is initially at rest. Higher pressure and density are maintained inside Ω_L than outside. The interior and exterior states are given by

$$\begin{bmatrix} \rho_L \\ \mathbf{v}_L \\ p_L \end{bmatrix} = \begin{bmatrix} 2.0 \\ 0.0 \\ 15.0 \end{bmatrix}, \quad \begin{bmatrix} \rho_R \\ \mathbf{v}_R \\ p_R \end{bmatrix} = \begin{bmatrix} 1.0 \\ 0.0 \\ 1.0 \end{bmatrix}.$$

The abrupt removal of the membrane at $t = 0$ gives rise to a radially expanding shock wave driven by the pressure difference. The challenge of this test is to capture the moving discontinuities while preserving the radial symmetry of the solution.

All computations are performed using linear finite elements on unstructured meshes constructed via regular subdivision of the coarse mesh depicted in Fig. 2. As explained in Sect. 10, it is advisable to initialize the numerical solution in a

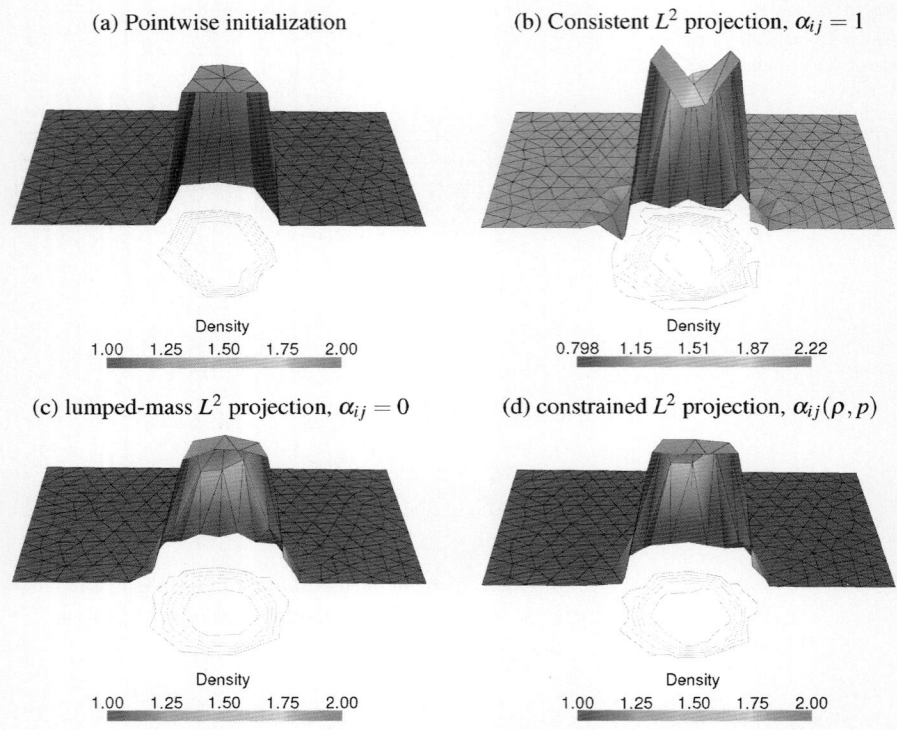

Fig. 3 Radially symmetric Riemann problem: initial density ρ_h^0 on the coarse mesh

conservative manner. The total mass and energy of the initial data are given by

$$\int_\Omega \rho \, d\mathbf{x} = 1 + (0.13)^2 \cdot \pi \approx 1.05309291584567,$$

$$\int_\Omega \rho E \, d\mathbf{x} = 2.5 + 35 \cdot (0.13)^2 \pi \approx 4.35825205459836.$$

Since the exact solution is discontinuous, the load vector (94) was assembled using adaptive cubature formulas [68]. The density profiles produced by 4 different initialization techniques are shown in Figs. 3(a)–(d). It can readily be seen that the consistent L^2 projection fails to preserve the bounds of the initial data, while its lumped counterpart gives rise to significant numerical diffusion. The synchronized FCT limiter (96) with $\alpha_{ij} = \alpha_{ij}(\rho, p)$ makes it possible to achieve a crisp resolution of the discontinuous initial profile without generating undershoots or overshoots.

Table 5 reveals that the pointwise initialization of nodal values is nonconservative. The consistent-mass L^2 projection preserves the total mass and energy but the initial density exhibits undershoots and overshoots of about 20%. Moreover, the initial pressure attains negative values, which results in an immediate crash of the code.

Table 5 Radially symmetric Riemann problem: constrained initialization on the coarse mesh

	$\int_\Omega \rho_h^0\,dx$	$\int_\Omega (\rho E)_h^0\,dx$	$\min(\rho_h^0)$	$\max(\rho_h^0)$	$\min(p_h^0)$	$\max(p_h^0)$
(a)	1.04799	4.17949	1.0	2.0	1.0	15.0
(b)	1.05309	4.35823	7.9845e−01	2.2239e+00	−1.8216e+00	1.8135e+01
(c), (d)	1.05309	4.35823	1.0	2.0	1.0	15.0

In contrast, the nodal values obtained with the lumped-mass L^2 projection and the flux-corrected version satisfy $1.0 \leq \rho_h^0 \leq 2.0$ and $1.0 \leq p_h^0 \leq 15.0$ as desired.

The evolution of the numerical solution initialized by the constrained L^2 projection was studied on the mesh obtained with 4 global refinements. The Crank-Nicolson time-stepping was employed with $\Delta t = 2 \cdot 10^{-3}$. Figures 4(a)–(d) display snapshots of the density (left) and pressure (right) at the final time $T = 0.13$. These solutions were obtained using the Roe tensorial dissipation and linearized FCT with the density-pressure limiter. Remarkably, both the low-order solution (top) and its flux-corrected counterpart (bottom) preserve the radial symmetry on the unstructured mesh. The symmetry plots shown in Figs. 4(e)–(f) show the nodal values $\rho_i = \rho_h(x_i, y_i)$ and $p_i = p_h(x_i, y_i)$ versus distance to the origin. The presented results are in a good agreement with the reference solutions computed using CLAW-PACK [38].

12.3 Double Mach Reflection

A more challenging test for the unsteady Euler equations is the double Mach reflection problem of Woodward and Colella [70]. In this benchmark, a Mach 10 shock impinges on a reflecting wall at the angle of 60° degrees. The computational domain is the rectangle $\Omega = (0, 4) \times (0, 1)$. The following pre-shock and post-shock values of the flow variables are used to define the initial and boundary conditions [3]

$$\begin{bmatrix} \rho_L \\ u_L \\ v_L \\ p_L \end{bmatrix} = \begin{bmatrix} 8.0 \\ 8.25\cos(30°) \\ -8.25\sin(30°) \\ 116.5 \end{bmatrix}, \quad \begin{bmatrix} \rho_R \\ u_R \\ v_R \\ p_R \end{bmatrix} = \begin{bmatrix} 1.4 \\ 0.0 \\ 0.0 \\ 1.0 \end{bmatrix}. \quad (113)$$

Initially, the post-shock values are prescribed in $\Omega_L = \{(x, y) \mid x < 1/6 + y/\sqrt{3}\}$ and the pre-shock values in $\Omega_R = \Omega \setminus \Omega_L$. The reflecting wall corresponds to $1/6 \leq x \leq 4$ and $y = 0$. No boundary conditions are required along the line $x = 4$. On the rest of the boundary, the post-shock conditions are prescribed for $x < 1/6 + (1 + 20t)/\sqrt{3}$ and the pre-shock conditions elsewhere [3]. The so-defined values along the top boundary describe the exact motion of the initial Mach 10 shock.

The density fields (30 isolines) depicted in Figs. 5, 6 and 7 were computed using bilinear finite elements on a sequence of structured meshes with equidistant grid spacings $h = 1/64, 1/128, 1/256$, and $1/512$. Integration in time was performed

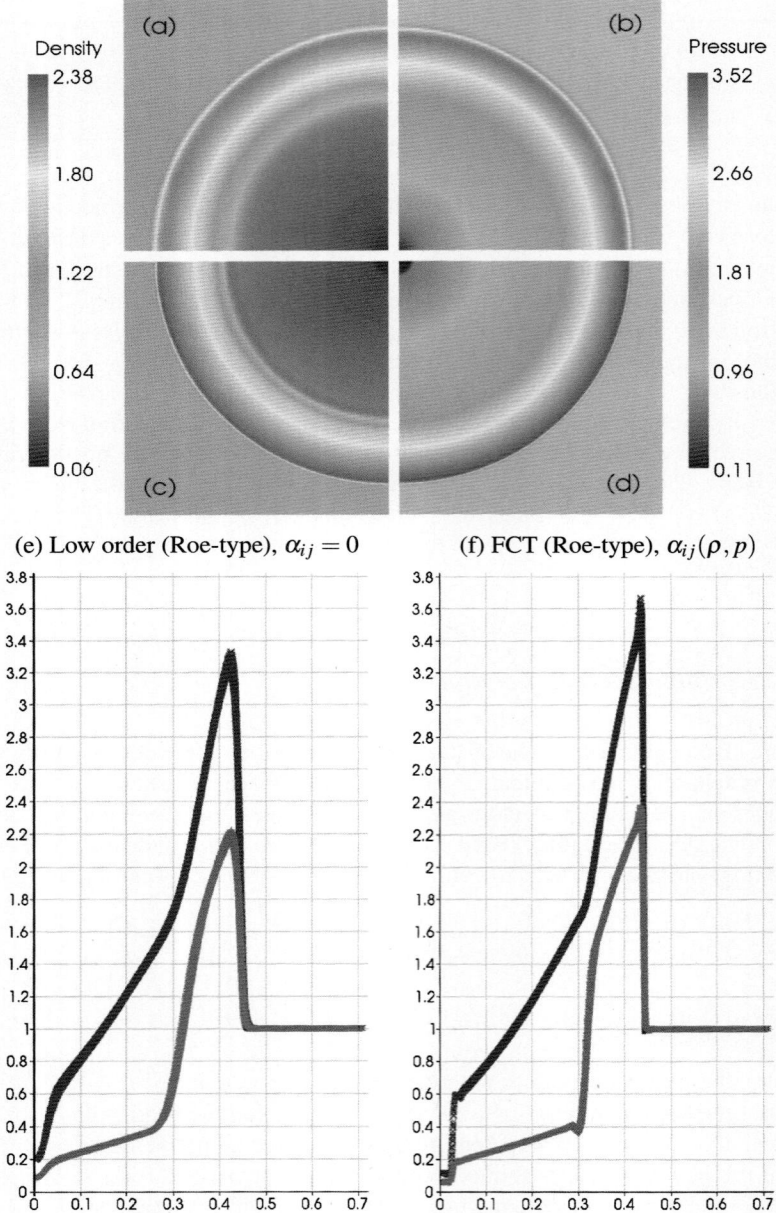

Fig. 4 Radially symmetric Riemann problem: density (*red*) and pressure (*blue*) at $T = 0.13$ (Color figure online)

Fig. 5 Double Mach reflection: isodensity contours at $T = 0.2$; low-order scheme, $\alpha_{ij} = 0$

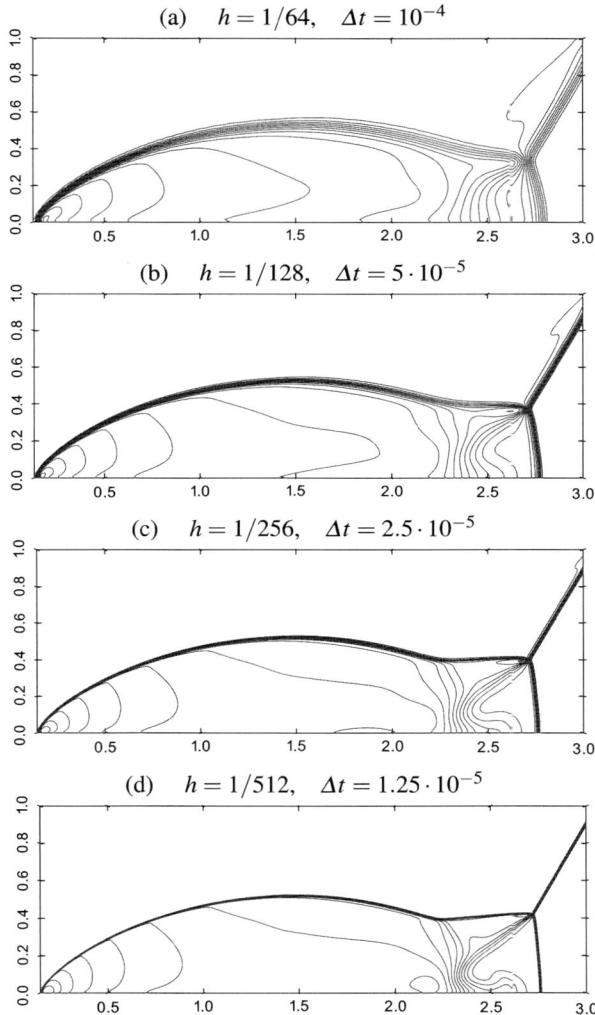

until $T = 0.2$ by the Crank-Nicolson scheme with the time step $\Delta t = 64h \cdot 10^{-4}$. The low-order solution displayed in Fig. 5 was calculated using the Roe-type artificial viscosity. Due to strong numerical diffusion, the complex interplay of incident, reflected, and Mach stem shock waves is resolved rather poorly, and so is the slipstream at the triple point. The use of FCT with synchronized limiting on primitive (Fig. 6) or conservative (Fig. 7) variables yields a marked improvement without producing 'staircase structures' or other artefacts observed by Woodward and Colella [70].

Fig. 6 Double Mach reflection: isodensity contours at $T = 0.2$; unsafe FCT^{Roe}, $\alpha_{ij}(\rho, p)$

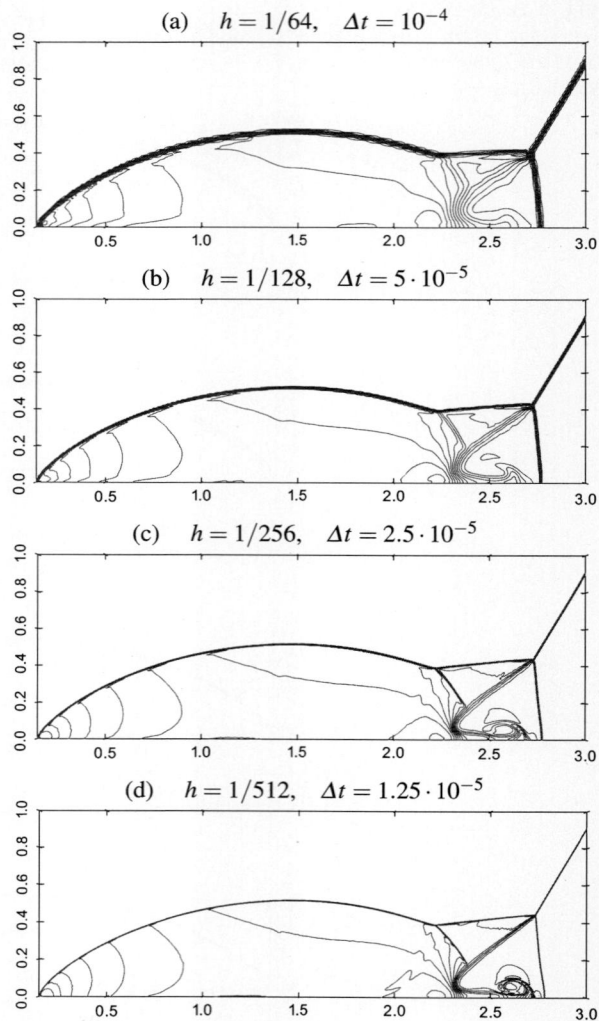

12.4 GAMM Channel

In the remainder of this section, we present the results of a numerical study for the stationary Euler equations. To begin with, we simulate the steady transonic flow in the GAMM channel with a 10% circular bump. For a detailed description of this popular benchmark, we refer to Feistauer et al. [13]. The gas enters the channel at free stream Mach number $M_\infty = 0.67$ and accelerates to supersonic velocities as it flows over the bump. The Mach number varies between approximately 0.22 and 1.41. An isolated shock wave forms in the local supersonic region. The inlet and outlet lie in the region of subsonic flow. Hence, the results are sensitive to the choice

Fig. 7 Double Mach reflection: isodensity contours at $T = 0.2$; unsafe FCT^{Roe}, $\alpha_{ij}(\rho, \rho E)$

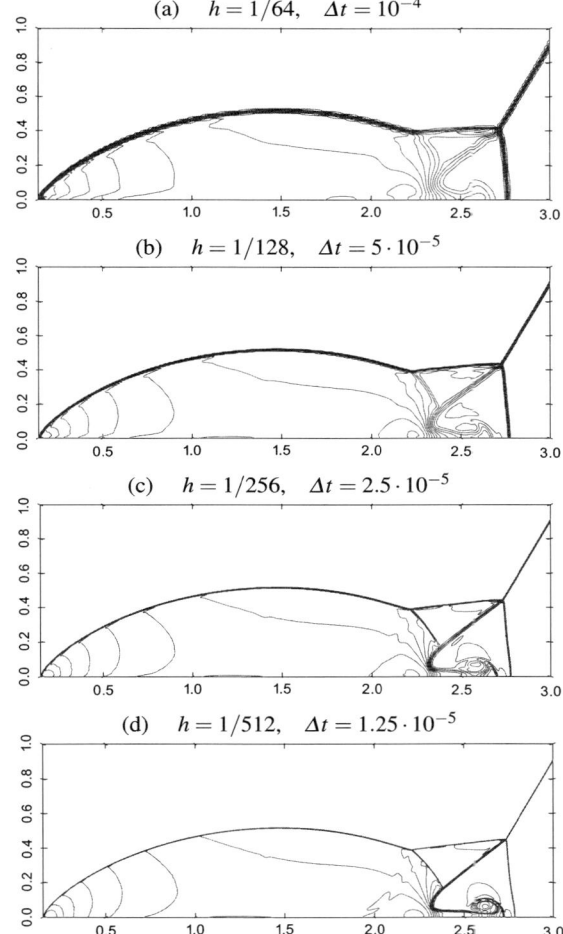

of physical and numerical boundary conditions. This makes the GAMM channel problem rather challenging when it comes to computing steady-state solutions.

Unless mentioned otherwise, the free stream boundary values for all stationary benchmark problems are given in the following dimensionless form [60]:

Variable	Free stream value
ρ_∞	1
u_∞	M_∞
v_∞	0
p_∞	$\frac{1}{\gamma}$
E_∞	$\frac{M_\infty^2}{2} + \frac{1}{\gamma(\gamma-1)}$

Fig. 8 GAMM channel: (**a**) stationary Mach number distribution and (**b**) the coarse grid

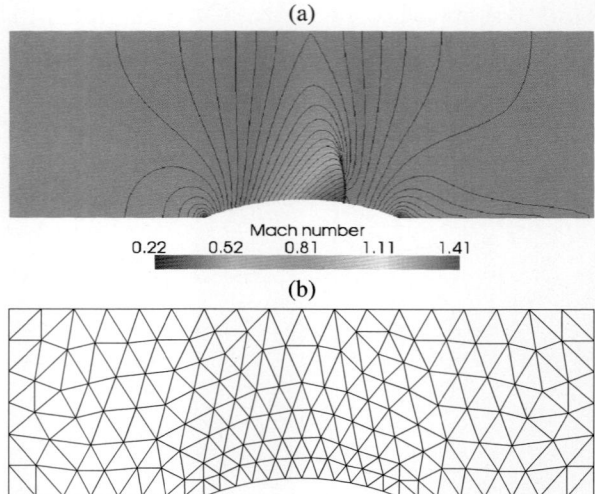

Table 6 GAMM channel: relative L^2 errors and grid convergence rates

Level	NVT	NEL	E_2^{Low}	p^{Low}	E_2^{Lim}	p^{Lim}
1	176	292	$5.47 \cdot 10^{-2}$	0.59	$3.05 \cdot 10^{-2}$	0.56
2	643	1168	$3.64 \cdot 10^{-2}$	0.64	$2.07 \cdot 10^{-2}$	1.04
3	2453	4672	$2.34 \cdot 10^{-2}$	0.59	$1.01 \cdot 10^{-2}$	0.99
4	9577	18688	$1.55 \cdot 10^{-2}$	0.61	$5.07 \cdot 10^{-3}$	1.45
5	37841	18688	$1.01 \cdot 10^{-2}$		$1.85 \cdot 10^{-3}$	
6	150433	299008				

The unstructured triangular mesh shown in Fig. 8(b) is successively refined to construct the computational mesh for the GAMM channel. Table 6 lists the number of vertices (NVT) and elements (NEL) for up to 5 quad-tree refinements. The stationary Mach number distribution computed with an algebraic flux correction scheme of TVD type [18, 19, 29] on mesh level 6 is presented in Fig. 8(a). It can readily be seen that the resolution of the shock wave is rather crisp and nonoscillatory.

The numerical solution to the Euler equations was initialized by the above free stream values and marched to the steady state using pseudo-time-stepping in conjunction with the semi-implicit linearization procedure (see Sect. 7). At the initial stage, we neglect the nonlinear antidiffusive term and begin with the inexpensive computation of a low-order predictor. When the residuals of the low-order scheme reach the prescribed tolerance, the limited antidiffusive correction is switched on, and the iteration process continues until convergence to a stationary solution.

During the startup phase, the pseudo-time-stepping scheme runs at the moderately large CFL number $\nu = 100$. When the relative residual falls below 10^{-2}, the

Algebraic Flux Correction II

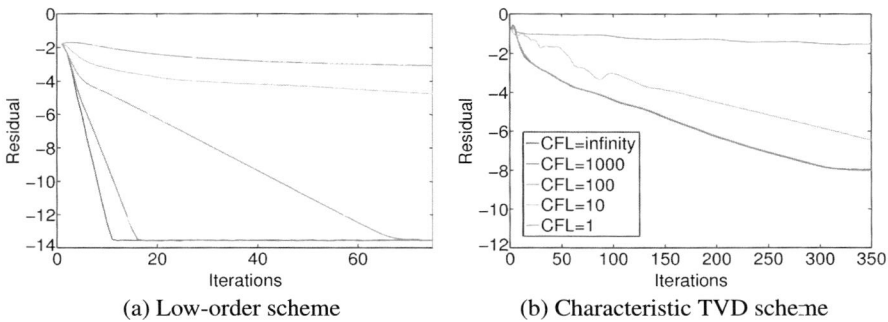

Fig. 9 GAMM channel: convergence history for various CFL numbers on mesh level 4

linearization becomes sufficiently accurate, and ν can be chosen arbitrarily large. In our experience, the semi-implicit algorithm converges even for $\nu = \infty$.

Figure 9 presents the convergence history for the flux-corrected Galerkin scheme and its low-order counterpart. In either case, the Neumann-type boundary conditions are imposed in a weak sense. The log-scale plots show the residual of the nonlinear system versus the number of pseudo-time steps for various CFL numbers. The employed mesh (refinement level 4) contains a total of 9, 577 vertices.

Remarkably, convergence to the steady-state solution accelerates as the CFL number increases. In the case of the low-order scheme, $\nu = \infty$ delivers the best convergence rates, whereby the norm of the residual falls below 10^{-12} after just ten iterations. Small values of the CFL number imply slow convergence, whereas fast and almost monotone error reduction is observed for large pseudo-time steps.

The flux-corrected Galerkin scheme exhibits a similar convergence behavior but requires a larger number of nonlinear iterations. As the CFL number is increased, the convergence rates improve until the threshold $\nu = 100$ is reached. A further increase of the pseudo-time step does not result in faster convergence. In contrast to the findings of Trépanier et al. [66], the rate of convergence does not deteriorate but stays approximately the same for all $\nu \geq 100$. However, the lagged treatment of the non-differentiable antidiffusive term and the oscillatory behavior of the correction factors produced by the limiter impose an upper bound on the rate of convergence. A better preconditioning of the discrete Jacobian operator and/or the use of convergence acceleration technique are likely to yield a further gain of efficiency.

The results of a grid convergence study for stationary solutions to the GAMM channel problem are presented in Table 6. The relative L^2 error defined as

$$E_2^{rel} = \frac{\|U_h - U\|_2}{\|U\|_2} \tag{114}$$

is calculated using the reference solution U computed on mesh level 6. The effective order of accuracy is $p \approx 0.6$ for the low-order predictor and $p \approx 1.0$ for the high-resolution scheme. The higher accuracy of the flux-corrected solution justifies the additional computational effort. The errors generated near the shock can be reduced using adaptive mesh refinement based on a goal-oriented error estimate [30].

Table 7 NACA 0012 airfoil: Test cases

Case	α	M_∞
I	2°	0.63
II	1.25°	0.8
III	1°	0.85

A proper implementation of boundary conditions is crucial for the overall accuracy of a numerical scheme for the Euler equations. Errors caused by an inappropriate boundary treatment may propagate into the interior of the domain and inhibit convergence to steady-state solutions. For an in-depth numerical study of the boundary conditions for the GAMM channel problem, we refer to Gurris et al. [18, 19]. It turns out that the fully implicit treatment of weakly imposed boundary conditions (see Sect. 11) leads to a much more robust and efficient implementation than the predictor-corrector algorithm described in the first edition of this chapter [32].

13 NACA 0012 Airfoil

In the next example, we simulate the steady gas flow past a NACA 0012 airfoil. The upper and lower surfaces are given by the function $f^\pm : [0, 1.00893] \mapsto \mathbb{R}$ with

$$f^\pm(x) = \pm 0.6\left(0.2969\sqrt{x} - 0.126x - 0.3516x^2 - 0.1015x^4\right). \tag{115}$$

We consider three test configurations labeled Case I–III. The corresponding values of the free stream Mach number M_∞ and inclination angle α are listed in Table 7.

The outer boundary of the computational domain is a circle of radius 10 centered at the tip of the airfoil. The unstructured coarse mesh and a zoom of the reference solution for Case II are displayed in Fig. 10. The stationary Mach number distribution is in a good agreement with the numerical results presented in [13, 23, 32, 52].

The low-order solution is initialized by the free stream values, and a few iterations with the CFL number $\nu = 10$ are performed before increasing the pseudo-time step. As before, the low-order predictor serves as an initial guess for the algebraic flux correction scheme equipped with the characteristic limiter of TVD type.

The nonlinear convergence history for mesh level 2 and the results of a grid convergence study for Case 2 are presented in Fig. 11 and Table 8, respectively. As in the previous example, the semi-implicit pseudo-time-stepping scheme converges faster as the CFL number is increased. In the case of $\nu = \infty$, the residual falls below 10^{-15} in 20 iterations. The high-resolution scheme exhibits similar convergence behavior, although the total number of iterations is much larger. It takes approximately 200 iterations for the residual to reach the tolerance 10^{-8}. Increasing the CFL numbers beyond the threshold $\nu = 100$ yields just a marginal improvement. The errors for $\nu = 100, 1000$, and ∞ are almost identical but considerably smaller than those

Fig. 10 NACA 0012 airfoil: coarse mesh and the Mach number distribution (zoom)

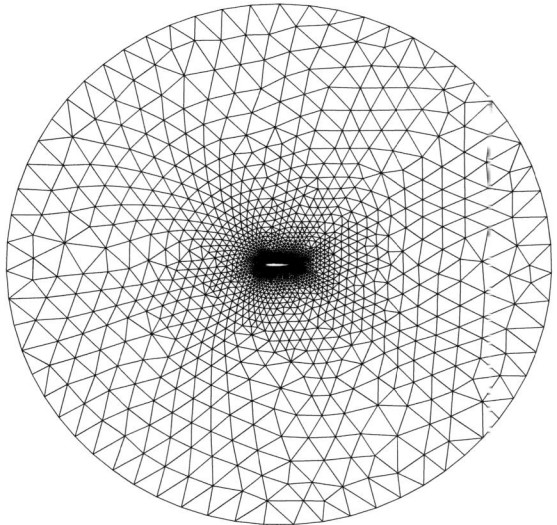

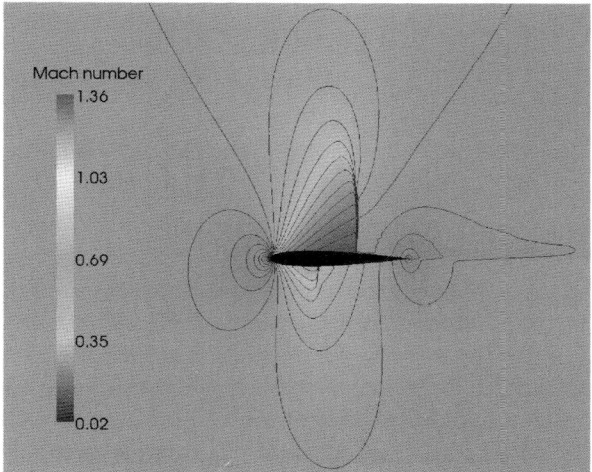

for $\nu = 1$ and 10. The effective order of accuracy is about 0.5 for the low-order scheme and 1.0 for the characteristic FEM-TVD scheme (see Table 8).

The drag and lift coefficients for all test cases are displayed in Table 9. They agree well with the available reference data [8, 15, 52], although the lift is slightly underestimated. This fact can be attributed to the relatively small size of the computational domain. It was shown in [8, 52] that the value of the lift coefficient tends to increase with the distance to the artificial far field boundary. The results presented therein were computed with far field distances of up to 2048 chords, while the far field boundary of our domain is located as few as 10 chords away from the airfoil.

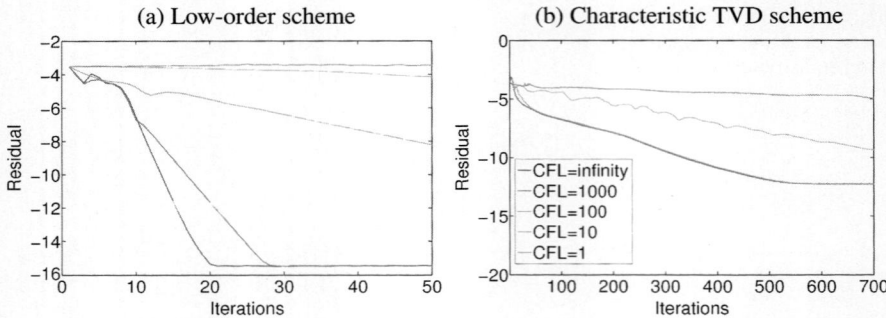

Fig. 11 NACA 0012 airfoil: convergence history for various CFL numbers on mesh level 2

Table 8 NACA 0012 airfoil: relative L^2 errors and grid convergence rates

Level	NVT	NEL	E_2^{Low}	p^{Low}	E_2^{Lim}	p^{Lim}
1	2577	4963	$4.08 \cdot 10^{-2}$	0.51	$1.68 \cdot 10^{-3}$	1.02
2	10117	19852	$2.86 \cdot 10^{-2}$	0.46	$8.27 \cdot 10^{-4}$	1.02
3	40086	79408	$2.08 \cdot 10^{-2}$		$2.68 \cdot 10^{-4}$	
4	159580	317632				

Table 9 NACA 0012 airfoil: drag and lift coefficients for all configurations

(a) Case I			(b) Case II			(c) Case III		
Level	C_D	C_L	Level	C_D	C_L	Level	C_D	C_L
1	$2.8194 \cdot 10^{-3}$	0.2791	1	$2.0043 \cdot 10^{-2}$	0.3065	1	$5.2434 \cdot 10^{-2}$	0.3205
2	$3.5473 \cdot 10^{-4}$	0.2977	2	$1.9198 \cdot 10^{-2}$	0.3169	2	$5.3217 \cdot 10^{-2}$	0.3400
3	$1.2927 \cdot 10^{-4}$	0.3071	3	$1.9501 \cdot 10^{-2}$	0.3199	3	$5.4087 \cdot 10^{-2}$	0.3441
4	$1.1355 \cdot 10^{-4}$	0.3120	4	$1.9933 \cdot 10^{-2}$	0.3200	4	$5.4636 \cdot 10^{-2}$	0.3436

14 Converging-Diverging Nozzle

In the last numerical example, we simulate the transonic flow in a converging-diverging nozzle. The free slip boundary condition (107) is prescribed on the upper and lower walls of the nozzle defined by the function $g^\pm : [-2, 8] \mapsto \mathbb{R}$ with [23]

$$g^\pm(x) = \begin{cases} \pm 1 & \text{if } -2 \leq x \leq 0, \\ \pm \frac{\cos(\frac{\pi x}{2})+3}{4} & \text{if } 0 < x \leq 4, \\ \pm 1 & \text{if } 4 < x \leq 8. \end{cases} \qquad (116)$$

At the subsonic inlet ($x = -2$, $-1 \leq y \leq 1$), the free stream Mach number equals $M_\infty = 0.3$. To facilitate comparison with the results presented by Hartmann and

Table 10 Converging-diverging nozzle: mesh properties

Level	NVT	NEL
1	33	20
2	105	80
3	369	320
4	1377	1280
5	5313	5120
6	20865	20480
7	82689	81920

Table 11 Grid convergence study: outflow boundary condition

Level	NVT	NEL	NVT$_{out}$	E_2^{out}	p^{out}
5	5313	5120	33	$2.50 \cdot 10^{-3}$	1.32
6	20865	20480	65	$1.00 \cdot 10^{-3}$	1.12
7	82689	81920	129	$4.62 \cdot 10^{-4}$	

Houston [23], we define the free stream pressure as $p_\infty = 1$ rather than $p_\infty = \frac{1}{\gamma}$. At the subsonic outlet ($x = 8$, $-1 \leq y \leq 1$), the exit pressure $p_{out} = \frac{2}{3}$ is prescribed as explained in Sect. 11.3.1. As the nozzle converges, the gas is accelerated to supersonic velocities. After entering the diverging part, the flow begins to decelerate and passes through a shock before becoming subsonic again [23].

A mesh of bilinear elements is generated from a structured coarse mesh using global refinements. The numbers of vertices and elements for 7 levels of refinement are listed in Table 10. Figure 12 displays the numerical solution computed on mesh level 7. There is a good agreement with the results obtained by Hartmann [22].

To assess the numerical error in the outlet boundary condition $p_{out} = \frac{2}{3}$, we present the pressure distribution at the outlet Γ_{out} in Fig. 13. The relative L^2 error

$$E_2^{out} = \frac{\|p - p_{out}\|_{2,\Gamma_{out}}}{\|p_{out}\|_{2,\Gamma_{out}}} \quad (117)$$

and the effective order of accuracy p^{out} for mesh levels 5–7 are listed in Table 11, where NVT$_{out}$ is the number of nodes at the outlet. It can be seen that the errors are very small, even on a relatively coarse mesh. As the mesh is refined, the errors shrink. This illustrates the consistency of the proposed boundary treatment.

15 Conclusions

This chapter sheds some light on the aspects of algebraic flux correction for systems of conservation laws. We extended the scalar limiting machinery to the compressible

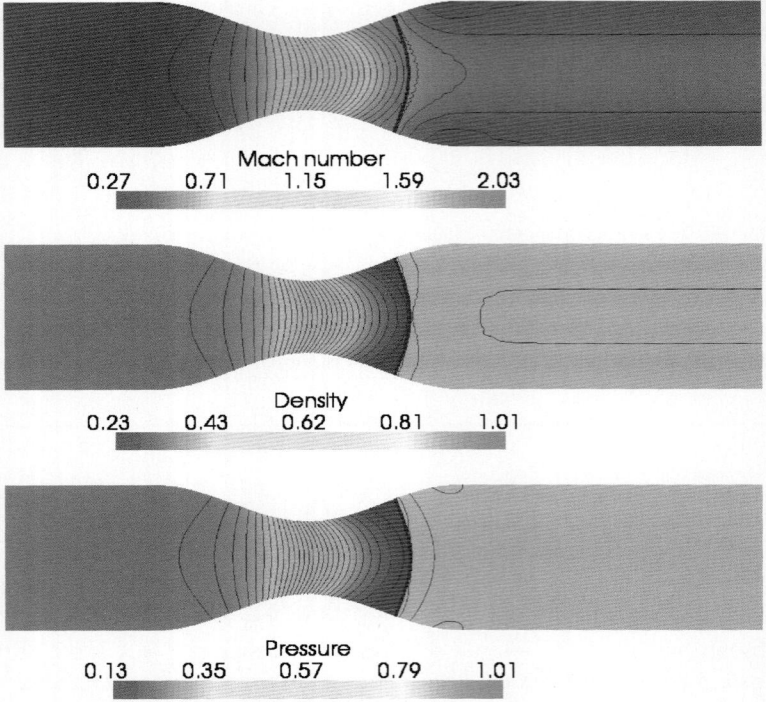

Fig. 12 Converging-diverging nozzle: FEM-TVD solution on mesh level 7

Fig. 13
Converging-diverging nozzle:
exit pressure distribution

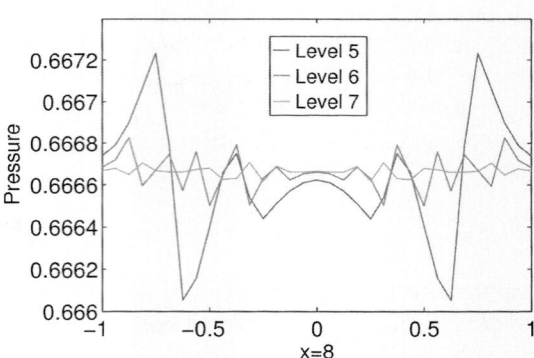

Euler equations and discussed various implementation details (initial and boundary conditions, linearization techniques, iterative solvers etc). Furthermore, we presented a new approach to constraining the primitive variables in synchronized FCT algorithms. It differs from other flux limiters for systems in that the transformation of variables is performed node-by-node rather than edge-by-edge. The generalized Zalesak limiter was equipped with a simple failsafe corrector designed to preserve the bounds of the low-order solution. A numerical study was performed to illustrate the practical utility of the proposed limiting techniques for the Euler equations.

In summary, flux limiting for hyperbolic systems may require (i) a careful choice of the variables to be controlled, (ii) a suitable synchronization of the correction factors, and (iii) a mechanism that makes it possible to 'undo' the antidiffusive correction whenever it turns out to be harmful. The accuracy and efficiency of the code depend on the employed linearizations. Moreover, the implementation of characteristic boundary conditions can make or break the numerical algorithm. All of these issues must be taken into account when it comes to solving real-life problems.

Acknowledgements The authors would like to thank Stefan Turek (Dortmund University of Technology), John Shadid (Sandia National Laboratories), and Mikhail Shashkov (Los Alamos National Laboratory) for many stimulating discussions and useful suggestions.

Appendix

In this appendix, we derive the artificial diffusion operator for the piecewise-linear Galerkin approximation to the one-dimensional Euler equations

$$\frac{\partial U}{\partial t} + \frac{\partial F}{\partial x} = 0. \tag{118}$$

In the 1D case, we have

$$U = \begin{bmatrix} \rho \\ \rho v \\ \rho E \end{bmatrix}, \quad F = \begin{bmatrix} \rho v \\ \rho v^2 + p \\ \rho H v \end{bmatrix}. \tag{119}$$

The differentiation of F by the chain rule yields the equivalent quasi-linear form

$$\frac{\partial U}{\partial t} + A \frac{\partial U}{\partial x} = 0, \tag{120}$$

where $A = \frac{\partial F}{\partial U}$ is the Jacobian matrix. It is easy to verify that

$$A = \begin{bmatrix} 0 & 1 & 0 \\ \frac{1}{2}(\gamma - 3)v^2 & (3 - \gamma)v & \gamma - 1 \\ \frac{1}{2}(\gamma - 1)v^3 - vH & H - (\gamma - 1)v^2 & \gamma v \end{bmatrix}. \tag{121}$$

The eigenvalues and right/left eigenvectors of A satisfy the system of equations

$$A\mathbf{r}_k = \lambda_k \mathbf{r}_k, \quad \mathbf{l}_k A = \lambda_k \mathbf{l}_k, \quad k = 1, 2, 3 \tag{122}$$

which can be written in matrix form as $AR = R\Lambda$ and $R^{-1}A = \Lambda R^{-1}$. Thus,

$$A = R\Lambda R^{-1}, \quad \Lambda = \mathrm{diag}\{v - c, v, v + c\} \tag{123}$$

in accordance with (9). The matrices of eigenvalues and eigenvectors are given by

$$\Lambda = \text{diag}\{v - c, v, v + c\}, \tag{124}$$

$$R = \begin{bmatrix} 1 & 1 & 1 \\ v - c & v & v + c \\ H - vc & \frac{1}{2}v^2 & H + vc \end{bmatrix} = [\mathbf{r}_1, \mathbf{r}_2, \mathbf{r}_3], \tag{125}$$

and

$$R^{-1} = \begin{bmatrix} \frac{1}{2}(b_1 + \frac{v}{c}) & \frac{1}{2}(-b_2 v - \frac{1}{c}) & \frac{1}{2}b_2 \\ 1 - b_1 & b_2 v & -b_2 \\ \frac{1}{2}(b_1 - \frac{v}{c}) & \frac{1}{2}(-b_2 v + \frac{1}{c}) & \frac{1}{2}b_2 \end{bmatrix} = \begin{bmatrix} \mathbf{l}_1 \\ \mathbf{l}_2 \\ \mathbf{l}_3 \end{bmatrix}, \tag{126}$$

where

$$b_1 = b_2 \frac{v^2}{2}, \qquad b_2 = \frac{\gamma - 1}{c^2}.$$

On a uniform mesh of linear finite elements, the coefficients of the lumped mass matrix M_L and of the discrete gradient operator C are given by

$$m_i = \Delta x, \qquad c_{ij} = \begin{cases} 1/2, & j = i + 1, \\ -1/2, & j = i - 1. \end{cases} \tag{127}$$

The lumped-mass Galerkin approximation is equivalent to the central difference scheme which can be written in the generic conservative form

$$\frac{d\mathbf{U}_i}{dt} + \frac{\mathbf{F}_{i+1/2} - \mathbf{F}_{i-1/2}}{\Delta x} = 0, \tag{128}$$

where

$$\mathbf{F}_{i+1/2} = \frac{\mathbf{F}_i + \mathbf{F}_{i+1}}{2}.$$

The numerical flux of the low-order scheme with $\mathbf{D}_{i+1/2}$ defined by (42) is

$$\mathbf{F}_{i+1/2} = \frac{\mathbf{F}_i + \mathbf{F}_{i+1}}{2} - \frac{1}{2}|\mathbf{A}_{i+1/2}|(\mathbf{U}_{i+1} - \mathbf{U}_i), \tag{129}$$

where $\mathbf{A}_{i+1/2}$ is the 1D Roe matrix. The so-defined approximation is known as Roe's approximate Riemann solver [54]. A detailed description of this first-order scheme can be found in many textbooks on gas dynamics [24, 37, 64]. Roe's method fails to recognize expansion waves and, therefore, may give rise to entropy-violating solutions (rarefaction shocks) in the neighborhood of sonic points. Hence, some additional numerical diffusion may need to be applied in regions where one of the characteristic speeds approaches zero [20, 21]. This trick is called an *entropy fix*.

The use of scalar dissipation (46) leads to a Rusanov-like low-order scheme with

$$\mathbf{F}_{i+1/2} = \frac{\mathbf{F}_i + \mathbf{F}_{i+1}}{2} - \frac{a_{i+1/2}}{2}(\mathbf{U}_{i+1} - \mathbf{U}_i), \tag{130}$$

where $a_{i+1/2}$ denotes the fastest characteristic speed. Zalesak [73] defines it as

$$a_{i+1/2} = \frac{|v_i| + |v_{i+1}|}{2} + \frac{c_i + c_{i+1}}{2}.$$

For reasons explained in [5], our definition of the Rusanov flux (130) is based on

$$a_{i+1/2} := \max\{|v_i| + c_i, |v_{i+1}| + c_{i+1}\}.$$

This formula yields a very robust and efficient low-order method for FCT [33].

References

1. Angrand, F., Dervieux, A.: Some explicit triangular finite element schemes for the Euler equations. Int. J. Numer. Methods Fluids **4**, 749–764 (1984)
2. Arminjon, P., Dervieux, A.: Construction of TVD-like artificial viscosities on 2-dimensional arbitrary FEM grids. INRIA Research Report 1111 (1989)
3. Athena test suite. http://www.astro.virginia.edu/VITA/ATHENA/dmr.html
4. Balakrishnan, N., Fernandez, G.: Wall boundary conditions for inviscid compressible flows on unstructured meshes. Int. J. Numer. Methods Fluids **28**, 1481–1501 (1998)
5. Banks, J.W., Henshaw, W.D., Shadid, J.N.: An evaluation of the FCT method for high-speed flows on structured overlapping grids. J. Comput. Phys. **228**, 5349–5369 (2009)
6. Barth, T.J.: Numerical aspects of computing viscous high Reynolds number flows on unstructured meshes. AIAA Paper 91-0721 (1991)
7. Behr, M.: On the application of slip boundary condition on curved boundaries. Int. J. Numer. Methods Fluids **45**, 43–51 (2004)
8. De Zeeuw, D., Powell, K.G.: An adaptive refined Cartesian mesh solver for the Euler equations. J. Comput. Phys. **104**, 56–68 (1993)
9. Dolejší, V., Feistauer, M.: A semi-implicit discontinuous Galerkin finite element method for the numerical solution of inviscid compressible flow. J. Comput. Phys. **198**, 727–746 (2004)
10. Donea, J., Selmin, V., Quartapelle, L.: Recent developments of the Taylor-Galerkin method for the numerical solution of hyperbolic problems. In: Numerical Methods for Fluid Dynamics III, pp. 171–185. Oxford University Press, Oxford (1988)
11. Engelman, M.S., Sani, R.L., Gresho, P.M.: The implementation of normal and/or tangential boundary conditions in finite element codes for incompressible fluid flow. Int. J. Numer. Methods Fluids **2**, 225–238 (1982)
12. Farrell, P.E., Piggott, M.D., Pain, C.C., Gorman, G.J., Wilson, C.R.: Conservative interpolation between unstructured meshes via supermesh construction. Comput. Methods Appl. Mech. Eng. **198**, 2632–2642 (2009)
13. Feistauer, M., Felcman, J., Straškraba, I.: Mathematical and Computational Methods for Compressible Flow. Clarendon Press, Oxford (2003)
14. Feistauer, M., Kučera, V.: On a robust discontinuous Galerkin technique for the solution of compressible flow. J. Comput. Phys. **224**, 208–231 (2007)
15. Fezoui, L., Stoufflet, B.: A class of implicit upwind schemes for Euler simulations with unstructured meshes. J. Comput. Phys. **84**, 174–206 (1989)
16. Fletcher, C.A.J.: The group finite element formulation. Comput. Methods Appl. Mech. Eng. **37**, 225–243 (1983)
17. Ghidaglia, J.-M., Pascal, F.: On boundary conditions for multidimensional hyperbolic systems of conservation laws in the finite volume framework. Technical Report, ENS de Cachan (2002)

18. Gurris, M.: Implicit finite element schemes for compressible gas and particle-laden gas flows. PhD Thesis, Dortmund University of Technology (2010)
19. Gurris, M., Kuzmin, D., Turek, S.: Implicit finite element schemes for the stationary compressible Euler equations. Int. J. Numer. Methods Fluids (2011). doi:10.1002/fld.2532
20. Harten, A.: On a class of high-resolution total-variation-stable finite-difference schemes. SIAM J. Numer. Anal. **21**, 1–23 (1984)
21. Harten, A., Hyman, J.M.: Self adjusting grid methods for one-dimensional hyperbolic conservation laws. J. Comput. Phys. **50**, 235–269 (1983)
22. Hartmann, R.: Homepage http://www.numerik.uni-hd.de/~hartmann/
23. Hartmann, R., Houston, P.: Adaptive discontinuous Galerkin finite element methods for the compressible Euler equations. J. Comput. Phys. **183**, 508–532 (2002)
24. Hirsch, C.: Numerical Computation of Internal and External Flows. Vol. II: Computational Methods for Inviscid and Viscous Flows. Wiley, Chichester (1990)
25. Krivodonova, L., Berger, M.: High-order accurate implementation of solid wall boundary conditions in curved geometries. J. Comput. Phys. **211**, 492–512 (2006)
26. Kuzmin, D.: Linearity-preserving flux correction and convergence acceleration for constrained Galerkin schemes. J. Comput. Appl. Math. (2012, to appear)
27. Kuzmin, D.: A Guide to Numerical Methods for Transport Equations. University Erlangen-Nuremberg, Erlangen (2010). http://www.mathematik.uni-dortmund.de/~kuzmin/Transport.pdf
28. Kuzmin, D.: Explicit and implicit FEM-FCT algorithms with flux linearization. J. Comput. Phys. **228**, 2517–2534 (2009)
29. Kuzmin, D.: Algebraic flux correction for finite element discretizations of coupled systems. In: Oñate, E., Papadrakakis, M., Schrefler, B. (eds.) Computational Methods for Coupled Problems in Science and Engineering II, pp. 653–656. CIMNE, Barcelona (2007)
30. Kuzmin, D., Möller, M.: Goal-oriented mesh adaptation for flux-limited approximations to steady hyperbolic problems. J. Comput. Appl. Math. **233**, 3113–3120 (2010)
31. Kuzmin, D., Möller, M.: Algebraic flux correction I. Scalar conservation laws. In: Kuzmin, D., et al. (eds.) Flux-Corrected Transport: Principles, Algorithms and Applications, pp. 155–206. Springer, Berlin (2005). Chapter 6 in the first edition of this book
32. Kuzmin, D., Möller, M.: Algebraic flux correction II. Compressible Euler equations. In: Kuzmin, D., et al. (eds.) Flux-Corrected Transport: Principles, Algorithms and Applications, pp. 207–250. Springer, Berlin (2005). Chapter 7 in the first edition of this book
33. Kuzmin, D., Möller, M., Shadid, J.N., Shashkov, M.: Failsafe flux limiting and constrained data projections for equations of gas dynamics. J. Comput. Phys. **229**, 8766–8779 (2010)
34. Kuzmin, D., Möller, M., Turek, S.: High-resolution FEM-FCT schemes for multidimensional conservation laws. Comput. Methods Appl. Mech. Eng. **193**, 4915–4946 (2004)
35. LeVeque, R.J.: Numerical Methods for Conservation Laws. Birkhäuser, Basel (1992)
36. LeVeque, R.J.: Simplified multi-dimensional flux limiting methods. In: Numerical Methods for Fluid Dynamics, vol. IV, pp. 175–190. Oxford University Press, Oxford (1993)
37. LeVeque, R.J.: Finite Volume Methods for Hyperbolic Problems. Cambridge University Press, Cambridge (2003)
38. LeVeque, R.J.: CLAWPACK—Conservation LAWs PACKage. http://www.amath.washington.edu/~claw/
39. Liska, R., Shashkov, M., Váchal, P., Wendroff, B.: Optimization-based synchronized Flux-Corrected Conservative interpolation (remapping) of mass and momentum for Arbitrary Lagrangian-Eulerian methods. J. Comput. Phys. **229**, 1467–1497 (2010)
40. Löhner, R.: Applied CFD Techniques: An Introduction Based on Finite Element Methods, 2nd edn. Wiley, New York (2008)
41. Löhner, R., Baum, J.D.: 30 years of FCT: Status and directions. In: Kuzmin, D., et al. (eds.) Flux-Corrected Transport: Principles, Algorithms and Applications, pp. 131–154. Springer, Berlin (2005). Chapter 5 in the first edition of this book
42. Löhner, R., Morgan, K., Peraire, J., Vahdati, M.: Finite element flux-corrected transport (FEM-FCT) for the Euler and Navier-Stokes equations. Int. J. Numer. Methods Fluids **7**, 1093–1109 (1987)

43. Löhner, R., Morgan, K., Zienkiewicz, O.C.: An adaptive finite element procedure for compressible high speed flows. Comput. Methods Appl. Mech. Eng. **51**, 441–465 (1985)
44. Luo, H., Baum, J.D., Löhner, R.: Numerical solution of the Euler equations for complex aerodynamic configurations using an edge-based finite element scheme. AIAA-93-2933 (1993)
45. Lyra, P.R.M., Morgan, K., Peraire, J., Peiro, J.: TVD algorithms for the solution of the compressible Euler equations on unstructured meshes. Int. J. Numer. Methods Fluids **19**, 827–847 (1994)
46. Lyra, P.R.M., Morgan, K.: A review and comparative study of upwind biased schemes for compressible flow computation. I: 1-D first-order schemes. Arch. Comput. Methods Eng. **7**(1), 19–55 (2000)
47. Lyra, P.R.M., Morgan, K.: A review and comparative study of upwind biased schemes for compressible flow computation. II: 1-D higher-order schemes. Arch. Comput. Methods Eng. **7**(3), 333–377 (2000)
48. Lyra, P.R.M., Morgan, K.: A review and comparative study of upwind biased schemes for compressible flow computation. III: Multidimensional extension on unstructured grids. Arch. Comput. Methods Eng. **9**(3), 207–256 (2002)
49. Möller, M.: Adaptive high-resolution finite element schemes. PhD thesis, Dortmund University of Technology (2008)
50. Möller, M.: Efficient solution techniques for implicit finite element schemes with flux limiters. Int. J. Numer. Methods Fluids **55**, 611–635 (2007)
51. Morgan, K., Peraire, J.: Unstructured grid finite element methods for fluid mechanics. Rep. Prog. Phys. **61**(6), 569–638 (1998)
52. Nejat, A.: A higher-order accurate unstructured finite volume Newton-Krylov algorithm for inviscid compressible flows. PhD Thesis, Vancouver (2007)
53. Peraire, J., Vahdati, M., Peiro, J., Morgan, K.: The construction and behaviour of some unstructured grid algorithms for compressible flows. In: Numerical Methods for Fluid Dynamics, vol. IV, pp. 221–239. Oxford University Press, Oxford (1993)
54. Roe, P.L.: Approximate Riemann solvers, parameter vectors and difference schemes. J. Comput. Phys. **43**, 357–372 (1981)
55. Rohde, A.: Eigenvalues and eigenvectors of the Euler equations in general geometries. AIAA Paper 2001-2609 (2001)
56. Selmin, V.: Finite element solution of hyperbolic equations. I. One-dimensional case. INRIA Research Report 655 (1987)
57. Selmin, V.: Finite element solution of hyperbolic equations. II. Two-dimensional case. INRIA Research Report 708 (1987)
58. Selmin, V.: The node-centred finite volume approach: bridge between finite differences and finite elements. Comput. Methods Appl. Mech. Eng. **102**, 107–138 (1993)
59. Selmin, V., Formaggia, L.: Unified construction of finite element and finite volume discretizations for compressible flows. Int. J. Numer. Methods Eng. **39**, 1–32 (1996)
60. Shapiro, R.A.: Adaptive Finite Element Solution Algorithm for the Euler Equations. Notes on Numerical Fluid Mechanics, vol. 32. Vieweg, Wiesbaden (1991)
61. Smith, T.M., Hooper, R.W., Ober, C.C., Lorber, A.A.: Intelligent nonlinear solvers for computational fluid dynamics. Conference Paper, Presentation at the 44th AIAA Aerospace Sciences Meeting and Exhibit, Reno, NV, January 2006
62. Smolarkiewicz, P.K., Grell, G.A.: A class of monotone interpolation schemes. J. Comput. Phys. **101**, 431–440 (1992)
63. Sod, G.: A survey of several finite difference methods for systems of nonlinear hyperbolic conservation laws. J. Comput. Phys. **27**, 1–31 (1978)
64. Toro, E.F.: Riemann Solvers and Numerical Methods for Fluid Dynamics. Springer, Berlin (1999)
65. Toro, E.F.: NUMERICA, A Library of Source Codes for Teaching, Research and Applications. Numeritek Ltd., http://www.numeritek.com (1999)
66. Trépanier, J.-Y., Reggio, M., Ait-Ali-Yahia, D.: An implicit flux-difference splitting method for solving the Euler equations on adaptive triangular grids. Int. J. Numer. Methods Heat Fluid Flow **3**, 63–77 (1993)

67. Váchal, P., Liska, R.: Sequential flux-corrected remapping for ALE methods. In: Bermudez de Castro, A., Gomez, D., Quintela, P., Salgado, P. (eds.) Numerical Mathematics and Advanced Applications (ENUMATH 2005), pp. 671–679. Springer, Berlin (2006)
68. Vogt, W.: Adaptive Verfahren zur numerischen Quadratur und Kubatur. Preprint No. M 1/06, IfMath TU Ilmenau (2006)
69. Wesseling, P.: Principles of Computational Fluid Dynamics. Springer, Berlin (2001)
70. Woodward, P.R., Colella, P.: The numerical simulation of two-dimensional fluid flow with strong shocks. J. Comput. Phys. **54**, 115–173 (1984)
71. Yee, H.C.: Construction of explicit and implicit symmetric TVD schemes and their applications. J. Comput. Phys. **43**, 151–179 (1987)
72. Yee, H.C., Warming, R.F., Harten, A.: Implicit Total Variation Diminishing (TVD) schemes for steady-state calculations. J. Comput. Phys. **57**, 327–360 (1985)
73. Zalesak, S.T.: The design of Flux-Corrected Transport (FCT) algorithms for structured grids. In: Kuzmin, D., et al. (eds.) Flux-Corrected Transport: Principles, Algorithms and Applications, pp. 29–78. Springer, Berlin (2005). Chapter 2 in the first edition of this book

Algebraic Flux Correction III

Incompressible Flow Problems

Stefan Turek and Dmitri Kuzmin

Abstract This chapter illustrates the use of algebraic flux correction in the context of finite element methods for the incompressible Navier-Stokes equations and related models. In the convection-dominated flow regime, nonlinear stability is enforced using algebraic flux correction. The numerical treatment of the incompressibility constraint is based on the 'Multilevel Pressure Schur Complement' (MPSC) approach. This class of iterative methods for discrete saddle-point problems unites fractional-step/operator-splitting methods and strongly coupled solution techniques. The implementation of implicit high-resolution schemes for incompressible flow problems requires the use of efficient Newton-like methods and optimized multigrid solvers for linear systems. The coupling of the Navier-Stokes system with scalar conservation laws is also discussed in this chapter. The applications to be considered include the Boussinesq model of natural convection, the k–ε turbulence model, population balance equations for disperse two-phase flows, and level set methods for free interfaces. A brief description of the numerical algorithm is given for each problem.

1 Introduction

One of the most fundamental models in fluid mechanics is the incompressible Navier-Stokes equations for the velocity $\mathbf{u}$ and pressure p of a Newtonian fluid

$$\frac{\partial \mathbf{u}}{\partial t} + \mathbf{u} \cdot \nabla \mathbf{u} - \nu \Delta \mathbf{u} + \nabla p = \mathbf{f}, \qquad (1)$$
$$\nabla \cdot \mathbf{u} = 0,$$

S. Turek (✉)
Institute of Applied Mathematics (LS III), TU Dortmund, Vogelpothsweg 87, 44227 Dortmund, Germany
e-mail: stefan.turek@math.tu-dortmund.de

D. Kuzmin
Applied Mathematics III, University Erlangen-Nuremberg, Cauerstr. 11, 91058 Erlangen, Germany
e-mail: kuzmin@am.uni-erlangen.de

where ν is the kinematic viscosity of the fluid and **f** is a given external force. In contrast to compressible flow models, there is no equation of state. The constant density ρ is "hidden" in the modified pressure p which adjusts itself instantaneously so as to render the velocity field **u** divergence-free. The solution to (1) is sought in a bounded domain $\Omega \subset \mathbb{R}^d$, $d = 2, 3$ on a finite time interval $(0, T]$. The choice of initial and boundary conditions depends on the particular application.

The Navier-Stokes equations (NSE) describe an amazing variety of fluid flows and represent a 'grand challenge' problem of profound importance to mathematicians, physicists, and engineers. It is not surprising that the NSE were among the seven *Millennium Problems* selected by the Clay Mathematics Institute in 2000. The associated $1,000,000 prize is to be awarded for "substantial progress toward a mathematical theory which will unlock the secrets hidden in the Navier-Stokes equations." During the first decade of the XXI century, no major breakthrough was achieved on the theoretical side of this enterprise. However, a lot of progress has been made in the development of numerical methods for the Navier-Stokes equations and their applications in Computational Fluid Dynamics (CFD).

Models based on the incompressible Navier-Stokes equations are widely used in applied mathematics and engineering sciences. The nonlinearity of the convective term, the incompressibility constraint, and the possible coupling of (1) with other equations make the numerical implementation of such models rather challenging. Numerical instabilities may be caused not only by the dominance of convective terms at high Reynolds numbers but also by the velocity-pressure coupling or by the numerical treatment of sources/sinks. In many applications, the flow is turbulent and takes place in a domain of complex geometrical shape. Additional difficulties are associated with the presence of moving boundaries, free interfaces, or unresolvable small-scale features. All peculiarities of a given model must be taken into account when it comes to the design of reliable and efficient numerical methods.

The performance of CFD software depends not only on the accuracy of the underlying discretization techniques but also on the choice of iterative solvers, data structures, and programming concepts. Explicit schemes are easy to implement and parallelize but give rise to severe time step restrictions. In the case of an implicit scheme, one has to solve sparse nonlinear systems for millions of unknowns at each time step. The computational cost can be reduced by using optimal preconditioners, multigrid solvers, local mesh refinement, and adaptive time step control. Last but not least, parallelization of the code is a must for many real-life applications.

The development of improved numerical algorithms for the incompressible Navier-Stokes equations has been actively pursued for more than 50 years. The number of publications on this topic is overwhelming. For a comprehensive overview, the reader is referred to the book by Gresho et al. [22]. In many cases, numerical solutions to the NSE are accurate enough to look realistic. The result of a 2D simulation for the laminar flow around a cylinder is shown in Fig. 1(a). The snapshot exhibits a remarkably good agreement with the experimental data in Fig. 1(b). However, a quantitative comparison of drag and lift coefficients produced by different codes reveals significant differences in their accuracy and efficiency [63].

A current trend in CFD is to combine the 'basic' Navier-Stokes equations (1) with more or less sophisticated engineering models for industrial applications. Ad-

Fig. 1 Flow around a cylinder: (**a**) numerical simulation with FEATFLOW [78], (**b**) experimental data (source: Van Dyke's 'Album of Fluid Motion' [89])

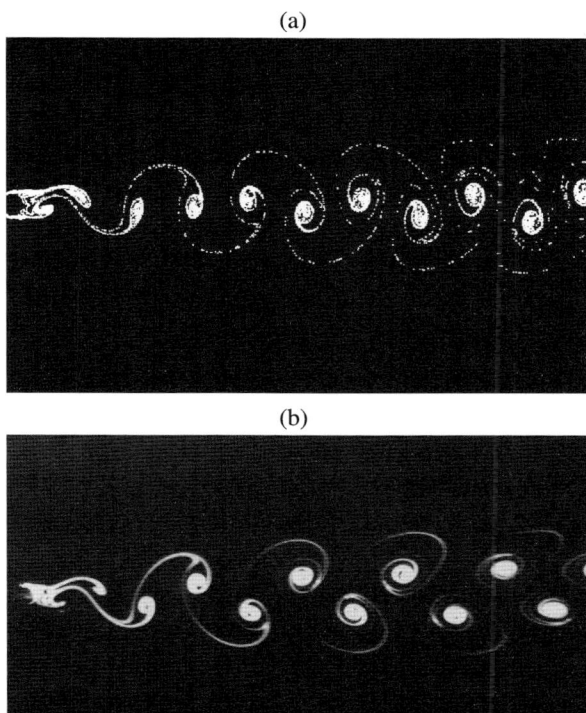

ditional equations are included to describe turbulence, nonlinear fluids, combustion, detonation, multiphase flow, free and moving boundaries, fluid-structure interaction, weak compressibility, and other effects. Some of these extensions will be discussed in the present chapter. All of them require a very careful choice of numerical approximations and iterative solution techniques. In summary, the main ingredients of a 'perfect' CFD code for a generalized Navier-Stokes model are as follows:

- *Discretization:* adaptive high-resolution schemes, discrete maximum principles;
- *Solvers:* robust and efficient iterative methods for linear and nonlinear systems;
- *Implementation:* optimal data structures, hardware-specific code, parallelization.

The availability and compatibility of these components would make it possible to attain high accuracy with a relatively small number of unknowns. Alternatively, discrete problems of the same size could be solved more efficiently. The marriage of accurate numerical methods and fast iterative solvers would make it possible to exploit the potential of modern computers to the full extent and improve the MFLOP/s rates of incompressible flow solvers by *orders of magnitude*. Hence, algorithmic aspects play an increasingly important role in contemporary CFD research.

This chapter begins with a brief review of the Multilevel Pressure Schur Complement (MPSC) approach to solving the incompressible Navier-Stokes equations at high and low Reynolds numbers. Next, the coupling of the basic flow model with additional transport equations is discussed in the context of the Boussinesq approximation for natural convection problems. Algebraic flux correction is shown to be a useful tool for enforcing positivity on unstructured meshes in 3D. In particular, a positivity-preserving implementation of the standard k–ε turbulence model is described. The application of the proposed algorithms to multiphase flow models is illustrated by a case study for population balance equations and free surface flows.

2 Discretization of the Navier-Stokes Equations

The incompressible Navier-Stokes equations are an integral part of all mathematical models to be considered in this chapter. First of all, we discretize (1) in space and time. For our purposes, it is convenient to begin with the time discretization. As a time-stepping method, we will use an implicit two-level θ-scheme (backward Euler or Crank-Nicolson) or the fractional-step θ-scheme proposed by Glowinski.

Let Δt denote the time step for advancing the solution from the time level t^n to the time level $t^{n+1} := t^n + \Delta t$. The value of Δt may be chosen adaptively. The semi-discrete version of (1) can be written in the following generic form [78]:

Given $\mathbf{u}(t^n)$ find $\mathbf{u} = \mathbf{u}(t^{n+1})$ and $p = p(t^{n+1})$ such that

$$[I + \theta \Delta t (\mathbf{u} \cdot \nabla - \nu \Delta)]\mathbf{u} + \Delta t \nabla p = \mathbf{g}, \quad \nabla \cdot \mathbf{u} = 0 \quad \text{in } \Omega, \qquad (2)$$

where

$$\mathbf{g} = [I - \theta_1 \Delta t (\mathbf{u}(t^n) \cdot \nabla - \nu \Delta)]\mathbf{u}(t^n) + \theta_2 \Delta t \mathbf{f}(t^{n+1}) + \theta_3 \Delta t \mathbf{f}(t^n). \qquad (3)$$

The values of the parameters θ and θ_i, $i = 1, 2, 3$ depend on the time-stepping scheme. For example, $\theta = \theta_2 = 1$, $\theta_1 = \theta_3 = 0$ for the backward Euler method.

Next, let us discretize the above problem in space using the finite element method (FEM). The algorithms to be presented in this chapter are also applicable to finite difference and finite volume approximations since the structure of the discrete problems is the same. We favor the finite element approach because the applications we have in mind require the use of high-order discretizations on unstructured meshes. Moreover, the FEM is backed by a solid mathematical theory that makes it possible to obtain rigorous a posteriori error estimates for adaptation in space and time.

The Galerkin finite element approximation to (2) is derived from a variational form of the semi-discretized Navier-Stokes equations. The discretization in space begins with the generation of a computational mesh $\mathscr{T}_h$ for the domain Ω. As usual, the subscript h refers to the local size of mesh cells (triangles or quadrilaterals in 2D, tetrahedra or hexahedra in 3D). Inside each cell, the numerical solution is defined in

Fig. 2 Nodal points of the nonconforming finite element pair $\tilde{Q}_1/Q_0$ in 2D

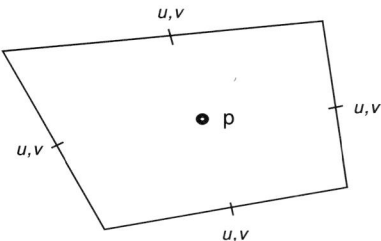

terms of polynomial basis functions. Let $\mathbf{V}_h$ and Q_h denote the finite-dimensional spaces for the velocity and pressure approximations, respectively. The discretization of (2) is stable if $\mathbf{V}_h$ and Q_h satisfy the *Babuška–Brezzi* (BB) condition [20]

$$\min_{q_h \in Q_h} \max_{\mathbf{v}_h \in \mathbf{V}_h} \frac{(q_h, \nabla \cdot \mathbf{v}_h)}{\|q_h\|_0 \|\nabla \mathbf{v}_h\|_0} \geq \gamma > 0 \qquad (4)$$

with a mesh–independent constant γ. If the use of equal-order interpolations is desired, additional stabilization terms must be included (see, e.g., [28]).

The lowest-order finite element approximations satisfying the above inf-sup condition are the nonconforming Crouzeix-Raviart ($\tilde{P}_1/P_0$) and Rannacher-Turek ($\tilde{Q}_1/Q_0$) elements [11, 62]. In either case, the degrees of freedom for the velocity are associated with edge/face mean values, whereas the pressure is approximated in terms of cell mean values. A sketch of the nodal points for a quadrilateral $\tilde{Q}_1/Q_0$ element is shown in Fig. 2. The benefits of using low-order nonconforming approximations include a relatively small number of unknowns and the availability of efficient multigrid solvers which are sufficiently robust in the whole range of Reynolds numbers, even on nonuniform and highly anisotropic meshes [64, 78]. Last but not least, algebraic flux correction is readily applicable to $\tilde{P}_1$ and $\tilde{Q}_1$ elements [41].

The most popular inf-sup stable approximation of higher order is the Taylor-Hood (P_2/P_1 or Q_2/Q_1) element. In our experience, the Q_2/P_1 element is a better choice for non-simplex meshes [12]. Since no algebraic flux correction schemes are currently available for higher-order finite elements, the Q_2/P_1 version of our Navier-Stokes solver is stabilized using continuous interior penalty techniques [56, 86].

The vectors of discrete nodal values for the velocity and pressure will also be denoted by $\mathbf{u}$ and p. The nonlinear discrete problem is formulated as follows:

Given $\mathbf{u}^n$ find $\mathbf{u} = \mathbf{u}^{n+1}$ and $p = p^{n+1}$ such that

$$A\mathbf{u} + \Delta t B p = \mathbf{g}, \qquad B^T \mathbf{u} = 0, \qquad (5)$$

where

$$\mathbf{g} = \left[M - \theta_1 \Delta t N(\mathbf{u}^n) \right] \mathbf{u}^n + \theta_2 \Delta t \mathbf{f}^{n+1} + \theta_3 \Delta t \mathbf{f}^n. \qquad (6)$$

Here M is the (consistent or lumped) mass matrix, B is the discrete gradient operator, and $-B^T$ is the discrete divergence operator. The matrix A is given by

$$A = M - \theta \Delta t N(\mathbf{u}), \tag{7}$$

where

$$N(\mathbf{u}) = K(\mathbf{u}) + \nu L,$$

$K(\mathbf{u})$ is the discrete transport operator and L is the viscous part of the stiffness matrix. The nonlinear operator $N(\mathbf{u})$ may also include artificial diffusion due to algebraic flux correction or other stabilization/shock-capturing techniques.

The discretization of the stationary Navier-Stokes equations also leads to a nonlinear system of the form (5). To use the same notation for steady and time-dependent flow problems, we replace (7) with the more general definition

$$A = \alpha M - \theta \Delta t N(\mathbf{u}). \tag{8}$$

The discrete evolution operator given by (7) corresponds to $\alpha = 1$. The steady-state approximation is defined by the parameter settings $\alpha = 0$, $\theta = 1$, $\Delta t = 1$.

The design of efficient iterative methods for the above discrete problem involves a linearization of $N(\mathbf{u})$ or iterative solution of the nonlinear system using fixed-point defect correction or Newton-like methods. Special techniques (explicit or implicit underrelaxation, line search, Anderson acceleration) may be implemented to achieve and speed up convergence. When it comes to the numerical treatment of the incompressibility constraint, one has a choice between a strongly coupled approach (simultaneous computation of $\mathbf{u}$ and p) and fractional-step algorithms (projection schemes [8, 91], pressure correction methods [16, 58]). The abundance of choices has generated a great variety of incompressible flow solvers that exhibit considerable differences in terms of their complexity, robustness, and efficiency.

The Multilevel Pressure Schur Complement (MPSC) formulation to be presented below makes it possible to put many existing solution algorithms into a common framework and combine their advantages. In particular, the iterative solver may be configured in an adaptive manner so as to achieve the best run-time characteristics for a given problem. For a more detailed presentation of the MPSC paradigm and additional numerical examples, we refer to the monograph by Turek [78].

3 Pressure Schur Complement Solvers

The linearized form of the fully discrete problem (5), as well as the linear systems to be solved at each iteration of a nonlinear scheme, can be written as

$$\begin{bmatrix} A & \Delta t B \\ B^T & 0 \end{bmatrix} \begin{bmatrix} \mathbf{u} \\ p \end{bmatrix} = \begin{bmatrix} \mathbf{g} \\ 0 \end{bmatrix}. \tag{9}$$

This is a typical saddle point problem in which the pressure p acts as the Lagrange multiplier for the discretized incompressibility constraint.

The *Schur complement* equation for the pressure can be derived using a formal elimination of the velocity unknowns. The discrete form of $\nabla \cdot \mathbf{u} = 0$ is

$$B^T \mathbf{u} = 0, \qquad (10)$$

where $\mathbf{u}$ is the solution to the discretized momentum equation, that is,

$$\mathbf{u} = A^{-1}(\mathbf{g} - \Delta t B p). \qquad (11)$$

Thus, an equivalent formulation of the discrete saddle-point problem (9) reads:

$$A\mathbf{u} = \mathbf{g} - \Delta t B p, \qquad (12)$$

$$B^T A^{-1} B p = \frac{1}{\Delta t} B^T A^{-1} \mathbf{g}. \qquad (13)$$

Since the right-hand side of (12) depends on the solution to (13), the two sub-problems should actually be solved in the reverse order:

1. Solve the pressure Schur complement (PSC) equation (13) for p.
2. Substitute p into the momentum equation (12) and compute $\mathbf{u}$.

In the fully nonlinear version, the Schur complement operator $S := B^T A^{-1} B$ depends on the solution to (12), so a number of outer iterations are performed.

The practical implementation of the two-step algorithm also requires a number of inner iterations. Since the matrix A^{-1} is full, the assembly and storage of S would be prohibitively expensive in terms of CPU time and memory requirements. Thus, it is imperative to solve the PSC equation in an iterative way. For instance, consider a preconditioned Richardson's method based on the following *basic iteration*

$$p^{(l)} = p^{(l-1)} + C^{-1}\left[\frac{1}{\Delta t} B^T A^{-1} \mathbf{g} - S p^{(l-1)}\right], \qquad (14)$$

where $l = 1, 2, \ldots, L$ is the iteration counter, C^{-1} is a suitable approximation to S^{-1}, and the expression in the brackets is the residual of the PSC equation.

By definition of S, an equivalent form of the pressure correction equation (15) is

$$p^{(l)} = p^{(l-1)} + C^{-1} \frac{1}{\Delta t} B^T A^{-1} \left[\mathbf{g} - \Delta t B p^{(l-1)}\right]. \qquad (15)$$

In practice, the matrices A and C are "inverted" by solving a linear system. Thus, the implementation of (15) can be split into the following basic tasks:

1. Given the pressure $p^{(l-1)}$, solve the discrete momentum equation

$$A\mathbf{u}^{(l)} = \mathbf{g} - \Delta t B p^{(l-1)}. \tag{16}$$

2. Given the velocity $\mathbf{u}^{(l)}$, solve the pressure correction equation

$$C q^{(l)} = \frac{1}{\Delta t} B^T \mathbf{u}^{(l)}. \tag{17}$$

3. Add the pressure increment $q^{(l)}$ to the current approximation

$$p^{(l)} = p^{(l-1)} + q^{(l)}. \tag{18}$$

The number of pressure correction cycles L can be fixed or variable. The iterative process may be terminated when the increments and residuals become small enough. Using $C := S$, one obtains the solution to (9) in one step ($L = 1$). The assembly of C can be avoided using a GMRES-like iterative solver that operates with matrix-vector products. The evaluation of $Cy = B^T A^{-1} B y$ would involve an iterative solution of the linear system $Ax = By$ followed by the matrix-vector multiplication $Cy := B^T y$. This procedure must be repeated as many times as necessary to reach the prescribed tolerance for the residual of the PSC equation. Hence, the computational cost per time step is likely to be very high even if multigrid acceleration is employed.

In many cases, the matrix-free 'inversion' of $C := S$ is impractical. In particular, the cost per time step is always the same, although a good initial guess is available when the time steps are small. In this case, the discrete evolution operator

$$A = M - \theta \Delta t N(\mathbf{u}) \approx M + \mathcal{O}(\Delta t) \tag{19}$$

represents a well-conditioned perturbation of the symmetric positive-definite mass matrix M. Hence, the discrete momentum equation can be solved efficiently for small Δt. However, the condition number of the PSC operator is given by

$$\text{cond}(S) = \text{cond}\big(B^T \big[M + \mathcal{O}(\Delta t)\big]^{-1} B\big) \approx \text{cond}(L) = \mathcal{O}\big(h^{-2}\big) \tag{20}$$

and does not improve when the time step is refined. The invariably high cost of solving the "elliptic" pressure Schur complement equation makes $C := S$ a poor choice when it comes to simulation of unsteady flows with small time steps.

A computationally efficient Schur complement preconditioner for time-dependent flow problems can be designed using approximations of the form

$$C := B^T \tilde{A}^{-1} B, \tag{21}$$

where $\tilde{A} \approx A$ is a matrix that can be 'inverted' in an efficient way. By (20), the condition number of the PSC operator is dominated by the elliptic part. Thus, the preconditioner can be defined using the symmetric positive definite matrix

$$\tilde{A} := M - \theta \Delta t \nu L. \qquad (22)$$

By (19), a usable preconditioner for high Reynolds number flows is given by

$$\tilde{A} := M. \qquad (23)$$

Replacing M with a lumped mass matrix M_L, one obtains a sparse approximation to the Schur complement operator. Another simple choice is the diagonal matrix

$$\tilde{A} := \mathrm{diag}(A). \qquad (24)$$

In general, the formula for $\tilde{A}$ should be as simple as possible but not simpler. Sparse approximations like $C := B^T M_L^{-1} B$ or $C := B^T \mathrm{diag}(A)^{-1} B$ rely on the diagonal dominance of A. The total number of iterations increases at large time steps, and convergence may fail if the off-diagonal part of A can no longer be neglected.

The preconditioning of (15) by a global matrix of the form (21) is called the *global pressure Schur complement* approach [78]. A typical implementation is based on the fractional-step algorithm (16)–(18). The well-known representatives of such segregated incompressible flow solvers include discrete projection schemes [15, 23, 60, 77], various modifications of the SIMPLE method, and Uzawa-like algorithms. For an overview of segregated methods, we refer to [16] and references therein.

An alternative to the sequential update of the velocity and pressure unknowns is the solution of small coupled subproblems. This solution strategy is recommended for steady-state computations and low Reynolds number flows. It should also be considered if the Navier-Stokes system is coupled with a RANS turbulence model or another set of convection-diffusion equations. If the variables are updated in a segregated manner, strong two-way coupling may result in slow convergence. In this case, it is worthwhile to replace (21) with a sum of local preconditioners

$$C^{-1} := \sum_i P_i^T S_i^{-1} P_i, \qquad (25)$$

where $S_i := B_i^T A_i^{-1} B_i$ is the Schur complement matrix for a local subproblem that corresponds to a small subdomain (a single element or a patch of elements) Ω_i. The multiplication by the transformation matrix P_i picks out the degrees of freedom associated with Ω_i, whereas the multiplication by P_i^T locates the global degrees of freedom to be updated after solving a local subproblem of the form $S_i x_i = P_i y$.

The basic iteration (15) preconditioned by (25) is called the *local pressure Schur complement* method [78]. The embedding of "local solvers" into an outer iteration loop of Jacobi or Gauss–Seidel type has a lot in common with domain decomposition methods but multilevel PSC preconditioners of the form (25) do not require a special treatment of interface conditions. A typical representative of such schemes is the Vanka smoother [90] which is widely used in the multigrid community.

As a matter of fact, it is possible to combine global PSC ("operator splitting") and local PSC ("domain decomposition") methods in a general-purpose CFD code. This can be accomplished by using *additive preconditioners* of the form

$$C^{-1} := \sum_i \alpha_i C_i^{-1}.$$

In what follows, we briefly discuss the design of such preconditioners and present the resulting algorithms. The convergence of these basic iteration schemes can be accelerated by using them as preconditioners for Krylov subspace methods (CG, BiCGStab, GMRES) or smoothers for a multigrid solver. The latter approach leads to a family of Multilevel pressure Schur complement (MPSC) methods that prove robust and efficient, as demonstrated by the benchmark computations in [63].

4 Global MPSC Approach

The construction of globally defined additive preconditioners for the Schur complement operator $S = B^T A^{-1} B$ is motivated by the following algebraic splitting

$$A = \alpha M + \beta K(\mathbf{u}) + \gamma L, \tag{26}$$

where $\beta = -\theta \Delta t$ and $\gamma = \nu \beta$. Consider $C := B^T \tilde{A}^{-1} B$, where $\tilde{A}$ is an approximation to A. The above decomposition of A into the reactive (M), convective (K), and viscous (L) part suggests the use of a similar splitting for C^{-1}. Let

- C_M be an approximation to the reactive part $B^T M^{-1} B$,
- C_K be an approximation to the convective part $B^T K^{-1} B$,
- C_L be an approximation to the viscous part $B^T L^{-1} B$.

The preconditioner C_M is well-suited for computations with small time steps. C_K is optimal for steady flows at high Reynolds numbers, and C_L is optimal for steady flows at low Reynolds numbers. Hence, a general-purpose PSC preconditioner can be defined as a suitable combination of the above. In particular, we consider

$$C^{-1} := \alpha' C_M^{-1} + \beta' C_K^{-1} + \gamma' C_L^{-1}, \tag{27}$$

where $\alpha' \in [0, \alpha]$, $\beta' \in [0, \beta]$, $\gamma' \in [0, \gamma]$ are parameters that can be used to activate, deactivate, and blend partial preconditioners depending on the flow regime.

To achieve the best overall performance, the meaning of 'optimality' has to be defined more precisely. Clearly, the most accurate preconditioner for each subproblem is the one that does not involve any approximations. In principle, even a full matrix of the form $B^T \tilde{A}^{-1} B$ can be "inverted" using a matrix-free iterative solver (see above). However, simpler partial preconditioners are likely be more efficient smoothers in the context of a multigrid method. The MPSC solver is well-designed if each subproblem can be solved efficiently and the convergence rates are not sensitive to the parameter settings and geometric properties of the mesh. Optimal preconditioners satisfying these criteria are introduced and analyzed in [78].

At high Reynolds numbers, the use of small time steps is dictated by the physical scales of flow motion. Thus, the lumped mass matrix M_L is a reasonable approximation to A, and the sparse matrix $C := B^T M_L^{-1} B$ may be used as a preconditioner for the basic iteration (15). The practical implementation of the PSC cycle

$$p^{(l)} = p^{(l-1)} + [B^T M_L^{-1} B]^{-1} \frac{1}{\Delta t} B^T A^{-1} [\mathbf{g} - \Delta t B p^{(l-1)}] \qquad (28)$$

is based on the fractional-step algorithm (16)–(18) and can be interpreted as a *discrete projection scheme* [15, 23, 60, 77]. An additional step is included to enforce the incompressibility constraint after the last iteration. The algorithm becomes:

1. Given the pressure $p^{(l-1)}$, solve the "viscous Burgers" equation

$$A\mathbf{u}^{(l)} = \mathbf{g} - \Delta t B p^{(l-1)}. \qquad (29)$$

2. Given the velocity $\mathbf{u}^{(l)}$, solve the "pressure Poisson" equation

$$B^T M_L^{-1} B q^{(l)} = \frac{1}{\Delta t} B^T \mathbf{u}^{(l)}. \qquad (30)$$

3. Add the pressure increment $q^{(l)}$ to the current approximation

$$p^{(l)} = p^{(l-1)} + q^{(l)}. \qquad (31)$$

To enforce $B^T \mathbf{u} = 0$, perform the divergence-free L^2 projection

$$\mathbf{u} = \mathbf{u}^{(l)} - \Delta t M_L^{-1} B q^{(l)}. \qquad (32)$$

The projection step is included because the intermediate velocity $\mathbf{u}^{(l)}$ is calculated using an approximate pressure $p^{(l-1)}$ and is generally not (discretely) divergence-free. Multiplying (32) by B^T and using (30), we obtain

$$B^T \mathbf{u} = B^T \mathbf{u}^{(l)} - \Delta t B^T M_L^{-1} B q^{(l)} = 0. \qquad (33)$$

It can be shown that $B^T M_L^{-1} B$ corresponds to a mixed discretization of the Laplacian operator [23]. If just one basic iteration is performed, algorithm (29)–(32) has the structure of a classical projection scheme for the time-dependent incompressible Navier-Stokes equations. In particular, a discrete counterpart of Chorin's method [8] is obtained with the trivial initial guess $p^{(0)} = 0$. The choice $p^{(0)} = p(t^n)$ leads to the discrete version of the second-order accurate van Kan scheme [91].

The derivation of continuous projection methods involves the use of operator splitting and the Helmholtz decomposition of the intermediate velocity [21, 60]. Replacing differential operators with matrices, one obtains a discrete projection scheme of the form (29)–(32). The advantages of the algebraic approach include

- applicability to discontinuous pressure approximations,
- consistent treatment of boundary conditions (no splitting),
- alleviation of spurious boundary layers for the pressure,
- convergence to the fully coupled solution as l increases,
- possibility of using other global PSC preconditioners.

On the negative side, discrete projection schemes lack inherent stabilization mechanisms, whereas the continuous Chorin and van Kan methods may be used with equal-order (P_1/P_1) interpolations if the time step is not too small [59].

The vectorizable global MPSC schemes are more efficient than coupled solvers in the high Reynolds number regime. If the discrete evolution operator A is dominated by the reactive part, it is sufficient to perform just one pressure Schur complement iteration per time step. The number of inner iterations for the viscous Burgers equation (29) can also be as small as 1 since $\mathbf{u}(t^n)$ is a good initial guess.

If an optimized multigrid method is used to solve the pressure Poisson problem (29), the total cost per time step is just a small fraction of that for a coupled solver. However, the sparse matrix $B^T M_L^{-1} B$ may become a poor approximation to $B^T A^{-1} B$ at large time steps. Therefore, the local MPSC approach presented in the next section is a better choice for low Reynolds number flows and steady-state computations.

5 Local MPSC Approach

In contrast to the global MPSC approach, local Schur complement preconditioners make it possible to update the velocity and pressure in a strongly coupled fashion. In this section, we explain the underlying design philosophy and practical implementation. As already mentioned, the basic idea is to solve small coupled subproblems associated with *patches* of degrees of freedom. We define a patch as a small subset of the vector of unknowns. The solutions to the local subproblems are used to correct the corresponding subsets of the global solution vector. The so-defined block-Jacobi or block-Gauß-Seidel iteration provides a very robust smoother for a multilevel solution strategy [13]. The local MPSC algorithm is amenable to a parallel implementation that exploits the fast cache of modern processors.

The coefficients of local subproblems for the multilevel "domain decomposition" method are extracted from the global matrices using a restriction matrix P_i that picks out the degrees of freedom associated with the i-th patch. We define

$$\begin{bmatrix} A_i & \Delta t B_i \\ B_i^T & 0 \end{bmatrix} := P_i \begin{bmatrix} A & \Delta t B \\ B^T & 0 \end{bmatrix} P_i^T. \qquad (34)$$

Thus, the 'boundary conditions' for subdomains are also taken from the global matrices. The local Schur complement matrix for the i-th subproblem is given by

$$S_i = B_i^T A_i^{-1} B_i. \qquad (35)$$

The block-Jacobi version of the local PSC method can be formulated as follows:

Given $\mathbf{u}^{(l-1)}$ and $p^{(l-1)}$, assemble the defect of the discrete problem (9)

$$\begin{bmatrix} \mathbf{r}^{(l-1)} \\ s^{(l-1)} \end{bmatrix} = \begin{bmatrix} \mathbf{g} \\ 0 \end{bmatrix} - \begin{bmatrix} A & \Delta t B \\ B^T & 0 \end{bmatrix} \begin{bmatrix} \mathbf{u}^{(l-1)} \\ p^{(l-1)} \end{bmatrix} \qquad (36)$$

and perform one basic iteration with the additive PSC preconditioner

$$\begin{bmatrix} \mathbf{u}^{(l)} \\ p^{(l)} \end{bmatrix} = \begin{bmatrix} \mathbf{u}^{(l-1)} \\ p^{(l-1)} \end{bmatrix} + \omega^{(l)} \sum_i P_i^T \begin{bmatrix} \tilde{A}_i & \Delta t B_i \\ B_i^T & 0 \end{bmatrix}^{-1} P_i \begin{bmatrix} \mathbf{r}^{(l-1)} \\ s^{(l-1)} \end{bmatrix}. \qquad (37)$$

The local stiffness matrix $\tilde{A}_i$ matrix is chosen to be an approximation to A_i. The default is $\tilde{A}_i := A_i$. The relaxation parameter $\omega^{(l)}$ can be fixed or chosen adaptively. The practical implementation of (37) begins with the solution of local problems

$$\begin{bmatrix} \tilde{A}_i & \Delta t B_i \\ B_i^T & 0 \end{bmatrix} \begin{bmatrix} \mathbf{v}_i^{(l)} \\ q_i^{(l)} \end{bmatrix} = P_i \begin{bmatrix} \mathbf{r}^{(l-1)} \\ s^{(l-1)} \end{bmatrix}. \qquad (38)$$

Next, the calculated local increments are inserted into the global vectors

$$\begin{bmatrix} \mathbf{v}^{(l)} \\ q^{(l)} \end{bmatrix} = \sum_i P_i^T \begin{bmatrix} \mathbf{v}_i^{(l)} \\ q_i^{(l)} \end{bmatrix}. \qquad (39)$$

Finally, the velocity and pressure approximations are updated thus:

$$\begin{bmatrix} \mathbf{u}^{(l)} \\ p^{(l)} \end{bmatrix} = \begin{bmatrix} \mathbf{u}^{(l-1)} \\ p^{(l-1)} \end{bmatrix} + \omega^{(l)} \begin{bmatrix} \mathbf{v}^{(l)} \\ q^{(l)} \end{bmatrix}. \qquad (40)$$

If some degrees of freedom are shared by two or more patches, a weighted average of the corresponding local increments is inserted into the global vector. The simplest strategy is to overwrite the contributions of previously processed patches or to calculate the arithmetic mean over all patch contributions.

The local subproblems (38) are so small that they can be solved using Gaussian elimination. A further reduction in the size of the linear system is offered by the Schur complement formulation of the local subproblem. The preconditioner

$$C_i^{-1} := \left[B_i \tilde{A}_i^{-1} B_i \right]^{-1} \qquad (41)$$

is a full matrix but its size depends on the number of pressure unknowns only. If the patch Ω_i contains just a moderate number of degrees of freedom, then the small matrix C_i is likely to fit into the processor cache. The local PSC problem can be solved very efficiently making use of hardware–optimized BLAS libraries. The corresponding velocity increment can be recovered as explained in Sect. 3.

In a sequential code, the block-Jacobi form of the basic iteration may be replaced with a block-Gauß-Seidel relaxation that calculates the local residuals using the latest solution values. Both versions are likely to perform well as long as there are no strong mesh anisotropies. However, severe convergence problems may occur on meshes with sharp angles and/or large aspect ratios. The local MPSC approach makes it possible to avoid the potential troubles by "hiding" the anisotropic mesh cells inside macroelements that have a regular shape. Several adaptive blocking strategies for generation of such macromeshes are described in [64, 78].

6 Multilevel Solution Strategy

The presented PSC schemes are particularly efficient if a *multilevel* solution strategy is adopted. To begin with, consider an abstract linear system of the form

$$A_N u_N = f_N. \quad (42)$$

The subscript N refers to the number of approximation levels. In geometric multigrid methods, these levels are characterized by the mesh size h. Let A_k and f_k denote the matrix and the right-hand side for the level number $k = 1, \ldots, N - 1$. The convergence of a basic iteration scheme on finer levels can be significantly accelerated by a few iterations on coarser levels. The multilevel solution algorithm can be interpreted as a hierarchical preconditioner for the slowly converging basic solver.

The main ingredients of a (geometric) multigrid method for solving (42) are:

- matrix–vector multiplication routines for the operators A_k, $k = 1, \ldots, N$,
- an inexpensive *smoother* (basic iteration scheme) and a *coarse grid solver*,
- *prolongation* I_{k-1}^k and *restriction* I_k^{k-1} operators for grid transfer.

Let u_k^0 denote the initial guess for the k-level iteration $MPSC(k, u_k^0, f_k)$. The so-defined *multigrid cycle* yields an approximate solution to the linear system

$$A_k u_k = f_k.$$

On the coarsest level, the number of unknowns is typically so small that the discrete problem $A_1 u_1 = f_1$ can be solved directly. The result is

$$MPSC(1, u_1^0, f_1) = A_1^{-1} f_1.$$

For all other levels of approximation ($k > 1$), the following algorithm is used [78]:

1. *Presmoothing*

    Given u_k^0, perform m basic iterations (smoothing steps) to obtain u_k^m.

2. *Coarse grid correction*

    Restrict the residual of the discrete problem to the coarse grid

    $$f_{k-1} = I_k^{k-1}(f_k - A_k u_k^m).$$

    Set $u_{k-1}^0 = 0$ and calculate u_{k-1}^i recursively for $i = 1, \ldots, p$

    $$u_{k-1}^i = MPSC(k-1, u_{k-1}^{i-1}, f_{k-1}).$$

3. *Relaxation and update*

    Correct u_k^m using a prolongation of the coarse grid solution

    $$u_k^{m+1} = u_k^m + \alpha_k I_{k-1}^k u_{k-1}^p.$$

4. *Postsmoothing*

    Given u_k^{m+1}, perform m smoothing steps to obtain u_k^{m+1+n}.

The relaxation parameter α_k may be fixed or chosen adaptively so as to minimize the error in a certain norm. Using the discrete energy norm, one obtains

$$\alpha_k = \frac{(f_k - A_k u_k^m, I_{k-1}^k u_{k-1}^p)_k}{(A_k I_{k-1}^k u_{k-1}^p, I_{k-1}^k u_{k-1}^p)_k}.$$

After sufficiently many cycles on level N, the above multigrid algorithm yields the converged solution to (42). An extension to the discrete saddle point problem (9) can be performed using a global or local pressure Schur complement approach.

The *global* MPSC approach corresponds to solving the generic system (42) with

$$A_N := B^T A^{-1} B, \qquad u_N := p, \qquad f_N := \frac{1}{\Delta t} B^T A^{-1} \mathbf{g}.$$

The basic iteration is given by (15). After solving the Schur complement equation for the pressure p, the velocity $\mathbf{u}$ is updated. The bulk of CPU time is spent on matrix-vector multiplications for smoothing, defect calculation, and adaptive coarse grid correction. The multiplication by $C = B^T \tilde{A}^{-1} B$ requires an iterative solution of a linear system, unless $\tilde{A}$ is a diagonal matrix. The choice $C = B^T M_L^{-1} B$ leads to a discrete projection scheme (16)–(18) that requires solving a viscous Burgers equation and a Poisson-like equation. Both subproblems can be solved efficiently using linear multigrid methods. For the reasons explained in Sect. 4, the global MPSC approach is recommended for unsteady flows at high Reynolds numbers.

The *local* MPSC approach corresponds to solving the generic system (42) with

$$A_N := \begin{bmatrix} A & \Delta t B \\ B^T & 0 \end{bmatrix}, \qquad u_N := \begin{bmatrix} \mathbf{u} \\ p \end{bmatrix}, \qquad f_N := \begin{bmatrix} \mathbf{g} \\ 0 \end{bmatrix}.$$

The basic iteration is the block-Jacobi method given by (37) or the block-Gauß-Seidel version of the local PSC method. The cost-intensive part is the smoothing step, as in the case of standard multigrid techniques for elliptic problems. Local MPSC schemes lead to very robust solvers for coupled problems. This solution strategy is recommended for flows at low and intermediate Reynolds numbers.

The presented MPSC solvers have been implemented in the open-source software package FEATFLOW [79]. The source code and documentation are available at http://www.featflow.de. Further algorithmic details (adaptive coarse grid correction, grid transfer operators, nonlinear iteration techniques, time step control, implementation of boundary conditions) can be found in the monograph by the first author [78]. Some programming strategies, data structures, and guidelines for the development of a hardware-oriented code are presented in [80–82, 84].

7 Coupling with Scalar Equations

In many practical applications, the Navier-Stokes equations are coupled with a system of conservation laws for scalar quantities transported with the flow. In the context of turbulence modeling, the additional variables may represent the turbulent kinetic energy k, its dissipation rate ε, or the components of the Reynolds stress tensor. The evolution of temperatures, concentrations, and volume fractions is also governed by convection-dominated transport equations with coefficients that depend on the solution to the basic flow model. The discrete maximum principle for these additional equations can be enforced using algebraic flux correction [38].

To explain the ramifications of a two-way coupling with scalar equations, we consider the Boussinesq model of natural convection. The weakly compressible flow induced by temperature gradients is described by the Navier-Stokes system

$$\frac{\partial \mathbf{u}}{\partial t} + \mathbf{u} \cdot \nabla \mathbf{u} + \nabla p = \nu \Delta \mathbf{u} + T \mathbf{e}_g, \qquad \nabla \cdot \mathbf{u} = 0, \tag{43}$$

where T is the temperature, and $\mathbf{e}_g$ stands for the unit vector directed opposite to the gravitational acceleration $\mathbf{g}$. The temperature equation is given by

$$\frac{\partial T}{\partial t} + \mathbf{u} \cdot \nabla T = d \Delta T. \tag{44}$$

In the nondimensional form of this model, the viscosity and diffusion coefficient

$$\nu = \sqrt{\frac{Pr}{Ra}}, \qquad d = \sqrt{\frac{1}{Ra\,Pr}}$$

depend on the Rayleigh number Ra and Prandtl number Pr. A detailed description of the Boussinesq model and the parameter settings for the *MIT benchmark problem* (natural convection in a differentially heated enclosure) can be found in [9].

7.1 Finite Element Discretization

Adding the buoyancy force and the temperature equation to the discretized Navier-Stokes equations, one obtains a nonlinear algebraic system of the form

$$A_u(\mathbf{u})\mathbf{u} + \Delta t M_T T + \Delta t B p = \mathbf{f}_u, \qquad (45)$$

$$B^T \mathbf{u} = 0, \qquad (46)$$

$$A_T(\mathbf{u})T = f_T. \qquad (47)$$

The subscripts u and T are used to distinguish between the evolution operators and right-hand sides of the momentum and temperature equations. As before, the matrices A_u and A_T can be decomposed into a reactive, convective, and diffusive part

$$A_u(\mathbf{v}) = \alpha_u M_u + \beta_u K_u(\mathbf{v}) + \gamma_u L_u, \qquad (48)$$

$$A_T(\mathbf{v}) = \alpha_T M_T + \beta_T K_T(\mathbf{v}) + \gamma_T L_T. \qquad (49)$$

The finite element spaces and discretization techniques for $\mathbf{u}$ and T may be chosen independently. For example, the temperature may be discretized with linear finite elements even if $\tilde{Q}_1/Q_0$ or Q_2/P_1 elements are employed for the Navier-Stokes part. Moreover, different stabilization techniques may be used for K_u and K_T.

The generic matrix form of the discretized Boussinesq model (45)–(47) reads

$$\begin{bmatrix} A_u(\mathbf{u}) & \Delta t M_T & \Delta t B \\ 0 & A_T(\mathbf{u}) & 0 \\ B^T & 0 & 0 \end{bmatrix} \begin{bmatrix} \mathbf{u} \\ T \\ p \end{bmatrix} = \begin{bmatrix} \mathbf{f}_u \\ f_T \\ 0 \end{bmatrix}. \qquad (50)$$

This generalization of (9) can be solved using a global or local MPSC algorithm.

7.2 Global MPSC Algorithm

In the case of unsteady buoyancy-driven flows, the equations of the Boussinesq model (50) can be solved in a segregated manner. A discrete projection method for the Navier-Stokes equations can be combined with an algebraic flux correction scheme for the temperature equation using outer iterations to update the unknown coefficients. The decoupled solution of the two subproblems makes it possible to develop software in a modular way making use of optimized multigrid solvers. Moreover, the time step can be chosen individually for each subproblem.

In the simplest implementation, one outer iteration per time step is performed. Given the velocity $\mathbf{u}^n$, temperature T^n, and pressure p^n at the time level t^n, the following fractional-step algorithm is used to advance the solution in time [87]:

1. Solve the viscous Burgers equation

$$A_u(\tilde{\mathbf{u}})\tilde{\mathbf{u}} = \mathbf{f}_u - \Delta t M_T T^n - \Delta t B p^n.$$

2. Solve the Pressure-Poisson equation

$$B^T M_L^{-1} B q = \frac{1}{\Delta t} B^T \tilde{\mathbf{u}}.$$

3. Correct the velocity and pressure

$$\mathbf{u}^{n+1} = \tilde{\mathbf{u}} - \Delta t M_L^{-1} B q,$$

$$p^{n+1} = p^n + q.$$

4. Solve the temperature equation

$$A_T(\mathbf{u}^{n+1}, T^{n+1}) T^{n+1} = f_T.$$

Since the matrix $A_u(\tilde{\mathbf{u}})$ depends on the unknown solution $\tilde{\mathbf{u}}$ to the discrete momentum equation, the system is nonlinear. We solve it using iterative defect correction or a Newton-like method. The discrete problem associated with the temperature equation is also nonlinear if algebraic flux correction is performed. Nonlinear solvers and convergence acceleration techniques for such systems are discussed in [38].

7.3 Local MPSC Algorithm

A generalization of the local MPSC approach can be used in situations when the above fractional-step algorithm proves insufficiently robust. The local problems are formulated using a restriction of the approximate Jacobian matrix associated with the nonlinear system (50). The structure of this matrix is as follows [64, 73]:

$$J(\sigma, \mathbf{u}^{(l)}) = \begin{bmatrix} A_u(\mathbf{u}^{(l)}) + \sigma R(\mathbf{u}^{(l)}) & \Delta t M_T & \Delta t B \\ \sigma R(T^{(l)}) & A_T(\mathbf{u}^{(l)}) & 0 \\ B^T & 0 & 0 \end{bmatrix}. \qquad (51)$$

The nonlinearity of the convective term gives rise to the 'reactive' part R which represents a solution-dependent mass matrix and may cause severe convergence problems. For this reason, we multiply R by an adjustable parameter σ. The choice $\sigma = 1$ corresponds to Newton's method. Setting $\sigma = 0$, one obtains the fixed-point defect correction scheme. In either case, the linearized problem is solved using a fully coupled multigrid solver equipped with a local MPSC smoother of 'Vanka' type [64]. The global matrix $J(\sigma, \mathbf{u}^{(l)})$ is decomposed into small blocks

$$J_i = P_i J P_i^T$$

associated with patches of regular shape. The smoothing of the global defect vector is performed patchwise by solving the corresponding local subproblems.

The size of the local matrices can be further reduced by using the Schur complement approach. For simplicity, consider the case $\sigma = 0$ (an extension to $\sigma > 0$ is straightforward). Using (47) to eliminate the temperature in (45), we obtain

$$A_u \mathbf{u} = \mathbf{f}_u - \Delta t M_T A_T^{-1} f_T - \Delta t B p. \qquad (52)$$

Next, we use (52) to eliminate the velocity in the discretized continuity equation

$$B^T \mathbf{u} = B^T A_u^{-1} [\mathbf{f}_u - \Delta t M_T A_T^{-1} f_T - \Delta t B p] = 0. \qquad (53)$$

Thus, the pressure Schur complement equation associated with (50) reads

$$B^T A_u^{-1} B p = B^T A_u^{-1} \left[\frac{1}{\Delta t} \mathbf{f}_u - M_T A_T^{-1} f_T \right]. \qquad (54)$$

At the local subproblem level, the matrix J_i is replaced with the Schur complement preconditioner C_i that has the same size as in the case of the basic Navier-Stokes system. After solving the local PSC equation and updating the pressure, the velocity and temperature increments are calculated and added to the global vectors.

The local MPSC algorithm is more difficult to implement than the fractional-step method presented in Sect. 7.2. However, the coupled solution strategy has a number of attractive features. Above all, steady-state solutions can be obtained without resorting to pseudo-time stepping. In the case of unsteady flows at low Reynolds numbers, the strongly coupled treatment of local subproblems makes it possible to use large time steps without any loss of robustness. On the other hand, the convergence behavior of multigrid solvers with Newton-type linearization may turn out to be unsatisfactory, and the computational cost per outer iteration is rather high compared to the global MPSC algorithm. The performance of both solution techniques is illustrated by the numerical study for the MIT benchmark problem [87].

8 Case Study: Turbulent Flows

Turbulence plays an important role in many incompressible flow problems. Since direct numerical simulation (DNS) of turbulent flows is unaffordable for Reynolds numbers of practical interest, eddy viscosity models based on the Reynolds Averaged Navier-Stokes (RANS) equations are commonly employed in CFD codes.

This section describes a numerical implementation of the k–ε model that has been in use since the 1970s. To model the effect of unresolved velocity fluctuations, the viscous part of the Navier-Stokes equations is replaced with

$$\nabla \cdot (\nu + \nu_T)[\nabla \mathbf{u} + (\nabla \mathbf{u})^T],$$

where ν_T is the turbulent eddy viscosity. In the standard k–ε model [50], ν_T depends on the turbulent kinetic energy k and its dissipation rate ε as follows:

$$\nu_T = C_\mu \frac{k^2}{\varepsilon}, \quad C_\mu = 0.09.$$

The evolution of k and ε is governed by the convection-diffusion-reaction equations

$$\frac{\partial k}{\partial t} + \nabla \cdot \left(\mathbf{u} k - \frac{\nu_T}{\sigma_k} \nabla k \right) = P_k - \varepsilon, \tag{55}$$

$$\frac{\partial \varepsilon}{\partial t} + \nabla \cdot \left(\mathbf{u} \varepsilon - \frac{\nu_T}{\sigma_\varepsilon} \nabla \varepsilon \right) = \frac{\varepsilon}{k}(C_1 P_k - C_2 \varepsilon), \tag{56}$$

where $P_k = \frac{\nu_T}{2}|\nabla \mathbf{u} + \nabla \mathbf{u}^T|^2$ is responsible for the production of k. The involved empirical constants are given by $C_1 = 1.44$, $C_2 = 1.92$, $\sigma_k = 1.0$, $\sigma_\varepsilon = 1.3$.

The above equations are nonlinear and strongly coupled, which makes them very sensitive to the choice of numerical algorithms. In particular, the discretization procedure must be positivity-preserving because negative values of the eddy viscosity would produce numerical instabilities and eventually result in a crash of the code.

8.1 Positivity-Preserving Linearization

In our implementation of k–ε model, the incompressible Navier-Stokes equations are discretized using the nonconforming $\tilde{Q}_1/Q_0$ element pair. Standard Q_1 elements are employed for k and ε. The discretization of (59)–(60) yields [41, 42]

$$A_k(\mathbf{u}, \nu_T)k = f_k, \tag{57}$$

$$A_\varepsilon(\mathbf{u}, \nu_T)\varepsilon = f_\varepsilon. \tag{58}$$

The use of algebraic flux correction for the convective terms is not sufficient for positivity preservation. Indeed, nonphysical negative values can also be produced by the right-hand sides f_k and f_ε. As shown by Patankar [58], a *negative slope linearization* of sink terms is required to maintain positivity.

To write the equations of the k–ε model in the desired form, we introduce

$$\gamma = \frac{\varepsilon}{k}.$$

The negative slope linearization of (59)–(60) is based on the representation [47]

$$\frac{\partial k}{\partial t} + \nabla \cdot \left(\mathbf{u} k - \frac{\nu_T}{\sigma_k} \nabla k \right) + \gamma k = P_k, \tag{59}$$

$$\frac{\partial \varepsilon}{\partial t} + \nabla \cdot \left(\mathbf{u} \varepsilon - \frac{\nu_T}{\sigma_\varepsilon} \nabla \varepsilon \right) + C_2 \gamma \varepsilon = \gamma C_1 P_k, \tag{60}$$

where ν_T and γ are evaluated using the solution from the last outer iteration [42].

After solving the linearized equations (59) and (60), the new values $k^{(l)}$ and $\varepsilon^{(l)}$ are used to calculate the linearization parameter $\gamma^{(l)}$ for the next outer iteration, if any. The associated eddy viscosity ν_T is bounded below by a certain fraction of the laminar viscosity $0 < \nu_{\min} \leq \nu$ and above by $\nu_{\max} = l_{\max} \sqrt{k}$, where $l_{\max}$ is the maximum admissible mixing length (the size of the largest eddies, e.g., the width of the domain). In our implementation, the limited mixing length

$$l_* = \begin{cases} C_\mu \frac{k^{3/2}}{\varepsilon}, & \text{if } C_\mu k^{3/2} < \varepsilon l_{\max}, \\ l_{\max}, & \text{otherwise} \end{cases} \tag{61}$$

is used to calculate the turbulent eddy viscosity by the formula

$$\nu_T = \max\{\nu_{\min}, l_* \sqrt{k}\}. \tag{62}$$

The corresponding linearization parameter γ is given by

$$\gamma = C_\mu \frac{k}{\nu_T}. \tag{63}$$

The above representation makes it possible to avoid division by zero and obtain bounded nonnegative coefficients without manipulating the values of k and ε.

8.2 Initial Conditions

It is not always easy to find reasonable initial values for the k–ε model. If the velocity is initialized by zero, it takes the flow some time to become turbulent. Therefore, we use a constant eddy viscosity ν_0 during a startup phase that ends at a certain time $t_* > 0$. The values to be assigned to k and ε at $t = t_*$ depend on the choice of ν_0 and on the mixing length $l_0 \in [l_{\min}, l_{\max}]$, where the threshold parameter $l_{\min}$ is related to the size of the smallest admissible eddies. Given v_0 and l_0, we define

$$k_0 = \left(\frac{v_0}{l_0}\right)^2, \qquad \varepsilon_0 = C_\mu \frac{k_0^{3/2}}{l_0}. \tag{64}$$

Alternatively, the initial values of k and ε can be estimated with a zero-equation turbulence model or defined using an extension of the boundary conditions.

8.3 Boundary Conditions

The k–ε model is very sensitive to the choice and numerical implementation of boundary conditions. In particular, an improper near-wall treatment can render the algorithm useless. The right choice of inflow values is also important. For this reason, we discuss the imposition of boundary conditions in some detail.

At the inflow boundary Γ_{in}, the values of all variables are commonly prescribed:

$$\mathbf{u} = \mathbf{g}, \quad k = c_\infty |\mathbf{u}|^2, \quad \varepsilon = C_\mu \frac{k^{3/2}}{l_0} \quad \text{on } \Gamma_{\text{in}}, \tag{65}$$

where $c_\infty \in [0.003, 0.01]$ and $|\mathbf{u}|$ stands for the magnitude of the velocity vector.

At the outlet Γ_{out}, the normal derivatives of all variables are set equal to zero

$$\mathbf{n} \cdot [\nabla \mathbf{u} + \nabla \mathbf{u}^T] = \mathbf{0}, \quad \mathbf{n} \cdot \nabla k = 0, \quad \mathbf{n} \cdot \nabla \varepsilon = 0 \quad \text{on } \Gamma_{\text{out}}. \tag{66}$$

In the context of finite element methods, the normal derivatives appear in the surface integrals that result from integration by parts in the variational form of the governing equations. These integrals do not need to be assembled if homogeneous Neumann ("do-nothing") boundary conditions of the form (66) are prescribed.

On a fixed solid wall Γ_w, the velocity must satisfy the no-penetration condition

$$\mathbf{n} \cdot \mathbf{u} = 0 \quad \text{on } \Gamma_w. \tag{67}$$

In laminar flow models, the tangential velocity is also set equal to zero, so that the *no-slip condition* $\mathbf{u} = \mathbf{0}$ holds on Γ_w. To avoid the need for resolving the viscous boundary layer in turbulent flow simulations, the boundary condition for the tangential direction is frequently given in terms of the wall shear stress

$$\mathbf{t}_w = \mathbf{n} \cdot \sigma - (\mathbf{n} \cdot \sigma \cdot \mathbf{n})\mathbf{n}, \quad \sigma = \nu[\nabla \mathbf{u} + \nabla \mathbf{u}^T]. \tag{68}$$

If $\mathbf{t}_w$ is prescribed on Γ_w, then (67) is called the *free slip condition* because the tangential velocity is defined implicitly and its value is generally unknown.

The practical implementation of the free-slip condition is nontrivial, unless the boundary of the domain is aligned with the axes of the Cartesian coordinate system. In contrast to the no-slip condition, (67) constrains a linear combination of several velocity components whose boundary values are unknown. Therefore, standard implementation techniques do not work. The free-slip condition can be implemented using element-by-element transformations to a local coordinate system aligned with the wall [17]. However, this strategy requires substantial modifications of the code. In our current implementation, we drive the normal velocities to zero in an iterative way using projections of the form $\mathbf{u} := \mathbf{u} - (\mathbf{n} \cdot \mathbf{u})\mathbf{n}$ [41]. Other implementation techniques are discussed in [40] in the context of compressible flow problems.

8.4 Wall Functions

To complete the problem statement, we still have to prescribe the tangential stress $\mathbf{t}_w$, as well as the boundary conditions for k and ε on the wall Γ_w. Note that the equations of the standard k–ε model are invalid in the near-wall region, where the Reynolds number is rather low and viscous effects are dominant. To bridge the gap between the no-slip boundaries and the region of turbulent flow, analytical solutions to the boundary layer equations are frequently used to determine the values of $\mathbf{t}_w$, k, and ε near the wall. The use of logarithmic wall laws leads to the following set of boundary conditions to be prescribed at a small distance y from the wall Γ_w

$$\mathbf{t}_w = -u_\tau^2 \frac{\mathbf{u}}{|\mathbf{u}|}, \qquad k = \frac{u_\tau^2}{\sqrt{C_\mu}}, \qquad \varepsilon = \frac{u_\tau^3}{\kappa y}, \tag{69}$$

where $\kappa = 0.41$ is the von Kármán constant. The friction velocity u_τ is given by

$$\frac{|\mathbf{u}|}{u_\tau} = \frac{1}{\kappa} \log y^+ + \beta, \qquad y^+ = \frac{u_\tau y}{\nu}. \tag{70}$$

The value of the parameter β depends on the wall roughness ($\beta = 5.2$ for smooth walls). The above logarithmic relationship is valid for $11.06 \leq y^+ \leq 300$.

The use of wall functions implies that a thin boundary layer of width y is removed, and the equations of the k–ε model should be solved in the reduced domain. Since the local Reynolds number y^+ is proportional to y, the wall distance should be chosen carefully. It is common to apply the wall laws (69) at the first internal node or integration point. However, the so-defined y depends on the mesh size and may fall into the viscous sublayer where (70) is invalid.

Another possibility is to adapt the mesh so that the location of boundary nodes always corresponds to a fixed value of y^+ which should be as small as possible for accuracy reasons. Taking the smallest value for which the logarithmic law still holds, one can neglect the width of the removed boundary layer and avoid mesh adaptation [25, 47]. In this case, the nodes located on the wall Γ_w should be treated as if they were shifted by the distance $y = \frac{y^+ \nu}{u_\tau}$ in the normal direction.

As explained in [25], the smallest wall distance for the definition of y^+ corresponds to the point where the logarithmic layer meets the viscous sublayer. At this point, the linear relation $y^+ = \frac{|\mathbf{u}|}{u_\tau}$ and the logarithmic law (70) must hold, whence

$$y^+ = \frac{1}{\kappa} \log y^+ + \beta. \tag{71}$$

This nonlinear equation can be solved iteratively. The resulting value of the parameter y^+ (for the default settings $\kappa = 0.41$, $\beta = 5.2$) is given by $y_*^+ \approx 11.06$.

The relationship between y_*^+ and the friction velocity u_τ becomes very simple:

$$u_\tau = \frac{|\mathbf{u}|}{y_*^+}. \tag{72}$$

On the other hand, the wall boundary condition for k implies that

$$u_\tau = C_\mu^{0.25} \sqrt{k}. \tag{73}$$

Following Grotjans and Menter [25], we use a combination of the above to define

$$\mathbf{t}_w = -\frac{u_\tau}{y_*^+} \mathbf{u}, \quad u_\tau = \max\left\{ C_\mu^{0.25} \sqrt{k}, \frac{|\mathbf{u}|}{y_*^+} \right\}. \tag{74}$$

This definition of $\mathbf{t}_w$ is consistent with (69) and prevents the momentum flux from going to zero at separation/stagnation points [25]. The natural boundary condition for the wall shear stress is used to evaluate the surface integral

$$\int_{\Gamma_w} \mathbf{t}_w \cdot \mathbf{w} \, ds = -\int_{\Gamma_w} \frac{u_\tau}{y_*^+} \mathbf{u} \cdot \mathbf{w} \, ds, \tag{75}$$

where $\mathbf{w}$ is the test function for the Galerkin weak form of the momentum equation.

By (69), the wall function for the turbulent eddy viscosity ν_T is given by

$$\nu_T = C_\mu \frac{k^2}{\varepsilon} = \kappa u_\tau y = \kappa y_*^+ \nu. \tag{76}$$

This relation is satisfied automatically if the wall functions for k and ε are implemented in the strong sense. However, the use of Dirichlet boundary conditions implies that the values of k and ε depend on $\mathbf{u}$ via the friction velocity $u_\tau = \frac{|\mathbf{u}|}{y_*^+}$ but there is no feedback. The result is an unrealistic one-way coupling.

To release the boundary values of k and ε and let them influence the tangential velocity via (74)–(75), the wall functions must be implemented in a weak sense. Differentiating (69), one obtains the Neumann boundary conditions [25]

$$\mathbf{n} \cdot \nabla k = -\frac{\partial k}{\partial y} = 0,$$

$$\mathbf{n} \cdot \nabla \varepsilon = -\frac{\partial \varepsilon}{\partial y} = \frac{u_\tau^3}{\kappa y^2} = \frac{\varepsilon}{y}. \tag{77}$$

Algebraic Flux Correction III

The unknown wall distance y can be expressed in terms of the turbulent eddy viscosity $\nu_T = \kappa u_\tau y$, which yields a natural boundary condition of Robin type

$$\mathbf{n} \cdot \nabla \varepsilon = \frac{\kappa u_\tau}{\nu_T} \varepsilon, \quad u_\tau = C_\mu^{0.25} \sqrt{k}. \tag{78}$$

The surface integrals associated with the Neumann boundary condition are given by

$$\int_{\Gamma_w} \frac{\nu_T}{\sigma_k} (\mathbf{n} \cdot \nabla k) w \, ds = 0, \tag{79}$$

$$\int_{\Gamma_w} \frac{\nu_T}{\sigma_\varepsilon} (\mathbf{n} \cdot \nabla \varepsilon) w \, ds = \int_{\Gamma_w} \frac{\kappa u_\tau}{\sigma_\varepsilon} \varepsilon w \, ds. \tag{80}$$

Alternatively, the strong form of the wall law $\varepsilon = \frac{u_\tau^3}{\kappa y} = \frac{u_\tau^4}{\kappa y_*^+ \nu}$ can be used to prescribe a Dirichlet boundary condition for ε or evaluate the right-hand side of (80).

If the wall functions for ε and/or k are prescribed in a weak sense, it is essential to calculate ν_T and P_k using the strong form of the wall law. That is, the correct value of the turbulent eddy viscosity is given by (76), while the production term

$$P_k = \frac{u_\tau^3}{\kappa y} = \frac{u_\tau^4}{\kappa y_*^+ \nu} \tag{81}$$

is in equilibrium with the dissipation rate. The friction velocity u_τ is defined by (74).

8.5 Chien's Low-Re k–ε Model

Logarithmic laws provide a reasonably accurate description model of the flow in the near-wall region avoiding the need for costly integration to the wall. The derivation is only valid for flat-plate boundary layers and developed flow conditions but wall functions of the form (69) are frequently used in more general settings with considerable success. An obvious drawback to this approach is the assumption that the viscous sublayer is very thin. Clearly, it is no longer safe to apply the wall functions on Γ_w if the wall distance associated with the constant y_*^+ becomes too large.

A robust, albeit costly, alternative to wall laws is the use of *damping functions* that provide a smooth transition from laminar to turbulent flow. In Chien's low-Reynolds number k–ε model [7], the turbulent eddy viscosity is redefined thus:

$$\nu_T = C_\mu f_\mu \frac{k^2}{\tilde{\varepsilon}}, \quad f_\mu = 1 - \exp(-0.0115 y^+), \tag{82}$$

$$\tilde{\varepsilon} = \varepsilon - 2\nu \frac{k}{y^2}. \tag{83}$$

This popular model is supported by the DNS results which indicate that the ratio $f_\mu = \frac{\nu_T \tilde{\varepsilon}}{C_\mu k^2}$ is not a constant but a function approaching zero at the wall.

The following modification of (59)–(60) is used in Chien's model [7]

$$\frac{\partial k}{\partial t} + \nabla \cdot \left(\mathbf{u}k - \frac{\nu_T}{\sigma_k}\nabla k\right) + \alpha k = P_k, \tag{84}$$

$$\frac{\partial \tilde{\varepsilon}}{\partial t} + \nabla \cdot \left(\mathbf{u}\tilde{\varepsilon} - \frac{\nu_T}{\sigma_\varepsilon}\nabla \tilde{\varepsilon}\right) + \beta \tilde{\varepsilon} = \gamma C_1 f_1 P_k, \tag{85}$$

where the coefficients and damping functions are given by

$$\alpha = \gamma + \frac{2\nu}{y^2}, \qquad \beta = C_2 f_2 \gamma + \frac{2\nu}{y^2}\exp(-0.5y^+),$$
$$\gamma = \frac{\tilde{\varepsilon}}{k}, \qquad f_1 = 1, \qquad f_2 = 1 - 0.22\exp\left(\frac{k^2}{6\nu\tilde{\varepsilon}}\right)^2. \tag{86}$$

In contrast to wall functions, the boundary conditions on Γ_w are very simple:

$$\mathbf{u} = 0, \quad k = 0, \quad \tilde{\varepsilon} = 0 \quad \text{on } \Gamma_w. \tag{87}$$

Note that the sink terms in (84) and (85) have positive coefficients, as required by Patankar's rule [58]. The value of y^+ is a function of the friction velocity:

$$y^+ = \frac{u_\tau y}{\nu}, \quad u_\tau = \max\{C_\mu^{0.25}\sqrt{k}, \sqrt{|\mathbf{t}_w|}\}. \tag{88}$$

The wall shear stress $\mathbf{t}_w$ is calculated using (68). Note that the computation of y^+ requires knowing the wall distance y. In the current implementation, we calculate it using a brute-force approach. More efficient techniques for computing distance functions can be found in the literature on level set methods (see Sect. 10).

8.6 Numerical Examples

To verify the above implementation the k–ε model, we perform a numerical study for two test problems. The first one is used to validate the code for Chien's low-Reynolds number k–ε (LRKE) model. In the second example, we use the LRKE solution to evaluate the results obtained with logarithmic wall functions implemented as Dirichlet (DIRBC) and Neumann (NEUBC) boundary conditions.

8.6.1 Channel Flow Problem

In the first example, we simulate the turbulent channel flow at $\text{Re}_\tau = 395$ based on the friction velocity u_τ, half of the channel width d, and kinematic viscosity ν. The reference data for this well-known benchmark problem are provided by the DNS results of Kim et al. [36]. In order to obtain the developed flow conditions required

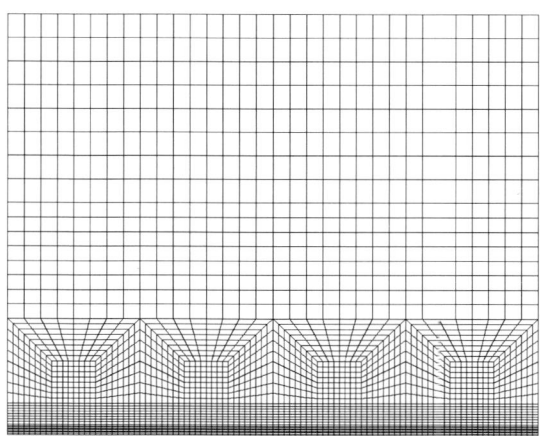

Fig. 3 Channel flow: local mesh refinement in the boundary layer

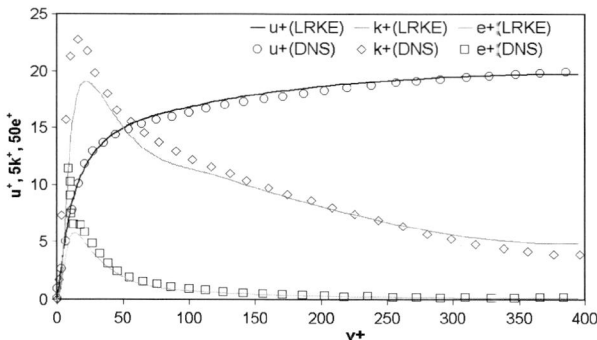

Fig. 4 Channel flow: LRKE solutions vs. Kim's DNS results for $Re_\tau = 395$

for validation, the inflow and outflow boundary conditions for the reduced domain were swapped repeatedly so as to emulate periodic boundary conditions.

The equations of the LRKE model are solved with the 3D code on a hexahedral mesh of 50,000 elements. Due to the need for high resolution, local mesh refinement is performed in the near-wall region, as shown in Fig. 3. The distance from the wall boundary to the nearest interior point corresponds to $y^+ \approx 2$. The numerical results for this test are presented in Fig. 4. The profiles of the nondimensional quantities

$$u^+ = \frac{u_x}{u_\tau}, \qquad k^+ = \frac{k}{u_\tau^2}, \qquad \varepsilon^+ = \frac{\varepsilon \nu}{u_\tau^4}$$

are in a good agreement with the DNS results [36] for this benchmark. The calculated profiles of u^+ and ε^+ are particularly close to the reference data.

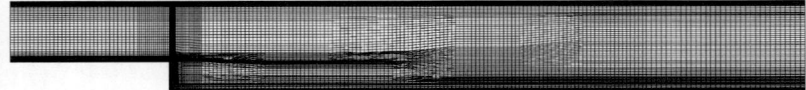

Fig. 5 Backward facing step: a 2D view of the computational mesh in the xy-plane

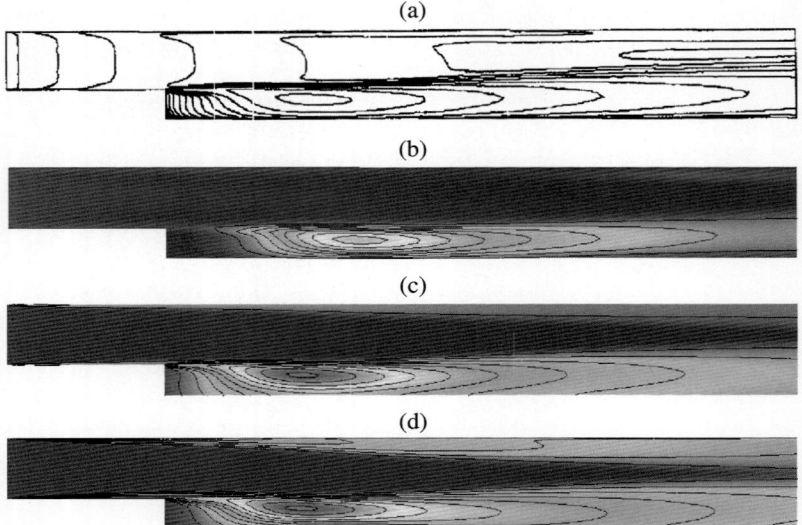

Fig. 6 Backward facing step: steady-state distribution of k for $Re = 47,625$. (**a**) reference solution [33], (**b**) DIRBC solution, (**c**) NEUBC solution, (**d**) LRKE solution

8.7 Backward Facing Step

In the second example, we simulate the turbulent flow past a backward facing step in 3D. The definition of the Reynolds number $Re = 47,625$ is based on the step height H, mean inflow velocity u_{mean}, and kinematic viscosity ν. The objective is to evaluate the performance of the k–ε model with three different kinds of near-wall treatment: LRKE vs. DIRBC and NEUBC implementation of wall functions.

All simulations are performed on the same mesh that consists of approximately 260,000 hexahedral elements. Local mesh refinement is performed in the near-wall region and behind the step (see Fig. 5). A comparison of the steady-state solutions for the turbulent kinetic energy k and eddy viscosity ν_T with the reference solution from [33] is presented in Figs. 6 and 7. Significant differences between the solutions computed using the strong and weak form of logarithmic wall functions are observed even in the "eyeball norm." DIRBC was found to produce disappointing results, whereas the accuracy of the NEUBC solution is similar to LRKE.

An important evaluation criterion for this popular test problem is the recirculation length defined as $L_R = x_r/H$. For the implementation based on wall functions

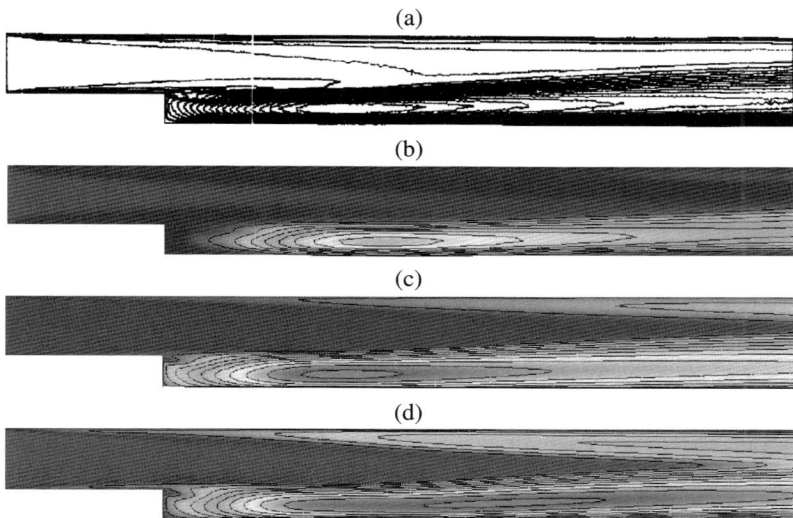

Fig. 7 Backward facing step: steady-state distribution of ν_T for $Re = 47{,}625$. (**a**) reference solution [33], (**b**) DIRBC solution, (**c**) NEUBC solution, (**d**) LRKE solution

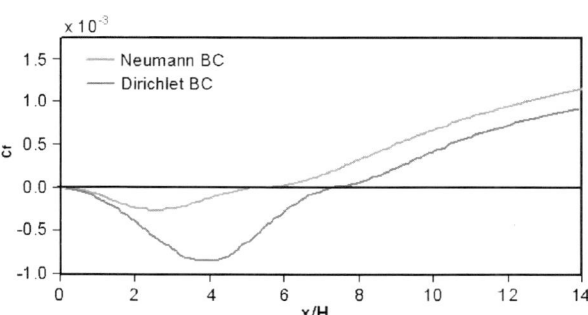

Fig. 8 Backward facing step: distribution of c_f along the lower wall, $Re = 47{,}625$

implemented as Dirichlet boundary conditions, this integral quantity can be readily inferred from the distribution of the skin friction coefficient

$$c_f = \frac{u_\tau^2}{u_{\text{mean}}^2} \frac{u_x}{|u_x|}$$

on the bottom wall (see Fig. 8). The recirculation length predicted by LRKE and NEUBC is underestimated ($L_R \approx 5.4$). The computational results published in the literature exhibit the same trend ($5.0 < L_R < 6.5$, see [25, 33, 74]). On the other hand, the implementation of wall functions in the strong sense yields $L_R \approx 7.1$, which matches the experimentally measured recirculation length ($L_R \approx 7.1$, see [35]). Unfortunately, this perfect agreement turns out to be a pure coincidence.

In Fig. 9, the calculated velocity profiles for 6 different distances from the step are compared to one another and to the experimental data from Kim's thesis [35].

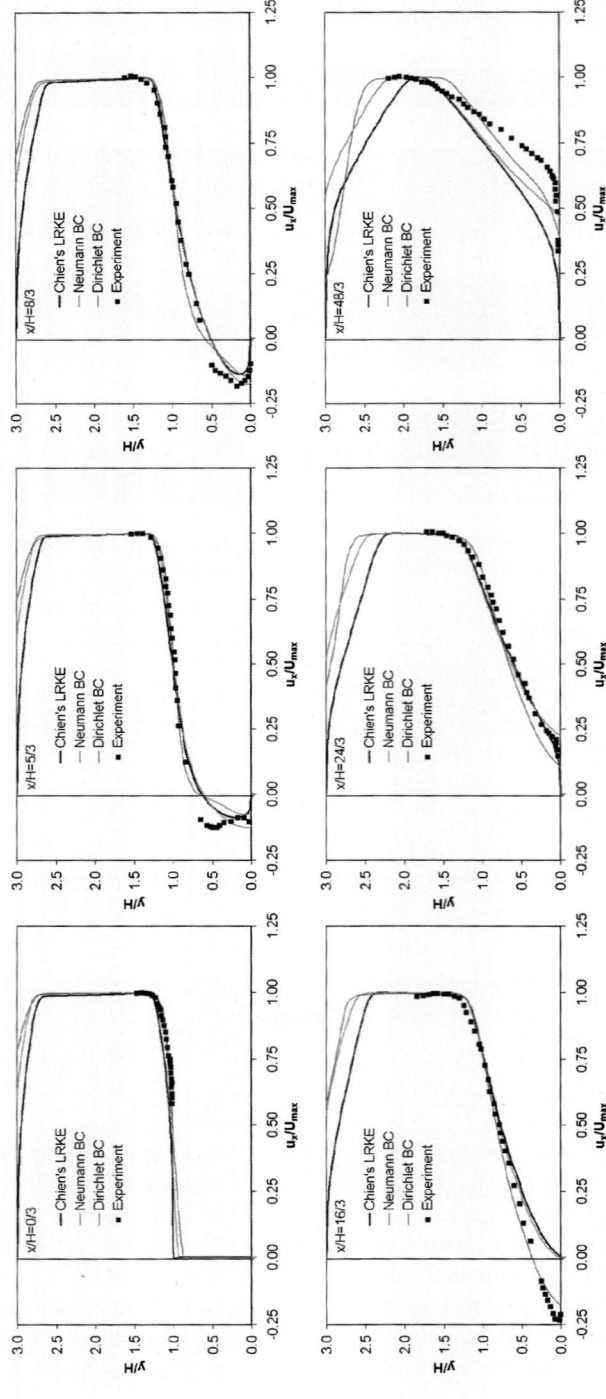

Fig. 9 Backward-facing step: profiles of u_x for 6 different distances x/H from the step

The corresponding profiles of k and ε are displayed in Fig. 10 and Fig. 11, respectively. This comparative study indicates that NEUBC yields essentially the same results as Chien's low-Reynolds number model, whereas the use of DIRBC leads to a significant discrepancy, especially at small distances from the step. It is also worth mentioning that the presented profiles of ε do not suffer from spurious undershoots which are frequently observed in other computations. This can be attributed to the positivity-preserving treatment of the convective terms and sinks in our algorithm.

9 Case Study: Population Balances

The hydrodynamic behavior of a polydisperse two-phase flow can be described by a RANS model for the continuous phase coupled with a *population balance* model for the size distribution of the disperse phase (bubbles, drops, or particles). Population balance equations (PBEs) describe crystallization processes, liquid-liquid extraction, gas-liquid dispersions, and polymerization, to name just a few important applications. The implementation of PBE models in CFD software adds an extra dimension to the problem, which increases the complexity of the code and incurs exorbitant computational costs. For this reason, examples of RANS-PBE multiphase flow models have been rare. In addition to our own work [3] to be presented here, we mention the Multiple Size Group (MUSIG) model [48] implemented in the commercial code ANSYS CFX and the recent publications by John et al. [26, 34] who used algebraic flux correction of FCT type to enforce positivity preservation.

9.1 Mathematical Model

The PBE for gas-liquid or liquid-liquid flows is an integro-differential transport equation for a *probability density function* f that depends on certain internal properties of the disperse phase. In the case of polydisperse bubbly flows, the internal coordinate of primary interest is the volume υ of the bubble, and $f(\mathbf{x}, t, \upsilon)$ is the probability that a bubble of volume υ will occupy location $\mathbf{x}$ at time t. The number density N_{ab} and volume fraction α_{ab} of bubbles with $\upsilon \in [\upsilon_a, \upsilon_b]$ are given by

$$N_{ab}(\mathbf{x}, t) = \int_{\upsilon_a}^{\upsilon_b} f(\mathbf{x}, t, \upsilon) \, d\upsilon, \tag{89}$$

$$\alpha_{ab}(\mathbf{x}, t) = \int_{\upsilon_a}^{\upsilon_b} f(\mathbf{x}, t, \upsilon) \upsilon \, d\upsilon. \tag{90}$$

The changes in the bubble size distribution are caused by convection in the physical space and by bubble-bubble interactions (breakage and coalescence) that change the profile of f along the internal coordinate. Let $\mathbf{u}_g(\mathbf{x}, t, \upsilon)$ denote the average velocity of bubbles that may be defined by adding an empirical slip velocity $\mathbf{u}_{\text{slip}}(m)$

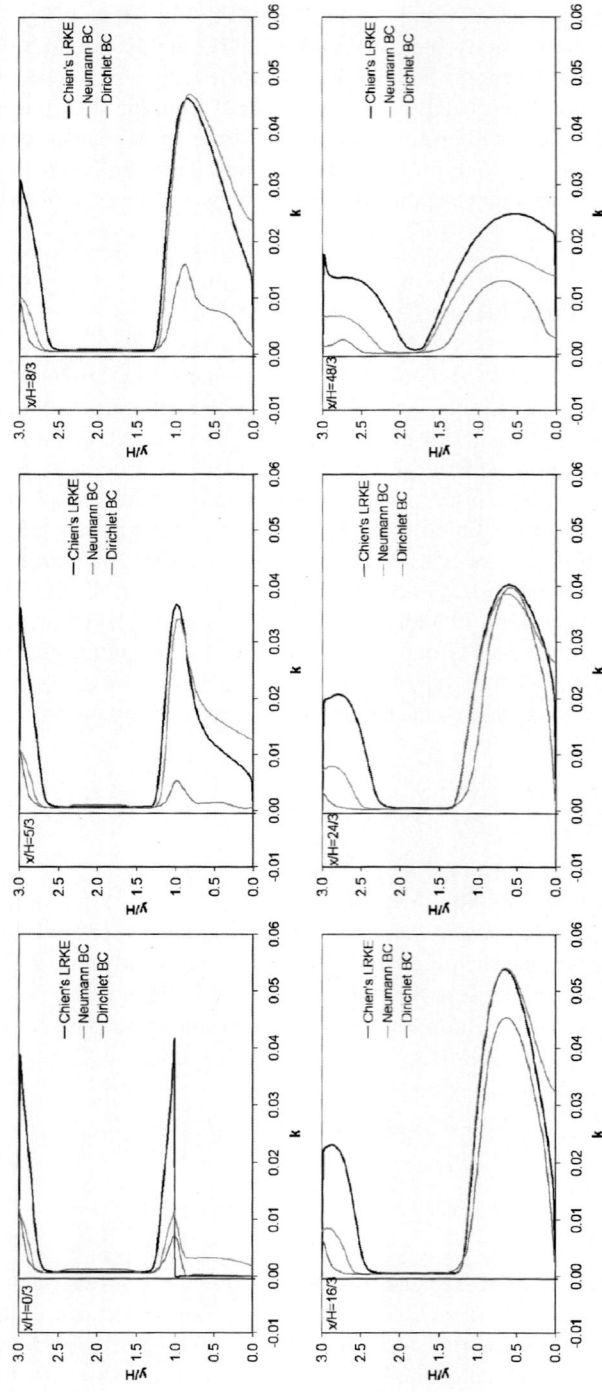

Fig. 10 Backward-facing step: profiles of k for 6 different distances x/H from the step

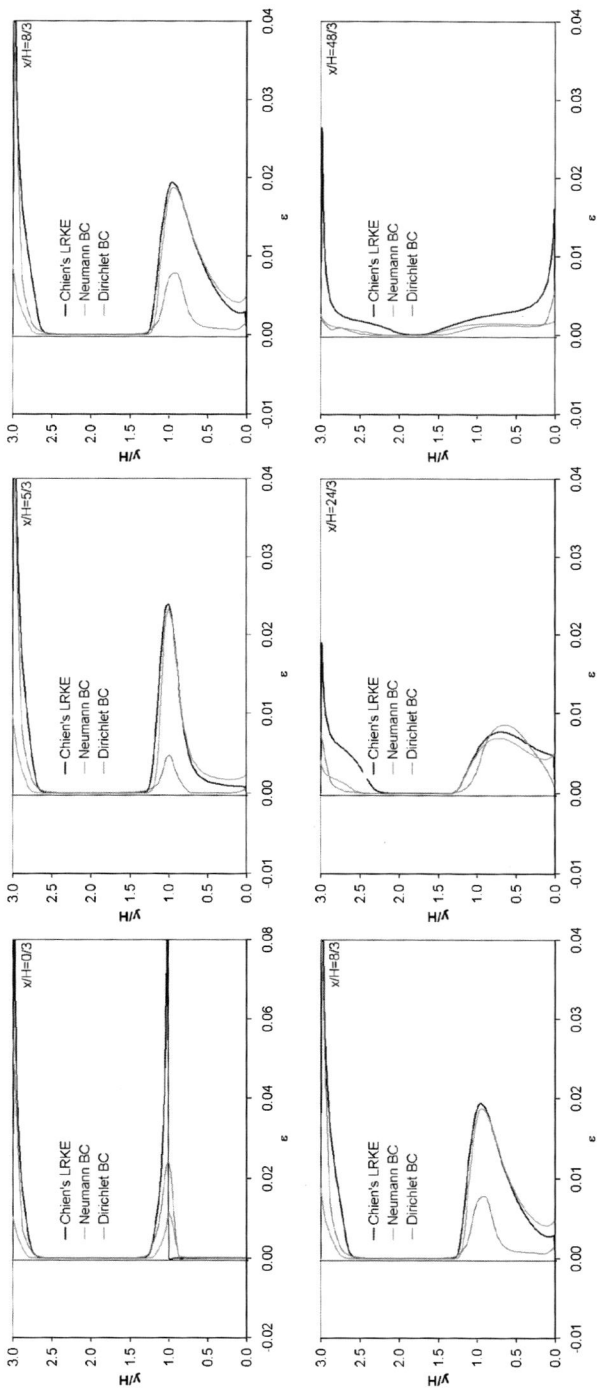

Fig. 11 Backward-facing step: profiles of ε for 6 different distances x/H from the step

to the solution $\mathbf{u}(\mathbf{x}, t)$ of the RANS model for the continuous phase. For simplicity, we assume that the slip velocity is constant, i.e., bubbles of all sizes are moving with the same velocity $\mathbf{u}_g$. The general form of population balance equation reads

$$\frac{\partial f}{\partial t} + \nabla \cdot \left(\mathbf{u}_g f - \frac{\nu_T}{\sigma_T} \nabla f \right) = B^+ + B^- + C^+ + C^-, \quad (91)$$

where ν_T is the turbulent eddy viscosity and σ_T is the turbulent Schmidt number.

The terms in the right-hand side of (91) describe the changes of f due to breakage (B) and coalescence (C) phenomena. The superscripts "+" and "−" are used to distinguish between sources and sinks. In this study, we use the models developed by Lehr et al. [44, 45] with some modifications proposed in [5]. Let r^B and r^C denote the *kernel functions* that describe the rates of breakage and coalescence, respectively. The modeling of $B^\pm$ and $C^\pm$ is based on the assumption that

- the probability that a parent bubble of volume υ will break up to form two daughter bubbles of volumes $\tilde{\upsilon}$ and $\upsilon - \tilde{\upsilon}$ is given by $r^B(\upsilon, \tilde{\upsilon}) f(\upsilon)$,
- the probability that two bubbles of volumes $\tilde{\upsilon}$ and $\upsilon - \tilde{\upsilon}$ will coalesce to form a bubble of volume υ is given by $r^C(\upsilon - \tilde{\upsilon}, \tilde{\upsilon}) f(\tilde{\upsilon}) f(\upsilon - \tilde{\upsilon})$.

Integrating the breakage and coalescence rates over all bubble sizes, one obtains

$$B^+ + B^- + C^+ + C^- = \int_\upsilon^\infty r^B(\upsilon, \tilde{\upsilon}) f(\tilde{\upsilon}) d\tilde{\upsilon} - \frac{f(\upsilon)}{\upsilon} \int_0^\upsilon \tilde{\upsilon} r^B(\tilde{\upsilon}, \upsilon) d\tilde{\upsilon}$$

$$+ \frac{1}{2} \int_0^\upsilon r^C(\tilde{\upsilon}, \upsilon - \tilde{\upsilon}) f(\tilde{\upsilon}) f(\upsilon - \tilde{\upsilon}) d\tilde{\upsilon}$$

$$- f(\upsilon) \int_0^\infty r^C(\tilde{\upsilon}, \upsilon) f(\tilde{\upsilon}) d\tilde{\upsilon}. \quad (92)$$

The model is closed by the choice of the kernel functions r^B and r^C, see [5, 44, 45].

9.2 Discretization of PBEs

In our algorithm [3], the population balance equation (92) is discretized using the *method of classes* which corresponds to a piecewise-constant approximation along the υ-coordinate. In the case of n classes, the *pivot volumes* are defined by

$$\upsilon_i = \upsilon_{\min} q^{i-1}, \quad i = 1, \ldots, n \quad (93)$$

where $\upsilon_{\min}$ is the volume of the smallest "resolved" class and q is a scaling factor.

The class width $\Delta \upsilon_i$ is defined as the length of the interval $[\upsilon_i^L, \upsilon_i^U]$, where [3]

$$\upsilon_i^U = \upsilon_i + \frac{1}{3}(\upsilon_{i+1} - \upsilon_i), \qquad \upsilon_i^L = \upsilon_i - \frac{2}{3}(\upsilon_i - \upsilon_{i-1}). \quad (94)$$

The method of classes transforms the integro-differential equation (92) into a system of n coupled transport equations for the class probability densities f_i

$$\frac{\partial f_i}{\partial t} + \nabla \cdot \left(\mathbf{u}_g f_i - \frac{\nu_T}{\sigma_T} \nabla f_i \right)$$

$$= \sum_{j=i}^{n} r^B_{i,j} f_j \Delta v_j - \frac{f_i}{v_i} \sum_{j=1}^{i} v_j r^B_{j,i} \Delta v_j + \frac{1}{2} \sum_{j=1}^{i} r^C_{j,k} f_j f_k \Delta v_j$$

$$- f_i \sum_{j=1}^{n} r^C_{j,i} f_j \Delta v_j, \quad i = 1, \ldots, n. \tag{95}$$

The number density and volume fraction of bubbles in the i-th class are given by

$$N_i = f_i \Delta v_i, \qquad \alpha_i = f_i v_i \Delta v_i = f_i N_i.$$

Multiplying (95) by $v_i \Delta v_i$, one obtains a system of transport equations for the class holdups α_i. This transformation leads to a conservative scheme such that the discretized source terms are balanced by the discretized sink terms, and the total holdup of the disperse phase is not affected by breakage or coalescence. We tacitly assume that the bubbles are incompressible so that the conservation of volume is equivalent to the conservation of mass. The number density is generally not conserved but the results of Buwa and Ranade [5] indicate that this inconsistency has hardly any influence on the specific interfacial area and the average bubble size.

The discretization of the bubble size distribution is conservative if a source in the equation for one class appears as a sink in the equation for another class. To verify this, consider a bubble of class i that breaks up into bubbles of classes j and k such that $v_i = v_j + v_k$. The increments to the three right-hand sides sum to zero:

$$i: \quad -\left(v_j r^B_{i,j} \Delta v_j \frac{f_i}{v_i} \right) v_i \Delta v_i - \left(v_k r^B_{i,k} \Delta v_k \frac{f_i}{v_i} \right) v_i \Delta v_i = -r^B_{i,j} \alpha_i \frac{v_j \Delta v_j}{v_i}$$

$$- r^B_{i,k} \alpha_i \frac{v_k \Delta v_k}{v_i},$$

$$j: \quad +\left(v_j r^B_{i,j} f_i \Delta v_i \right) v_j \Delta v_j = r^B_{i,j} \alpha_i \frac{v_j \Delta v_j}{v_i},$$

$$k: \quad +\left(v_k r^B_{i,k} f_i \Delta v_i \right) v_k \Delta v_k = r^B_{i,k} \alpha_i \frac{v_k \Delta v_k}{v_i}.$$

Next, suppose that bubbles of the j-th and k-th class coalesce to form a bubble of class i. The gains and losses in the three classes are as follows:

$$i: \quad +\frac{1}{2} \left(r^C_{j,k} f_j f_k \Delta v_j + r^C_{k,j} f_k f_j \Delta v_k \right) v_i \Delta v_i,$$

$$j: \quad -\left(f_j r^C_{j,k} f_k \Delta v_k \right) v_j \Delta v_j = -r^C_{j,k} \alpha_j f_k \Delta v_k,$$

$$k: \quad -\left(f_k r^C_{k,j} f_j \Delta v_j \right) v_k \Delta v_k = -r^C_{k,j} \alpha_k f_j \Delta v_j.$$

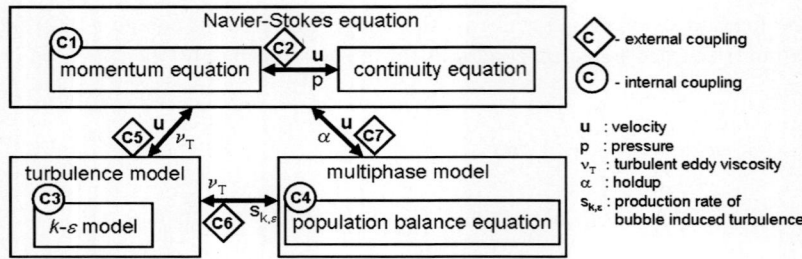

Fig. 12 Coupling of PBE with the turbulent flow model for the continuous phase

Suppose that all classes have the same width, that is, $\Delta v_i = \Delta v_j = \Delta v_k$. Using the fact that $v_i = v_j + v_k$, we obtain the following relationship

$$\frac{1}{2}\bigl(r^C_{j,k} f_j f_k \Delta v_j + r^C_{k,j} f_k f_j \Delta v_k\bigr)(v_j + v_k)\Delta v_i = r^C_{j,k}\alpha_j f_k \Delta v_k + r^C_{k,j}\alpha_k f_j \Delta v_j$$

which proves that the source and sink terms due to coalescence are also balanced.

In our implementation, the discretization of the internal coordinate is performed using nonuniform grids. To maintain the conservation of volume under coalescence, we calculate the sinks for every possible pair of classes and add their absolute values to the equation for the class that contains the emerging bubble. By this definition, the sources and sinks sum to zero, so that the total volume remains unchanged.

9.3 Integration of PBE in CFD Codes

The implementation of PBE in an existing CFD code calls for a block-iterative solution strategy. The diagram in Fig. 12 illustrates the coupling effects that arise when a PBE model is combined with the algorithm described in Sect. 8. In addition to the internal couplings within the Navier-Stokes system (C1 and C2), the k–ε model (C3), and the PBE transport equations (C4), the two-way couplings between these blocks must be taken into account (C5-C7). To reduce the computational cost, we currently neglect the influence of the disperse phase on the continuous phase and make a number of other simplifying assumptions (see below). The one-way coupling is a good approximation for flows driven by pressure and/or shear-induced turbulence. The numerical treatment of buoyancy-driven bubbly flows was addressed in [43] in the context of a drift-flux model with a two-way interphase coupling.

9.4 Numerical Examples

To our knowledge, there is no standard benchmark problem for population balance models coupled with the fluid dynamics of turbulent two-phase flows. In this section,

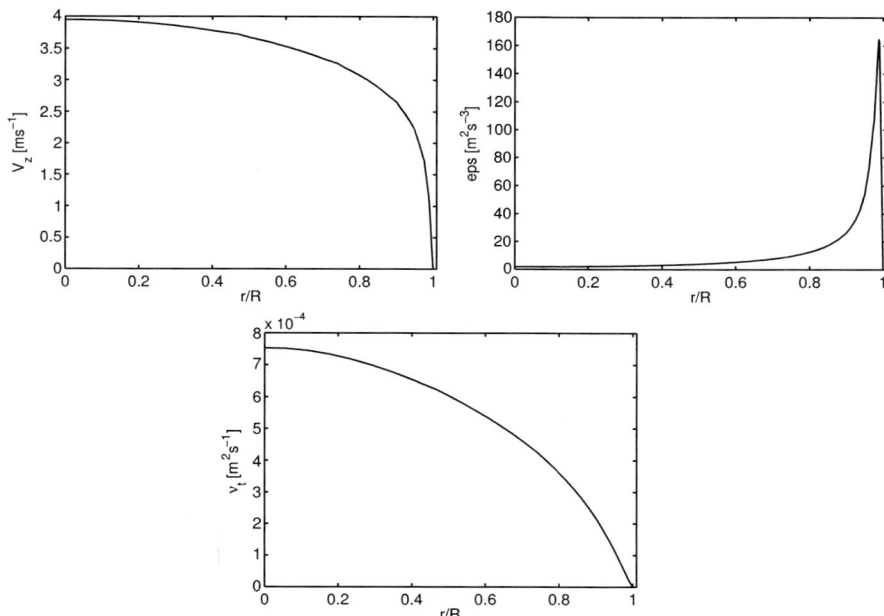

Fig. 13 Turbulent pipe flow: radial profiles of the axial velocity (*left*), turbulent dissipation rate (*middle*), and turbulent viscosity (*right*)

we study the influence of turbulence on the bubble size distribution in a turbulent 3D pipe flow. The main quantity of interest is the Sauter mean diameter d_{32} defined as the diameter of the sphere that has the same volume/surface area ratio as the entire ensemble. To show the potential of the CFD-PBE model in the context of an industrial application, we simulate the flow through a Sulzer static mixer SMV^{TM}. The results are compared to experimental data provided by Sulzer Chemtech Ltd.

9.4.1 Turbulent Pipe Flow

Turbulent pipe flow is well suited for testing population balance models with one spatial and one internal coordinate [27]. The preliminary validation of our algorithm was performed on a 3D version of this problem [3]. The continuous phase is water flowing through a 1 m long pipe of diameter $d = 3.8$ cm. The incompressible fluid that constitutes the droplets of the disperse phase has similar physical properties (density and viscosity). Due to this assumption, the interphase slip and buoyancy effects are neglected. That is, both phases are assumed to move with the mixture velocity which is calculated using the k–ε turbulence model. The Reynolds number for this simulation is $Re = \frac{dw}{v} = 114{,}000$, where w stands for the bulk velocity. The computational mesh is generated using a 2D to 3D extrusion of the mesh for the circular cross section. Each layer consists of 1,344 hexahedral elements.

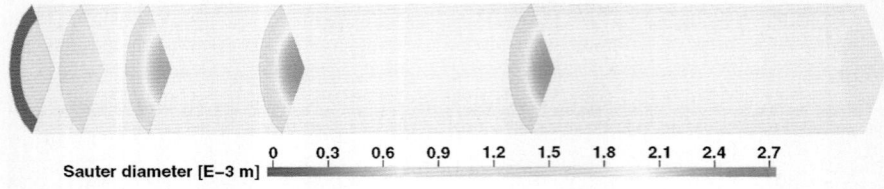

Fig. 14 Turbulent pipe flow: Sauter mean diameter d_{32} at $x = \{0, 0.06, 0.18, 0.33, 0.6\}$

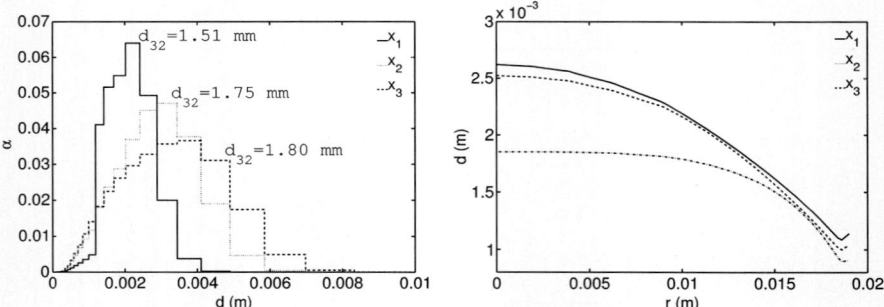

Fig. 15 Turbulent pipe flow: droplet size distribution (*left*) and radial variation of the Sauter mean diameter (*right*) at $x = \{0, 0.06, 0.18\}$

The calculated radial profiles of the axial velocity, turbulent dissipation rate, and eddy viscosity for the developed flow pattern are presented in Fig. 13. The results of the turbulent flow simulation determine the velocity and the breakage/coalescence rates for the population balance model. The CFD-PBE simulations are performed for 30 classes with nonuniform spacing that corresponds to the discretization factor $q = 1.7$. The feed stream is generated by a circular sparger of diameter 2.82 cm that produces droplets of diameter $d_{in} = 1.19$ mm. At the inlet, the volume fraction of droplets equals $\alpha_{in} = 0.55$. In the region of fully developed flow, the total holdup of the disperse phase has the constant value $\alpha_{tot} = 0.30$. Moreover, the droplet size distribution reaches an equilibrium under the developed flow conditions.

Figure 14 displays the distribution of the Sauter mean diameter d_{32} in five cross sections. For better visualization, the axis scaling $x : y : z = 10 : 1 : 1$ is employed in this diagram. Note that the equilibrium is attained at a short distance from the inlet. The distributions of the droplet size distribution and the radial profiles of the Sauter mean diameter for $x = \{0, 0.06, 0.18\}$ are presented in Fig. 15. The diagrams in Fig. 16 show the size distribution at the outlet and Sauter mean diameter along the x-axis for radii $r = \{0, R/3, 2R/3\}$. As expected, a high concentration of larger droplets is observed in the middle of the pipe, where the flow is fully turbulent and ε is relatively small. The concentration of smaller droplets is higher in the near-wall region, where ε is relatively large. The holdup distributions for three representative droplet classes (small, medium, and large) are presented in Fig. 17. The corresponding droplet diameters are given by 0.49 mm, 1.70 mm, and 4.90 mm.

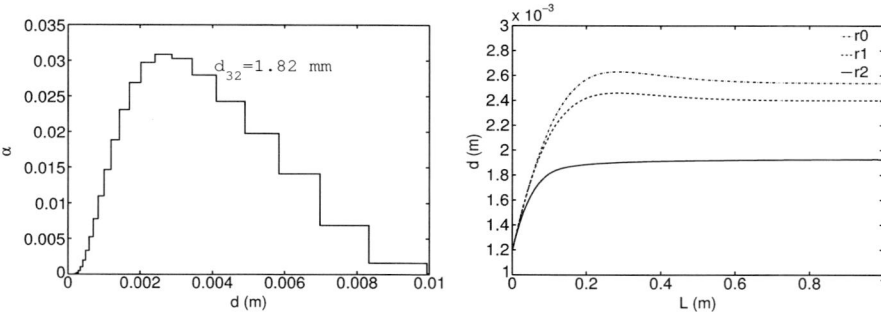

Fig. 16 Turbulent pipe flow: droplet size distribution at the outlet (*left*) and longitudinal variation of the Sauter mean diameter at $r = \{0, R/3, 2R/3\}$ (*right*)

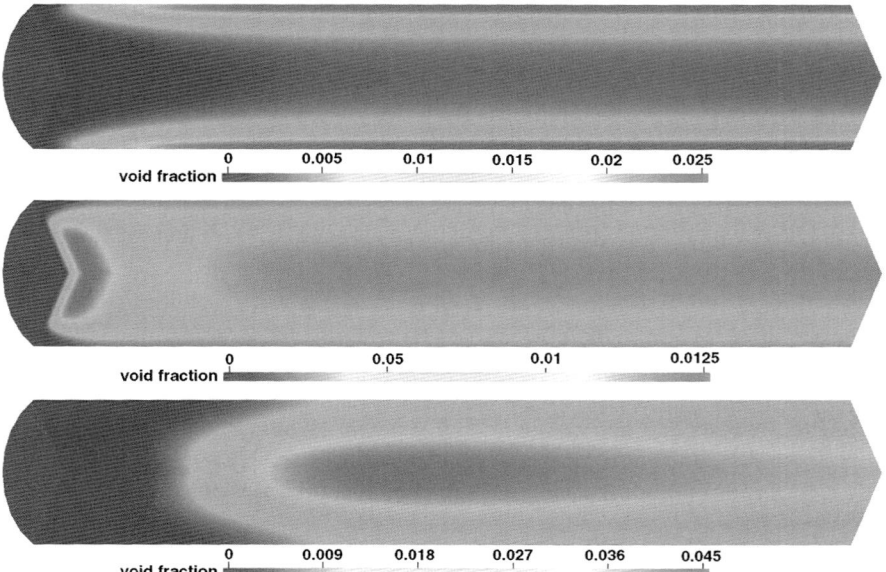

Fig. 17 Turbulent pipe flow: holdups of small (*top*), medium (*middle*), and large (*bottom*) droplets

9.4.2 Static Mixer SMV^{TM}

Static mixers are used in industry to disperse immiscible liquids as they flow around mixer elements rigidly installed in a tubular housing. The mechanical simplicity of static mixers makes them an attractive alternative to rotating impellers. Moreover, the dissipation of frictional energy in the packing is more uniform, and so is the resultant drop size distribution [61]. This homogeneity can be attributed to the stable flow pattern that depends on the geometry of the internal parts. The Sulzer SMV^{TM}

Fig. 18 Geometry of the SMV[TM] static mixer

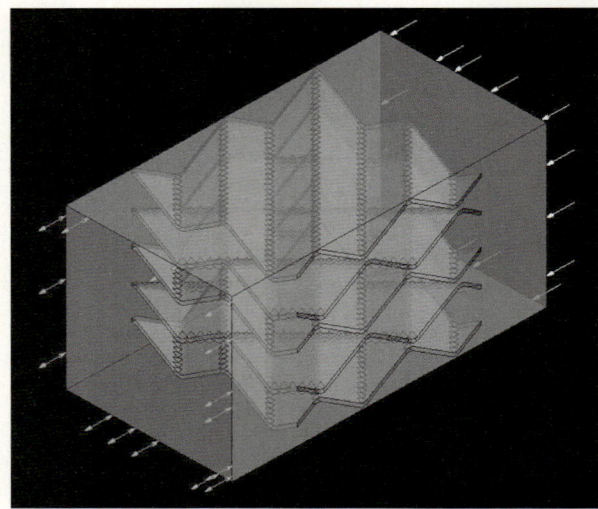

Table 1 Physical properties of the phases flowing in the SMV[TM] static mixer

	Water	Oil
ρ (kg m^{-3})	1000	847
ν (kg m^{-1}s^{-1})	1×10^{-3}	32×10^{-3}
σ (N m^{-1})	72×10^{-3}	21×10^{-3}

mixing elements consist of intersecting corrugated plates and channels. This design leads to fast and efficient dispersive mixing in the turbulent flow regime.

Many experimental and computational studies of laminar and turbulent static mixers can be found in the literature. For a detailed review, we refer to Thakur et al. [73]. Our interest in this industrial application is driven by the desire to explore the capabilities of the developed simulation tools. The complex geometry of the static mixer SMV[TM], as shown in Fig. 18 justifies the combination of a multidimensional flow model with PBEs. The inlet condition is that of a water-oil mixture with oil holdup $\alpha_{ij} = 0.1$, Sauter mean diameter $d_{32} = 10^{-3}$ m, and inflow speed $v_{in} = 1$ m/s. The physical properties of the two phases are listed in Table 1. The mixture is treated as a single fluid with density and viscosity defined as a weighted average of those for oil and water. The weights are given by the corresponding volume fractions.

Computations are performed on a mesh that consists of approximately 50,000 hexahedral elements. Due to the high computational cost, a one-way coupling between the flow and the PBEs is assumed. The simulation run begins with the computation of a steady-state solution for the turbulent flow field, see Fig. 19. The converged velocity and turbulent dissipation rate are used to solve the PBEs for 45 classes. The discretization constant equals $q = 1.4$ and the smallest droplets have

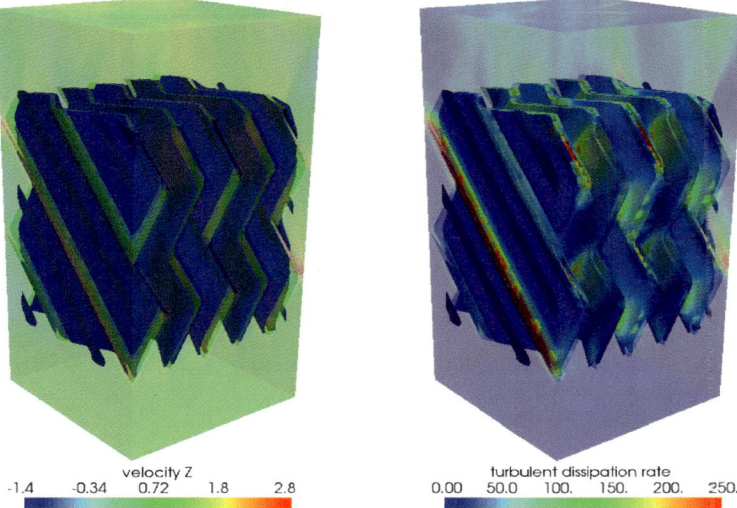

Fig. 19 The vertical velocity component (*left*) and turbulent dissipation rate (*right*)

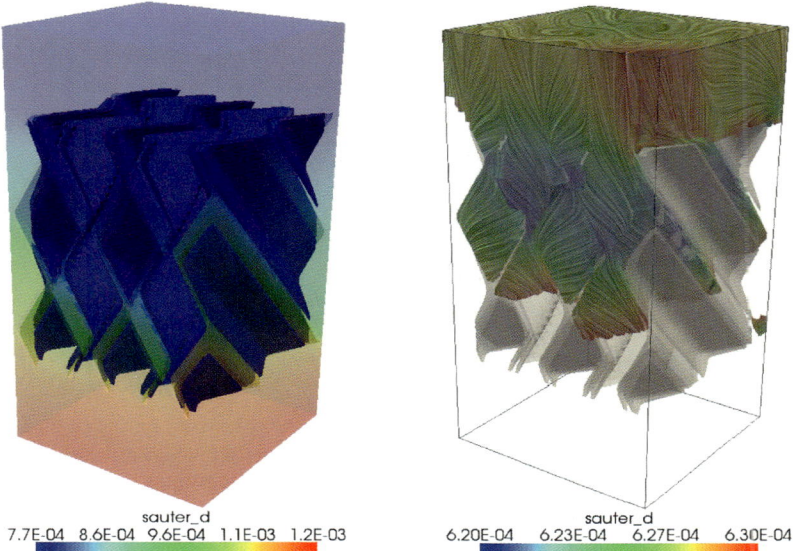

Fig. 20 Distribution of the Sauter mean diameter d_{32} for all classes (*left*) and droplet ensembles with $d_{32} \in [0.62, 0.63]$ mm (*right*)

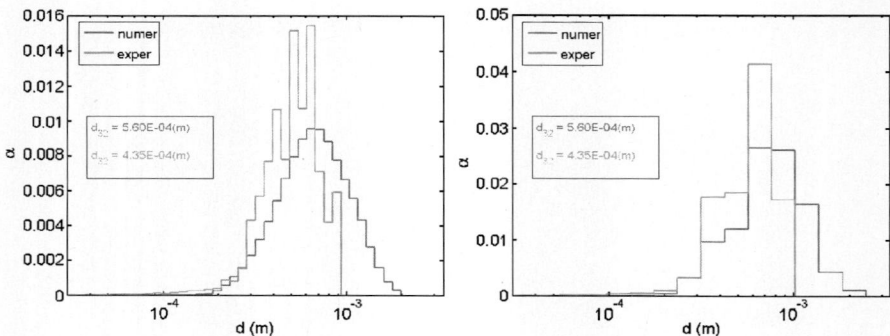

Fig. 21 Experimental and numerical results for the holdup with 45 (*left*) and 15 (*right*) classes

the diameter of 0.5 mm. The distributions of the Sauter mean diameter d_{32} and droplet ensembles with $d_{32} \in [0.62, 0.63]$ mm are displayed in Fig. 20.

For comparison purposes, we also present the experimental data provided by Sulzer Chemtech Ltd. The measurements are performed in the cross section right after the mixer element, and the detected droplets are assigned to the corresponding discrete classes. Since the number of classes for the numerical simulation is too large to obtain a representative number of droplets for each class, both numerical solutions and the measured data are mapped onto a size distribution with 15 classes, see Fig. 21. The results indicate that the CFD-PBE model provides a fairly good description of the population dynamics in turbulent mixtures. However, further effort is required to improve the accuracy of the model and of the numerical algorithms. This research will be continued in collaboration with Sulzer Chemtech Ltd.

10 Case Study: Interfacial Dynamics

Population balance models yield just a rough statistical estimate of the size distribution in gas-liquid and liquid-liquid dispersions. The position, shape, and size of individual drops or bubbles cannot be determined using such a model. To resolve the microscopic scales, the incompressible Navier-Stokes equations for the two immiscible fluids must be solved on subdomains separated by a moving boundary. The position of the interface is generally unknown and must be determined as a part of the problem. In this section, we describe *level set methods* that provide an implicit description of the interface and make it possible to solve a wide range of free boundary problems (deformation of drops/bubbles, breaking surface waves, slug flow, capillary microreactors, dendritic crystal growth) on fixed meshes.

10.1 The Level Set Method

The idea behind modern level set methods, as described in [55, 66, 67], is an implicit representation of the interface $\Gamma(t)$ in terms of a scalar variable $\varphi(\mathbf{x}, t)$ such that

$$\Gamma(t) = \{\mathbf{x} \mid \varphi(\mathbf{x}, t) = 0\}. \tag{96}$$

For practical purposes it is worthwhile to define φ as the *signed distance function*

$$\varphi(\mathbf{x}, t) = \pm \operatorname{dist}(\mathbf{x}, \Gamma(t)). \tag{97}$$

As a useful byproduct, one obtains the globally defined normal and curvature

$$\mathbf{n} = \frac{\nabla \varphi}{|\nabla \varphi|}, \qquad \kappa = -\nabla \cdot \mathbf{n}. \tag{98}$$

Since $|\varphi(\mathbf{x}, t)|$ is the (shortest) distance from $\mathbf{x}$ to $\Gamma(t)$, it may serve as an indicator of interface proximity for adaptive mesh refinement techniques [2, 37].

It can be shown that the evolution of φ is governed by the transport equation

$$\frac{\partial \varphi}{\partial t} + \mathbf{u} \cdot \nabla \varphi = 0. \tag{99}$$

The velocity field $\mathbf{u}$ is obtained by solving the generalized Navier-Stokes system

$$\rho \left(\frac{\partial \mathbf{u}}{\partial t} + \mathbf{u} \cdot \nabla \mathbf{u} \right) = -\nabla p + \nabla \cdot \left(\mu [\nabla \mathbf{u} + (\nabla \mathbf{u})^T] \right) + \mathbf{f}|_\Gamma, \tag{100}$$

$$\nabla \cdot \mathbf{u} = 0, \tag{101}$$

where $\mathbf{f}|_\Gamma$ is an interfacial force. The density ρ and viscosity μ are assumed to be constant in the interior of each phase and have a jump across Γ. We have

$$\rho(\mathbf{x}, t) = \rho_1 + (\rho_2 - \rho_1) H(\mathbf{x}, t), \tag{102}$$

$$\mu(\mathbf{x}, t) = \mu_1 + (\mu_2 - \mu_1) H(\mathbf{x}, t). \tag{103}$$

The value of the discontinuous Heaviside function H depends on the sign of φ

$$H(\varphi, \mathbf{x}, t) = \begin{cases} 1, & \text{if } \varphi(\mathbf{x}, t) > 0, \\ 0, & \text{if } \varphi(\mathbf{x}, t) < 0. \end{cases} \tag{104}$$

In numerical implementations, regularized approximations to H are employed.

In most existing level set codes, equations (99)–(101) are discretized using finite difference or finite volume approximations on structured meshes. However, the last decade has witnessed a lot of progress in the development of FEM-based level set algorithms [32, 46, 52, 57, 68, 75, 93]. In particular, discontinuous Galerkin methods have become popular in recent years [14, 24, 49]. The advantages of the finite element approach include the ease of mesh adaptation and the availability of a robust variational method for the numerical treatment of surface tension [1, 29].

10.2 Reinitialization

Even if the level set function φ is initialized using definition (97), it may cease to be a distance function as time evolves. In many situations, this is undesirable or unacceptable. First, nonphysical displacements of the interface and large conservation errors are likely to arise. Second, the lack of the distance function property has an adverse effect on the accuracy of numerical approximations to normals and curvatures. Third, if the gradients of φ become too steep, approximate solutions to (99) may be corrupted by spurious oscillations or excessive numerical diffusion.

The usual way to prevent a deterioration of the level set function is a postprocessing step known as 'reinitialization' or 'redistancing.' The purpose of this correction is to restore the distance function property of φ without changing its zero level set. Of course, it is possible to recalculate the distance from each mesh point to the interface. Such a 'direct' reinitialization is straightforward but computationally expensive, even if restricted to a narrow band around Γ. Alternatively, the distance function property can be enforced by solving the *Eikonal equation*

$$|\nabla \varphi| = 1 \tag{105}$$

subject to $\varphi = 0$ on $\Gamma(t) = \{\mathbf{x} \,|\, \tilde{\varphi}(\mathbf{x}, t) = 0\}$, where $\tilde{\varphi}$ is the level set function before reinitialization. The most popular techniques for solving (105) are *fast sweeping* methods [76], *fast marching* methods [65, 66], and the hyperbolic PDE approach [72]. In the latter method, equation (105) is treated as the steady-state limit of

$$\frac{\partial \varphi}{\partial \tau} + \mathbf{w} \cdot \nabla \varphi = \mathrm{sign}(\tilde{\varphi}), \quad \mathbf{w} = \mathrm{sign}(\tilde{\varphi}) \frac{\nabla \varphi}{|\nabla \varphi|}. \tag{106}$$

The solution to this nonlinear equation is initialized by $\tilde{\varphi}$ and marched to the steady state. In practice, it is enough to restore the distance function property in a narrow band around the interface. Hence, a few pseudo-time steps are sufficient.

For stability reasons, the discontinuous sign function is typically replaced with a smooth approximation. This practice may result in a loss of accuracy and displacements of Γ. In the *interface local projection* method of Parolini [57], finite element techniques are employed to perform direct reinitialization in the interface region. The corrected values of φ provide the boundary conditions for the subsequent solution of (106) in a reduced domain, where $\mathrm{sign}(\tilde{\varphi})$ has no jumps.

To avoid the need for postprocessing, Ville et al. [93] replace (99) and (106) with a single transport equation. The so-defined 'convected' level set method leads to an elegant and efficient algorithm. We also subscribe to the viewpoint that convection and reinitialization should be combined as long as there is no fail-safe way to fix φ when the damage is already done. This has led us to develop a variational level set method in which the Eikonal equation (105) is treated as a constraint for the level set transport equation [39]. The nonlinear Lagrange multiplier term

$$\int_\Omega \lambda \nabla \varphi \cdot \nabla w \, \Delta \mathbf{x} \tag{107}$$

added to the weak form of (99) corrects the gradients by adding artificial diffusion ($\lambda > 0$) or antidiffusion ($\lambda < 0$) whenever $|\nabla \varphi| > 1$ or $|\nabla \varphi| < 1$, respectively. In our experience, no flux limiting is required since φ remains smooth. A detailed description of the Lagrange multiplier approach will be presented elsewhere [39].

10.3 Mass Conservation

A major drawback of level set algorithms is the lack of mass conservation. Indeed, $\rho(\varphi)$ given by (102) may fail to satisfy the nonlinear continuity equation

$$\frac{\partial \rho(\varphi)}{\partial t} + \nabla \cdot \big(\mathbf{u}\rho(\varphi)\big) = 0. \tag{108}$$

As an alarming consequence, the volume of incompressible fluids may change in an unpredictable manner. In particular, this is likely to happen when evolving interfaces undergo topological changes such as coalescence or breakup.

Both transport and redistancing may be responsible for mass conservation errors in level set algorithms. To some extent, these errors can be reduced by using more accurate numerical schemes and adaptive mesh refinement techniques [53]. Many tricks for improving the conservation properties of level set algorithms have been proposed in recent years [14, 46, 68, 71, 88]. Again, the usual approach relies on the use of postprocessing techniques designed to preserve the total volume

$$V(t) = \int_\Omega H(\varphi, \mathbf{x}, t) \Delta \mathbf{x} = V(0), \quad \forall t \geq 0, \tag{109}$$

where H is the Heaviside function defined by (104). Smolianski [71] enforces this constraint by adding a constant c_φ to the nonconservative approximation

$$\bar{\varphi} = \varphi + c_\varphi, \quad \int_\Omega H(\varphi + c_\varphi, \mathbf{x}, t) \Delta \mathbf{x} = V(0). \tag{110}$$

This level correction ensures global mass conservation but there is a danger that the lost mass will reappear in a wrong place. If one fluid consists of multiple disconnected components, global conservation does not ensure that the mass/volume of each component is conserved. Clearly, manipulations of the form (110) are inappropriate in such situations. In our opinion, an incorrect distribution of mass is more harmful than (readily identifiable) mass conservation errors.

Lesage and Dervieux [46] proposed a localized mass corrector in which the constant c_φ is multiplied by the nodal residual of a *dual level set* equation. If the mass is conserved in a control volume around node i, then the value of φ_i remains unchanged. However, the corrections to other nodes depend on the global constant c_φ, which implies that the distribution of the lost mass may still be incorrect.

In the conservative level set method of Olsson and Kreiss [54], φ is replaced with a regularized Heaviside function. This definition makes the algorithm akin to the phase field (diffuse interface) method. Due to the presence of a steep front and the absence of Cahn-Hilliard terms, the use of flux limiting is a must. A finite difference TVD scheme is used to solve the transport equation in the original publication [54]. In the context of a finite element approximation, the conservative level set method can be implemented using algebraic flux correction of FCT or TVD type.

10.4 Surface Tension

The overall accuracy of level set algorithms depends not only on the computation of φ but also on the numerical treatment of the surface tension force

$$\mathbf{f}|_\Gamma(\mathbf{x}, t) = \sigma \kappa \mathbf{n} \delta(\mathbf{x}, t), \tag{111}$$

where σ is a surface tension coefficient and δ is the Dirac delta function localizing the effect of $\mathbf{f}|_\Gamma$ to Γ. The normal $\mathbf{n}$ and curvature κ are given by (98).

In a finite element code, the values of $\mathbf{n}$ and κ can be obtained using variational recovery techniques [30]. A better approach to the numerical treatment of surface tension effects is based on the following fact from differential geometry:

$$\kappa \mathbf{n} = \underline{\Delta} \mathrm{id}_\Gamma,$$

where id_Γ is the identity mapping on Γ and $\underline{\Delta}$ is the Laplace-Beltrami operator

$$\underline{\Delta} f := \underline{\nabla} \cdot (\underline{\nabla} f), \qquad \underline{\nabla} f := \nabla f - (\mathbf{n} \cdot \nabla f) \mathbf{n}.$$

The contribution of (111) to the weak form of the momentum equation (100) is calculated using the definition of $\delta(\mathbf{x}, t)$ and integration by parts [1, 29, 30]

$$\int_\Omega \mathbf{f}|_\Gamma \cdot \mathbf{w} \, \Delta \mathbf{x} = -\int_\Gamma \sigma \underline{\nabla} \mathbf{x} \cdot \underline{\nabla} \mathbf{w} \, ds. \tag{112}$$

Since a fully explicit treatment of this term leads to a *capillary time step restriction*, we follow the semi-implicit approach proposed by Bänsch [1] in the context of a front-tracking method. Plugging $\mathbf{x}^{n+1} = \mathbf{x}^n + \Delta t \mathbf{u}^{n+1}$ into (112), we obtain

$$\mathbf{f}_\sigma = -\int_{\Gamma^n} \sigma \underline{\nabla} \mathbf{x} \cdot \underline{\nabla} \mathbf{w} \, ds - \Delta t \int_{\Gamma^n} \sigma \underline{\nabla} \mathbf{u}^{n+1} \cdot \underline{\nabla} \mathbf{w} \, ds. \tag{113}$$

Note that the second term is linear in $\mathbf{u}^{n+1}$ and has the structure of a discrete diffusion operator. In contrast to the fully explicit approach, the discretization becomes more stable for large values of σ, as shown by the numerical study in [29, 30].

Following Hysing [29, 30], we evaluate $\mathbf{f}_\sigma$ using the continuum surface force (CSF) approximation [4]. By definition of the Dirac delta function, we have

$$\mathbf{f}_\sigma = -\int_\Omega \sigma \underline{\nabla} \mathbf{x} \cdot \underline{\nabla} \mathbf{w} \delta^n \, \Delta \mathbf{x} - \Delta t \int_\Omega \sigma \underline{\nabla} \mathbf{u}^{n+1} \cdot \underline{\nabla} \mathbf{w} \delta^n \, \Delta \mathbf{x}. \tag{114}$$

Since δ is singular, numerical integration is performed using a regularized delta function. Given an approximate distance function φ, we define

$$\delta_\varepsilon(\mathbf{x}) = \frac{\max\{0, \varepsilon - |\varphi|\}}{\varepsilon^2}, \qquad (115)$$

where ε is a small parameter. Note that there is no need to know the position of Γ that would be difficult to determine for bilinear and higher-order elements.

Sussman and Ohta [70] have recently found another promising way to achieve unconditional stability in a numerical implementation of stiff surface tension terms. Their algorithm is based on the concept of volume preserving motion by mean curvature. Reportedly, it offers a speed-up by a factor 3–5 for a given accuracy.

10.5 Putting It All Together

The above presentation of the level set method reveals that its practical implementation involves many choices and tradeoffs. The most important components are the solver for the Navier-Stokes equations with discontinuous coefficients, the numerical approximation of the level set transport equation, mechanisms for maintaining the distance function property and mass conservation, the method for computation of normals and curvatures, and the numerical treatment of surface tension

In the parallel 3D code developed by our group at the TU Dortmund, the incompressible Navier-Stokes equations are solved using a generalization of the discrete projection scheme described in Sect. 4. The velocity and pressure are discretized using $\tilde{Q}_1/Q_0$ or Q_2/P_1 elements. The level set equation is solved with a FEM-TVD scheme for continuous Q_1 elements [30, 32] or an upwind-biased P_1 discontinuous Galerkin (DG) method without any extra stabilization [85]. A variety of methods have been implemented to solve the Eikonal equation at the reinitialization step for the Q_1 version [31]. The DG approach makes it possible to reinitialize φ without displacing the free interface. The gradient of the piecewise-linear solution is constant inside each cell. To enforce $|\nabla\varphi| = 1$, we correct the slopes in elements crossed by the interface and solve (106) elsewhere, see [85] for details. The implementation of the surface tension force is based on the semi-implicit algorithm presented in Sect. 10.4. The option of solving contact angle problems is also provided.

10.6 Numerical Examples

In the absence of analytical solutions (which are very difficult to derive for interfacial two-phase flows) benchmarking is the only way to verify the developed method. Pure numerical benchmarks are of little help if no quantitative comparisons can be made. A visual inspection alone is rarely, if ever, sufficient for validation purposes. To illustrate this, consider the bubble shapes shown in Fig. 22. These shapes were

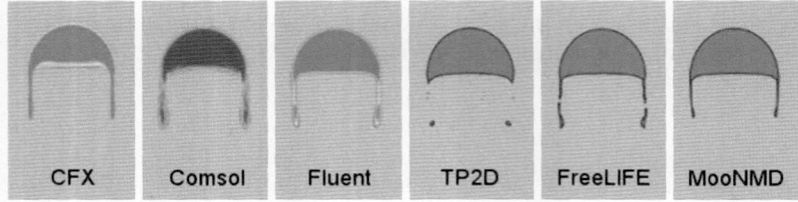

Fig. 22 Rising bubble simulation: numerical solutions produced by 6 codes

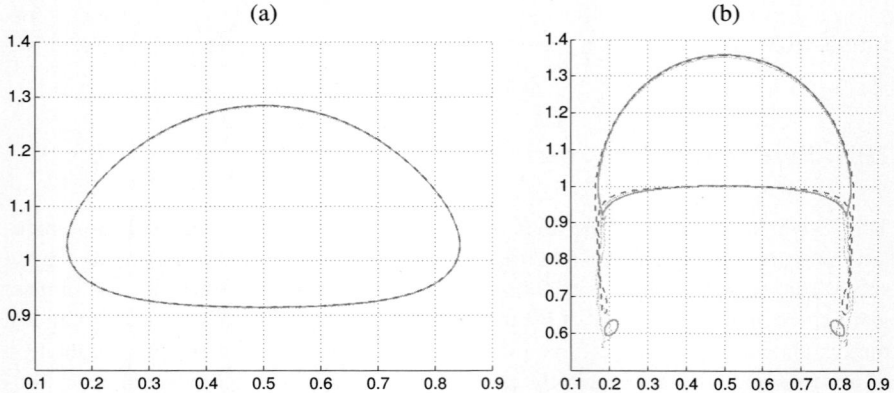

Fig. 23 Rising bubble benchmark: results for (**a**) Test 1 and (**b**) Test 2

calculated by six different codes with identical problem formulations. Ideally, the six solutions should be identical on fine meshes. Unfortunately, this is not the case. The shapes are quite similar but it is impossible to tell which solutions, if any, are really correct. In order to identify the good ones, one must replace the "eyeball norm" with some quantitative criteria for measuring the accuracy of simulation results.

10.6.1 Two-Dimensional Rising Bubble

In a recent paper [32], we proposed a new benchmark for interfacial two-phase flows. In collaboration with two other groups, we simulated a two-dimensional bubble rising in a liquid column. Two parameter constellations were considered. In the first test, the densities and viscosities of the two phases differ by a factor of 10, and the surface tension coefficient is chosen large enough to hold the bubble together. At the final time, the bubble assumes a typical ellipsoidal shape that was predicted very well by all codes under investigation, see Fig. 23(a). In the second test, the density and viscosity ratios are as large as 1000 and 100, respectively. Moreover, the value of the surface tension coefficient is reduced. The bubble shape falls into the skirted/dimpled ellipsoidal-cap regime, and a breakup occurs before the final time, see Fig. 23(b). The topological changes of the interface make this test rather

Table 2 3D rising bubble: empirical vs. simulated Reynolds numbers for Cases B, C, D

Case	Shape	Mo	Eo	Re_E	Re_S	$Re_{mA/2}$	$Re_{mA/3}$	$Re_{mB/3}$	$Re_{mB/4}$
B	Ellipsoidal	0.100	9.71	4.6	4.3	5.50	5.50	5.60	5.60
C	Skirted	0.971	97.1	20.0	18.0	17.7	18.0	18.0	18.0
D	Dimpled	1000	97.1	1.5	1.7	2.00	2.03	2.03	2.03

challenging. All computational details (geometry, initial and boundary conditions, parameter values) and the reference data for both cases are available online [6].

Since the publication of rising bubble benchmark, several other groups have contributed their results. It turned out that many different interface capturing techniques (level set, volume of fluid, phase field) produce very similar results. We remark that the rationale for developing a 2D test configuration was not an accurate prediction of physical reality (2D bubbles do not exist in nature) but the computation of reference solutions for evaluation of CFD software and underlying numerical methods.

10.6.2 Three-Dimensional Rising Bubble

The 3D version of our level set code has also been tested on a rising bubble problem [85]. The settings for this simulation correspond to test cases B, C, and D defined in the paper by van Sint Annaland et al. [92]. The proportions of the bubble diameter d and domain dimensions $a_x \times a_y \times a_z$ are $(d_b : a_x : a_y : a_z) = (3 : 10 : 10 : 20)$. The bubble undergoes significant deformations but does not break up. The densities and viscosities of the two immiscible fluids differ by a factor of 100. The values of the surface tension coefficient σ_{gl} and gravitational acceleration g_z are given in terms of the dimensionless Eötvös and Morton numbers defined as in [10]

$$\text{Eo} = \frac{g_z \Delta \rho_{gl} d_b^2}{\sigma_{gl}}, \quad \text{Mo} = \frac{g_z \mu_l^4 \Delta \rho_{gl}}{\rho_l^2 \sigma_{gl}}. \tag{116}$$

The Reynolds number associated with the terminal rise v_∞ velocity is defined by

$$\text{Re} = \frac{\rho_l v_\infty d_b}{\mu_l}. \tag{117}$$

In order to assess the dependence of the bubble shape and v_∞ on the mesh size, simulations were performed with two different meshes and two levels of refinements for each mesh (2, 3 for mesh A and 3, 4 for mesh B). The equilibrium bubble shapes shown in Fig. 24 indicate that the employed mesh resolution is sufficient, especially in the cases B and D. The measured and calculated values of the Reynolds number for all cases are listed in Table 2. The empirical data of Clift et al. [10] and simulation results of van Sint Annaland [92] are shown in the columns labeled Re_E and Re_S, respectively. The last 4 columns show our results obtained on meshes A and B for refinement levels 2–4. Although these results are essentially mesh-independent,

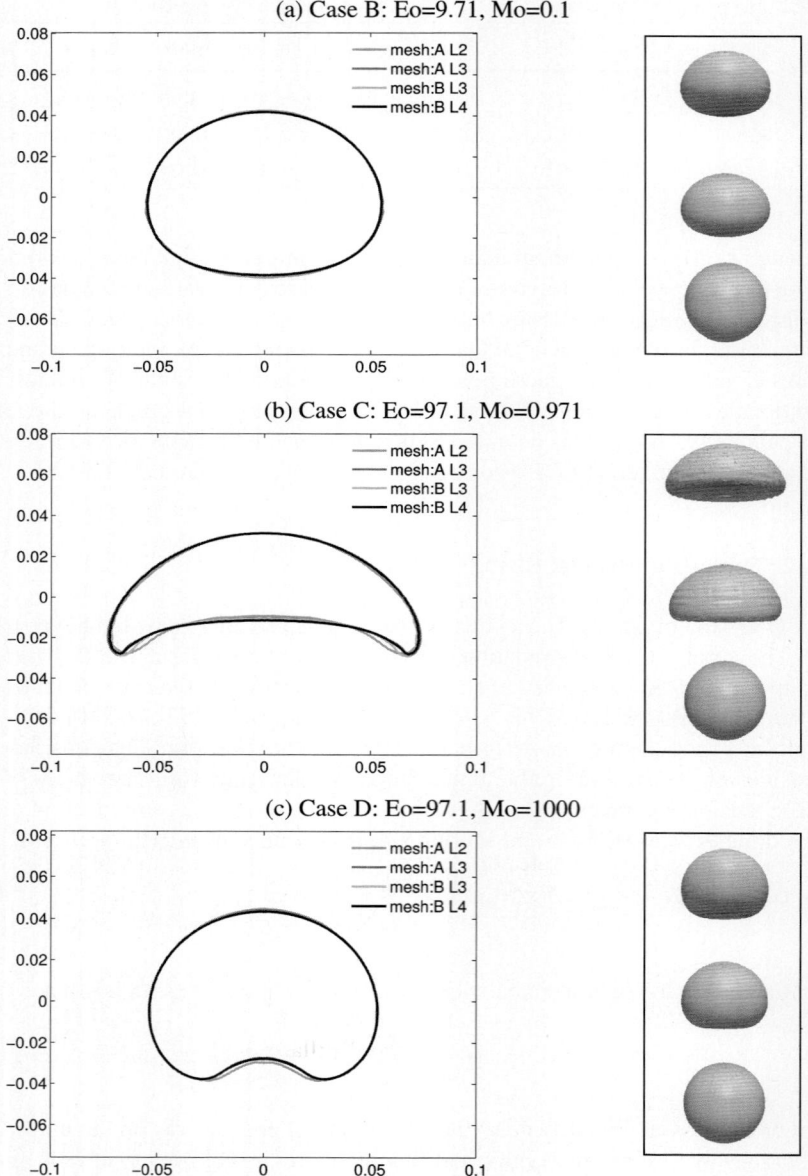

Fig. 24 3D rising bubble: equilibrium shapes (*left*) and snapshots of the deforming bubble (*right*)

Re_S exhibits a better correlation with Re_E. Since no grid convergence studies were performed in [92], it is unclear if the values of Re_S have also converged. This state

Fig. 25 Droplet dripping: a sketch of the domain around the capillary

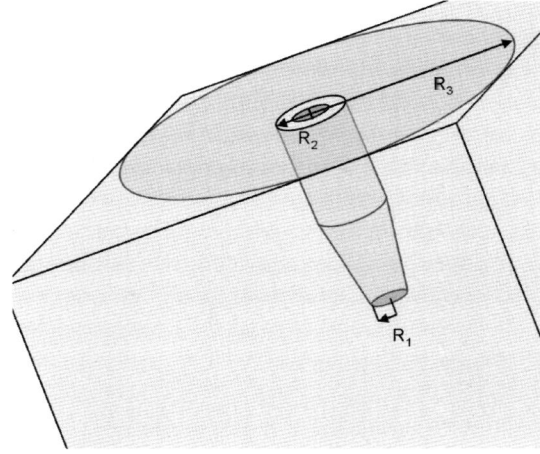

of affairs illustrates the urgent need for a collaborative research effort aimed at the development of a new 3D benchmark for interfacial two-phase flows.

10.6.3 Droplet Dripping

In the last numerical example, we simulate the process of droplet dripping in a liquid stream [85]. In the corresponding experimental setup, the continuous phase is a glucose-water mixture and the disperse phase is silicon oil. The dripping mode is characterized by relatively low volumetric flow rates and by the fact that the droplets are generated in the near vicinity of the capillary, so that the stream length is comparable to the size of the generated droplets. Since the temperature is kept at a constant value during the whole experiment, the densities and viscosities of the two phases are also constant. The experimental studies performed by the group of Prof. Walzel (BCI, TU Dortmund) provide the average values of target quantities like the droplet size, droplet generation frequency, and the stream length. These experimental data make it possible to validate the 3D simulation results to be presented below.

The geometry of the domain around the capillary is sketched in Fig. 25. The problem dimensions measured in decimeters (dm) are as follows:

domain dimensions	$0.3 \times 0.3 \times 1.2$
inner capillary radius	$R_1 = 0.015$
outer capillary radius	$R_2 = 0.030$
primary phase inlet radius	$R_3 = 0.15$

The physical properties of the continuous (C) and disperse (D) phase are given by

$$\rho_C = 1340 \text{ kg m}^{-3} = 1.34 \text{ kg dm}^{-3},$$
$$\rho_D = 970 \text{ kg m}^{-3} = 0.97 \text{ kg dm}^{-3},$$

$$\mu_C = \mu_D = 500 \text{ mPa s} = 0.050 \text{ kg dm s}^{-1},$$
$$g_z = -9.81 \text{ m s}^{-2} = -98.1 \text{ dm s}^{-2},$$
$$\sigma = 0.034 \text{ N m}^{-1} = 0.034 \text{ kg s}^{-2}.$$

The inflow boundary conditions are given in terms of the volumetric flow rates

$$\dot{V}_C = \int_{R_2}^{R_3} \left(2\pi r a_1 (R_3 - r)(r - R_2)\right) dr$$
$$= -2\pi a_1 \left[\frac{r^4}{4} - (R_2 + R_3)\frac{r^3}{3} + R_2 R_3 \frac{r^2}{2}\right]_{R_2}^{R_3}$$
$$= \frac{\pi a_1}{6}(R_2 + R_3)(R_3 - R_2)^3$$

and

$$\dot{V}_D = \int_0^{R_1} \left(2\pi r a_2 (R_1 - r)(R_1 + r)\right) dr = 2\pi a_2 \left[\frac{R_1^2 r^2}{2} - \frac{r^4}{4}\right]_0^{R_1} = \frac{\pi a_2}{2} R_1^4.$$

The parabolic velocity profile at the inflow boundary is defined by the formula

$$w = \begin{cases} a_2(R_1 - r)(R_1 + r), & \text{if } 0 < r < R_1, \\ a_1(R_3 - r)(r - R_2), & \text{if } R_2 < r < R_3, \\ 0, & \text{otherwise.} \end{cases}$$

The parameter values $a_1 = 10.14 \text{ dm}^{-1} \text{ s}^{-1}$, $a_2 = 763.7 \text{ dm}^{-1} \text{ s}^{-1}$ correspond to

$$\dot{V}_C = 99.04 \text{ ml min}^{-1} = 99.04 \text{ cm}^3 \text{ min}^{-1} = 99.04 \frac{10^{-3} \text{ dm}^3}{60 \text{ s}}$$
$$= 1.65 \cdot 10^{-3} \text{ dm}^3 \text{ s}^{-1},$$
$$\dot{V}_D = 3.64 \text{ ml min}^{-1} = 3.64 \text{ cm}^3 \text{ min}^{-1} = 3.64 \frac{10^{-3} \text{ dm}^3}{60 \text{ s}} = 6.07 \cdot 10^{-5} \text{ dm}^3 \text{ s}^{-1}.$$

The above operating conditions lead to a pseudo-steady dripping mode. The measured frequency of droplet formation is $f = 0.60$ Hz (cca 0.58 Hz^{exp}), the diameter of the generated droplets is $d = 0.058$ dm (cca 0.062 dm^{exp}), and the maximum stream length is $L = 0.102$ dm (cca 0.122 dm^{exp}). The process of droplet dripping is illustrated by the diagrams and photographs in Fig. 26. The agreement between the simulation results and physical reality is remarkably good. In this study, we used the $Q_2/P_1/P_1$ version of the 3D code. The total holdup of the disperse phase evolves as shown in Fig. 27. The slope of the lines that correspond to the experimental data is given by $q = 6.07 \cdot 10^{-5}$ dm^3 s^{-1}. The measured and simulated holdups follow the same trend, although the optional mass correction step was deactivated.

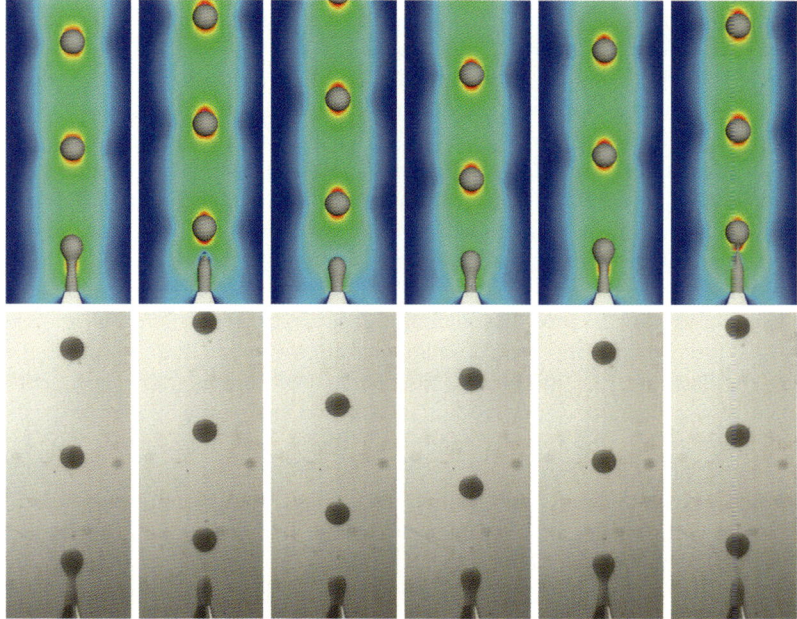

Fig. 26 Droplet dripping: 3D simulation (*top*) vs. experiment (*bottom*)

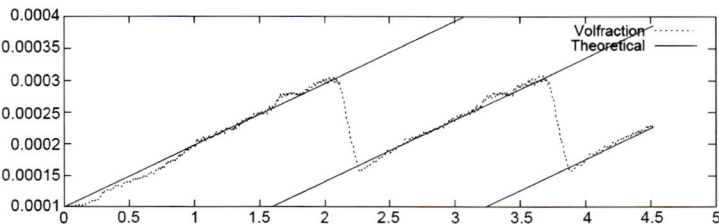

Fig. 27 Total holdup of the disperse phase: 3D simulation vs. experiment

11 Conclusions

In this chapter, we presented a family of multilevel pressure Schur complement methods for the incompressible Navier-Stokes equations. The coupling of the basic flow model with (systems of) scalar transport equations was illustrated by the case studies for the k–ε turbulence model, population balance equations, and level set algorithms. This survey covers a small but representative selection of incompressible problems that can be solved efficiently using the proposed tools. The current research activities of our groups cover a wide range of other applications such as particulate and granular flows [51, 56], viscoelastic fluids [12], computational hemodynamics [18], benchmarking for fluid-structure interaction [83], chemotaxis problems [69], and GPU computing [19, 82], to name just a few.

The design of professional CFD software for grand-challenge industrial problems requires an optimal interaction of discretization methods, iterative solvers, and software engineering aspects. The overall performance of the code depends on all of these components. Obtaining quantitatively accurate results in a computationally efficient manner is still an issue even for scalar convection-dominated transport problems and laminar flow models. The mathematical challenges of today include the extension of algebraic flux correction schemes to higher-order finite elements and tensor-valued transport operators, hp-adaptivity in space and time, rigorous a posteriori error estimation, and model-dependent improvements.

The optimization of iterative solvers for linear and nonlinear systems requires a further analysis of Newton-like methods, convergence acceleration techniques, monolithic multigrid solvers, and domain decomposition methods for parallel computing. Furthermore, the importance of benchmark computations and grid convergence studies cannot be overemphasized. We invite the reader to visit our CFD benchmarking site [6], get familiar with the test cases and propose new ones.

In addition to the above mathematical challenges, the growing demands of the CFD industry require a further investment in the development of hardware-oriented implementation techniques for modern computer architectures. The main bottleneck to high performance is not the actual data processing but slow memory access (see [80] for a critical discussion). For this reason, the actual MFLOP/s rates are typically very low compared to the theoretical peak performance. A major gain of efficiency can be achieved, for example, by using cache-based implementation techniques and exploiting the tensor product structure of stencils for block-structured grids. Such a hardware-oriented approach may yield an overall speedup factor of up to 1000 even on a single processor. On top of that, the use of optimal parallelization strategies may boost the performance of the code by further orders of magnitude.

In light of the above, the key to achieving optimal performance in the context of implicit finite element flow solvers lies in shifting the distribution of CPU times from costly memory access tasks (assembly of matrices/right-hand sides/residuals, adaptive mesh refinement/coarsening) toward more arithmetic-intensive work (solution of sparse linear systems). High-performance computing techniques based on this philosophy are already available and prove remarkably efficient [81].

In recent years, graphics processing units (GPUs) have become a popular tool for scientific computing. The contributions of our group include a GPU- and multicore-oriented implementation technique for geometric multigrid solvers [19]. Sparse matrix-vector multiplications are utilized throughout the multigrid pipeline: in the coarse-grid solver, in smoothers, and even in grid transfer operators. The current implementation can handle several low- and high-order finite element spaces in 2D and 3D. On a single GPU, we achieve speedups by nearly an order of magnitude compared to a multithreaded CPU code. We conclude that the practical implementation of a numerical algorithm may be as important as the choice of its mathematical components. This means that the methods of scientific computing will continue to evolve following the technological trends in computer architecture.

Acknowledgements The authors would like to thank Shu-Ren Hysing, Otto Mierka, and Evren Bayraktar (TU Dortmund) for contributing their results. The collaboration with Prof. Peter Walzel (TU Dortmund) and Sulzer Chemtech Ltd is gratefully acknowledged.

References

1. Bänsch, E.: Finite element discretization of the Navier-Stokes equations with a free capillary surface. Numer. Math. **88**, 203–235 (2001)
2. Barth, T.J., Sethian, A.: Numerical schemes for the Hamilton-Jacobi and level set equations on triangulated domains. J. Comput. Phys. **145**, 1–40 (1998)
3. Bayraktar, E., Mierka, O., Platte, F., Kuzmin, D., Turek, S.: Numerical aspects and implementation of population balance equations coupled with turbulent fluid dynamics. Comput. Chem. Eng. (2011). doi:10.1016/j.compchemeng.2011.04.001
4. Brackbill, J.U., Kothe, D.B., Zemach, C.: A continuum method for modeling surface tension. J. Comput. Phys. **100**, 335–354 (1992)
5. Buwa, V.V., Ranade, V.V.: Dynamics of gas-liquid flow in a rectangular bubble column: experiments and single/multi-group CFD simulations. Chem. Eng. Sci. **57**, 4715–4736 (2002)
6. CFD benchmarking site. http://www.featflow.de/en/benchmarks/cfdbenchmarking
7. Chien, K.-Y.: Predictions of channel and boundary-layer flows with a low-Reynolds number turbulence model. AIAA J. **20**, 33–38 (1982)
8. Chorin, A.J.: Numerical solution of the Navier–Stokes equations. Math. Comput. **22**, 745–762 (1968)
9. Christon, M.A., Gresho, P.M., Sutton, S.B.: Computational predictability of natural convection flows in enclosures. In: Bathe, K.J. (ed.) Proc. First MIT Conference on Computational Fluid and Solid Mechanics, pp. 1465–1468. Elsevier, Amsterdam (2001)
10. Clift, R., Grace, J.R., Weber, M.E.: Bubbles, Drops and Particles. Dover, New York (2005)
11. Crouzeix, M., Raviart, P.A.: Conforming and non-conforming finite element methods for solving the stationary Stokes equations. RAIRO **R–3**, 77–104 (1973)
12. Damanik, H.: Monolithic FEM techniques for viscoelastic fluids. PhD thesis, TU Dortmund (2011)
13. Damanik, H., Hron, J., Ouazzi, A., Turek, S.: Monolithic Newton-multigrid solution techniques for incompressible nonlinear flow models. Int. J. Numer. Methods Fluids (2012). doi:10.1002/fld.3656
14. Di Pietro, D.A., Lo Forte, S., Parolini, N.: Mass preserving finite element implementations of the level set method. Appl. Numer. Math. **56**, 1179–1195 (2006)
15. Donea, J., Giuliani, S., Laval, H., Quartapelle, L.: Finite element solution of the unsteady Navier-Stokes equations by a fractional step method. Comput. Methods Appl. Mech. Eng. **30**, 53–73 (1982)
16. Engelman, M.S., Haroutunian, V., Hasbani, I.: Segregated finite element algorithms for the numerical solution of large–scale incompressible flow problems. Int. J. Numer. Methods Fluids **17**, 323–348 (1993)
17. Engelman, M.S., Sani, R.L., Gresho, P.M.: The implementation of normal and/or tangential boundary conditions in finite element codes for incompressible fluid flow. Int. J. Numer. Methods Fluids **2**, 225–238 (1982)
18. Galdi, G., Rannacher, R., Robertson, A., Turek, S.: Hemodynamical Flows: Modelling, Analysis and Simulation. WS-Oberwolfach Seminars. Birkhäuser, Basel (2008). ISBN: 978-3-7643-7805-9
19. Geveler, M., Ribbrock, D., Göddeke, D., Zajac, P., Turek, S.: Towards a complete FEM-based simulation toolkit on GPUs: unstructured grid finite element geometric multigrid solvers with strong smoothers based on sparse approximated inverses. Comp. Fluids (2012). doi:10.1016/j.compfluid.2012.01.025. Special issue ParCFD'11

20. Girault, V., Raviart, P.A.: Finite Element Methods for Navier–Stokes Equations. Springer, Berlin (1986)
21. Glowinski, R.: Finite element methods for incompressible viscous flow. In: Ciarlet, P.G., Lions, J.L. (eds.) Numerical Methods for Fluids (Part 3). Handbook of Numerical Analysis, vol. IX, pp. 3–1176. North-Holland, Amsterdam (2003)
22. Gresho, P.M., Sani, R.L., Engelman, M.S.: Incompressible Flow and the Finite Element Method: Advection-Diffusion and Isothermal Laminar Flow. Wiley, New York (1998)
23. Gresho, P.M.: On the theory of semi–implicit projection methods for viscous incompressible flow and its implementation via a finite element method that also introduces a nearly consistent mass matrix, Part 1: Theory, Part 2: Implementation. Int. J. Numer. Methods Fluids **11**, 587–659 (1990)
24. Grooss, J., Hesthaven, J.S.: A level set discontinuous Galerkin method for free surface flows. Comput. Methods Appl. Mech. Eng. **195**, 3406–3429 (2006)
25. Grotjans, H., Menter, F.: Wall functions for general application CFD codes. In: ECCOMAS 98, Proceedings of the 4th Computational Fluid Dynamics Conference, pp. 1112–1117. Wiley, New York (1998)
26. Hackbusch, W., John, V., Khachatryan, A., Suciu, C.: A numerical method for the simulation of an aggregation-driven population balance system. Int. J. Numer. Methods Fluids (2011). doi:10.1002/fld.2656
27. Hu, B., Matar, O.K., Hewitt, G.F., Angeli, P.: Population balance modelling of phase inversion in liquid-liquid pipeline flows. Chem. Eng. Sci. **61**, 4994–4997 (2006)
28. Hughes, T.J.R., Franca, L.P., Balestra, M.: A new finite element formulation for computational fluid mechanics: V. Circumventing the Babuska–Brezzi condition: A stable Petrov–Galerkin formulation of the Stokes problem accommodating equal order interpolation. Comput. Methods Appl. Mech. Eng. **59**, 85–99 (1986)
29. Hysing, S.: A new implicit surface tension implementation for interfacial flows. Int. J. Numer. Methods Fluids **51**, 659–672 (2006)
30. Hysing, S.: Numerical simulation of immiscible fluids with FEM level set techniques. PhD thesis, TU Dortmund (2007)
31. Hysing, S., Turek, S.: The Eikonal equation: numerical efficiency vs. algorithmic complexity on quadrilateral grids. In: Proceedings of Algoritmy, pp. 22–31 (2005)
32. Hysing, S., Turek, S., Kuzmin, D., Parolini, N., Burman, E., Ganesan, S., Tobiska, L.: Quantitative benchmark computations of two-dimensional bubble dynamics. Int. J. Numer. Methods Fluids **60**, 1259–1288 (2009)
33. Ilinca, F., Hétu, J.-F., Pelletier, D.: A unified finite element algorithm for two-equation models of turbulence. Comput. Fluids **27**, 291–310 (1998)
34. John, V., Roland, M.: On the impact of the scheme for solving the higher-dimensional equation in coupled population balance systems. Int. J. Numer. Methods Eng. **82**, 1450–1474 (2010)
35. Kim, J.: Investigation of separation and reattachment of a turbulent shear layer: flow over a backward facing step. PhD thesis, Stanford University (1978)
36. Kim, J., Moin, P., Moser, R.D.: Turbulence statistics in fully developed channel flow at low Reynolds number. J. Fluid Mech. **177**, 133–166 (1987)
37. Kohno, H., Tanahashi, T.: Numerical analysis of moving interfaces using a level set method coupled with adaptive mesh refinement. Int. J. Numer. Methods Fluids **45**, 921–944 (2004)
38. Kuzmin, D.: Algebraic flux correction I. Scalar conservation laws. Chap. 6 in this book. doi:10.1007/978-94-007-4038-9_6
39. Kuzmin, D., Basting, C., Bänsch, E.: The Lagrange multiplier approach to maintaining the distance function property in level set algorithms. In preparation
40. Kuzmin, D., Möller, M., Gurris, M.: Algebraic flux correction II. Compressible Flow Problems. Chap. 7 in this book. doi:10.1007/978-94-007-4038-9_7
41. Kuzmin, D., Mierka, O., Turek, S.: On the implementation of the k–ε turbulence model in incompressible flow solvers based on a finite element discretization. Int. J. Comput. Sci. Math. **1**, 193–206 (2007)

42. Kuzmin, D., Turek, S.: Multidimensional FEM-TVD paradigm for convection-dominated flows. In: Proceedings of the IV European Congress on Computational Methods in Applied Sciences and Engineering (ECCOMAS 2004), vol. II (2004). ISBN:951-39-1869-6
43. Kuzmin, D., Turek, S.: Numerical simulation of turbulent bubbly flows. In: Proceedings of the 3rd International Symposium on Two-Phase Flow Modelling and Experimentation, Pisa, September 22–24, 2004
44. Lehr, F., Mewes, D.: A transport equation for interfacial area density applied to bubble columns. Chem. Eng. Sci. **56**, 1159–1166 (2001)
45. Lehr, F., Millies, M., Mewes, D.: Bubble size distribution and flow fields in bubble columns. AIChE J. **48**, 2426–2442 (2002)
46. Lesage, A.-C., Dervieux, A.: Conservation correction by dual level set. INRIA Report 7089 (November 2009)
47. Lew, A.J., Buscaglia, G.C., Carrica, P.M.: A note on the numerical treatment of the k-epsilon turbulence model. Int. J. Comput. Fluid Dyn. **14**, 201–209 (2001)
48. Lo, S.: Application of the MUSIG model to bubbly flows. AEAT-1096, AEA Technology (1996)
49. Marchandise, E., Geuzaine, P., Chevaugeon, N., Remacle, J.-F.: A stabilized finite element method using a discontinuous level set approach for solving two phase incompressible flows. J. Comput. Phys. **225**, 949–974 (2007)
50. Mohammadi, B., Pironneau, O.: Analysis of the k–Epsilon Turbulence Model. Wiley, New York (1994)
51. Münster, R., Mierka, O., Turek, S.: Finite element-fictitious boundary methods (FEM-FBM) for 3D particulate flow. Int. J. Numer. Methods Fluids (2011). doi:10.1002/fld.2558
52. Nagrath, S.: Adaptive stabilized finite element analysis of multi-phase flows using level set approach. PhD Thesis, Rensselaer Polytechnic Institute, New York (2004)
53. Nourgaliev, R.R., Wiri, S., Dinh, N.T., Theofanous, T.G.: On improving mass conservation of level set by reducing spatial discretization errors. Int. J. Multiph. Flow **31**, 1329–1336 (2005)
54. Olsson, E., Kreiss, G.: A conservative level set method for two phase flow. J. Comput. Phys. **210**, 225–246 (2005)
55. Osher, S., Fedkiw, R.: Level Set Methods and Dynamic Implicit Surfaces. Springer, New York (2003)
56. Ouazzi, A.: Finite element simulation of nonlinear fluids with application to granular material and powder. PhD thesis, TU Dortmund (2005)
57. Parolini, N.: Computational fluid dynamics for naval engineering problems. PhD thesis, EPFL Lausanne (2004)
58. Patankar, S.V.: Numerical Heat Transfer and Fluid Flow. McGraw-Hill, New York (1980)
59. Prohl, A.: Projection and Quasi-Compressibility Methods for Solving the Incompressible Navier-Stokes Equations. Advances in Numerical Mathematics. Teubner, Stuttgart (1997)
60. Quartapelle, L.: Numerical Solution of the Incompressible Navier-Stokes Equations. Birkhäuser, Basel (1993)
61. Rama Rao, N.V., Baird, M.H.I., Hrymak, A.N., Wood, P.E.: Dispersion of high-viscosity liquid-liquid systems by flow through SMX static mixer elements. Chem. Eng. Sci. **62**, 6885–6896 (2007)
62. Rannacher, R., Turek, S.: A simple nonconforming quadrilateral Stokes element. Numer. Methods Partial Differ. Equ. **8**, 97–111 (1992)
63. Schäfer, M., Turek, S. (with support of F. Durst, E. Krause, R. Rannacher): Benchmark computations of laminar flow around cylinder. In: Hirschel, E.H. (ed.) Flow Simulation with High-Performance Computers II. Notes on Numerical Fluid Mechanics, vol. 52, pp. 547–566. Vieweg, Wiesbaden (1996)
64. Schmachtel, R.: Robuste lineare und nichtlineare Lösungsverfahren für die inkompressiblen Navier-Stokes-Gleichungen. PhD thesis, University of Dortmund (2003)
65. Sethian, J.A.: A fast marching level set method for monotonically advancing fronts. Proc. Natl. Acad. Sci. USA **93**(4), 1591–1595 (1996)

66. Sethian, J.A.: Level Set Methods and Fast Marching Methods: Evolving Interfaces in Computational Geometry, Fluid Mechanics, Computer Vision, and Materials Science. Cambridge University Press, Cambridge (1999)
67. Sethian, J.A., Smereka, P.: Level set methods for fluid interfaces. Annu. Rev. Fluid Mech. **35**, 341–372 (2003)
68. Smolianski, A.: Numerical modeling of two-fluid interfacial flows. PhD thesis, University of Jyväskylä (2001)
69. Strehl, R., Sokolov, A., Kuzmin, D., Horstmann, D., Turek, S.: A positivity-preserving finite element method for chemotaxis problems in 3D. Ergebnisber. Angew. Math. 417, TU Dortmund (2010)
70. Sussman, M., Ohta, P.M.: A stable and efficient method for treating surface tension in incompressible two-phase flow. SIAM J. Sci. Comput. **31**, 2447–2471 (2009)
71. Sussman, M., Puckett, E.G.: A coupled level set and volume of fluid method for computing 3D and axisymmetric incompressible two-phase flows. J. Comput. Phys. **162**, 301–337 (2000)
72. Sussman, M., Smereka, P., Osher, S.: A level set approach for computing solutions to incompressible two-phase flow. J. Comput. Phys. **114**, 146–159 (1994)
73. Thakur, R.K., Vial, Ch., Nigam, K.D.P., Nauman, E.B., Djelveh, G.: Static mixers in the process industries—A review. Trans. IChemE **81**, 787–826 (2003)
74. Thangam, S., Speziale, C.G.: Turbulent flow past a backward-facing step: a critical evaluation of two-equation models. AIAA J. **30**, 1314–1320 (1992)
75. Tornberg, A.-K.: Interface tracking methods with applications to multiphase flows. PhD thesis, Royal Institute of Technology, Stockholm (2000)
76. Tsai, Y.R., Cheng, L.-T., Osher, S., Zhao, H.-K.: Fast sweeping algorithms for a class of Hamilton-Jacobi equations. SIAM J. Numer. Anal. **41**(2), 673–694 (2003)
77. Turek, S.: On discrete projection methods for the incompressible Navier-Stokes equations: An algorithmical approach. Comput. Methods Appl. Mech. Eng. **143**, 271–288 (1997)
78. Turek, S.: Efficient Solvers for Incompressible Flow Problems: An Algorithmic and Computational Approach. Lecture Notes in Computational Science and Engineering, vol. 6. Springer, Berlin (1999)
79. Turek, S., et al.: FEATFLOW: finite element software for the incompressible Navier-Stokes equations. User manual, University of Dortmund (2000). http://www.featflow.de
80. Turek, S., Becker, C., Kilian, S.: Hardware-oriented numerics and concepts for PDE software. Future **1095**, 1–23 (2003)
81. Turek, S., Becker, C., Kilian, S.: Some concepts of the software package FEAST. In: Palma, J.M., Dongarra, J., Hernandes, V. (eds.) VECPAR'98—Third International Conference for Vector and Parallel Processing. Lecture Notes in Computer Science. Springer, Berlin (1999)
82. Turek, S., Göddeke, D., Buijssen, S., Wobker, H.: Hardware-oriented multigrid finite element solvers on (GPU)-accelerated clusters. In: Kurzak, J., Bader, D.A., Dongarra, J. (eds.) Scientific Computing with Multicore and Accelerators, pp. 113–130. CRC Press, Boca Raton (2010). Chap. 6
83. Turek, S., Hron, J., Razzaq, M., Wobker, H., Schäfer, M.: Numerical Benchmarking of Fluid-Structure Interaction: A Comparison of Different Discretization and Solution Approaches. In: Bungartz, H.-J., Mehl, M., Schäfer, M. (eds.) Fluid Structure Interaction II: Modelling, Simulation, Optimization. Lecture Notes in Computational Science and Engineering, vol. 73, pp. 413–424. Springer, Berlin (2010)
84. Turek, S., Kilian, S.: An example for parallel ScaRC and its application to the incompressible Navier-Stokes equations. In: Proc. ENUMATH'97. World Scientific, Singapore (1998)
85. Turek, S., Mierka, O., Hysing, S., Kuzmin, D.: Numerical study of a high order 3D FEM-level set approach for immiscible flow simulation. Submitted to Proceedings of the ECCOMAS Thematic Conference on Computational Analysis and Optimization (June 9–11, 2011, Jyväskylä, Finland)
86. Turek, S., Ouazzi, A.: Unified edge-oriented stabilization of nonconforming FEM for incompressible flow problems: Numerical investigations. J. Numer. Math. **15**, 299–322 (2007)
87. Turek, S., Schmachtel, R.: Fully coupled and operator-splitting approaches for natural convection. Int. J. Numer. Methods Fluids **40**, 1109–1119 (2002)

88. van der Pijl, S.P., Segal, A., Vuik, C.: A mass-conserving level-set method for modelling of multi-phase flows. Int. J. Numer. Methods Fluids **47**, 339–361 (2005)
89. Van Dyke, M.: An Album of Fluid Motion. The Parabolic Press, Stanford (1982)
90. Vanka, S.P.: Implicit multigrid solutions of Navier–Stokes equations in primitive variables. J. Comput. Phys. **65**, 138–158 (1985)
91. Van Kan, J.: A second-order accurate pressure–correction scheme for viscous incompressible flow. SIAM J. Sci. Stat. Comput. **7**, 870–891 (1986)
92. Van Sint Annaland, M.S., Deen, N.G., Kuipers, J.A.M.: Numerical simulation of gas bubbles behaviour using a three-dimensional volume of fluid method. Chem. Eng. Sci. **60**, 2999–3011 (2005)
93. Ville, L., Silva, L., Coupez, T.: Convected level set method for the numerical simulation of fluid buckling. Int. J. Numer. Methods Fluids **66**, 324–344 (2011)

Algebraic Flux Correction and Geometric Conservation in ALE Computations

Guglielmo Scovazzi and Alejandro López Ortega

Abstract In this chapter, we describe the important role played by the so-called Geometric Conservation Law (GCL) in the design of Flux-Corrected Transport (FCT) methods for Arbitrary Lagrangian-Eulerian (ALE) applications. We propose a conservative synchronized remap algorithm applicable to arbitrary Lagrangian-Eulerian computations with nodal finite elements. Unique to the proposed method is the direct incorporation of the geometric conservation law (GCL) in the resulting numerical scheme. We show how the geometric conservation law allows the proposed method to inherit the positivity preserving and local extrema diminishing (LED) properties typical of FCT schemes for pure transport problems. The extension to systems of equations which typically arise in meteorological and compressible flow computations is performed by means of a synchronized strategy. The proposed approach also complements and extends the work of the first author on nodal-based methods for shock hydrodynamics, delivering a fully integrated suite of Lagrangian/remap algorithms for computations of compressible materials under extreme load conditions. Numerical tests in multiple dimensions show that the method is robust and accurate in typical computational scenarios.

1 Introduction

Arbitrary Lagrangian-Eulerian (ALE) algorithms (see Hirt et al. [20], Donea et al. [10]) utilize computational grids (or meshes) which are neither fixed (Eulerian) nor tied to the motion of the deformable medium (Lagrangian). ALE strategies are often used in applications which involve fluid/structure interaction problems (see, e.g., Donea et al. [10], Lesoinne and Farhat [35], Forster et al. [17]), shock hydrodynamics (see instead Benson [2, 4]), and, more recently, transport problems for

G. Scovazzi (✉)
Numerical Analysis and Applications, Sandia National Laboratories, MS-1319, Albuquerque, NM 87185-1319, USA
e-mail: gscovaz@sandia.gov

A. López Ortega
Graduate Aerospace Laboratories, California Institute of Technology, MC 105-50, Pasadena, CA 91125, USA

climate/meteorological simulations as discussed by Smolarkiewicz and Margolin [59] and references therein.

There are fundamentally two different approaches in ALE methods. The first is to solve in a single stage (monolithically) the ALE equations (see, e.g., Masud and Hughes [48] or Lesoinne and Farhat [35]). The second is instead to adopt an operator splitting procedure (see Donea et al. [10], Benson [3]): A pure Lagrangian computation is initially performed, then the nodes of the Lagrangian mesh are repositioned to improve the overall grid quality (rezoning), and finally the numerical solution is transferred from the old to the new grid. This last step is usually termed remap, and, more specifically, consists in transferring the numerical solution from a given mesh to a similar mesh with the same connectivity but different node locations.

Here we specifically focus on local continuous remapping, that is remapping that takes place at each time step in ALE computations, and is based on the solution of local problems under the assumption that mesh displacements are small. An alternative remap strategy (not discussed here) is global remapping, in which no assumption is made on the magnitude of mesh displacements and a global, conservative (constrained) interpolation problem is solved (see, e.g., Kuzmin et al. [34]).

In this chapter, we describe the important role played by the so-called Geometric Conservation Law (GCL) in the design of Flux-Corrected Transport (FCT) methods for ALE applications. The concept of a GCL, originally introduced by Lesoinne and Farhat [35] in the context of finite element/volume methods, has been recognized of great importance for the robustness and accuracy of ALE computations. In simple terms, the GCL is a consistency relationship between the change of volume of moving computational cells and the discrete divergence of the mesh velocity field.

Margolin and Shashkov [46, 47] pointed out that precise connections between advection algorithms for transport problems and continuous remap algorithms had yet to be completely understood. We show here how the GCL is the missing link necessary to bridge this theoretical gap, and that it is possible to cast a continuous remap problem as a pure transport problem, provided that the GCL is exactly satisfied in the discrete equations.

We would also like to point out that the GCL has recently being recognized as a fundamental ingredient in the proof of *a priori* error estimates for high-order finite element discretizations in ALE computations of the advection-diffusion equation by Bonito et al. [6].

One unique aspect of the approach discussed here is the explicit incorporation of the GCL into the discrete FCT equations, a feature which enables the overall algorithm to inherit conservation and local extremum diminishing (LED) properties typical of FCT schemes for pure transport problems. The scope of the present chapter is to present a thorough overview of a new class of GCL-compatible algorithms, stemming from the recent work of López Ortega and Scovazzi [40].

We propose a synchronized remap strategy based on flux-corrected transport algorithms, and in particular *algebraic* FCT methods (originally developed by Kuzmin and Turek [30], Kuzmin et al. [31–34], Kuzmin [29]).

We point out, however, that the main concepts presented can easily be extended to the case of monolithic ALE (as in the work of Boiarkine et al. [5]) and general FCT methods based on finite volume (FV) or finite difference (FD) discretizations (such as the ones originally developed by Boris and Book [7–9], Zalesak [62], Löhner et al. [37–39]).

The proposed method also complements the work in Scovazzi et al. [54, 55], Scovazzi [51], on Lagrangian shock hydrodynamics with nodal-based finite elements, and shows the feasibility of a suite of computational methods for shock hydrodynamics on nodal-based discretizations.

The exposition is organized as follows: Section 2 introduces a novel (to the authors' knowledge) continuum mechanics interpretation of ALE remap and the continuous version of the GCL. Section 3 describes the basic discretization of the continuum equations. Sections 4 and 5 present the proposed methodology first in the case of scalar equations and then in the case of systems of equations, respectively. Results from numerical tests in one, two, and three dimensions are presented in Sect. 6 and conclusions are summarized in Sect. 7.

2 General Concepts in Arbitrary Lagrangian-Eulerian Remap

ALE remap is the algorithmic procedure by which, on a fixed computational domain, data associated with a given computational grid (or mesh) is transferred onto a different grid with the same connectivity and improved quality. Remap is usually combined with Lagrangian algorithms to provide flexible and robust computational tools for fluid transport, as in Smolarkiewicz and Margolin [59], or shock hydrodynamics, as in Donea et al. [10], Benson [3], Dukowicz and Baumgardner [11], Dukowicz and Kodis [12], Dukowicz and Padial [13], Maire et al. [45], Kucharik et al. [28], Loubère et al. [42, 43], Galera et al. [19].

2.1 A Continuum Mechanics View on Remap

At the continuum level, a differential equation can be associated to the remap process. For this purpose a few definitions from continuum mechanics have to be introduced. Let the open set $\Omega_\chi \subset \mathbb{R}^{n_d}$ (where n_d is the number of spatial dimensions) denote the initial configuration of the mesh domain to be remapped, and let $\chi \in \Omega_\chi$ denote the position vector of a *particle* associated with the initial configuration of the mesh. For example, nodes of the mesh, barycenters of elements, etc., can be considered as particles. We can also define an open *portion* $\omega_\chi \subset \Omega_\chi$ as a subset of the mesh domain. For example, the interior of a single element or of a cluster of elements can define a portion.

In many situations of practical interest the remap procedure does not change the morphology of the boundary $\partial \Omega_\chi$ of Ω_χ. This happens every time the boundary motion is prescribed as *pure slip*, with no material inflow/outflow. On the contrary, an interior portion ω_χ will in general change its position as a result of the remap.

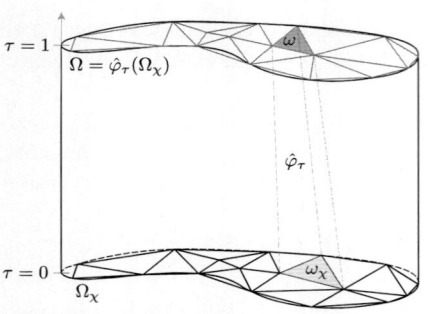
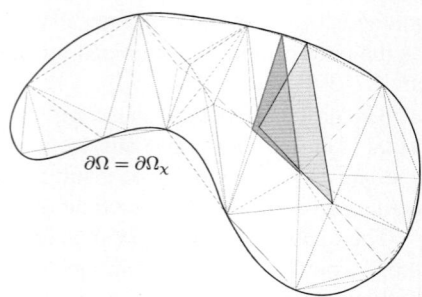

(a) Parametrization $\hat{\boldsymbol{\varphi}}_\tau$ of the mesh motion. (b) Projection along the τ-axis.

Fig. 1 Interpretation of the mesh remap motion $\hat{\boldsymbol{\varphi}}_\tau$. In (**a**), $\hat{\boldsymbol{\varphi}}_\tau$ is characterized as a family of maps parametrized with τ. An element portion ω_χ is remapped into $\omega = \hat{\boldsymbol{\varphi}}_\tau(\omega_\chi)$. In general, mesh particles do not need to deform along straight paths, but indeed they do when linear interpolation is used to represent $\hat{\boldsymbol{\varphi}}_\tau$. (**b**) shows the initial and final configurations of the mesh. *Black lines* represent the initial state of the mesh, and *red lines* the final state. The initial and final configurations of a mesh portion (a single element in this case) are represented by the *shaded areas in light blue and light red*, respectively (Color figure online)

Using the classical notation of continuum mechanics, the *mesh motion* (or remap motion) can be defined as

$$\hat{\boldsymbol{\varphi}} : \omega_\chi \to \omega = \hat{\boldsymbol{\varphi}}(\omega_\chi), \qquad \forall \omega_\chi \subset \Omega_\chi, \qquad (1)$$

$$\boldsymbol{\chi} \mapsto \boldsymbol{x} = \hat{\boldsymbol{\varphi}}_\tau(\boldsymbol{\chi}) = \hat{\boldsymbol{\varphi}}(\boldsymbol{\chi}, \tau), \quad \forall \boldsymbol{\chi} \in \omega_\chi, \ \tau \in [0,1]. \qquad (2)$$

The scalar τ is a parameter used to span the family of mesh deformations $\hat{\boldsymbol{\varphi}}_\tau$, and the notations $\hat{\boldsymbol{\varphi}}_\tau(\boldsymbol{\chi})$ and $\hat{\boldsymbol{\varphi}}(\boldsymbol{\chi}, \tau)$ can be used interchangeably. For $\tau = 0$ the mesh is in its initial configuration (i.e., $\hat{\boldsymbol{\varphi}}_0 = \boldsymbol{I}$, or $\hat{\boldsymbol{\varphi}}_0(\omega_\chi) = \omega_\chi$). For $\tau = 1$, the mesh is in its final configuration (i.e., $\hat{\boldsymbol{\varphi}}_1(\omega_\chi) = \omega$). Values of τ within the open interval $(0,1)$ indicate intermediate configurations of the mesh. Figure 1 shows a graphical interpretation of the concepts presented so far. In particular,

$$\hat{\boldsymbol{u}} = \left.\frac{\partial \hat{\boldsymbol{\varphi}}_\tau(\boldsymbol{\chi})}{\partial \tau}\right|_\chi \qquad (3)$$

indicates the mesh displacement for a mesh particle initially located at $\boldsymbol{\chi}$. We can also define the *mesh deformation gradient*

$$\hat{\boldsymbol{F}} = \nabla_\chi\big(\hat{\boldsymbol{\varphi}}_\tau(\boldsymbol{\chi})\big), \qquad (4)$$

and the *mesh deformation Jacobian determinant*

$$\hat{J} = \det(\hat{\boldsymbol{F}}) > 0, \qquad (5)$$

which represents the relative change in volume undergone by an infinitesimal portion of the mesh domain under the mesh map $\hat{\boldsymbol{\varphi}}_\tau$.

Algebraic Flux Correction and Geometric Conservation

Remark 1 In light of the above definition of $\hat{\boldsymbol{\varphi}}_\tau$, it would be wrong to interpret τ as time. There is in fact no concept of time in the context of ALE remap.

2.2 The Differential Remap Equation

Using the previous definitions, the remap differential equation for a scalar field $\rho(\boldsymbol{x}) = \rho(\hat{\boldsymbol{\varphi}}_\tau(\boldsymbol{\chi}))$ reads:

$$\left.\frac{\partial \rho}{\partial \tau}\right|_x = 0. \tag{6}$$

This simple equation expresses the fact that the mesh deformation should leave ρ unchanged, if observed from a reference frame fixed in space (Eulerian). This fact is also implicitly expressed by the dependency of ρ only on $\boldsymbol{x}$. Equation (6) is meaningful for sufficiently smooth mesh deformations, and is the basis of the proposed numerical approach. Using the chain rule of differentiation, the following classical identity holds:

$$\left.\frac{\partial \rho(\hat{\boldsymbol{\varphi}}(\boldsymbol{\chi},\tau))}{\partial \tau}\right|_\chi = \left.\frac{\partial \rho}{\partial \tau}\right|_x + \hat{\boldsymbol{u}} \cdot \nabla_x \rho, \tag{7}$$

where ∇_x is the spatial (Eulerian) gradient. This formula is a generalization of the relation between the Eulerian and Lagrangian derivatives of a field. Equation (6) can therefore be recast as

$$\left.\frac{\partial \rho(\boldsymbol{x},\tau)}{\partial \tau}\right|_\chi - \hat{\boldsymbol{u}} \cdot \nabla_x \rho = 0, \tag{8}$$

and its interpretation becomes clear looking at Fig. 1. Suppose an observer is "sitting" on a node initially located at position $\boldsymbol{\chi}$ and is transported by the node as the mesh reaches its final configuration. Then the measured rate of change of the field ρ at that node is related to the spatial gradient of ρ by how much the node displaces. This concept serves as the ground for a loose interpretation of the remap stage as advection. In reality, remap and advection are not equivalent, as will become clear momentarily.

Remark 2 Equation (8) is a general relationship, which is always true in the case of smooth fields, even for very large displacements of the mesh. It will be clear subsequently that specific discretizations of (8) may impose some limitations on the magnitude of the mesh displacements. Nonetheless, as long as the mesh motion map is smooth and well-posed (invertible), (8) is always correct.

2.3 Geometric Conservation Law

In its continuum (differential) form, the *geometric conservation law* expresses the compatibility between the rate of relative volume change $\hat{J}^{-1}\partial \hat{J}/\partial \tau|_\chi$ and the divergence of the mesh displacement field $\hat{u}$:

$$\left.\frac{\partial \hat{J}}{\partial \tau}\right|_\chi = \hat{J}\, \nabla_x \cdot \hat{u}. \qquad (9)$$

This relation can be derived from standard calculus identities (for complete details, see, e.g., López Ortega and Scovazzi [40]). A more intuitive integral form is obtained by integrating (9) over a portion ω_χ of the domain Ω_χ, using the Gauss theorem and the fact that $\hat{J}d\omega_\chi = d\omega$:

$$\frac{d}{d\tau}\left(\int_\omega d\omega\right) = \int_{\omega_\chi} \left.\frac{\partial \hat{J}}{\partial \tau}\right|_\chi d\omega_\chi = \int_\omega \nabla_x \cdot \hat{u}\, d\omega = \int_{\partial\omega} \hat{u}\cdot n\, d(\partial\omega), \qquad (10)$$

where n denotes the outward-pointing normal along the boundary $\partial\Omega$ of Ω. Thus, (10) relates the change in volume of the remapped portion ω of the initial domain Ω_χ to the engulfment of new regions of the domain due to the displacement of the remapped mesh.

2.4 Conservative Remap Equations

In general, remap algorithms need to preserve some basic conservation properties. Now (6) or (8) are not written in conservative form and their direct discretization with a finite difference/element/volume method would not be globally conservative. Assume that ρ is a conserved variable, meaning that the integral of ρ over Ω must be conserved through the remap procedure (e.g., ρ may indicate the mass density of a material, and its integral, the total mass of the mechanical system). Summing the product of (9) and ρ with the product of $\hat{J}$ and (8) leads to

$$\left.\frac{\partial(\hat{J}\rho)}{\partial \tau}\right|_\chi - \hat{J}\nabla_x \cdot (\rho\hat{u}) = 0, \qquad (11)$$

which is a conservation statement for remap. This is evident by integration of (11) over ω_χ, namely,

$$\int_{\omega_\chi} \left.\frac{\partial(\hat{J}\rho)}{\partial \tau}\right|_\chi d\omega_\chi - \int_{\omega_\chi} \hat{J}\nabla_x \cdot (\rho\hat{u})d\omega_\chi = 0. \qquad (12)$$

Then, recalling ω_χ is independent of τ and applying the Gauss divergence theorem,

$$\frac{d}{d\tau}\left(\int_\omega \rho\, d\omega\right) = \frac{d}{d\tau}\left(\int_{\omega_\chi} \hat{J}\rho\, d\omega_\chi\right) = \int_\omega \nabla_x \cdot (\rho\hat{u})d\omega = \int_{\partial\omega} \rho\hat{u}\cdot n\, d(\partial\omega). \qquad (13)$$

Equation (13), whenever the boundary integral vanishes, leads to a global conservation statement for the integral of ρ.

Remark 3 We reiterate that equations (6) (or (8)) do not lead to discrete integral conservation statements. Moreover, the integral conservation statement results from a combination of (6) (or (8)) and the geometric conservation law (9).

2.5 Boundary Conditions

Equation (11) needs to be complemented with appropriate boundary conditions. For the sake of simplicity, and without loss of generality, only the case in which the boundary does not deform will be considered, that is

$$\hat{u} \cdot n = 0 \quad \text{on } \partial \Omega_\chi. \tag{14}$$

In this case, boundary conditions may not be specified. Alternatively, it is possible to impose Dirichlet boundary conditions of the type

$$\rho\big(\hat{\varphi}_\tau^{-1}(x)\big) = \bar{\rho}(\chi) \quad \text{on } \partial \Omega_\chi. \tag{15}$$

This amounts to enforce that the solution does not change under remap at the boundary of the computational domain.

Remark 4 In the case of ALE shock hydrodynamics, computations are performed as a sequence of alternating Lagrangian and remap stages (continuous remap). Then condition (14) is equivalent to impose that the boundary of the domain Ω undergoes pure Lagrangian motion. In this case, no boundary conditions are typically imposed for the density and internal energy, while strong Dirichlet conditions of the type (15) are imposed for the velocity component normal to the boundary.

3 A Geometrically-Conservative Second-Order Remap

The sections that follow are devoted to the presentation of the discrete approximation to the remap equations using a geometrically-conservative FCT method.

3.1 A Conservative Nodal-Based Finite Element Formulation

A conservative finite element formulation can be developed by testing equation (11) on a space of appropriate *variations*. In particular, the rest of the paper is focused on piecewise-linear globally-continuous approximations. Consider the scalar equation

(11), with Dirichlet boundary conditions given as in Sect. 2.5. In this case, $\partial \Omega \equiv \partial \Omega_\chi$, and consequently, the trial space $\mathscr{S}^h$ and test space $\mathscr{V}^h$ are given by

$$\mathscr{S}^h = \{ \psi^h \in C^0(\Omega_\chi) : \psi^h \in \mathscr{P}_1(\omega_{\chi;e}), \ \psi^h = \bar{\rho}(\chi) \text{ on } \partial \Omega_\chi \}, \quad (16)$$

$$\mathscr{V}^h = \{ \psi^h \in C^0(\Omega_\chi) : \psi^h \in \mathscr{P}_1(\omega_{\chi;e}), \ \psi^h = 0 \text{ on } \partial \Omega_\chi \}, \quad (17)$$

where $\mathscr{P}_1(\omega_{\chi;e})$ is the space of piecewise-linear functions over the element domain $\omega_{\chi;e}$, and $\bar{\rho}$ is the Dirichlet boundary condition. If no boundary conditions are specified, then the trial/test spaces coincide and take the form

$$\mathscr{S}^h = \mathscr{V}^h = \{ \psi^h \in C^0(\Omega_\chi) : \psi^h \in \mathscr{P}_1(\omega_{\chi;e}) \}. \quad (18)$$

Incidentally, (16)–(17) and (18) lead to the same variational statement, which is readily obtained testing (11) against a member of $\mathscr{V}^h$. Namely, given a mesh displacement field $\hat{\boldsymbol{u}}$, seek $\rho^h \in \mathscr{S}^h$ such that, $\forall \psi^h \in \mathscr{V}^h$,

$$\int_{\Omega_\chi} \psi^h(\chi) \frac{\partial (\hat{J} \rho^h(\hat{\boldsymbol{\varphi}}_\tau(\chi)))}{\partial \tau} \bigg|_\chi d\Omega_\chi = \frac{d}{d\tau} \int_{\Omega_\chi} \psi^h \rho^h \hat{J} d\Omega_\chi$$

$$= \int_{\Omega_\chi} \psi^h \nabla_x \cdot (\rho^h \hat{\boldsymbol{u}}) \hat{J} d\Omega_\chi. \quad (19)$$

The last equality in (19) can be used to derive an integration procedure in the fashion of space-time integration algorithms (see, e.g., Scovazzi et al. [54], Scovazzi [49–51], Scovazzi and Love [53], Hulme [21], Jamet [25], Aziz and Monk [1] and references therein). Integrating along τ and changing reference frames in the left-hand-side yields

$$\int_\Omega \psi^h(\hat{\boldsymbol{\varphi}}_1^{-1}(\boldsymbol{x})) \rho^h(\boldsymbol{x}) d\Omega - \int_{\Omega_\chi} \psi^h(\chi) \rho^h(\chi) d\Omega$$

$$= -\int_0^1 \left(\int_{\Omega_\chi} \nabla_x \psi^h(\chi) \cdot (\rho^h \hat{\boldsymbol{u}}) \hat{J} d\Omega_\chi \right) d\tau, \quad (20)$$

where integration by parts has been performed on the right-hand-side, accounting for boundary conditions. This equation represents the foundation of our algorithmic approach, and many remap algorithms found in the literature fall under this general framework (see, e.g., Fressmann and Wriggers [18], Benson [2–4], Donea et al. [10], Dukowicz and Baumgardner [11], Dukowicz and Kodis [12], Dukowicz and Padial [13], Maire et al. [45], Margolin and Shashkov [46, 47], Vàchal et al. [61], Vàchal and Liska [60], Liska et al. [36]). The spatial discretization has already been chosen by defining the variational spaces $\mathscr{V}^h$ and $\mathscr{S}^h$, the remaining step is the choice of an appropriate approximation space and quadrature formulas along τ.

Remark 5 (Conservation) The variational formulation (20) is globally conservative. In fact, when homogeneous Neumann conditions are assumed (no normal flux of the solution ρ across the boundary), the unit constant function over the entire domain Ω_χ is in the test space. Consequently, for this test function choice, the integral on the right-hand-side of (20) vanishes, leading to the global conservation statement

$$\int_\Omega \rho^h(x)\,\mathrm{d}\Omega = \int_{\Omega_\chi} \rho^h(\chi)\,\mathrm{d}\Omega. \tag{21}$$

3.2 Approximation of the Mesh Motion and Mesh Displacements

The map $\hat{\varphi}$ is discretized making use of a piecewise linear discretization in space and (pseudo-)time, namely

$$\hat{\varphi}_\tau^h(\chi) = \sum_{A=1}^{n_{np}} \left(\tau \hat{\varphi}_1(\chi_A) + (1-\tau) \hat{\varphi}_0(\chi_A) \right) N_A(\chi), \tag{22}$$

where $N_A(\chi)$ is the finite element shape function centered at node A, and n_{np} is the total number of nodal mesh points. In particular N_A is continuous, piecewise-linear and of compact support, that is $N_A(\chi_B) = \delta_{AB}$ (the Kronecker delta tensor, such that $\delta_{AB} = 1$ if $A = B$, and $\delta = 0$ if $A \neq B$). Furthermore, the shape function basis has the partition of unity property, that is $\sum_A N_A = 1$. By definition (3), equation (22) also implies

$$\hat{u}(\chi) = \sum_{A=1}^{n_{np}} \left(\hat{\varphi}_1(\chi_A) - \hat{\varphi}_0(\chi_A) \right) N_A(\chi) = \sum_{A=1}^{n_{np}} \hat{u}(\chi_A) N_A(\chi), \tag{23}$$

independent of the parameter τ. Recall that $\hat{u}$ is a *datum* in the remap problem, since the new mesh positioning is chosen based on mesh quality considerations.

3.3 A Group Finite Element Approach

Effective remap algorithms need to maintain a number of important algorithmic properties. Global conservation has already been mentioned as a fundamental requirement. In addition, second-order accuracy is deemed necessary in practical applications. In order to satisfy these requirements a simple approach is to adopt an algorithm similar to the space-time discretizations presented in Scovazzi et al. [54, 55], Scovazzi [49–51], Scovazzi and Love [53] (see also references therein), with appropriate modifications to take advantage of the particular structure of the remap problem. Specifically, the test space will be given by the tensor product of spatial functions in $\mathscr{V}^h$ with piecewise-constant functions in τ, while the trial (solution) space will be given by the tensor product of spatial functions in $\mathscr{S}^h$ with piecewise-linear, continuous functions in τ. Consequently, a natural discretization for ρ^h is analogous to the one for mesh motion, that is,

$$\rho^h(\hat{\boldsymbol{\varphi}}_\tau(\boldsymbol{\chi})) = \sum_{A=1}^{n_{np}} (\tau\rho_1(\boldsymbol{\chi}_A) + (1-\tau)\rho_0(\boldsymbol{\chi}_A))N_A(\boldsymbol{\chi}), \qquad (24)$$

with $\rho_\tau(\boldsymbol{\chi}_A)$ the nodal value of ρ^h for $\tau = 0, 1$. This definition can be combined with (23) to discretize the product $\rho^h \hat{\boldsymbol{u}}$.

However, as will appear clear in what follows, the so-called *group finite element formulation* of Fletcher [14] is more advantageous in obtaining a discretely conservative algorithm, particularly when adopting an FCT method for a system of equations. This amounts to choose:

$$\rho^h(\hat{\boldsymbol{\varphi}}_\tau(\boldsymbol{\chi}))\hat{\boldsymbol{u}}(\boldsymbol{\chi}) = \sum_{A=1}^{n_{np}} \big(\tau\rho(\hat{\boldsymbol{\varphi}}_1(\boldsymbol{\chi}_A))$$
$$+ (1-\tau)\rho(\hat{\boldsymbol{\varphi}}_0(\boldsymbol{\chi}_A))\big)\hat{\boldsymbol{u}}(\boldsymbol{\chi}_A)N_A(\hat{\boldsymbol{\varphi}}_\tau(\boldsymbol{\chi})). \qquad (25)$$

Needless to say, our goal is to obtain a computationally efficient algorithm. With this purpose in mind, and *only* for the integral term in the right-hand-side of (20), it is possible to further approximate ρ^h by replacing the linear interpolation with a τ-average reminiscent of a trapezoidal integration rule

$$\rho^h(\hat{\boldsymbol{\varphi}}_\tau(\boldsymbol{\chi}))\hat{\boldsymbol{u}}(\boldsymbol{\chi}) = \sum_{A=1}^{n_{np}} \frac{\rho^h(\hat{\boldsymbol{\varphi}}_0(\chi_A)) + \rho^h(\hat{\boldsymbol{\varphi}}_1(\chi_A))}{2}\hat{\boldsymbol{u}}(\boldsymbol{\chi}_A)N_A(\hat{\boldsymbol{\varphi}}_\tau(\boldsymbol{\chi})). \qquad (26)$$

Using the original definition (24) is more involved computationally, but does not yield an improvement in order of accuracy. Combining all previous discretization choices, we obtain the following discrete system:

$$\mathbf{V}_1\boldsymbol{\rho}_1 - \mathbf{V}_0\boldsymbol{\rho}_0 = \bar{\mathbf{K}}\left(\frac{\boldsymbol{\rho}_1 + \boldsymbol{\rho}_0}{2}\right), \qquad (27)$$

where

$$\mathbf{V}_{(\cdot)} = [\mathbf{V}_{(\cdot);AB}], \qquad (28)$$

$$\mathbf{V}_{(\cdot);AB} = \int_{\Omega_\chi} N_A(\boldsymbol{\chi})N_B(\boldsymbol{\chi})\hat{J}(\hat{\boldsymbol{\varphi}}_{(\cdot)}(\boldsymbol{\chi}))\mathrm{d}\Omega_\chi$$

$$= \int_{\hat{\boldsymbol{\varphi}}_{(\cdot)}(\Omega_\chi)} N_A(\hat{\boldsymbol{\varphi}}_{(\cdot)}(\boldsymbol{\chi}))N_B(\hat{\boldsymbol{\varphi}}_{(\cdot)}(\boldsymbol{\chi}))\mathrm{d}\Omega, \qquad (29)$$

$$\bar{\mathbf{K}} = \int_0^1 \mathbf{K}(\tau)\mathrm{d}\tau = \left[\int_0^1 \mathbf{K}_{AB}(\tau)\mathrm{d}\tau\right] \qquad (30)$$

$$\mathbf{K}_{AB}(\tau) = \mathbf{C}_{AB}(\tau) \cdot \hat{\mathbf{u}}_B \quad \text{(no sum)}, \tag{31}$$

$$\mathbf{C}_{AB}(\tau) = -\int_{\Omega_\chi} \nabla_x N_A(\chi) N_B(\chi) \hat{J}(\hat{\boldsymbol{\varphi}}_\tau(\chi)) \mathrm{d}\Omega_\chi$$

$$= -\int_{\hat{\boldsymbol{\varphi}}_\tau(\Omega_\chi)} \nabla_x N_A N_B \mathrm{d}\Omega, \tag{32}$$

$$\boldsymbol{\rho}_{(\cdot)} = \{\rho_{(\cdot);A}\}, \tag{33}$$

with $\hat{\mathbf{u}}_B = \hat{u}(\chi_B)$ and $\rho_{(\cdot);A} = \rho^h(\hat{\boldsymbol{\varphi}}_{(\cdot)}(\chi_A))$.

Remark 6 The partition-of-unity property of the finite element shape functions, that is $\sum_A N_A = 1$, directly implies $\sum_A \mathbf{C}_{AB} = \mathbf{0}$.

Remark 7 It can be proven (see, e.g., López Ortega and Scovazzi [40]), that if A is not a boundary node, $\mathbf{C}_{AB}$ has the skew-symmetry property, that is $\mathbf{C}_{AB} = -\mathbf{C}_{BA}$, which in turn implies $\sum_B \mathbf{C}_{AB} = \mathbf{0}$. As a consequence, $\mathbf{C}_{AA} = \mathbf{0}$ for interior nodes, which also implies that the diagonal entries of $\bar{\mathbf{K}}$ are zero (this fact can also be proved directly using Gauss theorem along a single coordinate direction, see again López Ortega and Scovazzi [40]).

3.4 Discretization of the Geometric Conservation Law

It is also important to realize that substituting into (27) a constant solution ρ^h in space and time, we obtain the discretization of the geometric conservation law (10), namely,

$$\mathbf{V}_1^L - \mathbf{V}_0^L = \Sigma[\bar{\mathbf{K}}], \tag{34}$$

where $\mathbf{V}_1^L$, $\mathbf{V}_0^L$, $\Sigma[\bar{\mathbf{K}}]$, are lumped, row-sum matrices obtained from $\mathbf{V}_1$, $\mathbf{V}_C$, and $\bar{\mathbf{K}}$. Respectively:

$$\mathbf{V}_{(\cdot)}^L = \left[\mathrm{diag}\left(\sum_{B=1}^{n_{np}} \mathbf{V}_{(\cdot);AB}\right)\right], \tag{35}$$

and, for a given matrix $\mathbf{M} = [M_{AB}]$, the operator $\Sigma[\cdot]$ computes the row-sum diagonal matrix

$$\Sigma[\mathbf{M}] = \left[\mathrm{diag}\left(\sum_{B=1}^{n_{np}} M_{AB}\right)\right]. \tag{36}$$

Note in particular that, due to the boundary conditions on $\hat{u}$,

$$\sum_{B=1}^{n_{np}} \mathbf{V}_{(\cdot);AB} = \sum_{B=1}^{n_{np}} \int_{\Omega_\chi} N_A(\chi) N_B(\chi) \hat{J}(\hat{\boldsymbol{\varphi}}_{(\cdot)}(\chi)) \mathrm{d}\Omega_\chi = \int_{\hat{\boldsymbol{\varphi}}_{(\cdot)}(\Omega_\chi)} N_A \mathrm{d}\Omega, \tag{37}$$

$$\sum_{B=1}^{n_{np}} \bar{\mathsf{K}}_{AB} = \sum_{B=1}^{n_{np}} \left(\int_0^1 \mathbf{C}_{AB}(\tau) \cdot \hat{\mathbf{u}}_B \mathrm{d}\tau \right) = -\left(\int_0^1 \left(\int_{\Omega_\chi} \nabla_x N_A \cdot \hat{\mathbf{u}} \hat{J} \mathrm{d}\Omega_\chi \right) \mathrm{d}\tau \right)$$

$$= \left(\int_0^1 \left(\int_{\hat{\varphi}_\tau(\Omega_\chi)} N_A \nabla_x \cdot \hat{\mathbf{u}} \mathrm{d}\Omega \right) \mathrm{d}\tau \right). \tag{38}$$

Remark 8 $\Sigma[\bar{\mathsf{K}}] = \sum_B \bar{\mathsf{K}}_{AB}$ can be interpreted as a weak variational projection of the divergence of the displacement field onto the nodal shape function space.

Remark 9 An algorithm which embeds the geometric conservation law is always capable of representing a constant solution, independently of the mesh motion chosen. Algorithms which do not abide the geometric conservation law would instead introduce spurious numerical distributions of sources/sinks in the computational domain, causing the numerical solution to oscillate around the constant state. It will be shown momentarily that the geometric conservation law is essential in guaranteeing local extremum diminishing (LED) properties for flux-corrected transport remap.

4 Flux Corrected Synchronized Remap for Scalar Fields

Equations (27) and (34) are the basis for a synchronized, nodal-based, flux corrected remap (FCR) method described next. The derivations which follow show how the step-by-step integration of the discrete GCL (34) in previous work on algebraic FCT methods by Kuzmin [29], Kuzmin et al. [31, 32, 34], Kuzmin and Turek [30], leads to an conservative, accurate and LED-preserving FCR method.

We would also like to mention that Vàchal and Liska [60] have applied the ideas of FCT to classical staggered discretization arising in traditional shock hydrodynamics algorithms. In that case, thermodynamic variables, such as density and internal energy, are approximated as piecewise-constant fields and the linear momentum is collocated at the nodes of the computational grid. This results in different discrete remap operators needed to transfer momentum and thermodynamic quantities between meshes. In our approach, instead, all variables are collocated at the nodes of the mesh, and all conserved variables are remapped using the *same* discrete operators, with clear advantages from the point of view of computational efficiency. Our approach also encompasses the *repair* paradigm of Margolin and Shashkov [46, 47] by means of the *fail safe* approach described in Kuzmin et al. [34], of much simpler implementation.

The discussion is initially focused on the scalar case, and will later extend to systems of conservation laws. Equation (27) is a linear system and can be directly solved:

$$\boldsymbol{\rho}_1 = \left(\mathsf{V}_1 - \frac{1}{2} \bar{\mathsf{K}} \right)^{-1} \left(\mathsf{V}_0 + \frac{1}{2} \bar{\mathsf{K}} \right) \boldsymbol{\rho}_0. \tag{39}$$

However, because (39) is derived from a Bubnov-Galerkin formulation of the variational remap problem, that is a non-monotone centered discretization, oscillations are to be expected. Additionally, the inversion of a linear system may result computationally inefficient for large scale application of remap strategies in conjunction with explicit dynamics algorithms. We propose to address the former issue using an FCR approach, and the latter using a predictor/multi-corrector strategy.

4.1 Low-Order Discretization

The first stage in the development of the FCR algorithm is to assemble a conservative low-order scheme which provides positivity preserving and local extrema diminishing (LED) properties (see Jameson [22–24]) at the expense of lowering numerical accuracy. In addition, the low-order scheme must be compatible with the discretization (stencil) of the Galerkin high-order scheme it is derived from. As shown by Kuzmin and Turek [30] and in Chap. 6 of the book by Kuzmin et al. [33], such low-order scheme can be derived from (27) by row-sum lumping the matrices $\mathbf{V}_0$ and $\mathbf{V}_1$, and augmenting the matrix $\bar{\mathbf{K}}$ by an appropriate *algebraic* diffusion matrix $\bar{\mathbf{D}}$. Namely, the low-order scheme can be expressed as:

$$\mathbf{V}_1^L \rho_1^{low} - \mathbf{V}_0^L \rho_0 = \bar{\mathbf{L}} \rho_{1/2}^{low}, \tag{40}$$

where $\bar{\mathbf{L}} = \bar{\mathbf{K}} + \bar{\mathbf{D}}$, $\mathsf{V}_{(\cdot);AA}^L = \sum_B \mathsf{V}_{AB}$, $\rho_{1/2}^{low} = 1/2(\rho_1^{low} + \rho_0)$, and the superscript *low* stands for low order.

4.2 Positivity Properties and the Algebraic Diffusion Matrix

Positivity is a very valuable property in algorithms used to simulate systems of conservation laws, in which some solution variables (e.g., density, pressure, internal energy) must always preserve a positive sign. The low-order scheme can be written as:

$$\mathbf{A} \rho_1^{low} = \mathbf{B} \rho_0, \tag{41}$$

with $\mathbf{A} = \mathbf{V}_1^L - 1/2\bar{\mathbf{L}}$ and $\mathbf{B} = \mathbf{V}_0^L + 1/2\bar{\mathbf{L}}$. Assuming ρ_0 is an array with positive entries, the same property is desirable for the remapped array ρ_1^{low}. This is equivalent to require that $\mathbf{A}^{-1}\mathbf{B}$ is a positive matrix (a matrix with positive entries), condition which is satisfied if $\mathbf{A}$ is a so-called M-matrix and $\mathbf{B}$ is positive ($\mathsf{B}_{AB} \geq 0$). Sufficient conditions for $\mathbf{A}$ to be an M-matrix are:

$$\mathsf{A}_{AA} \geq 0, \tag{42a}$$

$$\mathsf{A}_{AB} \leq 0, \quad \text{for } B \neq A, \tag{42b}$$

$$\mathsf{A}_{AA} \geq -\sum_{B \neq A} \mathsf{A}_{AB}. \tag{42c}$$

Condition (42c) is usually referred as *diagonal dominance*, and must hold either for all rows or, in the case **A** is irreducible, for at least one row.

Following Kuzmin and Turek [30] or Chap. 6 of Kuzmin et al. [33], we design a positive scheme by constructing a matrix $\bar{\mathbf{D}}$ so that it behaves as an algebraic diffusion operator. In addition, the proposed approach yields a conservative, local extrema diminishing algorithm. $\bar{\mathbf{D}} = [\bar{D}_{AB}]$ being a diffusion matrix implies symmetry (i.e., $\bar{D}_{AB} = \bar{D}_{BA}$). Conservation requires that $\sum_B \bar{D}_{AB} = \sum_A \bar{D}_{AB} = 0$, that is $\bar{\mathbf{D}}$ should not perturb any discrete global conservation statements. In particular, the previous requirements entail $\bar{D}_{AA} = -\sum_{B \neq A} \bar{D}_{AB}$, where $\sum_{F \neq G}$ is shorthand notation for $\sum_{F=1, F \neq G}^{nnp}$. A simple approach to deriving a matrix $\bar{\mathbf{D}}$ from $\bar{\mathbf{K}}$ is to enforce $\bar{D}_{AB} = \max(-\bar{K}_{AB}, -\bar{K}_{BA}, 0)$, for $A \neq B$, and $\bar{D}_{AA} = -\sum_{A \neq B} \bar{D}_{AB} = -\sum_{B \neq A} \bar{D}_{AB}$ (see Kuzmin and Turek [30]).

Remark 10 By construction, $\bar{D}_{AB} \geq 0$, for $B \neq A$, and, consequently, $\bar{D}_{AA} \leq 0$. Again, by construction, $\bar{L}_{AB} \geq 0$ for $B \neq A$.

As shown in full detail in López Ortega and Scovazzi [40], the previous definition of the matrix $\bar{\mathbf{D}}$ yields a matrix **A** which satisfies exactly (42a)–(42c). With regard to the matrix **B**, the condition $B_{AB} \geq 0$ is clearly satisfied when $B \neq A$, since the off-diagonal entries of **B** are just the off-diagonal entries of $\bar{\mathbf{L}}$, while the condition $B_{AA} \geq 0$ leads to a Courant-Friedrichs-Lewy (CFL) condition for positivity, namely $\min_A (V^L_{0;AA} - 1/2 \bar{D}_{AA}) > 0$, which restricts the magnitude of the mesh displacements.

4.3 Discrete Geometric Conservation Law

It is now important to appreciate the role played by the geometric conservation law (34) in the context of the proposed algebraic FCR approach. The discussion in this and the following sections differs from and extends the classical work on algebraic FCT schemes of Kuzmin and Turek [30], Kuzmin et al. [31–34], Kuzmin [29]. Consider the difference between (40) and the product of $\rho_{1/2}^{low}$ and (34):

$$\mathbf{V}_1^L \rho_1^{low} - \mathbf{V}_0^L \rho_0 - \rho_{1/2}^{low}(\mathbf{V}_1^L - \mathbf{V}_0^L) = (\bar{\mathbf{K}} - \Sigma[\bar{\mathbf{K}}] + \bar{\mathbf{D}}) \rho_{1/2}^{low}, \tag{43}$$

or, after a few simple algebraic manipulations,

$$\mathbf{V}_{1/2}^L (\rho_1^{low} - \rho_0) = \hat{\mathbf{L}} \rho_{1/2}^{low}, \tag{44}$$

where $\mathbf{V}_{1/2}^L = 1/2(\mathbf{V}_1^L + \mathbf{V}_0^L)$ and $\hat{\mathbf{L}} = \hat{\mathbf{K}} + \bar{\mathbf{D}}$ with $\hat{\mathbf{K}} = \bar{\mathbf{K}} - \Sigma[\bar{\mathbf{K}}]$.

Remark 11 By definition, $\hat{\mathbf{K}}$ and $\hat{\mathbf{L}}$ have zero row sum. In addition, the off-diagonal entries of $\hat{\mathbf{L}}$ are non-negative, by construction of $\bar{\mathbf{D}}$.

Algebraic Flux Correction and Geometric Conservation 313

It is at this point crucial to observe that if $\mathbf{V}_1^L$ is computed using (35) and satisfies (34) exactly, then (40) and (44) are exactly equivalent. Linear simplex-type finite elements (i.e., line elements in one dimension, triangular elements in two dimensions and tetrahedral elements in three dimensions) exactly satisfy (34), provided that one and two quadrature points are used along the τ coordinate for evaluating (38) in two and three dimensions, respectively. For a detailed proof of this statement, see Lesoinne and Farhat [35], Formaggia and Nobile [15, 16] or Scovazzi and Hughes [52] and references therein.

Because the spatial integrals in (38) cannot be integrated exactly for general mesh geometries in the case of quadrilateral or hexahedral linear elements, we will always make the choice of computing $\mathbf{V}_1^L$ using (34) rather than (35). This approach ensures that the lumped volume matrix $\mathbf{V}_1^L$ is always compatible with a geometric conservation law irrespective of the type of finite element adopted.

Remark 12 Note that the geometric conservation law somewhat supersedes the condition on diagonal dominance on $\mathbf{A}$ discussed in Sect. 4.2. In fact, in the context of the present method, condition (42c) translates into:

$$\mathsf{V}^L_{1;AA} - \frac{1}{2}(\bar{\mathsf{K}}_{AA} + \bar{\mathsf{D}}_{AA}) \geq \frac{1}{2} \sum_{B \neq A} (\bar{\mathsf{K}}_{AB} + \bar{\mathsf{D}}_{AB}), \tag{45}$$

that is,

$$\mathsf{V}^L_{1/2;AA} = \mathsf{V}^L_{1;AA} - \frac{1}{2}\sum_B \bar{\mathsf{K}}_{AB} \geq \frac{1}{2}\sum_B \bar{\mathsf{D}}_{AB} = 0. \tag{46}$$

Now $\mathsf{V}^L_{1/2;AA} \geq 0$ by definition, and, consequently, the geometric conservation law implies $\mathbf{A}$ is diagonally dominant.

4.3.1 Geometric Conservation and Local Extremum Diminishing Properties

A numerical scheme is LED if it prevents the creation of new local extrema with respect to the distribution of extrema in the initial solution as discussed by Jameson [22–24], Kuzmin and Turek [30], Kuzmin et al. [33]. It is very important to realize at this point that the change in mesh geometry during remap prevents the direct application of the standard FCT methodology for pure transport problems based on the premise that $\mathbf{V}_1^L = \mathbf{V}_0^L$ (as developed, e.g., in Kuzmin and Turek [30] and Chap. 6 of Kuzmin et al. [33]).

The forthcoming discussion shows that satisfaction of the geometric conservation law allows the proposed FCR scheme to inherit the LED properties of the FCT schemes it is derived from. Typically, the LED properties are discussed in the semi-discrete context (see again Kuzmin and Turek [30] and Chap. 6 of Kuzmin et al. [33]). For a transport problem, this means that the equations are discretized in time but not in space. In the case of remap, this is equivalent to applying discretization in space but not along the τ coordinate. Namely,

$$\partial_\tau \left(\mathbf{V}_\tau^L \boldsymbol{\rho}^{low} \right) = \mathbf{L} \boldsymbol{\rho}^{low}, \qquad (47)$$

where $\mathbf{L} = \mathbf{K} + \mathbf{D}$, with $D_{AB} = \max(-K_{AB}, -K_{BA}, 0)$. $\mathbf{K}$ is defined in (31) and differs from $\bar{\mathbf{K}}$, as its entries are not integrated along τ. Note however that $\mathbf{D}$ and $\bar{\mathbf{D}}$ and $\mathbf{L}$ and $\bar{\mathbf{L}}$ share the same properties, as integration along τ does not change the sign of matrix entries. Analogously, the semi-discrete version of the geometric conservation law reads

$$\partial_\tau \mathbf{V}_\tau^L = \Sigma[\mathbf{K}]. \qquad (48)$$

Proceeding as in Sect. 4.3, by taking the difference of (47) and the product of $\boldsymbol{\rho}^{low}$ and (48), and rearranging terms, the semi-discrete version of (44) is obtained:

$$\partial_\tau \boldsymbol{\rho}^{low} = \left(\mathbf{V}_\tau^L \right)^{-1} \mathbf{L} \boldsymbol{\rho}^{low}, \qquad (49)$$

or, in components,

$$\partial_\tau \rho_A^{low} = \frac{1}{V_{\tau;AA}^L} \sum_{B \neq A} \mathsf{L}_{AB} \left(\rho_B^{low} - \rho_A^{low} \right), \qquad (50)$$

where we have used the fact that $\mathsf{L}_{AA} = -\sum_{B \neq A} \mathsf{L}_{AB}$, since $\mathbf{L}$ has zero row sum. By construction, the off-diagonal entries of $\mathbf{L}$ are non-negative, and, in addition, the sparsity pattern of $\mathbf{L}$ is given by the local compact support of the linear shape functions N_A's. In particular, the only non-zero entries of $\mathbf{L}$ are in correspondence of neighboring nodes (i.e., nodes connected by a common element). In conclusion, if the solution ρ^{low} has a maximum at node A, then $\partial_\tau \rho_A^{low} \leq 0$, since $\rho_B^{low} - \rho_A^{low} \leq 0$ for every node B in the neighborhood of A. The case of a minimum is analogous, and this concludes the classical proof of LED properties in Jameson [22–24].

In the fully-discrete context of the proposed FCR approach, the previous argument can be applied to (44) using (34), namely,

$$\rho_{1;A}^{low} - \rho_{0;A}^{low} = \frac{1}{V_{1/2;A}^L} \sum_{A=1}^{n_{np}} \hat{\mathsf{L}}_{AB} \rho_{1/2;B}^{low}$$

$$= \frac{1}{V_{1/2;A}^L} \sum_{B \neq A} \hat{\mathsf{L}}_{AB} \left(\rho_{1/2;B}^{low} - \rho_{1/2;A}^{low} \right). \qquad (51)$$

Remark 13 It is fundamental to observe that if the geometric conservation law is not respected, then (40) is not equivalent to (44), and, as a result, (51) does not hold.

4.3.2 Geometric Conservation and Accuracy

When the matrix $\mathbf{D}$ is set to zero, the higher order method is recovered. Using again the geometric conservation law, (43) and (44) lead to the two equivalent discrete equations for the higher-order scheme:

$$\mathbf{V}_1^L \rho_1 - \mathbf{V}_0^L \rho_0 = \bar{\mathbf{K}} \rho_{1/2}, \qquad (52)$$

$$\mathbf{V}_{1/2}^L (\rho_1 - \rho_0) = \hat{\mathbf{K}} \rho_{1/2}, \qquad (53)$$

where the last equation is in advective form. To be more specific, because $\hat{\mathbf{K}}$ has zero sum, it is analogous to a matrix obtained in the case of advection due to a divergence-free convective field. Hence, the theory developed in Kuzmin et al. [35] directly applies. Specifically, since piecewise-linear finite elements are used in obtaining (53), the theory of finite element indicates that the proposed high-order method is second-order accurate, in terms of truncation error.

4.4 An Efficient Predictor/Multi-corrector Scheme

The simplest possible predictor/multi-corrector version of (40) reads:

$$\mathbf{V}_1^L \rho_1^{low;(i+1)} - \mathbf{V}_0^L \rho_0 = (\bar{\mathbf{K}} + \bar{\mathbf{D}}) \rho_{1/2}^{low;(i)}, \qquad (54)$$

where $(i+1)$ and (i) indicate the current and previous iterate of the solution, $\rho_{1/2}^{low;(i)} = 1/2(\rho_1^{low;(i)} + \rho_0)$, and $\mathbf{V}_1^L$ is computed according to (34). In practice, the first pass of the predictor/multi-corrector scheme is a classical explicit Euler step, while successive corrections allow the scheme to reach second-order accuracy in τ. The positivity properties of the scheme are easily derived by observing that the first step of (54) corresponds to (41) with $\mathbf{A} = \mathbf{V}_1^L$ and $\mathbf{B} = \mathbf{V}_0^L + \bar{\mathbf{L}}$. The analysis of LED properties is more delicate, as the equation corresponding to (43) reads:

$$\mathbf{V}_1^L \rho_1^{low;(i+1)} - \mathbf{V}_0^L \rho_0 - \rho_{1/2}^{low;(i)}(\mathbf{V}_1^L - \mathbf{V}_0^L) = (\bar{\mathbf{K}} - \Sigma[\bar{\mathbf{K}}] + \bar{\mathbf{D}}) \rho_{1/2}^{low;(i)}, \qquad (55)$$

that is

$$\mathbf{V}_{1/2}^L (\rho_1^{low;(i+1)} - \rho_0) = \hat{\mathbf{L}} \rho_{1/2}^{low;(i)} - (\rho_{1/2}^{low;(i+1)} - \rho_{1/2}^{low;(i)})(\mathbf{V}_1^L - \mathbf{V}_0^L). \qquad (56)$$

This means that only in the limit of a large number of iterations the predictor/multi-corrector shares the same LED properties of the original algorithms from which it is derived. However, the presence of the term $(\rho_{1/2}^{low;(i+1)} - \rho_{1/2}^{low;(i)})(\mathbf{V}_1^L - \mathbf{V}_0^L)$ did not produce any spurious maxima/minima to be formed in any of the numerical simulations performed, no matter how few iterations were performed. The computations presented subsequently were performed with one predictor and two corrector passes.

4.5 High-Order Anti-diffusive Fluxes

As shown in the previous section, the low-order solution obtained from (40) prevents the over/undershoots possibly generated by the original high-order scheme, but it is in general overly diffusive. In FCT methods, an appropriate local convex combination of the low- and high-order solutions is taken to obtain the best compromise between accuracy and robustness (monotonicity). The first step is to write the so-called anti-diffusive fluxes, which are the numerical fluxes that yield the high-order scheme when added to the low-order scheme. Namely,

$$\mathbf{P} = \{\mathsf{P}_A\}, \tag{57}$$

$$\mathsf{P}_A = \sum_{B \neq A} \mathsf{f}_{AB}, \tag{58}$$

$$\mathsf{f}_{AB} = -\bar{\mathsf{D}}_{AB}(\rho_{1/2;B} - \rho_{1/2;A})$$
$$-\mathsf{V}_{1;AB}(\rho_{1;B} - \rho_{1;A}) + \mathsf{V}_{0;AB}(\rho_{0;B} - \rho_{0;A}) \quad \text{(no sum)}, \tag{59}$$

where we have used the fact that the $\bar{\mathsf{D}}$ has zero sum ($\bar{\mathsf{D}}_{AA} = -\sum_{B \neq A} \bar{\mathsf{D}}_{AB}$) and the property of the row-sum lumped volume matrices, for which $\mathsf{V}^L_{(\cdot);AA} = \sum_B \mathsf{V}_{(\cdot);AB}$, so that $\mathsf{V}_{(\cdot);AA} - \mathsf{V}^L_{(\cdot);AA} = \sum_{B \neq A} \mathsf{V}_{(\cdot);AB}$. In the context of finite element approximations, the term f_{AB} can be interpreted as an *internodal anti-diffusive flux*, that is the anti-diffusive contribution of node B on the change of value in the solution at node A.

Remark 14 Note that, by construction, $\mathsf{f}_{AB} = -\mathsf{f}_{BA}$, ensuring that the high-order anti-diffusive fluxes maintain the overall conservation properties of the resulting algorithm.

An appropriate convex combination of low- an high-order fluxes is then taken by weighting each internodal flux f_{AB} with a coefficient $\alpha_{AB} \in [0, 1]$, so that a higher-order non-oscillatory solution is obtained. The choice of α_{AB} is performed using a limiting strategy due to Zalesak [62].

4.6 Zalesak's Limiter

The aim of the Zalesak's limiting strategy is to preserve the positivity and LED properties of the low-order fluxes while increasing the overall order of accuracy of the method when the solution is smooth. To this end, we first split the total flux P_A in positive and negative contributions:

$$\mathsf{P}^{\pm}_A = \sum_{B \neq A} \genfrac{}{}{0pt}{}{\max}{\min} \{0, \mathsf{f}_{AB}\}. \tag{60}$$

In practice, the flux $\mathsf{P}^{\pm}_A$ represents the total negative/positive contribution to the solution at node A, which, in turn, may be responsible for undershoots/overshoots. One then uses the terms

$$\mathsf{Q}^+_A = \mathsf{V}_A(\rho^{\max}_A - \rho_A), \tag{61}$$

$$Q_A^- = V_A(\rho_A^{\min} - \rho_A) \tag{62}$$

to estimate a distance from the solution and the local maxima/minima. For the iterative algorithm, monotonicity is enforced by estimating $Q_A^\pm$ using the maximum and minimum of the previous iterate (i) at $\tau = 1$ and the initial solution at $\tau = 0$. This procedure slightly extends the approach presented in Chap. 6 of Kuzmin et al. [33], in which the terms $Q_A^\pm$ are computed using only the low-order scheme. To estimate the maximum and minimum values $\rho_A^{\max}$ and $\rho_A^{\min}$, we just have to look at the neighboring nodes because the shape functions have compact support. Note also that, by definition, $P_A^+ \geq 0$, $P_A^- \leq 0$, $Q_A^+ \geq 0$, and $Q_A^- \leq 0$. We then define the portion of the flux $P_A^\pm$ which can be accepted in or out of a node without producing overshoots or undershoots:

$$R_A^\pm = \begin{cases} \min\{1, Q_A^\pm / P_A^\pm\} & \text{if } P_A^\pm \neq 0, \\ 1 & \text{if } P_A^\pm = 0. \end{cases} \tag{63}$$

In order to prevent overshoots or undershoots, the following weights on the fluxes f_{AB} must be used:

$$\alpha_{AB} = \begin{cases} \min\{R_A^+, R_B^-\} & \text{if } f_{AB} \geq 0, \\ \min\{R_A^-, R_B^+\} & \text{if } f_{AB} < 0, \end{cases} \tag{64}$$

where the minima help enforce the most restrictive condition between the inflow flux at node A and outflow flux at node B (and vice versa). Note that, by definition (64), $\alpha_{AB} = \alpha_{BA}$, which implies global conservation of the overall scheme, since the low-order scheme is conservative and the limited anti-diffusive fluxes are conservative (antisymmetric).

Remark 15 (Prelimiting) As an initialization step before performing Zalesak's limiting, all fluxes satisfying $f_{AB}(\rho_A - \rho_B) < 0$ are set to 0, to prevent the nominally anti-diffusive fluxes to be dissipative. This is a standard approach in FCT methods for pure transport problems (see Chap. 6 of Kuzmin et al. [33] for details in the context of finite element formulations and Zalesak [62] for the original conceptualization).

4.7 Algorithmic Implementation

The proposed scheme stems from the work of Kuzmin et al. [31], but with the important difference that the geometric conservation law is incorporated at each stage of the computational procedure. Here is the sequence of steps:

1. Use the discrete GCL (34) to update the nodal volumes:

$$V_1^L - V_0^L = \Sigma[\bar{K}]. \tag{65}$$

2. Solve iteratively (54):

$$\mathsf{V}_1^L \rho_1^{low;(i+1)} - \mathsf{V}_0^L \rho_0 = (\bar{\mathbf{K}} + \bar{\mathbf{D}}) \rho_{1/2}^{low;(i)}. \qquad (66)$$

Note that unless otherwise stated, numerical computations were performed with one predictor pass and two corrector passes. A single corrector pass would be sufficient to achieve second-order accuracy, but an additional corrector pass was found beneficial in further improving the quality of the solution.

3. Compute the antidiffusive fluxes (59), and the parameters α_{AB} using Zalesak's limiter.
4. Add antidiffusive fluxes to the low order solution computed in step 1:

$$\mathsf{V}_{1;AA}^L \rho_{1;A} = \mathsf{V}_{1;AA}^L \rho_{1;A}^{low} + \sum_{B \neq A} \alpha_{AB} \mathfrak{f}_{AB}. \qquad (67)$$

Remark 16 The proposed implementation is not the only possible choice. A different implementation using an iterated remap strategy is presented in full detail by López Ortega and Scovazzi [40].

5 Flux Corrected Synchronized Remap for Systems of Equations

The proposed FCR iterative algorithm can be extended to a multivariable system, and the main focus will be on systems of equations arising in shock hydrodynamics applications. Let us denote by $\boldsymbol{U} = [\rho, \rho \boldsymbol{v}, \rho E]^T$ the vector of conserved variables, where ρ indicates the material density, $\boldsymbol{v}$ the velocity, and $E = e + \boldsymbol{v} \cdot \boldsymbol{v}/2$ the total energy, with e the internal energy. In this case, the conservative form of the remap equations simply reads

$$\left. \frac{\partial (\hat{J} \boldsymbol{U})}{\partial \tau} \right|_\chi - \hat{J} \nabla_x \cdot (\boldsymbol{U} \otimes \hat{\boldsymbol{u}}) = 0, \qquad (68)$$

where $\boldsymbol{U} \otimes \hat{\boldsymbol{u}}$ is the generalized flux associated with the mesh displacement. A group finite element approximation can be constructed as

$$\boldsymbol{U}(\hat{\boldsymbol{\varphi}}(\chi)) = \sum_{A=1}^{n_{np}} \boldsymbol{U}_A N_A(\hat{\boldsymbol{\varphi}}(\chi)), \qquad (69)$$

where $\boldsymbol{U}_A = \boldsymbol{U}(\hat{\boldsymbol{\varphi}}(\chi_A))$. Consequently, the previous discussion generalizes naturally to the case of a system of equations, by simply applying the flux-corrected remap strategy to each individual component of the solution vector $\boldsymbol{U}$.

There are only a number of specific issues related to the synchronization of the FCR procedure among the various equations and the interpretation of LED properties. In general applications, one is never interested in the LED properties for linear momentum and total energy, which are guaranteed by a trivial, component-wise extension of the scalar FCR procedure. Rather, one is interested in the LED properties for a set of *primitive* variables, which are typically the density, the velocity and the internal energy. Of these three LED requirements, the one on the velocity is usually not crucial, since the sign of the velocity components is not in general predetermined in complex flow computations. However, the density and internal energy have physical and thermodynamical meaning only if they are positive. The aim of this section is the development of a reliable FCR strategy preserving positivity and LED properties for density and internal energy. We also note that the discussion here is simply aimed at remap, and is fundamentally different from Kuzmin et al. [31], in which the full solution of Euler systems is pursued with FCT approaches in the Eulerian context.

5.1 Low-Order Scheme

The low-order procedure extends the approach presented in the scalar case. In this case, we generate a single diffusion matrix $\bar{D}$ and we then apply it to each of the conservative equations to prevent the oscillations of conserved variables. We found in practice that the action of this diffusion operator was also sufficient to prevent the oscillations of the primitive variables. Because the low-order scheme is a trivial vectorization of the scalar case, we leave the details to the reader.

5.2 Anti-diffusive Correction and Zalesak's Limiter for Systems of Conservation Laws

Let f_{AB}^{ρ}, $f_{AB}^{\rho v}$, and $f_{AB}^{\rho E}$ denote the anti-diffusive fluxes computed for density, linear momentum, and total energy. Taking inspiration from Kuzmin et al. [34] and using linearization (by means of Frechét differentiation), we can obtain expressions for the velocity and internal energy fluxes in terms of the momentum and total energy fluxes, namely:

$$\mathbf{f}_{AB}^{v} = \frac{\mathbf{f}_{AB}^{\rho v} - \mathbf{v}_A f_{AB}^{\rho}}{\rho_A}, \tag{70}$$

$$f_{AB}^{e} = \frac{f_{AB}^{\rho E} - \mathbf{v}_A \cdot \mathbf{f}_{AB}^{\rho v} + 1/2 \mathbf{v}_A \cdot \mathbf{v}_A f_{AB}^{\rho} - e_A f_{AB}^{\rho}}{\rho_A}, \tag{71}$$

where $\mathbf{v}_A = v(\hat{\varphi}(\chi_A))$ and $e_A = e(\hat{\varphi}(\chi_A))$. Note that the primitive variables fluxes do not necessarily maintain the anti-symmetric property of the conserved variables fluxes. In fact, given the previous definitions, $f_{AB}^{e} \neq -f_{BA}^{e}$. This is not a problem however, as long as the definition of $P_A^{e;\pm}$ is modified appropriately,

$$P_A^{e;\pm} = \sum_{B \neq A} {\max \atop \min} \{0, f_{AB}^e, -f_{BA}^e\}, \tag{72}$$

and similarly for each of the components of the velocity cumulative flux $P_A^{v;\pm}$. Analogously,

$$Q_A^{e;+} = V_A(e_A^{\max} - e_A), \tag{73}$$

$$Q_A^{e;-} = V_A(e_A^{\min} - e_A), \tag{74}$$

and similarly for $Q_A^{v;\pm}$. Finally, in order to synchronize Zalesak's limiting procedure for the system of equations, the limiting weights α_{AB} are defined as $\alpha_{AB} = \min\{\alpha_{AB}^\rho, \alpha_{AB}^e\}$, implementing the most restrictive condition between the density and internal energy equations. The weights α_{AB} are then used to limit the *conservative* antidiffusive fluxes f_{AB}^ρ, $f_{AB}^{\rho v}$, and $f_{AB}^{\rho E}$. This approach was first explored by Löhner et al. [38] and produced numerical results of very good quality.

Remark 17 (Frame invariance) The choice of not incorporating the weights associated with the velocity equation in the computation of the final weights α_{AB} is made to preserve frame invariance properties of the overall formulation. In fact, directly limiting each component of the velocity field, as often done in the literature (see, e.g., Kuzmin et al. [33]), would result in destroying the frame invariance properties of the resulting formulation. While in pure Eulerian computations this aspect can be overlooked, in arbitrary Eulerian-Lagrangian computations it may lead to erratic behavior of the numerical discretization, as documented by Scovazzi [49, 50], Scovazzi and Love [53]. In the proposed algorithm, the weights α_{AB} are applied to the entire vector of momentum anti-diffusive fluxes, and preserve the vector structure of the equations. The reader can easily check that even if a change of reference frame is performed, the nodal low-order and the anti-diffusive fluxes will remain unchanged. This is in fact the key to the overall frame invariance of the proposed remap algorithm.

In summary, our proposed solution strategy proceeds as follows: First the geometric conservation law is updated, followed by the density, momentum and total energy, using the α_{AB}'s derived using the vector Zalesak's limiting procedure. From the nodal momentum and the nodal density, the nodal velocity is computed. Using the total energy, the density and the velocity, the nodal internal energy is finally obtained. This approach is conservative and proved reliable in computations.

5.3 Fail-Safe Approach

The linearization performed clearly induces some approximation in the limiting procedure and overshoots or undershoots may be still present (although usually fairly small). Kuzmin et al. [34] proposed a strategy to correct this issue as follows: If an overshoot/undershoot occurs at a node A, then the weights of the anti-diffusive fluxes are reduced at that node and its neighbors B's. In particular, the weights α_{AB}'s are iteratively reduced by a certain fraction of the initial weight (one third in our computations), until the overshoot/undershoot disappears. In the worst case scenario, the weight is reduced to zero. More details can be found in Kuzmin et al. [34]. We note that in most tests the fail-safe procedure was not required. We also point out that the in the proposed context, the fail-safe approach has some relation to the repair algorithm of Shashkov and Wendroff [57], Margolin and Shashkov [47], but is of much simpler implementation.

6 Numerical Results

Extensive remap tests for the proposed FCR algorithm are presented next. The numerical tests are performed in the case of scalar and vector quantities in one, two, and three dimensions, and have been collected in the vast literature of remap algorithms, and in particular from Loubère and Shashkov [41], Shashkov and Lipnikov [58], Kucharik et al. [27], Knupp et al. [26], Maire et al. [45], Margolin and Shashkov [46, 47], Vàchal et al. [61], Vàchal and Liska [60], and Liska et al. [36]. As already mentioned in Sect. 4.7, unless otherwise stated, all tests were performed solving iteratively (54) with one predictor pass and two corrector passes.

6.1 One-Dimensional Tests

In the first battery of tests presented remap is performed for a single scalar field in one dimension, and comprehensive error convergence studies are included.

6.1.1 Rezoning Strategy

The mesh motion is based on a cyclic rezone strategy described in Liska et al. [36], and Margolin and Shashkov [46, 47]. Each of the mesh nodes displaces according to

$$x(\varepsilon, t) = x_{min} + (x_{max} - x_{min})\tilde{\varepsilon}(\varepsilon, \tau), \tag{75}$$

$$\tilde{\varepsilon}(\varepsilon, \tau) = \big(1 - \alpha(\tau)\big)\varepsilon + \alpha(\tau)\varepsilon^3, \tag{76}$$

$$\alpha(\tau) = \frac{\sin 4\pi \tau}{2}, \tag{77}$$

for $0 \le \varepsilon \le 1$ and $0 \le \tau \le 1$. The node positions are $x_A^k = x(\varepsilon_A, \tau^k)$ where $\varepsilon_A = (A-1)/n_{el}$ and $\tau^k = k/k_{max}$, with $A = 1, 2, \ldots, n_{el} + 1 = n_{np}$, and $k =$

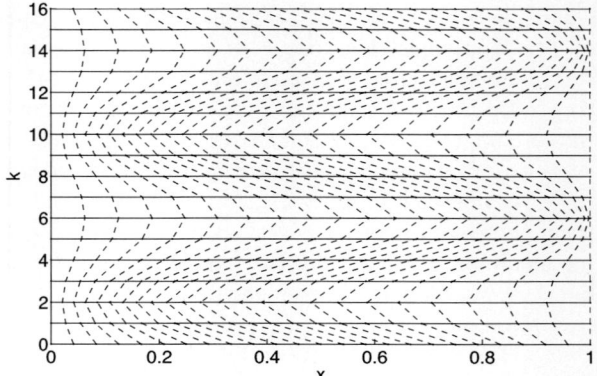

Fig. 2 Mesh motion for $n_{el} = 24$ and $k_{max} = 16$. Each element completes two sinusoidal displacement cycles

$0, 1, \ldots, k_{max}$. Here $n_{np} = n_{el} + 1$ is the number of nodes, n_{el} is the number of elements, and k_{max} the number of pseudo-time steps. Tests are run for $n_{el} = 64, 128, 256$ with $k_{max} = 5n_{el}$, so that the ratio between the mesh spacing and the increment in τ remains unchanged (equivalently, the mesh CFL number is constant under refinement). In Fig. 2 the mesh motion is shown for $n_{el} = 24$ and $k_{max} = 16$.

Additional tests were also performed with the randomly perturbed mesh motion suggested in Margolin and Shashkov [46, 47]. These tests show the same qualitative and quantitative behavior of the cyclic rezoning, and are not reported here for the sake of brevity. In particular, the proposed algorithm matches very well in performance the schemes proposed in Margolin and Shashkov [46, 47], and Liska et al. [36].

Remark 18 In general, randomized mesh motion is valuable in assessing convergence rates, as they prevent error cancellations that may lead to unexpected superconvergence results. However, it is the opinion of the authors that the randomized mesh motion proposed in Margolin and Shashkov [46, 47], and Liska et al. [36] prevents nodes to undergo very large displacements (compared with the cyclic mesh motion proposed in the same references), and is less effective in testing the overall robustness of remap algorithms. The cyclic remap utilized here was chosen because it prevents error cancellations while still allowing large cumulative mesh displacements.

6.1.2 Remap of Scalar Fields

Four scalar tests were initially performed. As shown in Fig. 3, the first three tests consist, respectively, in remapping a discontinuous (square), a triangular (peak), and a Gaussian (smooth) profiles, all in the range $[0, 1]$. The fourth test involves remapping a sinusoidal function, with range in the interval $[-1, 1]$. In Fig. 3, the low-order and the FCR schemes are qualitatively compared with the exact solutions of the four tests for a grid with $n_{el} = 128$ and $k_{max} = 640$. Detailed error comparison and convergence rates are presented in Table 1. The scheme compares well

Algebraic Flux Correction and Geometric Conservation

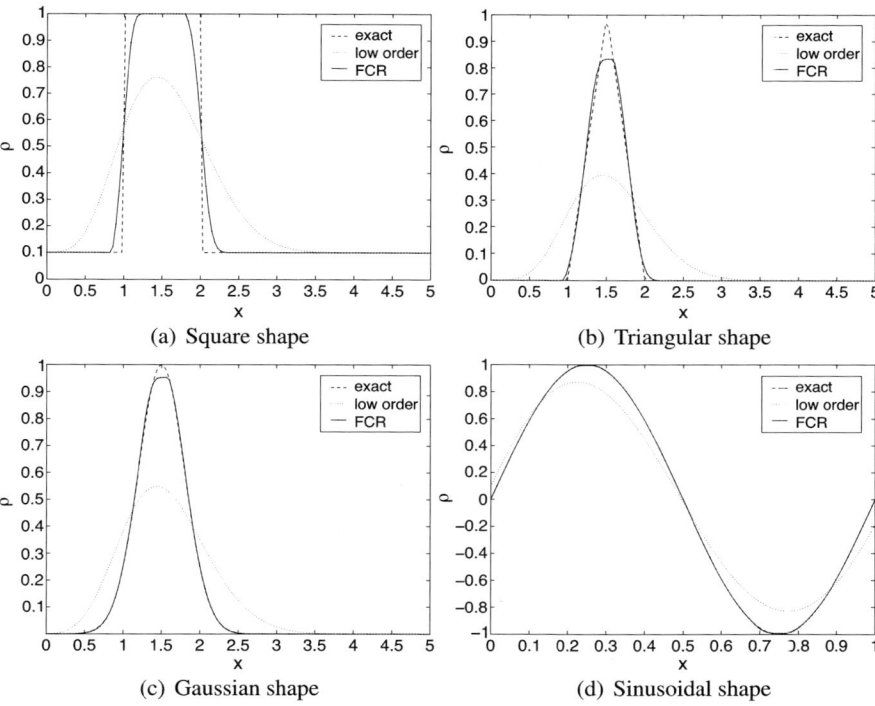

Fig. 3 Remapped solutions for an initial square, triangular, Gaussian and sinusoidal shapes. The FCR solution is compared with the low-order solution

Table 1 L^1-error, L^2-error and convergence rates for the various tests shown in Fig. 3. Convergence rates are computed using the errors for $n_{el} = 512$ and $n_{el} = 1024$

n_{el}	Mesh							
	Square		Triangle		Gaussian		Sine	
	L^1	L^2	L^1	L^2	L^1	L^2	L^1	L^2
64	2.103e−1	2.364e−1	8.863e−2	1.063e−1	7.412e−2	7.303e−2	4.236e−3	6.892e−3
128	1.292e−1	1.859e−1	3.680e−2	0.472e−1	1.435e−2	1.779e−2	0.900e−3	1.824e−3
256	0.790e−1	1.456e−1	1.497e−2	0.225e−1	0.287e−2	0.460e−2	0.194e−3	0.504e−3
512	0.484e−1	1.139e−1	0.562e−2	0.105e−1	0.063e−2	0.130e−2	0.043e−3	0.145e−3
1024	0.298e−1	0.891e−1	0.210e−2	0.050e−1	0.014e−2	0.038e−2	0.010e−3	0.043e−3
Rate	0.6997	0.3543	1.4150	1.0704	2.1531	1.7581	2.1031	1.7459

to (and sometimes surpasses) current state-of-the-art implementations by Margolin and Shashkov [46, 47], Maire et al. [45], and Liska et al. [36] in terms of error magnitude and convergence rates.

Fig. 4 Remapped FCR solution for the scalar exponential jump test for different levels of mesh refinement

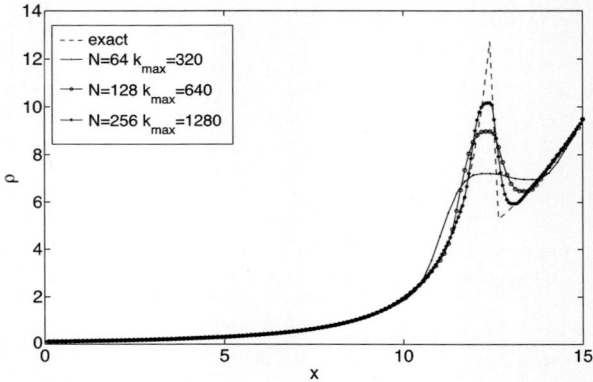

Table 2 L^1-error, L^2-error, and convergence rates for the scalar exponential shock problem. Convergence rates are computed between the current and lower level of refinement (e.g., the rate for $n_{el} = 128$ is computed comparing errors at $n_{el} = 64$ and $n_{el} = 128$)

n_{el}	Mesh			
	L^1-error		L^2-error	
	Error	Rate	Error	Rate
64	4.684e0		3.412e0	
128	3.161e0	0.5672	2.952e0	0.2090
256	1.757e0	0.8473	2.126e0	0.4733
512	1.027e0	0.7750	1.649e0	0.3665
1024	0.599e0	0.7762	1.278e0	0.3677

A more demanding test found in the literature is the exponential shock (see Liska et al. [36], Margolin and Shashkov [46, 47], and Vàchal and Liska [60]), in which the initial solution profile is given by an exponential increase followed by a shock and loosely represents the density profile due to a blast. Convergence of the remapped solution for the FCR algorithm can be compared in Fig. 4 for three pairs of (n_{el}, k_{max}). Results show that this algorithm avoids terracing of the solution, which is a common phenomenon in FCT-type algorithms in combination with Zalesak's limiter (see Kuzmin et al. [31]). Numerical errors are presented in Table 2.

6.1.3 Remap of Vector Fields

Two one-dimensional multivariable tests are described in Liska et al. [36] and Vàchal and Liska [60]. In the first test, the initial condition vector is given by a discontinuity in density, velocity and internal energy. In the second test, a vector extension of the scalar exponential jump is proposed, in which the density jump is complemented by discontinuous linear functions for the velocity and internal energy. More details can be found in Liska et al. [36], and Vàchal and Liska [60].

Table 3 L^1-error, L^2-error and convergence rates for the remap of a discontinuous vector field. Convergence rates are computed using the errors for $n_{el} = 512$ and $n_{el} = 1024$

n_{el}	Mesh					
	L^1			L^2		
	ρ	v	e	ρ	v	e
64	8.161e−2	3.556e−2	2.322e−2	2.669e−1	1.295e−1	8.945e−2
128	5.013e−2	2.184e−2	1.425e−2	2.096e−1	1.015e−1	7.005e−2
256	3.082e−2	1.344e−2	0.888e−2	1.644e−1	0.795e−1	5.490e−2
512	1.902e−2	0.833e−2	0.546e−2	1.289e−1	0.624e−1	4.311e−2
1024	1.184e−2	0.523e−2	0.346e−2	1.011e−1	0.490e−1	3.396e−2
Rate	0.6872	0.6746	0.6521	0.3505	0.3488	0.3422

Table 4 L^1-error, L^2-error and convergence rates for the vector exponential shock problem. Convergence rates are computed using the errors for $n_{el} = 512$ and $n_{el} = 1024$

n_{el}	Mesh					
	L^1			L^2		
	ρ	v	e	ρ	v	e
64	4.541e0	3.769e−1	2.097e−1	3.201e0	2.606e−1	1.552e−1
128	3.233e0	1.939e−1	1.007e−1	2.784e0	1.993e−1	1.119e−1
256	1.815e0	1.116e−1	0.564e−1	2.029e0	1.570e−1	0.864e−1
512	1.028e0	0.675e−1	0.333e−1	1.585e0	1.261e−1	0.676e−1
1024	0.599e0	0.409e−1	0.198e−1	1.236e0	1.009e−1	0.531e−1
Rate	0.7771	0.7228	0.7500	0.4060	0.3568	0.3817

The numerical results are obtained with Zalesak's limiting only applied to density and internal energy. No limiting is performed on the velocity components, as it would violate the invariance properties of the algorithm under general rotations, and, at the same time, would increase more than twice the computational cost in three-dimensions. The failsafe algorithm is used to avoid possible overshoots and undershoots.

Figure 5 and Table 3 show results for the discontinuous field. This test confirms that limiting the density and internal energy is sufficient for the monotonicity preservation of all the variables of the system.

Figure 6 shows results for the vector exponential shock test. The algorithm produces smooth results where the initial solution has an exponential behavior and an good resolution close to the peak. Table 4 contains the relative error in the L^1 and L^2-norms.

The conclusion drawn from this first set of one-dimensional tests is that the FCR algorithm shows very good robustness and accuracy in all problems tested and avoids the terracing phenomenon that is observed in other FCT-type algorithms.

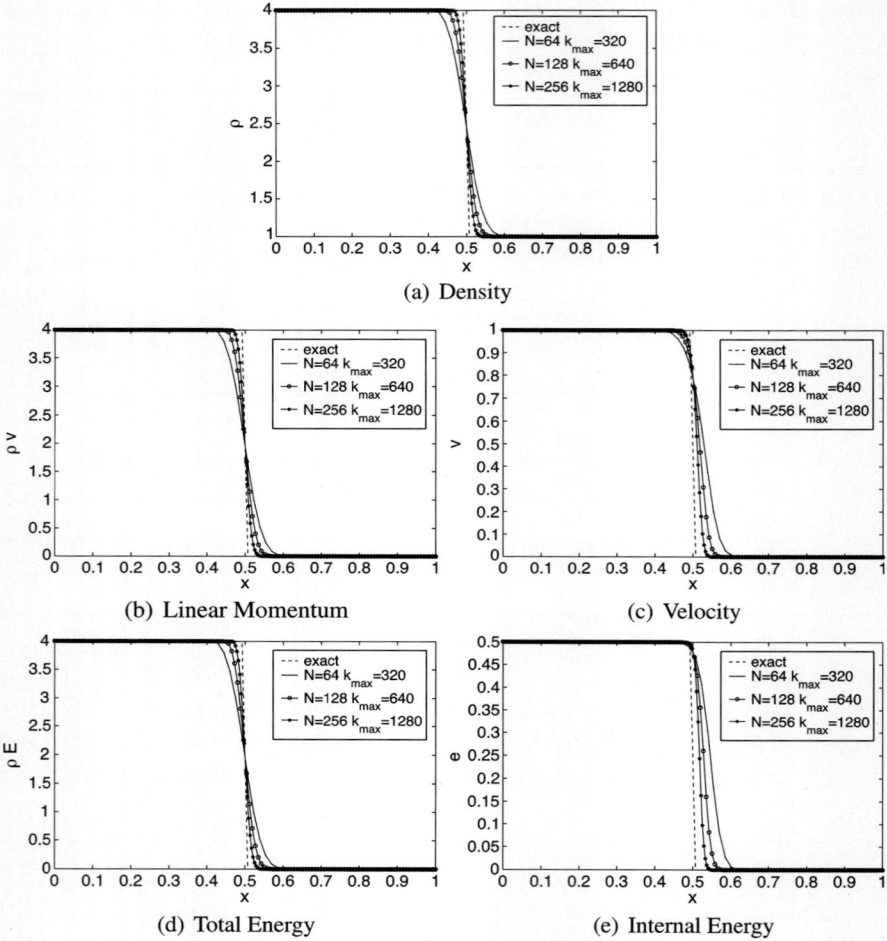

Fig. 5 Remap of discontinuous vector field. Density, linear momentum and total energy are conserved. Note the monotonicity of all variables, in spite of the fact that only internal energy and density are limited

6.2 Multi-dimensional Tests

In this section, we extend up to three dimensions many of the one-dimensional tests of the previous sections.

6.2.1 Rezoning Strategies

Computations were performed using three rezoning strategies, denoted R1, R2, and R3, respectively.

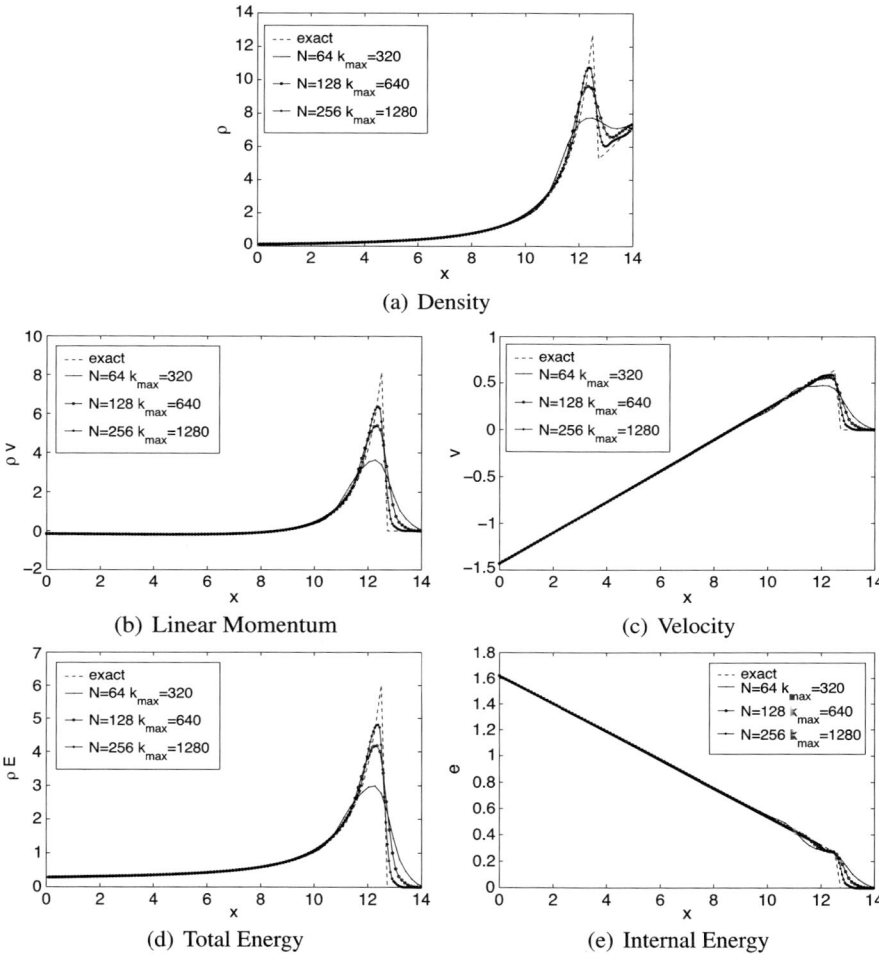

Fig. 6 Remap of the vector version of the exponential shock with FCR. Constraining of multiple variables does not seem to influence the peak value for the density when compared to the previous results for a single scalar variable

R1: *Orthogonal decoupled mesh motion.* This first strategy was suggested by Margolin and Shashkov [46, 47] and is only used in two-dimensional tests. It is based on taking a tensor product of one-dimensional displacement fields. Namely, for $\tau \in [0, 1]$, and $\xi_1, \xi_2 \in [0, 1]$:

$$x_1(\xi_1, \tau) = x_{1;min} + (x_{1;max} - x_{1;min})\tilde{\xi}_1(\xi_1, \tau), \tag{78a}$$

$$x_2(\xi_2, \tau) = x_{2;min} + (x_{2;max} - x_{2;min})\tilde{\xi}_2(\xi_2, \tau), \tag{78b}$$

$$\tilde{\xi}_1(\xi_1, \tau) = \bigl(1 - \alpha(\tau)\bigr)\xi_1 + \alpha(\tau)\xi_1^3, \tag{78c}$$

Fig. 7 Nodal mesh displacements associated with the R1, R2 and R3 rezoning strategies, for a 64 × 64 grid

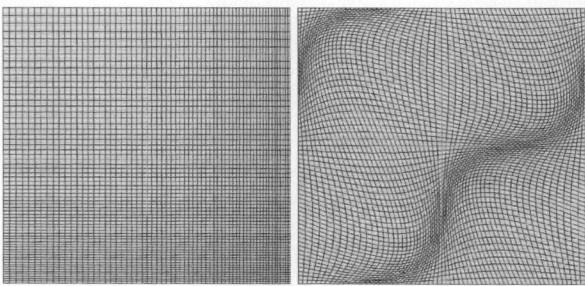

(a) R1 rezoning at $\tau = 0.15$. (b) R2 rezoning at $\tau = 0.5$.

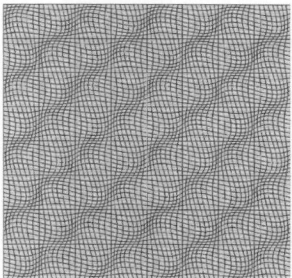

(c) R3 rezoning at $\tau = 0.5$.

$$\tilde{\xi}_2(\xi_2, \tau) = (1 - \alpha(\tau))\xi_2 + \alpha(\tau)\xi_2^2, \tag{78d}$$

$$\alpha(\tau) = \frac{\sin 4\pi \tau}{2}, \tag{78e}$$

where x_i and ξ_i ($i = 1, 2$) are components along the ith axis of the physical and parent reference frame, respectively, and are defined analogously to the one-dimensional case. Observe that orthogonality of mesh edges is preserved throughout rezoning. Also, note that the initial and final configurations of the mesh coincide. A plot of the deformed mesh at some $\tau = 0.15$ is shown in Fig. 7(a).

R2: *Coupled sinusoidal rezoning.* This second approach, also proposed by Margolin and Shashkov [46, 47], is based on coupled non-orthogonal displacements of the mesh by means of tensor products of sinusoidal functions. Namely, for $\tau \in [0, 1]$ and $\xi_i \in [0, 1]$:

$$x_i(\xi_i, \tau) = x_{i;min} + (x_{i;max} - x_{i;min})\tilde{\xi}_i(\boldsymbol{\xi}, \tau), \tag{79a}$$

$$\tilde{\xi}_i(\boldsymbol{\xi}, \tau) = \xi_i + \alpha(\tau)\left(\prod_{m=1}^{n_d} \sin(2\pi\lambda\xi_m)\right), \tag{79b}$$

$$\alpha(\tau) = \begin{cases} \tau/\Delta & \text{if } 0 \leq \tau \leq 0.5, \\ (1-\tau)/\Delta & \text{if } 0.5 < \tau \leq 1, \end{cases} \tag{79c}$$

with $\lambda = 1.0$ and $\Delta = 5.0$. Also, in this case, the initial and final configurations of the mesh coincide. A plot of the deformed mesh at $\tau = 0.5$ is shown in Fig. 7(b).

R3: *A variation on R2.* The third rezoning strategy is very similar to the second. The same mesh motion is used, with $\lambda = 4.0$ and $\Delta = 30.0$. A plot of the deformed mesh at $\tau = 0.5$ is shown in Fig. 7(c).

6.2.2 Two- and Three-Dimensional Remap of Scalar Fields

We propose two/three-dimensional extensions of the scalar one-dimensional tests, given by the following initial conditions for the scalar field ρ in the domain $[-0.5, 0.5] \times [-0.5, 0.5]$:

$$\rho_1(\boldsymbol{x}) = \begin{cases} 1, & \text{if } r(\boldsymbol{x}) \leq r_0, \\ 0, & \text{otherwise,} \end{cases} \tag{80}$$

$$\rho_2(\boldsymbol{x}) = \begin{cases} 1 - r(\boldsymbol{x})/r_0, & \text{if } r(\boldsymbol{x}) \leq r_0, \\ 0, & \text{otherwise,} \end{cases} \tag{81}$$

where

$$r(\boldsymbol{x}) = \|\boldsymbol{x}\|_2 = \sqrt{\boldsymbol{x} \cdot \boldsymbol{x}} = \sqrt{\sum_{m=1}^{n_d} x_i^2}, \quad \text{and} \quad r_0 = 0.25.$$

Additionally,

$$\rho_3(\boldsymbol{x}) = \prod_{m=1}^{n_d} \sin(2\pi x_i). \tag{82}$$

These initial conditions are stated so that they apply in both the two- and three-dimensional context. In what follows, we will refer to ρ_1 as the "top hat" function, to ρ_2 as the "cone" function, and to ρ_3 as the "sine" function. Similar notation will be used in the three-dimensional tests.

Starting with the two-dimensional tests, the top hat and cone initial conditions have been remapped using the R1 rezoning strategy, while the sinusoidal function has been remapped using the R2 rezoning strategy. Results are shown in Tables 5 and 6, and Fig. 8. We tested the algorithms on two types of grids. Grids of the first type are made of regular structured quadrilateral elements (squares, in particular) at various levels of resolution. Grids of the second type are obtained by subdividing each of the square elements of grids of the first type into two triangular elements, without a preferential direction.

This battery of tests shows that the convergence rates closely match the corresponding one-dimensional tests, and that the proposed remap approach is insensitive to the choice of quadrilateral or triangular elements.

Table 5 L^1-error and convergence rate of the FCR method for the two-dimensional tests on top hat, cone, and tensor product of sinusoidal functions. Convergence rates are computed using the errors for $n_{el} = 128 \times 128$ and $n_{el} = 256 \times 256$. Note also that "Q4" stands for four-node quadrilateral elements and "T3" stands for three-node triangular elements. The top hat and cone initial conditions have been remapped using R1 rezoning, while the sinusoidal function has been remapped using the R2 rezoning

n_{el}	Mesh					
	Top hat		Cone		Sine	
	Q4	T3	Q4	T3	Q4	T3
64×64	3.715e−2	3.807e−2	2.473e−3	2.560e−3	1.207e−3	1.509e−3
128×128	2.290e−2	2.288e−2	0.884e−3	0.879e−3	0.275e−3	0.365e−3
256×256	1.397e−2	1.396e−2	0.315e−3	0.314e−3	0.066e−3	0.093e−3
Rate	0.7132	0.7124	1.4886	1.4857	2.0560	1.9789

Table 6 L^2-error and convergence rate of the FCR method for two-dimensional tests on top hat, cone, and tensor-product of sinusoidal functions. Convergence rates are computed using the errors for $n_{el} = 128 \times 128$ and $n_{el} = 256 \times 256$. Note also that "Q4" stands for four-node quadrilateral elements and "T3" stands for three-node triangular elements. The top hat and cone initial conditions have been remapped using R1 rezoning, while the sinusoidal function has been remapped using the R2 rezoning

n_{el}	Mesh					
	Top hat		Cone		Sine	
	Q4	T3	Q4	T3	Q4	T3
64×64	1.067e−1	1.081e−1	6.900e−3	7.110e−3	2.334e−3	2.516e−3
128×128	0.836e−1	0.836e−1	2.993e−3	2.979e−3	0.787e−3	0.853e−3
256×256	0.652e−1	0.651e−1	1.343e−3	1.338e−3	0.276e−3	0.295e−3
Rate	0.3596	0.3601	1.1566	1.1544	1.5138	1.5306

For the sake of completeness, three-dimensional tests were also performed, on the cubic domain $[-0.5, 0.5] \times [-0.5, 0.5] \times [-0.5, 0.5]$. All computations utilized the R2 rezoning, with meshes at various levels of resolution. Due to the computational burden, we used coarser meshes with respect to the two-dimensional computations.

Results are presented in Tables 7 and 8, and compare well to the results in one and two dimensions, possibly with the exception of the top hat function tests. The explanation for this behavior is due to the fact that the finest grid used ($80 \times 80 \times 80$) is not sufficient to enter the asymptotic convergence range of the numerical errors. Hence, we cannot expect the convergence rates of this case to exactly match the two-dimensional case, computed on comparatively much finer grids.

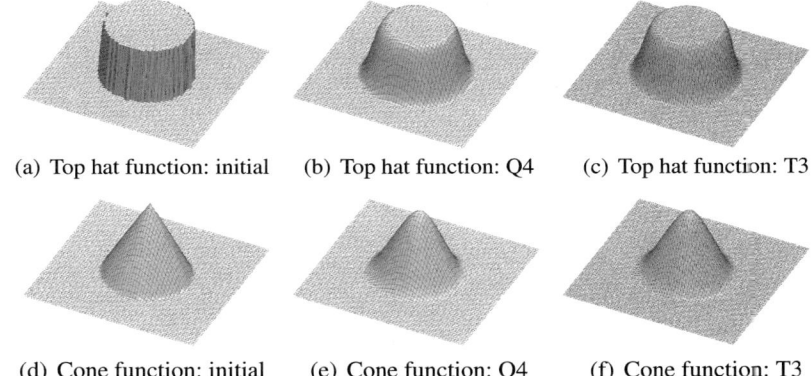

(a) Top hat function: initial (b) Top hat function: Q4 (c) Top hat function: T3

(d) Cone function: initial (e) Cone function: Q4 (f) Cone function: T3

Fig. 8 Solutions for the remap of the top hat and cone functions using the R1 rezone approach on the 64 × 64-element mesh. "Q4" stands for four-node quadrilateral elements and "T3" stands for three-node triangular elements

Table 7 L^1-error and convergence rate for three-dimensional tests on top hat, cone, and tensor-product sinusoidal functions. Convergence rates are computed using the errors for $n_{el} = 40 \times 40 \times 40$ and $n_{el} = 80 \times 80 \times 80$. All solutions are obtained using the R2 rezoning. Note also that "H8" stands for eight-node hexahedral elements and "T4" stands for four-node tetrahedral elements

n_{el}	Mesh					
	Top hat		Cone		Sine	
	H8	T4	H8	T4	H8	T4
10 × 10 × 10	1.452e−2	1.555e−2	1.179e−3	1.260e−3	3.028e−2	3.887e−2
20 × 20 × 20	1.034e−2	1.035e−2	0.854e−3	0.816e−3	1.083e−2	1.175e−2
40 × 40 × 40	0.796e−2	0.786e−2	0.345e−3	0.356e−3	0.278e−2	0.327e−2
80 × 80 × 80	0.532e−2	0.528e−2	0.120e−3	0.133e−3	0.065e−2	0.084e−2
Rate	0.5811	0.5753	1.5169	1.4186	2.1071	1.9578

6.2.3 Two- and Three-Dimensional Remap of Vector Fields

A two-dimensional version of the vector discontinuity remap test can be defined using the following initial distributions of density velocity and internal energy, in the domain $[0, 1] \times [0, 1]$:

$$\rho(x, y) = e(x, y) = \begin{cases} 2, & \text{if } y > (x - 0.4)/0.3, \\ 1, & \text{if } y \leq (x - 0.4)/0.3, \end{cases} \tag{83}$$

$$v(x, y) = \begin{cases} (2, -0.6), & \text{if } y > (x - 0.4)/0.3, \\ (1, -0.3), & \text{if } y \leq (x - 0.4)/0.3. \end{cases} \tag{84}$$

Table 8 L^2-error and convergence rate for three-dimensional tests on top hat, cone, and tensor-product sinusoidal functions. Convergence rates are computed using the errors for $n_{el} = 40 \times 40 \times 40$ and $n_{el} = 80 \times 80 \times 80$. All solutions obtained using the R2 rezoning. Note also that "H8" stands for eight-node hexahedral elements and "T4" stands for four-node tetrahedral elements

n_{el}	Mesh					
	Top hat		Cone		Sine	
	H8	T4	H8	T4	H8	T4
$10 \times 10 \times 10$	5.898e−2	6.177e−2	6.190e−3	6.319e−3	4.905e−2	5.927e−2
$20 \times 20 \times 20$	5.154e−2	5.075e−2	3.800e−3	3.658e−3	1.615e−2	1.740e−2
$40 \times 40 \times 40$	4.751e−2	4.714e−2	1.802e−3	1.814e−3	0.442e−2	0.504e−2
$80 \times 80 \times 80$	3.954e−2	3.946e−2	0.793e−3	0.801e−3	0.135e−2	0.158e−2
Rate	0.2650	0.2566	1.1841	1.1803	1.7103	1.6733

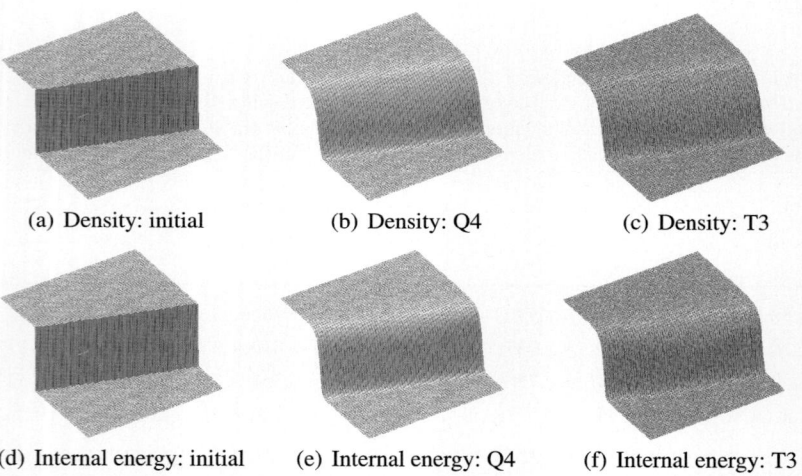

(a) Density: initial (b) Density: Q4 (c) Density: T3

(d) Internal energy: initial (e) Internal energy: Q4 (f) Internal energy: T3

Fig. 9 Density and internal energy solution for the two-dimensional vector shock discontinuity test. The mesh motion has been performed using the R1 rezoning

Results from this test using the R1 rezone strategy are shown in Fig. 9 and Tables 9 and 10, for the same computational grids used in the previous two-dimensional tests. Also in this case the error convergence rates compare well with the one-dimensional case. This test was extended also to the three dimensional domain $[0, 1] \times [0, 1] \times [0, 1]$, using the following initial conditions

$$\rho(x, y, z) = e(x, y, z) = \begin{cases} 2, & \text{if } y > (1.2x + z - 1.1)/0.31, \\ 1, & \text{if } y \leq (1.2x + z - 1.1)/0.31, \end{cases} \tag{85}$$

Table 9 L^1-relative error and convergence rates for the two-dimensional vector shock problem. Convergence rates are computed using the errors for $n_{el} = 128 \times 128$ and $n_{el} = 256 \times 256$. The mesh motion has been performed using the R1 rezoning

n_{el}	Mesh					
	ρ		v		e	
	Q4	T3	Q4	T3	Q4	T3
64 × 64	4.554e−2	5.771e−2	5.895e−2	7.457e−2	4.733e−2	5.968e−2
128 × 128	2.904e−2	3.023e−2	3.730e−2	3.879e−2	2.977e−2	3.092e−2
256 × 256	1.850e−2	1.955e−2	2.357e−2	2.485e−2	1.869e−2	1.965e−2
Rate	0.6504	0.6286	0.6622	0.6425	0.6714	0.6537

Table 10 L^2-relative error and convergence rates for the two-dimensional vector shock problem. Convergence rates are computed using the errors for $n_{el} = 128 \times 128$ and $n_{el} = 256 \times 256$. The mesh motion has been performed using the R1 rezoning

n_{el}	Mesh					
	ρ		v		e	
	Q4	T3	Q4	T3	Q4	T3
64 × 64	1.192e−1	1.355e−1	1.452e−1	1.645e−1	1.632e−1	1.844e−1
128 × 128	0.954e−1	0.975e−1	1.164e−1	1.189e−1	1.305e−1	1.331e−1
256 × 256	0.765e−1	0.789e−1	0.933e−1	0.961e−1	1.042e−1	1.072e−1
Rate	0.3193	0.3062	0.3191	0.3063	0.3240	0.3123

$$\boldsymbol{v}(x, y, z) = \begin{cases} (2, -0.6, -0.3), & \text{if } y > (1.2x + z - 1.1)/0.31, \\ (1, -0.3, 0.0), & \text{if } y \leq (1.2x + z - 1.1)/0.31, \end{cases} \quad (86)$$

also depicted in Fig. 10. Numerical errors and convergence rates are presented in Tables 11 and 12, and show good agreement with the one- and two-dimensional tests.

We conclude this section with a two-dimensional extension of the vector exponential shock problem. This test mimics the remapping of a Sedov blast flow and a similar version of it, limited however to only the density and momentum equations was presented in Liska et al. [36]. The domain is given by the square $[-0.5, 0.5] \times [-0.5, 0.5]$, with initial fields specified as:

$$\rho(x, y) = \begin{cases} \varepsilon + 6\left(\frac{r}{r_0}\right)^8, & \text{if } r < r_0, \\ 1 + \varepsilon, & \text{otherwise,} \end{cases} \quad (87)$$

$$\boldsymbol{v}(x, y) = \begin{cases} 0.83x, & \text{if } r < r_0, \\ 0, & \text{otherwise,} \end{cases} \quad (88)$$

Table 11 L^1-relative error and convergence rates for the three-dimensional vector shock problem. Convergence rates are computed using the errors for $n_{el} = 40 \times 40 \times 40$ and $n_{el} = 80 \times 80 \times 80$

n_{el}	Mesh					
	ρ		v		e	
	H8	T4	H8	T4	H8	T4
$10 \times 10 \times 10$	3.666e−2	3.627e−2	6.252e−2	6.327e−2	4.396e−2	4.502e−2
$20 \times 20 \times 20$	2.901e−2	2.942e−2	4.831e−2	4.874e−2	3.309e−2	3.320e−2
$40 \times 40 \times 40$	1.996e−2	2.032e−2	3.285e−2	3.330e−2	2.221e−2	2.243e−2
$80 \times 80 \times 80$	1.277e−2	1.309e−2	2.095e−2	2.137e−2	1.411e−2	1.432e−2
Rate	0.6443	0.6345	0.6487	0.6401	0.6548	0.6467

Table 12 L^2-relative error and convergence rates for the three-dimensional vector shock problem. Convergence rates are computed using the errors for $n_{el} = 40 \times 40 \times 40$ and $n_{el} = 80 \times 80 \times 80$

n_{el}	Mesh					
	ρ		v		e	
	H8	T4	H8	T4	H8	T4
$10 \times 10 \times 10$	9.328e−2	9.402e−2	1.241e−1	1.261e−1	1.426e−1	1.460e−1
$20 \times 20 \times 20$	8.820e−2	8.970e−2	1.149e−1	1.164e−1	1.298e−1	1.311e−1
$40 \times 40 \times 40$	7.504e−2	7.625e−2	0.967e−1	0.981e−1	1.084e−1	1.095e−1
$80 \times 80 \times 80$	6.074e−2	6.180e−2	0.779e−1	0.791e−1	0.869e−1	0.880e−1
Rate	0.3051	0.3030	0.3133	0.3107	0.3183	0.3153

Fig. 10 Three-dimensional vector shock discontinuity remap test. Color plot of the initial density solution on a unit cube domain. It is clearly visible the outline of the discontinuity separating the domain in two regions where the density has values $\rho = 2$ (*red*) and $\rho = 1$ (*blue*), respectively. An identical initial condition is used for the internal energy (Color figure online)

Table 13 L^1-relative error and convergence rates for the two-dimensional vector exponential shock problem. Convergence rates are computed using the errors for $n_{el} = 512 \times 512$ and $n_{el} = 1024 \times 1024$. The mesh motion has been performed using the R3 rezoning

n_{el}	Mesh						
	ρ		v		e	
	Q4	T3	Q4	T3	Q4	T3
64 × 64	5.539e−2	5.122e−2	7.340e−3	6.855e−3	9.469e−3	9.185e−3
128 × 128	3.690e−2	3.618e−2	4.466e−3	4.241e−3	3.519e−3	3.492e−3
256 × 256	2.417e−2	2.368e−2	2.741e−3	2.623e−3	1.819e−3	1.802e−3
512 × 512	1.534e−2	1.501e−2	1.660e−3	1.617e−3	1.061e−3	1.057e−3
1024 × 1024	0.949e−2	0.880e−2	0.100e−3	0.092e−3	0.064e−3	0.059e−3
Rate	0.6928	0.7702	0.7281	0.8215	0.7347	0.8406

Table 14 L^2-relative error and convergence rates for the two-dimensional vector exponential shock problem. Convergence rates are computed using the errors for $n_{el} = 512 \times 512$ and $n_{el} = 1024 \times 1024$. The mesh motion has been performed using the R3 rezoning

n_{el}	Mesh						
	ρ		v		e	
	Q4	T3	Q4	T3	Q4	T3
64 × 64	2.290e−1	2.185e−1	2.574e−2	2.468e−2	2.781e−2	2.688e−2
128 × 128	2.028e−1	2.019e−1	2.196e−2	2.111e−2	2.009e−2	1.947e−2
256 × 256	1.737e−1	1.719e−1	1.808e−2	1.755e−2	1.596e−2	1.556e−2
512 × 512	1.436e−1	1.420e−1	1.449e−2	1.427e−2	1.260e−2	1.243e−2
1024 × 1024	0.1150e−1	0.1102e−1	1.1408e−2	1.090e−2	0.985e−2	0.944e−2
Rate	0.3207	0.3653	0.3448	0.3885	0.3550	0.3968

$$e(x, y) = \begin{cases} 0.25 + 3\left(1 - \frac{r}{r_0}\right), & \text{if } r < r_0, \\ 0, & \text{otherwise,} \end{cases} \tag{89}$$

with $r_0 = 0.375$, and $\varepsilon = 10^{-3}$. Note that $\varepsilon > 0$ is introduced to avoid the occurrence of a zero density at the origin of the domain, which will cause divisions by zero in the algorithm. The mesh was displaced using the R3 rezoning strategy. Convergence rates, which closely match the one-dimensional case, are presented in Tables 13 and 14. Plots of the remapped densities and energies are presented in Fig. 11. The numerical results of this last test confirm the trends already outlined in previous sections, and demonstrates the feasibility of the proposed approach in more complex realistic scenarios.

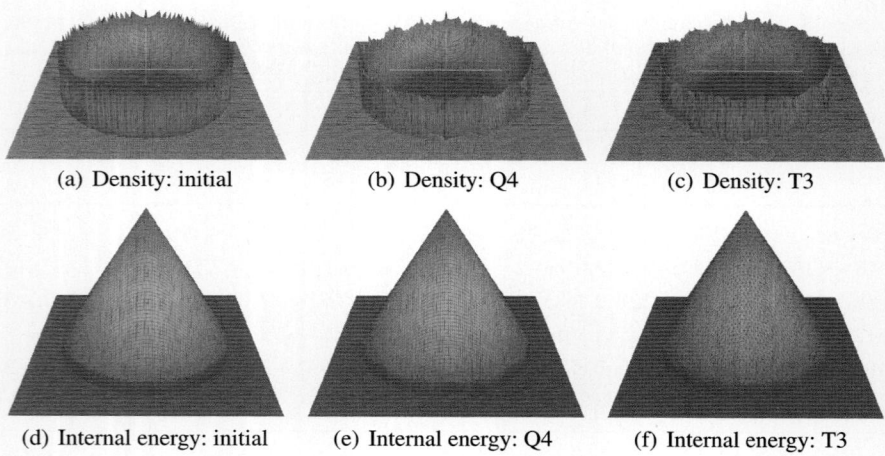

Fig. 11 Density and internal energy solution for the two-dimensional exponential shock test. The mesh motion has been performed using the R3 rezoning. Note that the energy peak value is preserved as the mesh does not displace at the origin

6.3 Arbitrary Lagrangian Eulerian Computations of Two-Dimensional Sedov Test

We conclude the presentation of numerical computations using the proposed remap strategy in the context of ALE mesh movement, to test performance with regard to preservation of symmetry and directionality of vector fields.

A complete study of performance in the ALE case is beyond the scope of the present chapter, and will be the object of future developments. Here, we show the results of a two-dimensional blast test (see Sedov [56]) performed comparing the Lagrangian method described in Scovazzi et al. [55] and an arbitrary Lagrangian-Eulerian (ALE) approach. The ALE method is obtained by applying the continuous remap described here after each step of the Lagrangian method described in Scovazzi et al. [55].

The mesh is repositioned using a simple iterative approach that has the purpose of improving mesh quality by smoothing. Let us denote by $\mathbf{x}_1^{(j)} = \mathbf{x}^{(j)}(\tau = 1)$ the jth iterate of the new nodal coordinates of the mesh. Likewise $\mathbf{V}_1^{L;(i)}$ and $\mathbf{V}_1^{(i)}$, respectively, are the lumped and consistent mass matrices associated with the basis of piecewise linear nodal finite element shape functions, constructed over the ith iterate of the mesh configuration. The mesh is repositioned as follows:

$$\begin{cases} \mathbf{V}_1^{L;(i)} \mathbf{x}_1^{(i+1)} = \mathbf{V}_1^{(i)} \mathbf{x}_1^{(i)}, \\ \mathbf{x}_1^{(0)} = \mathbf{x}_0, \end{cases} \quad (90)$$

Fig. 12 Velocity vector plots for Lagrangian and ALE simulations of a Sedov test on a 45 × 45 mesh of quadrilaterals

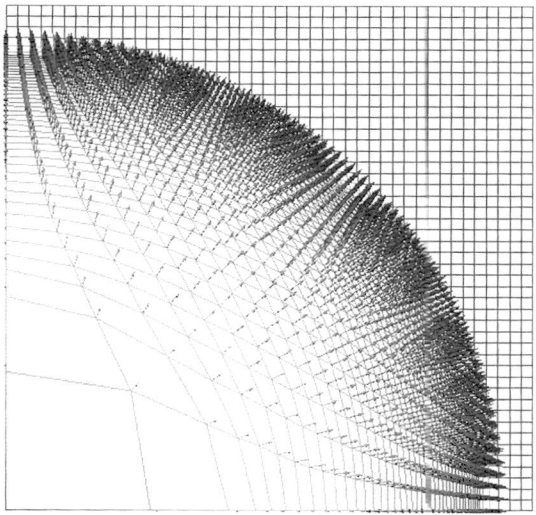

(a) Lagrangian, quadrilaterals

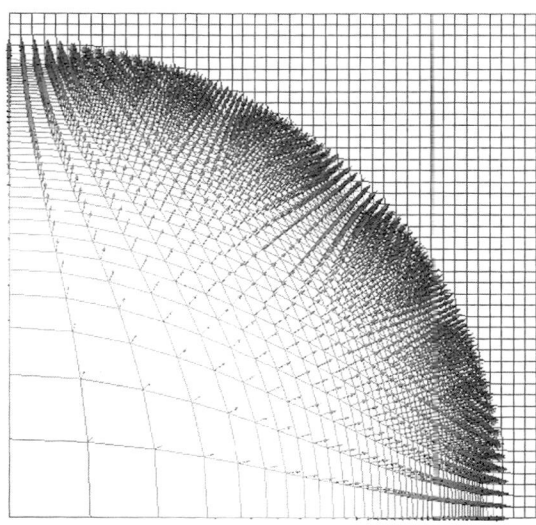

(b) ALE, quadrilaterals

where $\mathbf{x}_0 = \mathbf{x}(\tau = 0)$. Rewriting (90) as

$$\mathbf{x}_1^{(i+1)} = \mathbf{x}_1^{(i)} + \left(\mathbf{V}_1^{(i)}\right)^{-1}\left(\mathbf{V}_1^{(i)} - \mathbf{V}_1^{L;(i)}\right)\mathbf{x}_1^{(i+1)} \qquad (91)$$

it is easy to verify the well-known fact that the term $\mathbf{V}_1^{(i)} - \mathbf{V}_1^{L;(i)}$ has the discrete structure of a Laplacian viscosity operator. Hence, the proposed rezoning method acts as a smoothing on the mesh nodal positions. Three rezoning iterations were performed after each Lagrangian step in the computations.

Fig. 13 Velocity vector plots for Lagrangian and ALE simulations on a 45 × 45 mesh of quadrilaterals with every element subdivided into four triangles

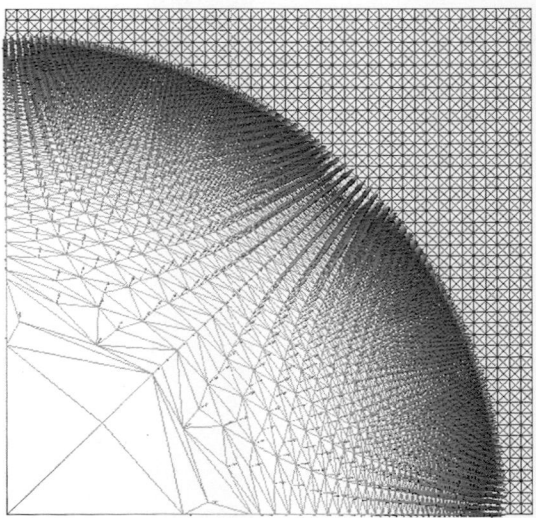

(a) Lagrangian, triangles

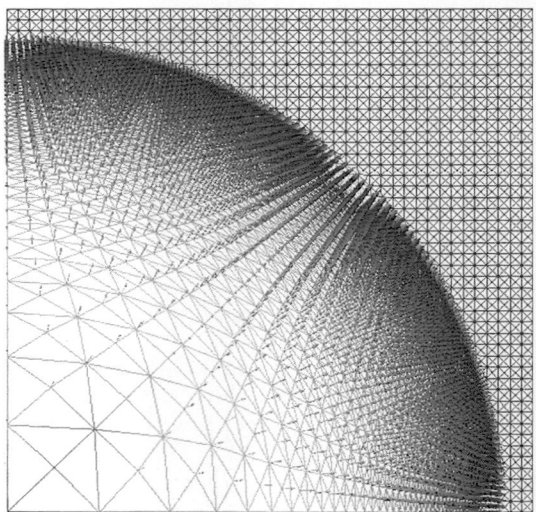

(b) ALE, triangles

In Figs. 12 and 13, velocity vector plots are presented for quadrilateral and triangular meshes. Quantitative results are shown in Figs. 14 and 15.

The solution of the Sedov test has radial symmetry, and, in principle, the tangential component of the velocity should be zero. Due to numerical errors, the small values of the tangential velocity components are present in Lagrangian computations (red dots). It is evident in Fig. 15 that the remap step (blue dots) does not amplify such errors.

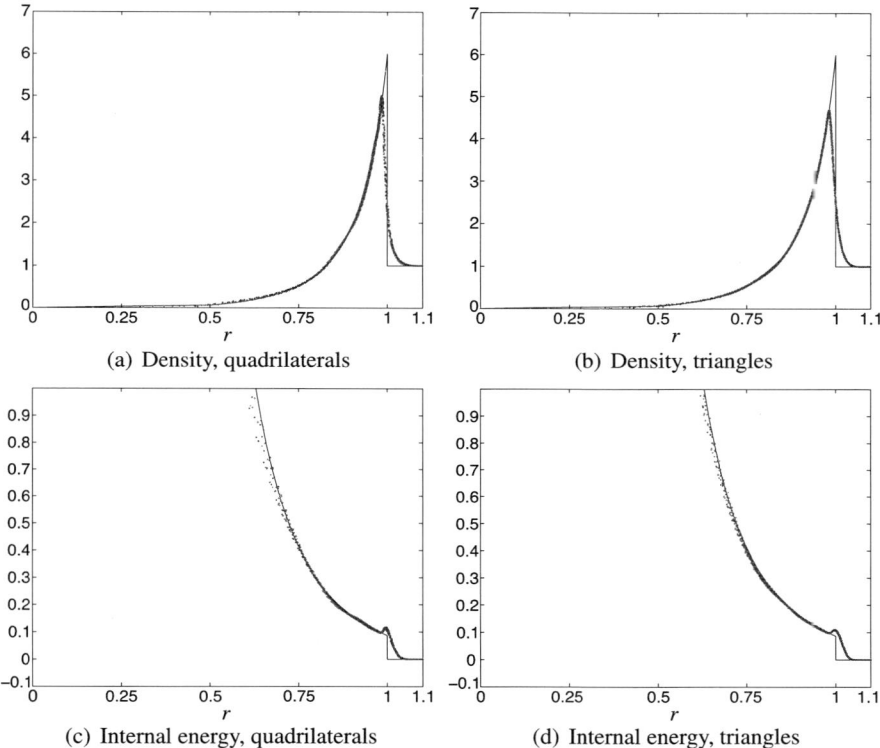

Fig. 14 Lagrangian (*red dots*) versus ALE (*blue dots*) computations of the classical Sedov test: Density and internal energy. The *blue and red dots* represent the solution angularly projected along the radial coordinate. The exact solution is shown as a *black continuous line* (Color figure online)

Note also that due to the symmetry of the mesh, the tangential component of the velocity must be symmetric with respect to the horizontal axis. This property is also preserved discretely through remap. It is also worthwhile to observe that the error in global angular momentum is within machine accuracy (the global angular momentum should be conserved in the Sedov test). The importance of angular momentum preservation in shock hydrodynamics computations is discussed in Love and Scovazzi [44].

7 Summary

We have developed a conservative synchronized ALE remap approach for computations on nodal finite elements. The proposed method leverages the geometric conservation property, with important implications on stability, accuracy, and local extremum diminishing properties. The method complements very well the work of

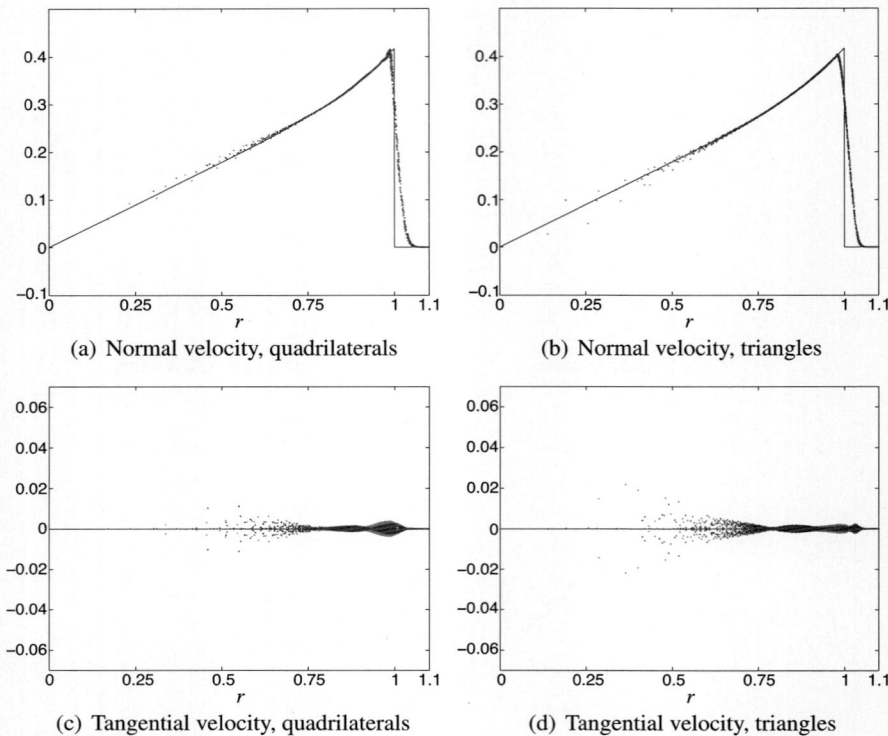

Fig. 15 Lagrangian (*red dots*) versus ALE (*blue dots*) computations of the classical Sedov test: Velocity components. The *blue and red dots* represent the solution angularly projected along the radial coordinate. The exact solution is shown as a *black continuous line* (Color figure online)

the first author on Lagrangian shock hydrodynamics with nodal-based finite elements (Scovazzi et al. [54, 55], Scovazzi [51]), and shows the feasibility of a suite of computational methods for shock hydrodynamics on nodal-based discretizations. Extensive testing in one, two, and three dimensions has been presented to evaluate the performance of the proposed algorithm.

Acknowledgements The authors would like to acknowledge, for very valuable discussions, Dr. M. Möller at University of Dortmund, Professor D. Kuzmin at the University of Erlangen-Nuremberg, Professor A. Bonito at Texas A&M University, Professor R.H. Nochetto at the University of Maryland, Dr. E. Love and Dr. J. Shadid at Sandia National Laboratories, and Dr. M. Shashkov at Los Alamos National Laboratory. A. López Ortega would also like to acknowledge his advisors D.I. Pullin and D.I. Meiron at the Graduate Aerospace Laboratories at California Institute of Technology.

A. López Ortega is supported by the PSAAP program, funded by the Department of Energy National Nuclear Security Administration under Award Number DE-FC52-08NA29613.

G. Scovazzi would like to acknowledge the continuing support of Dr. J. Stewart and Dr. S. Domino through Computer Science Research Foundation Grants at Sandia National Laboratories. Sandia National Laboratories is a multi-program laboratory operated by Sandia Corporation, a wholly owned subsidiary of Lockheed Martin Corporation, for the U.S. Department of Energy's National Nuclear Security Administration under contract DE-AC04-94AL85000.

References

1. Aziz, A.K., Monk, P.: Continuous finite elements in space and time for the heat equation. Math. Comput. **144**, 70–97 (1998)
2. Benson, D.J.: An efficient, accurate, simple ALE method for nonlinear finite element programs. Comput. Methods Appl. Mech. Eng. **72**, 305–350 (1989)
3. Benson, D.J.: Computational methods in Lagrangian and Eulerian hydrocodes. Comput. Methods Appl. Mech. Eng. **99**, 235–394 (1992)
4. Benson, D.J.: Momentum advection on a staggered grid. J. Comput. Phys. **100**, 143–162 (1992)
5. Boiarkine, O., Kuzmin, D., Čanić, S., Guidoboni, G., Mikelić, A.: A positivity-preserving ALE finite element scheme for convection-diffusion equations in moving domains. J. Comput. Phys. **230**, 2896–2914 (2011)
6. Bonito, A., Kyza, I., Nochetto, R.H.: A priori error analysis of time discrete higher-order ALE formulations (2011, in preparation)
7. Boris, J.P., Book, D.L.: Flux-corrected transport. I. SHASTA, a fluid transport algorithm that works. J. Comput. Phys. **11**, 38–69 (1973)
8. Boris, J.P., Book, D.L.: Flux-corrected transport. II. Generalization of the method. J. Comput. Phys. **18**, 248–283 (1975)
9. Boris, J.P., Book, D.L.: Flux-corrected transport. III. Minimal error FCT algorithms. J. Comput. Phys. **20**, 397–431 (1976)
10. Donea, J., Giuliani, S., Halleux, J.P.: An arbitrary Lagrangian-Eulerian finite element method for transient dynamic fluid-structure interactions. Comput. Methods Appl. Mech. Eng. **33**, 689–723 (1982)
11. Dukowicz, J.K., Baumgardner, J.: Incremental remapping as a transportation/advection algorithm. J. Comput. Phys. **160**, 318–335 (2000)
12. Dukowicz, J.K., Kodis, J.W.: Accurate conservative remapping (rezoning) for arbitrary Lagrangian-Eulerian computations. SIAM J. Sci. Stat. Comput. **8**, 305–321 (1987)
13. Dukowicz, J.K., Padial, N.: REMAP3D: A conservative three-dimensional remapping code. LA-12136-MS Report, Los Alamos National Laboratory, Los Alamos, NM, USA (1991)
14. Fletcher, C.A.J.: The group finite element formulation. Comput. Methods Appl. Mech. Eng. **37**, 225–244 (1983). doi:10.1016/0045-7825(83)90122-6
15. Formaggia, L., Nobile, F.: A stability analysis for the arbitrary Lagrangian-Eulerian formulation with finite elements. East-West J. Numer. Math. **7**(2), 105–131 (1999)
16. Formaggia, L., Nobile, F.: Stability analysis of second-order time accurate schemes for ALE-FEM. Comput. Methods Appl. Mech. Eng. **193**, 4097–4116 (2004)
17. Forster, C., Wall, W.A., Ramm, E.: Artificial added mass instabilities in sequential staggered coupling of nonlinear structures and incompressible viscous flows. Comput. Methods Appl. Mech. Eng. **196**, 1278–1293 (2007)
18. Fressmann, D., Wriggers, P.: Advection approaches for single- and multi-material arbitrary Lagrangian-Eulerian finite element procedures. Comput. Mech. **225**, 153–190 (2007). doi:10.1007/s00466-005-0016-7
19. Galera, S., Maire, P.-H., Breil, J.: A two-dimensional unstructured cell-centered multi-material ALE scheme using VOF interface reconstruction. J. Comput. Phys. **229**, 5755–5787 (2010). doi:10.1016/j.jcp.2010.04.019
20. Hirt, C.W., Amsden, A.A., Cook, J.L.: An arbitrary Lagrangian-Eulerian computing method for all flow speeds. J. Comput. Phys. **14**, 227–253 (1974)
21. Hulme, B.L.: Discrete Galerkin and related one-step methods for ordinary differential equations. Math. Comput. **26**(120), 881–891 (1972)

22. Jameson, A.: Computational algorithms for aerodynamic analysis and design. Appl. Numer. Math. **13**, 743–776 (1993)
23. Jameson, A.: Analysis and design of numerical schemes for gas dynamics 1. Artificial diffusion, upwind biasing, limiters and their effect on accuracy and multigrid convergence. Int. J. Comput. Fluid Dyn. **4**, 171–218 (1995)
24. Jameson, A.: Computational algorithms for aerodynamic analysis and design. Int. J. Numer. Methods Fluids **20**, 743–776 (1995)
25. Jamet, P.: Stability and convergence of a generalized Crank-Nicolson scheme on a variable mesh for the heat equation. SIAM J. Numer. Anal. **17**, 530–539 (1980)
26. Knupp, P., Margolin, L.G., Shashkov, M.J.: Reference Jacobian optimization-based rezone strategies for arbitrary Lagrangian-Eulerian methods. J. Comput. Phys. **176**, 93–128 (2002)
27. Kucharik, M., Shaskov, M., Wendroff, B.: An efficient linearity-and-bound-preserving remapping method. J. Comput. Phys. **188**, 462–471 (2003)
28. Kucharik, M., Breil, J., Maire, P.-H., Berndt, M., Shashkov, M.J.: Hybrid remap for multi-material ALE. Comput. Fluids (2010). doi:10.1016/j.compfluid.2010.08.004
29. Kuzmin, D.: Explicit and implicit FEM-FCT algorithms with flux linearization. J. Comput. Phys. **228**, 2517–2534 (2009)
30. Kuzmin, D., Turek, S.: Flux correction tools for finite elements. J. Comput. Phys. **175**, 525–538 (2002). doi:10.1006/jcph.2001.6955
31. Kuzmin, D., Möller, M., Turek, S.: High-resolution FEM-FCT schemes for multidimensional conservation laws. Comput. Methods Appl. Mech. Eng. **193**, 4915–4946 (2003)
32. Kuzmin, D., Möller, M., Turek, S.: Multidimensional FEM-FCT schemes for arbitrary time-stepping. Int. J. Numer. Methods Fluids **42**, 265–295 (2003)
33. Kuzmin, D., Löhner, R., Turek, S. (eds.): Flux-Corrected Transport: Principles, Algorithms and Applications. Springer, Berlin (2005)
34. Kuzmin, D., Möller, M., Shadid, J.N., Shashkov, M.J.: Failsafe flux limiting and constrained data projections for systems of conservation laws. J. Comput. Phys. **229**, 8766–8779 (2010)
35. Lesoinne, M., Farhat, C.: Geometric conservation laws for flow problems with moving boundaries and deformable meshes, and their impact on aeroelastic computations. Comput. Methods Appl. Mech. Eng. **134**, 71–90 (1996)
36. Liska, R., Shashkov, M.J., Vàchal, P., Wendroff, B.: Optimization-based synchronized flux-corrected conservative interpolation (remapping) of mass and momentum for arbitrary Lagrangian-Eulerian methods. J. Comput. Phys. **225**, 1467–1497 (2010)
37. Löhner, R., Morgan, K., Zienkiwicz, O.C.: The solution of non-linear hyperbolic equation systems by the finite element method. Int. J. Numer. Methods Fluids **4**, 1043–1063 (1984). doi:10.1002/fld.1650041105
38. Löhner, R., Morgan, K., Peraire, J., Vahdati, M.: Finite element flux-corrected transport (FEM-FCT) for the Euler and Navier-Stokes equations. Int. J. Numer. Methods Fluids **7**, 1093–1109 (1987). doi:10.1002/fld.1650071007
39. Löhner, R., Morgan, K., Vahdati, M., Boris, J.P., Book, D.L.: FEM-FCT: Combining unstructured grids with high resolution. Commun. Appl. Numer. Methods **4**, 717–729 (1988). doi:10.1002/cnm.1530040605
40. López Ortega, A., Scovazzi, G.: A geometrically-conservative, synchronized, flux-corrected remap for arbitrary Lagrangian-Eulerian computations with nodal finite elements. J. Comput. Phys. (2011). doi:10.1016/j.jcp.2011.05.005
41. Loubère, R., Shashkov, M.J.: A subcell remapping method on staggered polygonal grids for arbitrary-Lagrangian Eulerian methods. J. Comput. Phys. **209**, 105–138 (2005)
42. Loubère, R., Maire, P.-H., Shashkov, M.J.: ReALE: A reconnection arbitrary Lagrangian-Eulerian method in cylindrical coordinates. Computers and Fluids (2010). doi:10.1016/j.compfluid.2010.08.024
43. Loubère, R., Maire, P.-H., Shashkov, M.J., Breil, J., Galera, S.: ReALE: A reconnection-based arbitrary Lagrangian-Eulerian method. J. Comput. Phys. **229**, 4724–4761 (2010). doi:10.1016/j.jcp.2010.03.011

44. Love, E., Scovazzi, G.: On the angular momentum conservation and incremental objectivity properties of a predictor/multi-corrector method for Lagrangian shock hydrodynamics. Comput. Methods Appl. Mech. Eng. **198**, 3207–3213 (2009)
45. Maire, P.-H., Breil, J., Galera, S.: A cell-centered arbitrary Lagrangian-Eulerian (ALE) method. Int. J. Numer. Methods Fluids **56**, 1161–1166 (2008)
46. Margolin, L.G., Shashkov, M.: Second-order sign-preserving conservative interpolation (remapping) on general grids. J. Comput. Phys. **184**(1), 266–298 (2003)
47. Margolin, L.G., Shashkov, M.: Remapping, recovery and repair on a staggered grid. Comput. Methods Appl. Mech. Eng. **193**(39–41), 4139–4155 (2004)
48. Masud, A., Hughes, T.J.R.: A space-time Galerkin/least-squares finite element formulation of the Navier-Stokes equations for moving domain problems. Comput. Methods Appl. Mech. Eng. **146**, 91–126 (1997)
49. Scovazzi, G.: A discourse on Galilean invariance and SUPG-type stabilization. Comput. Methods Appl. Mech. Eng. **196**, 1108–1132 (2007)
50. Scovazzi, G.: Galilean invariance and stabilized methods for compressible flows. Int. J. Numer. Methods Fluids **54**, 757–778 (2007)
51. Scovazzi, G.: Stabilized shock hydrodynamics: II. Design and physical interpretation of the SUPG operator for Lagrangian computations. Comput. Methods Appl. Mech. Eng. **196**, 966–978 (2007)
52. Scovazzi, G., Hughes, T.J.R.: Lecture notes on continuum mechanics on arbitrary moving domains. Technical Report SAND-2007-6312P, Sandia National Laboratories (2007)
53. Scovazzi, G., Love, E.: A generalized view on Galilean invariance in stabilized compressible flow computations. Int. J. Numer. Methods Fluids **64**, 1065–1083 (2010). doi:10.1002/fld.2417
54. Scovazzi, G., Christon, M.A., Hughes, T.J.R., Shadid, J.N.: Stabilized shock hydrodynamics: I. A Lagrangian method. Comput. Methods Appl. Mech. Eng. **196**, 923–966 (2007)
55. Scovazzi, G., Shadid, J.N., Love, E., Rider, W.J.: A conservative nodal variational multiscale method for Lagrangian shock hydrodynamics. Comput. Methods Appl. Mech. Eng. **199**, 3059–3100 (2010). doi:10.1016/j.cma.2010.03.027
56. Sedov, L.I.: Similarity and Dimensional Methods in Mechanics. Academic Press, New York (1959)
57. Shashkov, M., Wendroff, B.: The repair paradigm and application to conservation laws. J. Comput. Phys. **198**, 265–277 (2004)
58. Shashkov, M.J., Lipnikov, K.: The error-minimization-based rezone strategy for arbitrary Lagrangian-Eulerian methods. Commun. Comput. Phys. **1**, 53–81 (2005)
59. Smolarkiewicz, P., Margolin, L.G.: MPDATA: A finite-difference solver for geophysical ows. J. Comput. Phys. **140**, 459–480 (1998)
60. Vàchal, P., Liska, R.: Sequential flux-corrected remapping for ale methods. In: de Castro, A.B., Gómez, D., Quintela, P., Salgado, P. (eds.) Numerical Mathematics and Advanced Applications, ENUMATH 2005, pp. 671–679. Springer, Berlin (2006)
61. Vàchal, P., Garimella, R.V., Shashkov, M.J.: Untangling of 2D meshes in ALE simulations. J. Comput. Phys. **196**, 627–644 (2004)
62. Zalesak, S.T.: Fully multidimensional flux-corrected transport algorithms for fluids. J. Comput. Phys. **31**, 335–362 (1979)

Constrained-Optimization Based Data Transfer

A New Perspective on Flux Correction

Pavel Bochev, Denis Ridzal, Guglielmo Scovazzi, and Mikhail Shashkov

Abstract We formulate a new class of optimization-based methods for data transfer (remap) of a scalar conserved quantity between two close meshes with the same connectivity. We present the methods in the context of the remap of a mass density field, which preserves global mass (the integral of the density over the computational domain). The key idea is to formulate remap as a global inequality-constrained optimization problem for mass fluxes between neighboring cells. The objective is to minimize the discrepancy between these fluxes and the given high-order *target mass fluxes*, subject to constraints that enforce physically motivated bounds on the associated primitive variable. In so doing, we separate accuracy considerations, handled by the objective functional, from the enforcement of physical bounds, handled by the constraints. The resulting second-order, conservative, and bound-preserving optimization-based remap (OBR) formulation is applicable to general, unstructured, heterogeneous grids. Under some weak requirements on grid proximity we prove that the OBR algorithm preserves linear fields in one, two and three dimensions. The chapter also examines connections between the OBR and the flux-corrected remap (FCR), which can be interpreted as a modified version of OBR (M-OBR), with the same objective but a smaller feasible set. The feasible set for M-OBR (FCR) is given by simple box constraints derived by using a "worst-case" scenario approach,

---

P. Bochev (✉) · G. Scovazzi
Numerical Analysis and Applications, Sandia National Laboratories, MS-1320, Albuquerque, NM 87185-1320, USA
e-mail: pbboche@sandia.gov

G. Scovazzi
e-mail: gscovaz@sandia.gov

D. Ridzal
Optimization and Uncertainty Quantification, Sandia National Laboratories, MS-1320, Albuquerque, NM 87185-1320, USA
e-mail: dridzal@sandia.gov

M. Shashkov
XCP-4, Methods and Algorithms, Los Alamos National Laboratory, MS-F644, Los Alamos, NM 87545, USA
e-mail: shashkov@lanl.gov

which may result in loss of linearity preservation and ultimately accuracy for some grid motions. The optimality of the OBR solution means that, given a set of target fluxes and a distance measure, OBR finds the best possible approximations of these fluxes with respect to this measure, which also satisfy the physically motivated bounds. In this sense, OBR can serve as a natural benchmark for evaluating the accuracy of existing and future numerical methods for data transfer with respect to a given class of flux reconstruction methods and flux distance measures. In this context, we perform numerical comparisons between OBR, FCR and iFCR (a version of FCR which utilizes an iterative procedure to enhance the accuracy of FCR numerical fluxes).

1 Introduction

The problem of transferring data between computational grids under specific constraints arises in the computational sciences in many contexts (see, e.g., Laursen and Heinstein [13], Bochev and Day [4], Carey et al. [6]). Among the main applications, we focus on Arbitrary Lagrangian-Eulerian (ALE) methods (see Hirt et al. [9]) as the primary motivation for this work.

ALE methods based on so-called continuous remap involve three separate phases: (i) the Lagrangian update of the solution, including displacements of the computational grid; (ii) rezoning (repositioning) of the computational grid in order to reduce grid distortion accrued during the Lagrangian motion; and (iii) conservative interpolation (remap) of the Lagrangian solution onto the rezoned grid. Formally, it is possible to run ALE algorithms primarily in the Lagrangian mode with the occasional rezone/remap taking place only when the grid becomes too distorted. However, an alternative computational strategy that combines the best properties of Eulerian and Lagrangian methods is to perform rezoning and remapping at every time step (from which the terminology, continuous remapping).

An important property of the *continuous rezone* strategy is that individual grid movements can be limited to small perturbations of the Lagrangian (old) mesh, and, in turn, that conserved quantities are exchanged only between neighboring cells. In this case, the remap step is localized to neighborhoods of old mesh cells and eliminates expensive global search operations required to locate new cells in the old mesh. Note also that, since remap is performed at every time step, the accuracy of the continuous-rezone ALE strongly depends on the quality of the remap phase.

In what follows, we focus on the second-order conservative and bound-preserving remap of a scalar conserved quantity between two close meshes with the same connectivity. On each cell of the old mesh we are given the mean value of the primitive variable that is an otherwise unknown positive scalar function ("density"). The conserved variable is the product of this mean value and the cell volume ("mass"). The objective is to find an accurate approximation of the conserved variable on the new mesh such that the density, approximated by the remapped cell mass divided by the volume of the new cell, satisfies physically motivated bounds. In summary, we seek solutions to the remap problem which possess the following properties:

(P1) Conservation of total mass;
(P2) Preservation of linearity;
(P3) Preservation of local bounds for the primitive variable (namely, density).

Specifically, property (P1) is a fundamental requirement for remap, while property (P2) is a statement of accuracy. It requires the remap algorithm to recover exact masses in the new cells whenever the old masses correspond to a linear density function. Property (P3) accounts for the fact that physically motivated bounds are imposed on the primitive variable rather than on the conserved quantity. In the continuous rezone setting, every new cell is contained in the union of its Lagrangian prototype and its neighbors. The minimum and maximum mean density values on these Lagrangian cells provide natural lower and upper bounds for the mean density value on the new cell.

Conservation of total mass (P1) is guaranteed if the remap is discretely stated in mass flux form, as indicated by Margolin and Shashkov [17].

Two strategies are commonly used in existing remappers to fulfill (P2) and (P3). The first one employs slope-limited bound-preserving reconstruction of the primitive variable, as presented in Dukowicz and Kodis [8], Jones [10], Miller et al. [19]. This first approach suffers from two main drawbacks: On the one hand, many of the slope limiters in wide use today are not linearity-preserving on irregular grids, as shown in Berger et al. [2]; on the other hand slope limiters usually impose geometric restrictions on the mesh (e.g., cell alignment, logically structured grids, etc.) The second strategy relaxes the bound-preserving requirement in the reconstruction, and in turn the geometric conditions on the mesh. The approach then proceeds with a mass re-distribution to satisfy (P3), see e.g., Kucharik et al. [11], Margolin and Shashkov [18], Loubere and Shashkov [15], Loubere et al. [16]. Unfortunately, both bound-preserving reconstruction and mass "repair" tend to obscure the sources of discretization errors and make the analysis of accuracy more complex.

The alternative approach pursued here relies as well on the mass flux form of remap to provide (P1), but achieves (P2) and (P3) without bound-preserving reconstruction or mass post-processing. This is because the remap step is rephrased as a *global* inequality-constrained optimization problem for mass fluxes between neighboring cells. The objective is to minimize the discrepancy between these fluxes and the given *target* mass fluxes, subject to constraints that enforce physically motivated bounds on the primitive variable (density).

This strategy is expected to be more robust, flexible and asymptotically accurate than the other two approaches mentioned for the following reasons. First, optimization-based remap (OBR) finds a global optimal solution from a feasible set defined by the local bounds, i.e. OBR always finds *the best possible, with respect to the target fluxes, remapped quantity that also satisfies these bounds*. Therefore, it does not rely on local "worst-case" assumptions, which can reduce the accuracy, as both bound-preserving reconstruction and mass redistribution.

Second, OBR can be easily adapted to different problems by choosing the most appropriate target fluxes and discrepancy measures (norms) for these problems.

Third, OBR enforces the local bounds (P3) by a set of linear inequalities, which are completely impervious to the shape of the cells in the mesh. Therefore, in principle, OBR can be applied to arbitrary grids, including grids comprising of polygons or polyhedra.

It is important to mention at this point that Rider and Kothe [20] and Berger et al. [2] used constrained optimization in lieu of standard limiters to define a bound-preserving reconstruction method on general cells. In that work the least-squares gradient recovery on a cell is constrained by the local minimum and maximum of the data, i.e., the limiting remains based on local "worst-case" assumptions. In contrast, we pose the entire remap problem as a globally constrained minimization problem in which all bounds are considered simultaneously. This possibility was first brought up in Liska et al. [14]. Using ideas from flux-corrected transport (FCT, see e.g. Kuzmin et al. [12]) these authors developed a flux-corrected remap (FCR) algorithm. Then, they interpreted FCR as "a process of replacing a global constrained optimization problem by series of local constrained optimization problems by considering the worst case scenario". Liska et al. [14] did not examine in detail this connection, and left open the question about the preservation of linearity in FCR.

The material that follows is aimed at presenting the key components of the proposed approach in detail, and to ultimately demonstrate that the global inequality-constrained optimization strategy leads to robust, accurate and efficient remappers. For this reason, we use the Euclidean norm to measure the flux discrepancy and define the target fluxes using density reconstruction that is exact for linear functions. While not the only possible choices, the former leads to differentiable objectives and the latter provides the preservation of linearity (P2).

Furthermore, we show that under some fairly weak requirements on mesh proximity OBR satisfies (P2) on arbitrary unstructured grids in one, two and three dimensions, including grids with polygonal or polyhedral cells.

We also clarify the intuitive interpretation of FCR given in Liska et al. [14]. We show that the FCR solution coincides with the solution of a modified version of OBR (M-OBR), which has the same objective but a simpler set of box constraints derived from the OBR constraints by using a "worst-case" scenario. FCR is then viewed as an approximate solution procedure for OBR, which seeks minimizers in a reduced feasible set. Because M-OBR (FCR) has a smaller feasible set, preservation of linearity may be lost and accuracy may suffer for some grid configurations.

Numerical studies confirm these conjectures, showing that for certain types of grids FCR defaults to a first-order accurate scheme, while OBR achieves the theoretically best possible accuracy (second order) for a linearity-preserving scheme. We also present examples of grids in one and two dimensions for which OBR is linearity preserving when FCR is not, and grids for which OBR preserves (P3) when FCR does not. These trends also extend to the case of *iterated* FCR (iFCR), a recursive algorithm derived from standard FCR, in which the low-order remap fluxes are sequentially updated using the most recent FCR monotone iterate. The iFCR

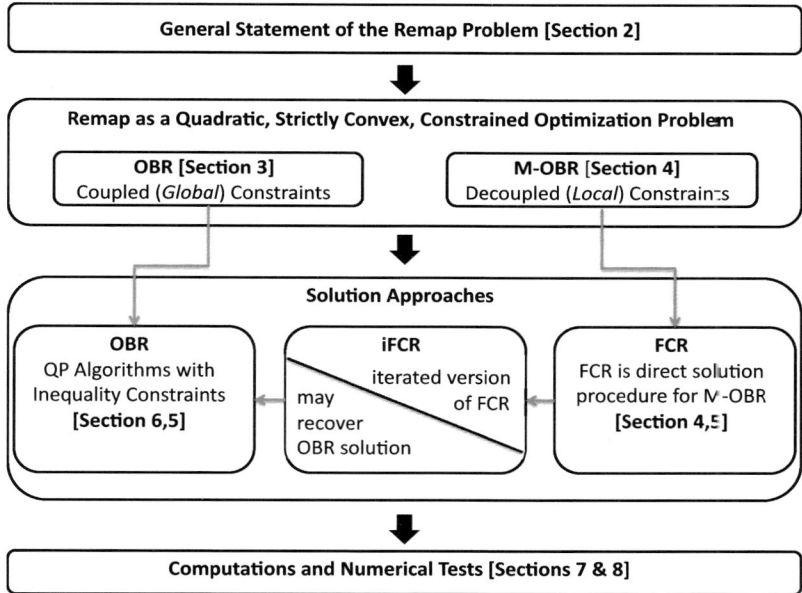

Fig. 1 Outline of the contents of the chapter and the main flow of the presentation

algorithm is clearly more expensive than the simple FCR algorithm, but provides a more challenging benchmark for testing the accuracy of OBR.

Our analysis also explains why the FCR fluxes are required to be *convex* combinations of low and high-order fluxes, without appealing to analogies with FCT. We show that the convexity requirement is introduced implicitly when the OBR constraints are approximated by simpler box constraints. This restricts the optimal solution of the global M-OBR problem to convex combinations of low-order and high-order fluxes. Because FCR is a solution procedure for the M-OBR problem, the convexity requirement becomes part of the "formula" for the optimal solution.

The chapter is organized as follows (see also Fig. 1 for a roadmap of the presentation of the material). Notation and a formal statement of the remap problem is presented in Sect. 2, and the new optimization-based formulation of remap is developed in Sect. 3. There we also establish sufficient conditions for the preservation of linearity in OBR. Connections between OBR, FCR, and iFCR are examined in Sect. 4. Sections 5 and 6 discuss implementation details of OBR and FCR. Section 7 presents three instructive computational examples, and Sect. 8 focuses on numerical estimates of convergence rates and assessment of the CBR performance.

2 The Remap Problem

2.1 Notation

In what follows $\Omega \subset \mathbb{R}^d$, $d = 1, 2, 3$, denotes an open bounded domain with a Lipschitz continuous boundary $\partial \Omega$. Bold face lower case Roman symbols denote points in the computational domain with $\mathbf{x} \in \Omega$ reserved for the independent variable. The symbol $K_h(\Omega)$ stands for a conforming partition of Ω into K cells κ_i, $i = 1, \ldots, K$, with volumes and barycenters given by

$$V(\kappa_i) = \int_{\kappa_i} dV \quad \text{and} \quad \mathbf{b}_i = \frac{\int_{\kappa_i} \mathbf{x}\, dV}{V(\kappa_i)}, \qquad (1)$$

respectively. $S(K_h)$ is the set of all sides in the mesh $K_h(\Omega)$, and $S(\kappa_i)$ is the subset of $S(K_h)$ associated with cell κ_i. A side can be oriented in two different ways, which we refer to as positive and negative. We assume that each side $\sigma_i \in S(K_h)$ is endowed with a unique positive or negative orientation ω_i. It is convenient to associate ω_i with the numeric values $+1$ and -1, for positively and negatively oriented sides, respectively. We recall that conforming partitions of Ω consist of cells that cover the domain without gaps or overlaps. The partition $K_h(\Omega)$ can be uniform or nonuniform, and the cells are not required to have the same shape or to be convex. For instance, in two dimensions $K_h(\Omega)$ can contain triangles, quadrilaterals and convex and non-convex polygons. This makes our approach applicable to a wide range of grids and methodologies. For example, we can think of a two-dimensional AMR grid (see, e.g., Berger and Colella [3]) as consisting of quadrilaterals and (degenerate) polygons, while in three dimensions (see, e.g., Bell et al. [1]) such grids will contain cubes and polyhedra.

We assume that Ω is endowed with two different grid partitions $K_h(\Omega)$ and $\widetilde{K}_h(\Omega)$ having the same connectivity. In the context of ALE methods we refer to $K_h(\Omega)$ as the old or Lagrangian grid and $\widetilde{K}_h(\Omega)$ as the new or rezoned[1] grid. Quantities defined on the new grid will have the tilde accent, e.g. $\widetilde{f}$, whereas the quantities on $K_h(\Omega)$ will have no accent. The cells on the new grid are denoted by $\widetilde{\kappa}_i$, with barycenters $\widetilde{\mathbf{b}}_i$, $i = 1, \ldots, K$. Because $K_h(\Omega)$ and $\widetilde{K}_h(\Omega)$ have the same connectivity, it is convenient to assume that the new cells are numbered in the same order as the old cells. Therefore, the Lagrangian prototype of the rezoned cell $\widetilde{\kappa}_i$ is the cell κ_i.

The neighborhood $N(\kappa_i)$ of κ_i comprises of the cell κ_i itself and all its neighbors, i.e. those cells in $K_h(\Omega)$ that share a vertex (in 1D), vertex or edge (in 2D) and vertex, edge or face (in 3D) with κ_i. The remap problem is stated under the assumption that the rezoned grid satisfies the *locality condition*

$$\widetilde{\kappa}_i \subset N(\kappa_i), \quad \text{for all } i = 1, \ldots, K, \qquad (2)$$

---

[1] Typically, in a continuous rezone ALE the rezoned grid is close to the Lagrangian but has better geometric quality.

that is, each rezoned cell $\widetilde{\kappa}_i$ is contained in $N(\kappa_i)$, the neighborhood of its Lagrangian prototype. Here the relation $\widetilde{\kappa}_i \subset N(\kappa_i)$ is interpreted geometrically (in contrast to its set-relational definition).[2] In the context of ALE methods, assumption (2) corresponds to using the continuous rezone strategy. Finally, $\mathscr{I}$ denotes the operator that returns the index of a cell, i.e. $\mathscr{I}(\kappa_i) = \mathscr{I}(\widetilde{\kappa}_i) = i$. The extension of this operator to sets of cells is natural, e.g.

$$\mathscr{I}(N(\kappa_i)) = \{\mathscr{I}(\kappa_i) | \kappa_i \in N(\kappa_i)\}$$

is the set of all indices of the cells in $N(\kappa_i)$.

For completeness, we review the specialization of some notation to one-dimensional domains $\Omega = [a, b]$ where $a < b$ are real numbers. In this case $K_h(\Omega)$ is defined by a set of $K + 1$ points $a = x_0 < x_1 < \cdots < x_{K-1} < x_K = b$, the Lagrangian cells are the intervals $\kappa_i = [x_{i-1}, x_i]$ and their volumes are $V(\kappa_i) = h_i = x_i - x_{i-1}$. The new grid $\widetilde{K}_h(\Omega)$ comprises of rezoned cells $\widetilde{\kappa}_i = [\widetilde{x}_{i-1}, \widetilde{x}_i]$ such that $a = \widetilde{x}_0 < \widetilde{x}_1 < \cdots < \widetilde{x}_{K-1} < \widetilde{x}_K = b$. In one dimension, (2) assumes a particularly simple form:

$$\widetilde{\kappa}_i \subset (\kappa_{i-1} \cup \kappa_i \cup \kappa_{i+1}) \quad \text{for } i = 2, \ldots, K - 1,$$

$$\widetilde{\kappa}_1 \subset (\kappa_1 \cup \kappa_2) \quad \text{and} \quad \widetilde{\kappa}_K \subset (\kappa_{K-1} \cup \kappa_K),$$

or

$$\widetilde{\kappa}_i \subset [x_{i-2}, x_{i+1}] \quad \text{for } i = 2, \ldots, K - 1,$$

$$\widetilde{\kappa}_1 \subset [a, x_2] \quad \text{and} \quad \widetilde{\kappa}_K \subset [x_{K-2}, b].$$

An equivalent form of the locality condition is given by

$$x_{i-1} \leq \widetilde{x}_i \leq x_{i+1}, \quad i = 1, \ldots, K - 1. \tag{3}$$

Material in this chapter also requires some notation for Euclidean spaces $\mathbb{R}^n$. We use Roman and Greek symbols with an arrow accent, and bold face Roman capitals for vectors and matrices, respectively, e.g., $\vec{c} \in \mathbb{R}^n$, $\vec{F} \in \mathbb{R}^n$, $\vec{\lambda} \in \mathbb{R}^n$, and $\mathbf{A} \in \mathbb{R}^{n \times m}$. The superscript $(\cdot)^\mathsf{T}$ indicates vector and matrix transposition. The *Euclidean inner product*, $\langle \cdot, \cdot \rangle : \mathbb{R}^N \to \mathbb{R}$, is $\langle \vec{a}, \vec{b} \rangle = \vec{a}^\mathsf{T} \vec{b}$, and the *Euclidean norm* $\|\vec{a}\|_2^2 = \langle \vec{a}, \vec{a} \rangle = \vec{a}^\mathsf{T} \vec{a}$. We use the Euclidean space notation to state algebraic forms of the optimization problems and for various coefficient vectors.

---

[2] In this chapter, we use the set-relational definitions and the corresponding geometric interpretations of $\subset$, $\subseteq$, $\cup$, $\cap$, $\setminus$ and $\in$ interchangeably. Their meaning will be clear from the context. In particular, relations between entities defined on $\widetilde{K}_h(\Omega)$ and those defined on $K_h(\Omega)$ only make sense when interpreted geometrically relative to the common domain Ω.

2.2 Statement of the Remap Problem

We recall the formal statement of mass-density remap following Margolin and Shashkov [17], Liska et al. [14]. We assume that there is a positive function $\rho(\mathbf{x}) > 0$, referred to as *density*, that is defined on Ω and whose values on the boundary $\partial\Omega$ are known. The only information given about $\rho(\mathbf{x})$ in the interior of Ω is its mean value on the old cells:

$$\rho_i = \frac{\int_{\kappa_i} \rho(\mathbf{x}) dV}{V(\kappa_i)}.$$

Equivalently, we can write

$$\rho_i = \frac{m_i}{V(\kappa_i)} \quad \text{or} \quad m_i = \rho_i V(\kappa_i) \tag{4}$$

where

$$m_i = \int_{\kappa_i} \rho(\mathbf{x}) dV$$

is the (old) cell mass. Here we have implicitly assumed that the initial distribution of $\rho(\mathbf{x})$ is known exactly, and that the previous integral represents the exact mass associated with cell i. The total mass is

$$M = \int_{\Omega} \rho(\mathbf{x}) dV = \sum_{i=1}^{K} \int_{\kappa_i} \rho(\mathbf{x}) dV = \sum_{i=1}^{K} m_i = \sum_{i=1}^{K} \rho_i V(\kappa_i).$$

For further reference we note that the mean density on every Lagrangian cell κ_i trivially satisfies the bounds

$$\rho_i^{\min} \leq \rho_i \leq \rho_i^{\max}, \tag{5}$$

where

$$\rho_i^{\min} = \begin{cases} \min_{j \in \mathscr{I}(N(\kappa_i))} \{\rho_j\} & \text{if } \kappa_i \cap \partial\Omega = \emptyset, \\ \min\left\{ \min_{j \in \mathscr{I}(N(\kappa_i))} \{\rho_j\}, \min_{\mathbf{x} \in N(\kappa_i) \cap \partial\Omega} \rho(\mathbf{x}) \right\} & \text{if } \kappa_i \cap \partial\Omega \neq \emptyset \end{cases} \tag{6}$$

and

$$\rho_i^{\max} = \begin{cases} \max_{j \in \mathscr{I}(N(\kappa_i))} \{\rho_j\} & \text{if } \kappa_i \cap \partial\Omega = \emptyset, \\ \max\left\{ \max_{j \in \mathscr{I}(N(\kappa_i))} \{\rho_j\}, \max_{\mathbf{x} \in N(\kappa_i) \cap \partial\Omega} \rho(\mathbf{x}) \right\} & \text{if } \kappa_i \cap \partial\Omega \neq \emptyset. \end{cases} \tag{7}$$

In words, for cells that do not intersect the boundary $\partial\Omega$, the values of $\rho_i^{\min}$ and $\rho_i^{\max}$ give the smallest and the largest mean densities in the neighborhood of κ_i,

Constrained-Optimization Based Data Transfer

respectively. For cells adjacent to the boundary, $\rho_i^{\min}$ is the smaller of the smallest mean cell density in the cell neighborhood and the smallest density on the boundary segment $N(\kappa_i) \cap \partial\Omega$; $\rho_i^{\max}$ is defined analogously. Bounds for the cell masses follow from (4) and (5):

$$\rho_i^{\min} V(\kappa_i) = m_i^{\min} \leq m_i \leq m_i^{\max} = \rho_i^{\max} V(\kappa_i) \quad \forall \kappa_i \in K_h(\Omega). \tag{8}$$

A formal statement of the mass-density remap problem is as follows

Definition 1 (Remapping of mass-density) Given mean density values ρ_i on the *old* grid cells κ_i, find accurate approximations $\widetilde{m}_i$ for the masses of the *new* cells $\widetilde{\kappa}_i$,

$$\widetilde{m}_i \approx \widetilde{m}_i^{\text{ex}} = \int_{\widetilde{\kappa}_i} \rho(\mathbf{x}) dV; \quad i = 1, \ldots, K, \tag{9}$$

such that the following conditions hold:

(R1) The total mass is conserved:

$$\sum_{i=1}^{K} \widetilde{m}_i = \sum_{i=1}^{K} m_i = M.$$

(R2) If the exact density $\rho(\mathbf{x})$ is a linear function on all of Ω, then the remapped masses are exact:

$$\widetilde{m}_i = \widetilde{m}_i^{\text{ex}} = \int_{\widetilde{\kappa}_i} \rho(\mathbf{x}) dV; \quad i = 1, \ldots, K. \tag{10}$$

(R3) Given approximate masses $\widetilde{m}_i$ on the new cells, define $\widetilde{\rho}_i = \widetilde{m}_i / V(\widetilde{\kappa}_i)$. Let $\rho_i^{\min}$ and $\rho_i^{\max}$ be the quantities defined in (6)–(7). Then the bounds

$$\rho_i^{\min} \leq \widetilde{\rho}_i \leq \rho_i^{\max}$$

and

$$\rho_i^{\min} V(\widetilde{\kappa}_i) = \widetilde{m}_i^{\min} \leq \widetilde{m}_i \leq \widetilde{m}_i^{\max} = \rho_i^{\max} V(\widetilde{\kappa}_i) \tag{11}$$

hold on every new cell $\widetilde{\kappa}_i$. □

Requirements (R1)–(R3) in Definition 1 are derived from the desired remap properties (P1)–(P3). (R1) and (R2) are formal statements of (P1) and (P2), whereas (R3) follows from the bounds in (5) and (8), and the locality assumption (2). Therefore, the last requirement is specific to a continuous rezone strategy and may have to be modified for other settings. Such a modification is beyond the scope of this chapter.

3 A Constrained Optimization Formulation of the Remap Problem

In this section we develop an inequality-constrained optimization formulation of remap that satisfies requirements (R1)–(R3). The conservation of total mass (R1) is the simplest one. For any two grids that satisfy the locality assumption (2), the new cells have the following representation (cf. Margolin and Shashkov [17, equation (3.9)]):

$$\widetilde{\kappa}_i = \left(\kappa_i \cup \bigcup_{j \in \mathscr{I}(N(\kappa_i))} \widetilde{\kappa}_i \cap \kappa_j \right) \Big\backslash \left(\bigcup_{j \in \mathscr{I}(N(\kappa_i))} \kappa_i \cap \widetilde{\kappa}_j \right). \tag{12}$$

Using (12) we can express the exact masses of the new cells in *flux form*

$$\widetilde{m}_i^{\text{ex}} = m_i + \sum_{j \in \mathscr{I}(N(\kappa_i))} F_{ij}^{\text{ex}}, \tag{13}$$

where the (exact) fluxes are (cf. Margolin and Shashkov [17, equation (3.12)])

$$F_{ij}^{\text{ex}} = \int_{\widetilde{\kappa}_i \cap \kappa_j} \rho(\mathbf{x}) dV - \int_{\kappa_i \cap \widetilde{\kappa}_j} \rho(\mathbf{x}) dV. \tag{14}$$

Formula (14) implies that the exact mass fluxes are antisymmetric: $F_{ij}^{\text{ex}} = -F_{ji}^{\text{ex}}$. Assume that F_{ij} are approximate mass fluxes that are also antisymmetric

$$F_{ij} = -F_{ji}. \tag{15}$$

Using these fluxes in (16) yields a formula for the approximation of the new cell masses

$$\widetilde{m}_i = m_i + \sum_{j \in \mathscr{I}(N(\kappa_i))} F_{ij}, \tag{16}$$

which preserves the total mass, i.e. satisfies (R1) in Definition 1. To satisfy (R2) we introduce the notion of *high-order target* mass fluxes

$$F_{ij}^H = \int_{\widetilde{\kappa}_i \cap \kappa_j} \rho_j^H(\mathbf{x}) dV - \int_{\kappa_i \cap \widetilde{\kappa}_j} \rho_i^H(\mathbf{x}) dV, \tag{17}$$

where $\rho_i^H(\mathbf{x})$ is a density reconstruction on κ_i that is exact for linear functions. If $\rho(\mathbf{x})$ is linear, then $F_{ij}^H = F_{ij}^{\text{ex}}$, i.e., the target fluxes coincide[3] with the exact fluxes for linear functions. In this case, using (16) with the target fluxes gives the exact new cell masses, i.e., (R2) holds. However, if $\rho(\mathbf{x})$ is not linear, using F_{ij}^H in (16) will likely lead to violation of (R3), especially when $\rho(\mathbf{x})$ is not smooth. We then

---

[3] In practice, this also means that the integrals in (17) should be approximated by quadratures that are exact for linear functions.

constrain the set of approximate fluxes F_{ij} introduced in (15)–(16) by the global system of linear inequalities

$$\widetilde{m}_i^{\min} \leq m_i + \sum_{j \in \mathscr{I}(N(\kappa_i))} F_{ij} \leq \widetilde{m}_i^{\max}; \quad i = 1, \ldots, K, \tag{18}$$

obtained by substituting the approximate mass in (11) with the flux form formula (16). By construction, any F_{ij} that solves (18) produces new cell masses that satisfy (R3). To summarize,

- using the flux form (16) guarantees the conservation of total mass (R1);
- using (16) with the target fluxes F_{ij}^H ensures preservation of linearity (R2);
- using (16) with fluxes F_{ij} which solve (18) secures the preservation of local bounds (R3).

We use optimization to reconcile the last two properties. Let us regard the fluxes F_{ij} as the unknowns, the inequalities (18) as the constraints, and the minimization of the Euclidean distance[4] between the target and the unknown fluxes as the objective. The resulting constrained optimization problem reads

$$\begin{cases} \min_{F_{ij}} \sum_{i=1}^{K} \sum_{j \in \mathscr{I}(N(\kappa_i))} (F_{ij} - F_{ij}^H)^2 \\ \text{subject to} \\ \quad F_{ij} = -F_{ji}, \quad\quad\quad\quad\quad\quad\quad i = 1, \ldots, K, \ j \in \mathscr{I}(N(\kappa_i)) \\ \quad \widetilde{m}_i^{\min} \leq m_i + \sum_{j \in \mathscr{I}(N(\kappa_i))} F_{ij} \leq \widetilde{m}_i^{\max}, \quad i = 1, \ldots, K. \end{cases} \tag{19}$$

Explicit enforcement of the antisymmetry constraint by using only the fluxes F_{pq} for which $p < q$ simplifies the optimization problem:

$$\begin{cases} \min_{F_{ij}} \sum_{i=1}^{K} \sum_{\substack{j \in \mathscr{I}(N(\kappa_i)) \\ i < j}} (F_{ij} - F_{ij}^H)^2 \\ \text{subject to} \\ \quad \widetilde{m}_i^{\min} - m_i \leq \sum_{\substack{j \in \mathscr{I}(N(\kappa_i)) \\ i < j}} F_{ij} - \sum_{\substack{j \in \mathscr{I}(N(\kappa_i)) \\ i > j}} F_{ji} \leq \widetilde{m}_i^{\max} - m_i, \quad i = 1, \ldots, K, \end{cases} \tag{20}$$

where we have also moved m_i to the left and right of the chain of inequalities. Any feasible point of (20) satisfies (R1) and (R3) by construction.

We proceed to show that (20) has a non-empty feasible set, i.e., there is always a non-trivial optimal solution, and that the optimal solution preserves linear densities.

---

[4] The Euclidean distance is used for simplicity. The objective can be defined using any valid distance function (or, equivalently, norm).

Theorem 1 *Assume that $K_h(\Omega)$ and $\widetilde{K}_h(\Omega)$ are such that the locality condition* (2) *holds. For any given set of masses m_i and associated densities $\rho_i = m_i/V(\kappa_i)$ on $K_h(\Omega)$ there exist antisymmetric fluxes $\{F_{ij}\}$ which satisfy the inequality constraints in* (20), *resp* (19).

Proof We need to show that there are antisymmetric fluxes F_{ij} such that

$$\rho_i^{\min} V(\widetilde{\kappa}_i) \leq \rho_i V(\kappa_i) + \sum_{\kappa_j \in N_i} F_{ij} \leq \rho_i^{\max} V(\widetilde{\kappa}_i).$$

Fix a cell index $1 \leq i \leq K$, and choose $\widehat{\rho}_j$, for $\kappa_j \in N_j$ according to

$$\rho_i^{\min} \leq \widehat{\rho}_j \leq \rho_i^{\max} \quad \text{for } j \neq i \quad \text{and} \quad \widehat{\rho}_i = \rho_i. \tag{21}$$

The representation formula (12) motivates the following definition:

$$F_{ij} = \widehat{\rho}_j V(\widetilde{\kappa}_i \cap \kappa_j) - \widehat{\rho}_i V(\kappa_i \cap \widetilde{\kappa}_j). \tag{22}$$

Clearly, $F_{ij} = -F_{ji}$. Using the fluxes (22)

$$\rho_i V(\kappa_i) + \sum_{\kappa_j \in N_i} F_{ij} = \rho_i \left[V(\kappa_i) - \sum_{j \neq i} V(\kappa_i \cap \widetilde{\kappa}_j) \right] + \sum_{j \neq i} \widehat{\rho}_j V(\widetilde{\kappa}_i \cap \kappa_j)$$

$$= \rho_i V(\widetilde{\kappa}_i \cap \kappa_i) + \sum_{j \neq i} \widehat{\rho}_j V(\widetilde{\kappa}_i \cap \kappa_j) = \sum_{\kappa_j \in N_i} \widehat{\rho}_j V(\widetilde{\kappa}_i \cap \kappa_j).$$

From $\widetilde{\kappa}_i = \bigcup_{\kappa_j \in N_i}(\widetilde{\kappa}_i \cap \kappa_j)$ and the bounds in (21) it follows that

$$\sum_{\kappa_j \in N_i} \widehat{\rho}_j V(\widetilde{\kappa}_i \cap \kappa_j) \leq \rho_i^{\max} \sum_{\kappa_j \in N_i} V(\widetilde{\kappa}_i \cap \kappa_j) = \rho_i^{\max} V(\widetilde{\kappa}_i);$$

$$\sum_{\kappa_j \in N_i} \widehat{\rho}_j V(\widetilde{\kappa}_i \cap \kappa_j) \geq \rho_i^{\min} \sum_{\kappa_j \in N_i} V(\widetilde{\kappa}_i \cap \kappa_j) = \rho_i^{\min} V(\widetilde{\kappa}_i),$$

which proves the theorem. $\square$

Preservation of linearity (R2) requires the target fluxes F_{ij}^H to be in the feasible set of (20) whenever $\rho(\mathbf{x})$ is linear, i.e.,

$$\widetilde{m}_i^{\min} - m_i \leq \sum_{\substack{j \in \mathscr{I}(N(\kappa_i)) \\ i<j}} F_{ij}^H - \sum_{\substack{j \in \mathscr{I}(N(\kappa_i)) \\ i>j}} F_{ji}^H \leq \widetilde{m}_i^{\max} - m_i, \quad i = 1, \ldots, K. \tag{23}$$

The proof of this fact requires a simple technical result.

Lemma 1 *Let $n > 0$ be an integer and let $\vec{c} \in \mathbb{R}^n$ be an arbitrary fixed vector. For any closed and bounded set of points $P \subset \mathbb{R}^n$*

$$\min_{\mathbf{x} \in P}(\vec{c}^T \mathbf{x}) = \min_{\mathbf{x} \in \mathscr{H}(P)}(\vec{c}^T \mathbf{x}) \quad \text{and} \quad \max_{\mathbf{x} \in P}(\vec{c}^T \mathbf{x}) = \max_{\mathbf{x} \in \mathscr{H}(P)}(\vec{c}^T \mathbf{x}), \tag{24}$$

where $\mathscr{H}(P)$ is the convex hull of P.

Proof The real-valued function $\vec{c}^T \mathbf{x}$ is continuous on $\mathbb{R}^n$. The set P is closed and bounded, which implies that $\vec{c}^T \mathbf{x}$ attains its minimum and maximum over P. Since the convex hull of a closed and bounded set is closed and bounded, see Rockafellar [21, Theorem 17.2], the same is true for $\mathscr{H}(P)$.[5]

The function $\vec{c}^T \mathbf{x}$ is linear, hence both convex and concave. The claim of the lemma follows from a standard result on the supremum of convex (infimum of concave) functions, see e.g. Rockafellar [21, Theorem 32.2]. □

The following theorem provides sufficient conditions on mesh movement for (23) to hold.

Theorem 2 *Assume the locality condition (2) and suppose that the exact density $\rho(\mathbf{x})$ is linear in all of Ω. Let B_i denote the set of barycenters of the Lagrangian cells in $N(\kappa_i)$,*

$$B_i = \{\mathbf{b}_j \mid j \in \mathscr{I}(N(\kappa_i))\},$$

and let $\widetilde{\mathbf{b}}_i$ be the barycenter of the rezoned cell $\widetilde{\kappa}_i$. Sufficient conditions for the target fluxes to be in the feasible set of (20), that is for (23) to hold, are

$$\widetilde{\mathbf{b}}_i \in \mathscr{H}(B_i) \qquad \text{if } \kappa_i \cap \partial \Omega = \emptyset, \qquad (25)$$

$$\widetilde{\mathbf{b}}_i \in \mathscr{H}\big(B_i \cup (N(\kappa_i) \cap \partial \Omega)\big) \quad \text{if } \kappa_i \cap \partial \Omega \neq \emptyset, \qquad (26)$$

where $\mathscr{H}(\cdot)$ denotes the convex hull.

Proof Because $\rho(\mathbf{x})$ is linear and the density reconstruction is exact for linear functions it follows that the remapped mass equals the exact mass on every rezoned cell $\widetilde{\kappa}_i$:

$$\widetilde{m}_i = m_i + \sum_{\substack{j \in \mathscr{I}(N(\kappa_i)) \\ i < j}} F_{ij}^H - \sum_{\substack{j \in \mathscr{I}(N(\kappa_i)) \\ i > j}} F_{ji}^H = m_i + \sum_{\substack{j \in \mathscr{I}(N(\kappa_i)) \\ i < j}} F_{ij}^{\text{ex}} - \sum_{\substack{j \in \mathscr{I}(N(\kappa_i)) \\ i > j}} F_{ji}^{\text{ex}} = \widetilde{m}_i^{\text{ex}}.$$

Therefore, proving that (23) holds reduces to showing that

$$\widetilde{m}_i^{\min} \leq \widetilde{m}_i^{\text{ex}} \leq \widetilde{m}_i^{\max} \quad \text{for all } i = 1, \ldots, K. \qquad (27)$$

Recalling $\rho(\mathbf{x}) = c_0 + \vec{c}^T \mathbf{x}$ and using the barycenter formula (1) yields

$$\widetilde{m}_i^{\text{ex}} = \int_{\widetilde{\kappa}_i} (c_0 + \vec{c}^T \mathbf{x}) dV = c_0 V(\widetilde{\kappa}_i) + \vec{c}^T \left[\int_{\widetilde{\kappa}_i} \mathbf{x} \, dV\right]$$

---

[5]This guarantees that taking min and max in (24) is well-defined. Otherwise, the correct statement of this result should involve inf and sup.

$$= c_0 V(\widetilde{\kappa}_i) + \vec{c}^{\mathsf{T}} \left[\frac{\int_{\widetilde{\kappa}_i} \mathbf{x} \, dV}{V(\widetilde{\kappa}_i)} \right] V(\widetilde{\kappa}_i) = \left(c_0 + \vec{c}^{\mathsf{T}} \widetilde{\mathbf{b}}_i \right) V(\widetilde{\kappa}_i).$$

We consider two cases, $\kappa_i \cap \partial\Omega = \emptyset$ and $\kappa_i \cap \partial\Omega \neq \emptyset$.

Case 1: Suppose $\kappa_i \cap \partial\Omega = \emptyset$. Using

$$\rho_i^{\min} = \min_{j \in \mathscr{I}(N(\kappa_i))} \{\rho_j\} \quad \text{and} \quad \rho_i^{\max} = \max_{j \in \mathscr{I}(N(\kappa_i))} \{\rho_j\},$$

the barycenter formula yields

$$\widetilde{m}_i^{\min} = \min_{j \in \mathscr{I}(N(\kappa_i))} \left[\frac{\int_{\kappa_j}(c_0 + \vec{c}^{\mathsf{T}}\mathbf{x})dV}{V(\kappa_j)} \right] V(\widetilde{\kappa}_i) = \min_{\mathbf{b}_j \in B_i} \left(c_0 + \vec{c}^{\mathsf{T}}\mathbf{b}_j \right) V(\widetilde{\kappa}_i)$$

for the lower bound and

$$\widetilde{m}_i^{\max} = \max_{j \in \mathscr{I}(N(\kappa_i))} \left[\frac{\int_{\kappa_j}(c_0 + \vec{c}^{\mathsf{T}}\mathbf{x})dV}{V(\kappa_j)} \right] V(\widetilde{\kappa}_i) = \max_{\mathbf{b}_j \in B_i} \left(c_0 + \vec{c}^{\mathsf{T}}\mathbf{b}_j \right) V(\widetilde{\kappa}_i)$$

for the upper bound in (27). From Lemma 1 it follows that

$$\min_{\mathbf{b}_j \in B_i} \left(c_0 + \vec{c}^{\mathsf{T}}\mathbf{b}_j \right) = \min_{\mathbf{x} \in \mathscr{H}(B_i)} \left(c_0 + \vec{c}^{\mathsf{T}}\mathbf{x} \right) \tag{28}$$

and

$$\max_{\mathbf{b}_j \in B_i} \left(c_0 + \vec{c}^{\mathsf{T}}\mathbf{b}_j \right) = \max_{\mathbf{x} \in \mathscr{H}(B_i)} \left(c_0 + \vec{c}^{\mathsf{T}}\mathbf{x} \right). \tag{29}$$

Consequently, whenever $\kappa_i \cap \partial\Omega = \emptyset$, (27) is equivalent to

$$\min_{\mathbf{x} \in \mathscr{H}(B_i)} \left(c_0 + \vec{c}^{\mathsf{T}}\mathbf{x} \right) \leq \left(c_0 + \vec{c}^{\mathsf{T}}\widetilde{\mathbf{b}}_i \right) \leq \max_{\mathbf{x} \in \mathscr{H}(B_i)} \left(c_0 + \vec{c}^{\mathsf{T}}\mathbf{x} \right). \tag{30}$$

A sufficient condition for (30) is given by (25), see Fig. 2.

Case 2: Suppose $\kappa_i \cap \partial\Omega \neq \emptyset$. We have

$$\rho_i^{\min} = \min \left\{ \min_{j \in \mathscr{I}(N(\kappa_i))} \{\rho_j\}, \min_{\mathbf{x} \in N(\kappa_i) \cap \partial\Omega} \left(c_0 + \vec{c}^{\mathsf{T}}\mathbf{x} \right) \right\}$$

and

$$\rho_i^{\max} = \max \left\{ \max_{j \in \mathscr{I}(N(\kappa_i))} \{\rho_j\}, \max_{\mathbf{x} \in N(\kappa_i) \cap \partial\Omega} \left(c_0 + \vec{c}^{\mathsf{T}}\mathbf{x} \right) \right\}.$$

Using again the barycenter formula, we obtain

$$\rho_i^{\min} = \min \left\{ \min_{\mathbf{x} \in B_i} \left(c_0 + \vec{c}^{\mathsf{T}}\mathbf{x} \right), \min_{\mathbf{x} \in N(\kappa_i) \cap \partial\Omega} \left(c_0 + \vec{c}^{\mathsf{T}}\mathbf{x} \right) \right\}$$

and

$$\rho_i^{\max} = \max \left\{ \max_{\mathbf{x} \in B_i} \left(c_0 + \vec{c}^{\mathsf{T}}\mathbf{x} \right), \max_{\mathbf{x} \in N(\kappa_i) \cap \partial\Omega} \left(c_0 + \vec{c}^{\mathsf{T}}\mathbf{x} \right) \right\}.$$

Constrained-Optimization Based Data Transfer

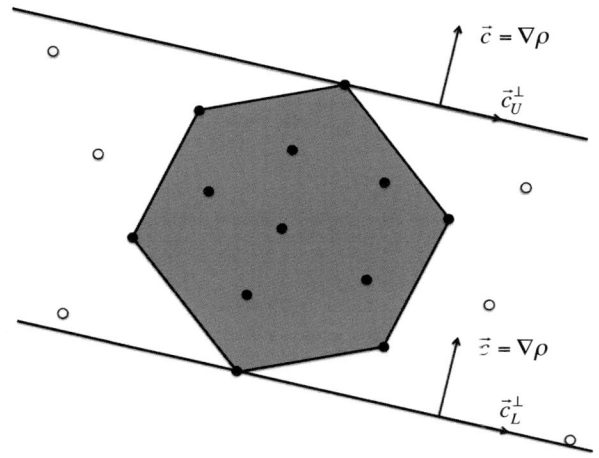

Fig. 2 The level sets of $\rho(\mathbf{x}) = c_0 + \vec{c}^T\mathbf{x}$ are perpendicular to $\nabla\rho(\mathbf{x}) = \vec{c}$ and the extrema of $\rho(\mathbf{x})$ are achieved along the parallel lines $\vec{c}_L^\perp$ and $\vec{c}_U^\perp$ shown in the plot. Therefore, inequality (30) holds for all points between the two lines, while (25) requires $\tilde{\mathbf{b}}_i$ to remain in the convex hull $\mathscr{H}(B_i)$ (the *gray hexagon*)

In other words,

$$\widetilde{m}_i^{\min} = \min_{\mathbf{x} \in B_i \cup (N(\kappa_i) \cap \partial\Omega)} \left(c_0 + \vec{c}^T\mathbf{x}\right) V(\widetilde{\kappa}_i)$$

and

$$\widetilde{m}_i^{\max} = \max_{\mathbf{x} \in B_i \cup (N(\kappa_i) \cap \partial\Omega)} \left(c_0 + \vec{c}^T\mathbf{x}\right) V(\widetilde{\kappa}_i).$$

Treating $B_i \cup (N(\kappa_i) \cap \partial\Omega)$ as a set of points in $\mathbb{R}^n$, another application of Lemma 1 gives

$$\widetilde{m}_i^{\min} = \min_{\mathbf{x} \in \mathscr{H}(B_i \cup (N(\kappa_i) \cap \partial\Omega))} \left(c_0 + \vec{c}^T\mathbf{x}\right) V(\widetilde{\kappa}_i)$$

and

$$\widetilde{m}_i^{\max} = \max_{\mathbf{x} \in \mathscr{H}(B_i \cup (N(\kappa_i) \cap \partial\Omega))} \left(c_0 + \vec{c}^T\mathbf{x}\right) V(\widetilde{\kappa}_i).$$

Therefore, whenever $\kappa_i \cap \partial\Omega \neq \emptyset$, a sufficient condition for (27) is given by (26). This concludes the proof. □

Remark 1 The sufficient condition (26) can be replaced by more restrictive conditions of the type

$$\tilde{\mathbf{b}}_i \in \mathscr{H}(B_i \cup S_i) \quad \text{if } \kappa_i \cap \partial\Omega \neq \emptyset,$$

where $S_i \subseteq (N(\kappa_i) \cap \partial\Omega)$, i.e. S_i is any (for example, finite) set of points taken from the boundary segment $N(\kappa_i) \cap \partial\Omega$.

Remark 2 The sufficient conditions (25) and (26) are not in any way dependent on the cell shape. As a result, the statement of Theorem 2 applies to general grids,

Using (32) and (34) we rewrite the constraints as follows:

$$\widetilde{m}_i = m_i + \sum_{\substack{j \in \mathscr{I}(N(\kappa_i)) \\ i<j}} F_{ij} - \sum_{\substack{j \in \mathscr{I}(N(\kappa_i)) \\ i>j}} F_{ji}$$

$$= m_i + \sum_{\substack{j \in \mathscr{I}(N(\kappa_i)) \\ i<j}} \left(F_{ij}^L + a_{ij} dF_{ij}\right) - \sum_{\substack{j \in \mathscr{I}(N(\kappa_i)) \\ i>j}} \left(F_{ji}^L + a_{ji} dF_{ji}\right)$$

$$= \left(m_i + \sum_{\substack{j \in \mathscr{I}(N(\kappa_i)) \\ i<j}} F_{ij}^L - \sum_{\substack{j \in \mathscr{I}(N(\kappa_i)) \\ i>j}} F_{ji}^L\right) + \sum_{\substack{j \in \mathscr{I}(N(\kappa_i)) \\ i<j}} a_{ij} dF_{ij} - \sum_{\substack{j \in \mathscr{I}(N(\kappa_i)) \\ i>j}} a_{ji} dF_{ji}$$

$$= \widetilde{m}_i^L + \sum_{\substack{j \in \mathscr{I}(N(\kappa_i)) \\ i<j}} a_{ij} dF_{ij} - \sum_{\substack{j \in \mathscr{I}(N(\kappa_i)) \\ i>j}} a_{ji} dF_{ji}.$$

From (33) it follows that

$$\widetilde{Q}_i^{\min} := \widetilde{m}_i^{\min} - \widetilde{m}_i^L \leq 0 \quad \text{and} \quad \widetilde{Q}_i^{\max} := \widetilde{m}_i^{\max} - \widetilde{m}_i^L \geq 0. \tag{35}$$

We write the transformed constraints using these quantities as

$$\widetilde{Q}_i^{\min} \leq \sum_{\substack{j \in \mathscr{I}(N(\kappa_i)) \\ i<j}} a_{ij} dF_{ij} - \sum_{\substack{j \in \mathscr{I}(N(\kappa_i)) \\ i>j}} a_{ji} dF_{ji} \leq \widetilde{Q}_i^{\max}, \quad i=1,\ldots,K. \tag{36}$$

In summary, after changing variables according to (34), the OBR problem (20) assumes the form

$$\begin{cases} \min_{a_{ij}} \sum_{i=1}^K \sum_{\substack{j \in \mathscr{I}(N(\kappa_i)) \\ i<j}} (1-a_{ij})^2 (dF_{ij})^2 \\ \text{subject to} \\ \widetilde{Q}_i^{\min} \leq \sum_{\substack{j \in \mathscr{I}(N(\kappa_i)) \\ i<j}} a_{ij} dF_{ij} - \sum_{\substack{j \in \mathscr{I}(N(\kappa_i)) \\ i>j}} a_{ji} dF_{ji} \leq \widetilde{Q}_i^{\max}, \quad i=1,\ldots,K. \end{cases} \tag{37}$$

Problems (20) and (37) are completely equivalent. For example, the global minimizer $a_{ij} = 1$ of (37), sans constraints, corresponds to $F_{ij} = F_{ij}^H$, which is the global minimizer of (20), sans constraints. Note also that the choice $a_{ij} = 0$ satisfies the constraints, due to (35). The sufficient conditions in Theorem 2 guarantee that $a_{ij} = 1$ are in the feasible set of (37) when the exact density $\rho(\mathbf{x})$ is a linear function in all of Ω.

4.2 The M-OBR Formulation

In this section we modify (37) to another inequality-constrained optimization problem, termed M-OBR, in which the same objective is minimized subject to a set of simple box constraints. The box constraints are sufficient for the original inequality constraints in (37) to hold and are derived by following the same reasoning as in Liska et al. [14]. To this end, we define the quantities

$$P_i^- = \sum_{\substack{j \in \mathscr{I}(N(\kappa_i)) \\ i<j}}^{dF_{ij} \leq 0} dF_{ij} - \sum_{\substack{j \in \mathscr{I}(N(\kappa_i)) \\ i>j}}^{dF_{ji} \geq 0} dF_{ji} \leq 0; \quad P_i^+ = \sum_{\substack{j \in \mathscr{I}(N(\kappa_i)) \\ i<j}}^{dF_{ij} \geq 0} dF_{ij} - \sum_{\substack{j \in \mathscr{I}(N(\kappa_i)) \\ i>j}}^{dF_{ji} \leq 0} dF_{ji} \geq 0; \tag{38}$$

$$D_i^- = \begin{cases} \frac{\widetilde{Q}_i^{\min}}{P_i^-} & \text{if } P_i^- < 0, \\ 0 & \text{if } P_i^- = 0 \end{cases} \quad \text{and} \quad D_i^+ = \begin{cases} \frac{\widetilde{Q}_i^{\max}}{P_i^+} & \text{if } P_i^+ > 0, \\ 0 & \text{if } P_i^+ = 0. \end{cases} \tag{39}$$

Using these quantities we reduce the constraints in (37) to a set of box constraints in three steps.

In the first step we replace the upper and lower bounds in the constraints of (37) by $D_i^- P_i^-$ and $D_i^+ P_i^+$, respectively:

$$D_i^- P_i^- \leq \sum_{\substack{j \in \mathscr{I}(N(\kappa_i)) \\ i<j}} a_{ij} dF_{ij} - \sum_{\substack{j \in \mathscr{I}(N(\kappa_i)) \\ i>j}} a_{ji} dF_{ji} \leq D_i^+ P_i^+, \quad i = 1, \ldots, K. \tag{40}$$

In the second step we split (40) into two parts, according to the signs of the flux differentials:

(a) $\displaystyle D_i^- P_i^- \leq \sum_{\substack{j \in \mathscr{I}(N(\kappa_i)) \\ i<j}}^{dF_{ij} \leq 0} a_{ij} dF_{ij} - \sum_{\substack{j \in \mathscr{I}(N(\kappa_i)) \\ i>j}}^{dF_{ji} \geq 0} a_{ji} dF_{ji} \leq 0,$

(b) $\displaystyle 0 \leq \sum_{\substack{j \in \mathscr{I}(N(\kappa_i)) \\ i<j}}^{dF_{ij} \geq 0} a_{ij} dF_{ij} - \sum_{\substack{j \in \mathscr{I}(N(\kappa_i)) \\ i>j}}^{dF_{ji} \leq 0} a_{ji} dF_{ji} \leq D_i^+ P_i^+,$

$i = 1, \ldots, K.$ (41)

Finally, using definition (38), we reduce (41) to a set of box constraints by applying the upper and the lower bounds componentwise:

(a) $\begin{cases} D_i^- dF_{ij} \leq a_{ij} dF_{ij} \leq 0 & \text{for } i < j, dF_{ij} \leq 0, \\ D_i^- dF_{ji} \geq a_{ji} dF_{ji} \geq 0 & \text{for } i > j, dF_{ji} \geq 0, \end{cases}$

(b) $\begin{cases} 0 \leq a_{ij} dF_{ij} \leq D_i^+ dF_{ij} & \text{for } i < j, dF_{ij} \geq 0, \\ 0 \geq a_{ji} dF_{ji} \geq D_i^+ dF_{ji} & \text{for } i > j, dF_{ji} \leq 0, \end{cases}$ (42)

$$i = 1, \ldots, K, j \in \mathscr{I}(N(\kappa_i)).$$

Using the box constraints (42) in lieu of the original set of inequalities in (37) yields the modified OBR problem (M-OBR)

$$\begin{cases} \min_{a_{ij}} \sum_{i=1}^{K} \sum_{\substack{j \in \mathscr{I}(N(\kappa_i)) \\ i<j}} (1 - a_{ij})^2 (dF_{ij})^2 \\ \text{subject to} \\ \text{(a)} \quad \begin{cases} D_i^- dF_{ij} \leq a_{ij} dF_{ij} \leq 0 & \text{for } i < j, dF_{ij} \leq 0, \\ D_i^- dF_{ji} \geq a_{ji} dF_{ji} \geq 0 & \text{for } i > j, dF_{ji} \geq 0, \end{cases} \\ \text{(b)} \quad \begin{cases} 0 \leq a_{ij} dF_{ij} \leq D_i^+ dF_{ij} & \text{for } i < j, dF_{ij} \geq 0, \\ 0 \geq a_{ji} dF_{ji} \geq D_i^+ dF_{ji} & \text{for } i > j, dF_{ji} \leq 0, \end{cases} \\ i = 1, \ldots, K, j \in \mathscr{I}(N(\kappa_i)). \end{cases} \quad (43)$$

We are now ready to study the connections of the global M-OBR formulation (43) with the OBR problem (37). The first result shows that (43) always has a solution.

Proposition 1 *The feasible set of the modified OBR problem* (43) *is non-empty.*

Proof The inequalities in (43) are always satisfied for $a_{ij} = 0$ because $D_i^- \geq 0$ and $D_i^+ \geq 0$ for all $i = 1, \ldots, K$. Therefore, the feasible set of (43) always contains at least one point. □

We note that $a_{ij} = 0$ results in $F_{ij} = F_{ij}^L$, which corresponds to a low-order mass remap or, using an advection parlance, to a "donor-cell" solution of the remap problem. Thus, at the least, the M-OBR problem admits the same solution as a conventional low-order local remapper.

The following theorem examines the relationship between M-OBR and OBR.

Theorem 3 *The feasible set of the M-OBR formulation* (43) *is a subset of the feasible set of the OBR formulation* (37).

Proof The feasible sets of the OBR and M-OBR problems are given by

$$\mathscr{U}_O = \{a_{ij} \in \mathbb{R} | (36) \text{ hold for } i = 1, \ldots, K \text{ and } j \in \mathscr{I}(N(\kappa_i))\},$$

and

$$\mathscr{U}_M = \{a_{ij} \in \mathbb{R} | (42) \text{ hold for } i = 1, \ldots, K \text{ and } j \in \mathscr{I}(N(\kappa_i))\},$$

respectively. To show that $\mathscr{U}_M \subseteq \mathscr{U}_O$ define the intermediate sets

$$\mathscr{U}_A = \{a_{ij} \in \mathbb{R} | (40) \text{ hold for } i = 1, \ldots, K \text{ and } j \in \mathscr{I}(N(\kappa_i))\},$$

and
$$\mathcal{U}_B = \{a_{ij} \in \mathbb{R} | (41) \text{ hold for } i = 1, \ldots, K \text{ and } j \in \mathcal{I}(N(\kappa_i))\},$$

corresponding to the first and the second stages in the transformation of the OBR constraints to the box constraints of M-OBR.

To prove the theorem we will show that

$$\mathcal{U}_M \subseteq \mathcal{U}_B \subseteq \mathcal{U}_A \subseteq \mathcal{U}_O.$$

Step 1: $\mathcal{U}_M \subseteq \mathcal{U}_B$. Let $\{a_{ij}\} \in \mathcal{U}_M$. Summing up the inequalities in (42) yields

$$\sum_{\substack{j \in \mathcal{I}(N(\kappa_i)) \\ i<j}}^{dF_{ij} \leq 0} D_i^- dF_{ij} - \sum_{\substack{j \in \mathcal{I}(N(\kappa_i)) \\ i>j}}^{dF_{ji} \geq 0} D_i^- dF_{ji} \leq \sum_{\substack{j \in \mathcal{I}(N(\kappa_i)) \\ i<j}}^{dF_{ij} \leq 0} a_{ij} dF_{ij} - \sum_{\substack{j \in \mathcal{I}(N(\kappa_i)) \\ i>j}}^{dF_{ji} \geq 0} a_{ji} dF_{ji} \leq 0,$$

$$0 \leq \sum_{\substack{j \in \mathcal{I}(N(\kappa_i)) \\ i<j}}^{dF_{ij} \leq 0} a_{ij} dF_{ij} - \sum_{\substack{j \in \mathcal{I}(N(\kappa_i)) \\ i>j}}^{dF_{ji} \geq 0} a_{ji} dF_{ji} \leq \sum_{\substack{j \in \mathcal{I}(N(\kappa_i)) \\ i<j}}^{dF_{ij} \leq 0} D_i^+ dF_{ij} - \sum_{\substack{j \in \mathcal{I}(N(\kappa_i)) \\ i>j}}^{dF_{ji} \geq 0} D_i^+ dF_{ji}.$$

From (38) we see that the left hand side in the first inequality equals $D_i^- P_i^-$ and the right hand side in the second inequality is $D_i^+ P_i^+$. Therefore, inequalities (41) hold for $\{a_{ij}\}$, i.e. $\{a_{ij}\} \in \mathcal{U}_B$. This proves the inclusion $\mathcal{U}_M \subseteq \mathcal{U}_B$.

Step 2: $\mathcal{U}_B \subseteq \mathcal{U}_A$. Assume that $\{a_{ij}\} \in \mathcal{U}_B$. Summing up inequalities (a) and (b) in (41) gives

$$D_i^- P_i^- \leq \sum_{\substack{j \in \mathcal{I}(N(\kappa_i)) \\ i<j}}^{dF_{ij} \leq 0} a_{ij} dF_{ij} - \sum_{\substack{j \in \mathcal{I}(N(\kappa_i)) \\ i>j}}^{dF_{ji} \geq 0} a_{ji} dF_{ji} + \sum_{\substack{j \in \mathcal{I}(N(\kappa_i)) \\ i<j}}^{dF_{ij} \geq 0} a_{ij} dF_{ij} - \sum_{\substack{j \in \mathcal{I}(N(\kappa_i)) \\ i>j}}^{dF_{ji} \leq 0} a_{ji} dF_{ji}$$

$$\leq D_i^+ P_i^+$$

from where it follows that (40) hold for $\{a_{ij}\}$, i.e. $\{a_{ij}\} \in \mathcal{U}_A$. This proves the inclusion $\mathcal{U}_B \subseteq \mathcal{U}_A$.

Step 3: $\mathcal{U}_B \subseteq \mathcal{U}_O$. Finally, let $\{a_{ij}\} \in \mathcal{U}_A$. Note that

$$\widetilde{Q}_i^{\min} \leq D_i^- P_i^- \quad \text{and} \quad D_i^+ P_i^+ \leq \widetilde{Q}_i^{\max}.$$

Therefore, inequalities (36) hold for $\{a_{ij}\}$, i.e. $\{a_{ij}\} \in \mathcal{U}_O$. This proves the inclusion $\mathcal{U}_A \subseteq \mathcal{U}_O$. □

Remark 3 Since the M-OBR feasible set is contained in the OBR feasible set due to Theorem 3, it follows that the OBR solution is always at least as accurate as the M-OBR solution.

4.3 FCR and M-OBR: Two Equivalent Algorithms

In this section we show that the M-OBR formulation is equivalent to the FCR algorithm in Liska et al. [14]. For convenience, below we summarize the FCR formulation for the mass-density remap. Full details can be found in [14, Sect. 3].

The original motivation for FCR is to replace a global optimization problem such as OBR by a series of local problems. To this end, FCR restricts the mass fluxes in (16) to *convex* combinations of the low-order and the high-order target fluxes, i.e.

$$F_{ij} = (1 - a_{ij})F_{ij}^L + a_{ij}F_{ij}^H = F_{ij}^L + a_{ij}dF_{ij}, \qquad (44)$$

where $a_{ij} = a_{ji}$ and $0 \le a_{ij} \le 1$. The convexity assumption is motivated by analogies with the FCT approach of Kuzmin et al. [12] for advection. Except for this requirement, formula (44) is identical to the change of variables in (34). In the FCR algorithm the approximate mass fluxes in (44) are computed using the following values for the unknown coefficients:

$$a_{ij} = \begin{cases} \min\{D_i^+, D_j^-, 1\} & \text{if } dF_{ij} > 0, \\ \min\{D_i^-, D_j^+, 1\} & \text{if } dF_{ij} < 0, \end{cases} \quad 1 \le i, j \le K \text{ and } i < j. \qquad (45)$$

For completeness, one can set $a_{ij} = 1$ whenever $dF_{ij} = 0$. In Liska et al. [14] it is shown that (45) is sufficient for the local mass-density bounds in (36) to hold.

We proceed to show that the solution of the global M-OBR problem is also given by (45). This fact establishes the equivalence of FCR and M-OBR and is a direct consequence of the following theorem.

Theorem 4 *The M-OBR formulation* (43) *is equivalent to the following set of independent, single-variable, constrained optimization problems: for $1 \le i, j \le K$ and $i < j$ solve*

$$\begin{cases} \min_{a_{ij}}(1 - a_{ij})^2(dF_{ij})^2 \\ \text{subject to} \\ \quad 0 \le a_{ij} \le \begin{cases} \min\{D_i^+, D_j^-\} & \text{if } dF_{ij} > 0, \\ \min\{D_i^-, D_j^+\} & \text{if } dF_{ij} < 0. \end{cases} \end{cases} \qquad (46)$$

Proof A flux differential dF_{ij}, $i < j$, can be negative, zero or positive. If $dF_{ij} = 0$, we denote the variable a_{ij} as *free*, because the box constraint in (42) holds for any value of a_{ij}. Note that the terms associated with free variables do not contribute to the objective, because $(1 - a_{ij})^2(dF_{ij})^2 = 0$. It follows that all free variables can be eliminated[6] from the optimization problem. Thus, without loss of generality we may assume that $dF_{ij} \ne 0$.

---

[6]For a complete match with FCR we can set all free variables to 1.

Constrained-Optimization Based Data Transfer

It is easy to see that whenever $dF_{ij} \neq 0$, the associated variable a_{ij} enters in exactly one constraint of type (a) and one constraint of type (b). Solving the inequalities for a_{ij} gives

$$0 \leq a_{ij} \leq D_i^+ \quad \text{and} \quad 0 \leq a_{ij} \leq D_j^-$$

for $i < j$ and $dF_{ij} > 0$, and

$$0 \leq a_{ij} \leq D_i^- \quad \text{and} \quad 0 \leq a_{ij} \leq D_j^+$$

for $i < j$ and $dF_{ij} < 0$. Succinctly,

$$0 \leq a_{ij} \leq \begin{cases} \min\{D_i^+, D_j^-\} & \text{if } dF_{ij} > 0, \\ \min\{D_i^-, D_j^+\} & \text{if } dF_{ij} < 0, \end{cases} \quad 1 \leq i, j \leq K \text{ and } i < j$$

is a new set of box constraints that is completely equivalent to (43). Because each of the terms in the objective functional depends on only one variable, it follows that (43) decouples into the set of independent, single-variable, constrained optimization problems given in (46). □

The equivalence of FCR and M-OBR easily follows.

Corollary 1 *The solution $\{a_{ij}\}$ of the M-OBR problem (43) is given by the FCR formula (45).*

Proof To find the solution of the M-OBR problem we set all free variables to 1. The rest of the variables are computed by solving the decoupled optimization problems in (46). For a given pair of indices $i < j$ let $D_{ij} \geq 0$ denote the upper bound in the constraint of the optimization problem for the variable a_{ij}. The cost functional $(1 - a_{ij})^2 (dF_{ij})^2$ in this problem represents a parabola with the vertex at $(1, 0)$. Therefore, the constrained minimum is achieved at the smaller of the two values $a_{ij} = 1$ or $a_{ij} = D_{ij}$. It follows that whenever $dF_{ij} \neq 0$, the solution of the optimization problem in (46) is given by formula (45). □

4.4 iFCR: An Iterative Extension of FCR

For the purpose of numerical comparisons, we introduce a variation of the standard FCR algorithm, called iterated FCR (iFCR), originally proposed by Schär and Smolarkiewicz [22]. The key idea of iFCR is that, by definition, FCR fluxes ensure monotonicity of the solution, and can be reused as base low-order fluxes for an additional FCR flux computation. This process can be repeated *ad infinitum*. The advantage of iFCR over FCR is mainly in accuracy, at the price of increased computational cost, as the FCR flux computation has to be repeated at each iteration of

Table 1 Outline of the iFCR algorithm

Initialize solution field with initial conditions.
Predictor: Compute FCR fluxes F_{ij} using (44) and (45).
 Define $F_{ij}^{(0)} = F_{ij}$ and $F_{ij}^{L;(0)} = F_{ij}^{L}$.
For $k = 0, \ldots, k_{\max}$ (*iFCR loop begins*)
 Replace $F_{ij}^{L;(k+1)} = F_{ij}^{(k)}$.
 Corrector: Compute $F_{ij}^{(k+1)}$ using (44) and (45).
End (*iFCR loop ends*)
Exit

---

the method. iFCR represents a more challenging benchmark in the analysis of performance of the OBR approach, and, of course—in the limit for a large number of iterations—may easily surpass in cost the OBR algorithm itself. The iFCR approach is described in Table 1.

5 Algorithms I: Exact Cell Intersection Versus Swept Region Flux Computations

Until now all our considerations were based on the exact cell intersection formula (12). This means that in order to implement the corresponding OBR and FCR algorithms we would have to find the intersections between the cells on the old and new meshes, which is computationally expensive. Instead, we implement the OBR and FCR algorithms using *swept regions* as in Margolin and Shashkov [17, Sect. 4]. These are the regions swept by the movement of the sides of the old cells. As a result, the swept regions are completely determined by the coordinates of the old and new nodes and do not require the computation of cell intersections.

Recall that $S(\kappa_i)$ is the set of all sides in cell κ_i. Each side σ_j has unique orientation $\omega_j = +1$, or -1, which induces orientation on the associated swept region Σ_j. The idea of the swept region approximation is to allow mass exchanges only between cells that share a side. In this case, the new cell masses can be approximated by the formula

$$\widetilde{m}_i = m_i + \sum_{j \in \mathscr{I}(S(\kappa_i))} \omega_j F_j, \qquad (47)$$

where summation is over the sides of the cell and F_j are mass fluxes corresponding to the (signed) swept regions Σ_j associated with side σ_j.

Our implementation of OBR and FCR uses (47) in lieu of the cell-intersection formula (16). Let Σ_j denote the swept region associated with side σ_j of cell κ_i. We define the target (high-order) fluxes as[7]

$$F_j^H = \int_{\Sigma_j} \rho_j^H(\mathbf{x}) dV, \qquad (48)$$

where ρ_j^H is a density reconstruction that is exact for linear functions. One can show that using formula (47) with the fluxes defined in (48) gives the exact cell masses whenever the density is linear (see Margolin and Shashkov [17]). This means that the preservation of linearity in OBR remains in full force when the method is implemented using swept regions, instead of exact cell intersections.

The situation with FCR is somewhat more complicated. In addition to the high-order fluxes (48) this method also uses the low order fluxes

$$F_j^L = \int_{\Sigma_j} \rho_j^L(\mathbf{x}) dV. \qquad (49)$$

It turns out that when the low-order approximations of the new cell masses are computed using (47) and (49), instead of (16) and (31), there is no guarantee that these masses will satisfy the bounds (33), see Margolin and Shashkov [17]. Additional restrictions on the mesh movement are required to ensure that these bounds hold. A sufficient condition for (33) is that the area of the old cell κ_i is greater than the sum of the absolute values of all negatively signed swept regions (see Margolin and Shashkov [17, p. 279]).

The fact that (33) can be violated when FCR is implemented using swept regions has important consequences. Without (33) holding, the two OBR formulations (20) and (37) are still equivalent. However, we cannot carry out the steps in Sect. 4.2, which reduced (37) to the M-OBR formulation (43). Therefore, violation of (33) invalidates Proposition 1, Theorems 3–4, and Corollary 1. What this means in practice is that the feasible set in (43) may become void, in which case the M-OBR problem has no solution. As a result, the FCR solution defined in (45) ceases to be connected to the global OBR optimization problem (20) and is not guaranteed to be in its feasible set. The practical dimension of this fact is that the FCR solution may violate the local bounds. Section 7.3 provides an instructive example in two dimensions that shows the loss of monotonicity when FCR is implemented using swept regions.

6 Algorithms II: Solution Techniques for the OBR Problem

We discuss optimization techniques for the solution of the OBR problem assuming a swept-region approximation. In compact matrix/vector notation problem (20) has

---

[7] Because side nodes can move in different directions swept regions are not simple extrusions of the sides, which can complicate the computation of integrals. Using Green's theorem, integrals of polynomials over swept regions can be replaced by integrals of higher-degree polynomials over the (lower-dimensional) boundaries of these regions, see Margolin and Shashkov [17], Dukowicz and Kodis [8]. This provides an efficient way to compute the fluxes, regardless of the shape of the swept regions.

the form

$$\min_{\vec{F} \in \mathbb{R}^M} \frac{1}{2}(\vec{F} - \vec{F}^H)^\top (\vec{F} - \vec{F}^H) \quad (50)$$
$$\text{subject to} \quad \vec{b}_{\min} \leq \mathbf{A}\vec{F} \leq \vec{b}_{\max},$$

where M denotes the number of unique flux variables, $\widetilde{F}_{ij}^h$. We also define $\vec{F} \in \mathbb{R}^M$, $\vec{F}^H \in \mathbb{R}^M$, $\vec{b}_{\min} \in \mathbb{R}^K$ and $\vec{b}_{\max} \in \mathbb{R}^K$ such that $\vec{F}_{\iota(i,j)} = \widetilde{F}_{ij}^h$, $\vec{F}_{\iota(i,j)}^H = \widetilde{F}_{ij}^T$, $(\vec{b}_{\min})_i = m_i^{\min} - \widetilde{m}_i$ and $(\vec{b}_{\max})_i = m_i^{\max} - \widetilde{m}_i$, respectively, where ι is an indexing function. Finally we let $\mathbf{A} \in \mathbb{R}^{K \times M}$ be a matrix with entries $-1, 0$ and 1 defining the inequality constraints in (20) or a related proxy (see swept-region approximation, Bochev et al. [5, Sects. 4.1, 4.2]). The matrix $\mathbf{A}$ is typically very sparse, with $M > K$ in 2D and 3D. We abbreviate the *nonnegative orthant* as $\mathbb{R}_+^m = \{\mathbf{x} \in \mathbb{R}^m : \mathbf{x} \geq 0\}$.

Rather than solving (50) directly, we focus on its dual formulation. This allows us to reformulate the problem into a simpler, *bound-constrained* optimization problem.

Theorem 5 *Given the definitions of* $\vec{F}^H \in \mathbb{R}^M$, $\vec{b}_{\min} \in \mathbb{R}^K$, $\vec{b}_{\max} \in \mathbb{R}^K$, *and* $\mathbf{A} \in \mathbb{R}^{K \times M}$ *from above, let us define* $J_p : \mathbb{R}^M \to \mathbb{R}$ *and* $J_d : \mathbb{R}^{2K} \to \mathbb{R}$ *as*

$$J_p(\vec{F}) = \frac{1}{2} \|\vec{F} - \vec{F}^H\|_2^2$$

and

$$J_d(\vec{\lambda}, \vec{\mu}) = \frac{1}{2} \|\mathbf{A}^\top \vec{\lambda} - \mathbf{A}^\top \vec{\mu}\|_2^2 - \langle \vec{\lambda}, \vec{b}_{\min} - \mathbf{A}\vec{F}^H \rangle - \langle \vec{\mu}, -\vec{b}_{\max} + \mathbf{A}\vec{F}^H \rangle.$$

Then, we have that

$$\min_{\vec{F} \in \mathbb{R}^M} \{J_p(\vec{F}) : \vec{b}_{\min} \leq \mathbf{A}\vec{F} \leq \vec{b}_{\max}\} = \min_{(\vec{\lambda}, \vec{\mu}) \in \mathbb{R}_+^{2K}} \{J_d(\vec{\lambda}, \vec{\mu})\}$$

where we call the first problem the primal and the second problem the dual. Furthermore,

$$\{\vec{F}^H + \mathbf{A}^\top (\vec{\lambda}^* - \vec{\mu}^*)\} = \arg \min_{\vec{F} \in \mathbb{R}^M} \{J_p(\vec{F}) : \vec{b}_{\min} \leq \mathbf{A}\vec{F} \leq \vec{b}_{\max}\}$$

whenever

$$(\vec{\lambda}^*, \vec{\mu}^*) \in \arg \min_{(\vec{\lambda}, \vec{\mu}) \in \mathbb{R}_+^{2K}} \{J_d(\vec{\lambda}, \vec{\mu})\}.$$

Proof We begin with the observation that J_p denotes a strictly convex, continuous function and that $\{\vec{F} \in \mathbb{R}^M : \vec{b}_{\min} \leq \mathbf{A}\vec{F} \leq \vec{b}_{\max}\}$ denotes a bounded, closed, convex set. Therefore, a unique minimum exists and is attained. Furthermore, since there exists an $\vec{F}$ such that $\vec{b}_{\min} < \mathbf{A}\vec{F} < \vec{b}_{\max}$, we satisfy Slater's constraint qualification. This tells us that strong duality holds, which implies that the Lagrangian dual exists and possesses the same optimal value as the original problem.

Based on this knowledge, we notice that

$$\min_{F \in \mathbb{R}^M} \left\{ J_p(\vec{F}) : \vec{b}_{\min} \leq \mathbf{A}\vec{F} \leq \vec{b}_{\max} \right\}$$
$$= \min_{F \in \mathbb{R}^M} \max_{(\vec{\lambda}, \vec{\mu}) \in \mathbb{R}_+^{2K}} \left\{ J_p(\vec{F}) - \langle \mathbf{A}\vec{F} - \vec{b}_{\min}, \vec{\lambda} \rangle - \langle \vec{b}_{\max} - \mathbf{A}F, \vec{\mu} \rangle \right\}$$
$$= \max_{(\vec{\lambda}, \vec{\mu}) \in \mathbb{R}_+^{2K}} \min_{F \in \mathbb{R}^M} \left\{ J_p(\vec{F}) - \langle \vec{F}, \mathbf{A}^\mathsf{T}(\vec{\lambda} - \vec{\mu}) \rangle + \langle \vec{b}_{\min}, \vec{\lambda} \rangle - \langle \vec{b}_{\max}, \vec{\mu} \rangle \right\}.$$

Next, we consider the function $J : \mathbb{R}^M \to \mathbb{R}$ where

$$J(\vec{F}) = J_p(\vec{F}) - \langle \vec{F}, \mathbf{A}^\mathsf{T}(\vec{\lambda} - \vec{\mu}) \rangle$$

and $(\vec{\lambda}, \vec{\mu}) \in \mathbb{R}^{2K}$ are fixed. We see that J is strictly convex. Therefore, it attains its unique minimum when $\nabla J = 0$. Specifically, when

$$\vec{F} - \vec{F}^H - \mathbf{A}^\mathsf{T}(\vec{\lambda} - \vec{\mu}) = 0,$$

which occurs if and only if

$$\vec{F} = \vec{F}^H + \mathbf{A}^\mathsf{T}(\vec{\lambda} - \vec{\mu}).$$

Therefore, we may find the optimal solution to our original problem with this equation when $(\vec{\lambda}, \vec{\mu})$ are optimal. In addition, we may use this knowledge to simplify our derivation of the dual. Let $\omega = \mathbf{A}^\mathsf{T}(\vec{\lambda} - \vec{\mu})$ and notice that

$$\max_{(\vec{\lambda}, \vec{\mu}) \in \mathbb{R}_+^{2K}} \min_{F \in \mathbb{R}^M} \left\{ J_p(\vec{F}) - \langle \vec{F}, \mathbf{A}^\mathsf{T}(\vec{\lambda} - \vec{\mu}) \rangle + \langle b_{\min}, \vec{\lambda} \rangle - \langle b_{\max}, \vec{\mu} \rangle \right\}$$
$$= \max_{(\vec{\lambda}, \vec{\mu}) \in \mathbb{R}_+^{2K}} \left\{ J_p(\vec{F}^H + \omega) - \langle \vec{F}^H + \omega, \omega \rangle + \langle b_{\min}, \vec{\lambda} \rangle - \langle b_{\max}, \vec{\mu} \rangle \right\}$$
$$= \max_{(\vec{\lambda}, \vec{\mu}) \in \mathbb{R}_+^{2K}} \left\{ \frac{1}{2} \|\omega\|_2^2 - \langle \vec{F}^H, \omega \rangle - \|\omega\|_2^2 + \langle b_{\min}, \vec{\lambda} \rangle - \langle b_{\max}, \vec{\mu} \rangle \right\}$$
$$= \max_{(\vec{\lambda}, \vec{\mu}) \in \mathbb{R}_+^{2K}} \left\{ -\frac{1}{2} \|\mathbf{A}^\mathsf{T}(\vec{\lambda} - \vec{\mu})\|_2^2 - \langle \mathbf{A}\vec{F}^H, \vec{\lambda} - \vec{\mu} \rangle + \langle b_{\min}, \vec{\lambda} \rangle - \langle b_{\max}, \vec{\mu} \rangle \right\}$$
$$= \min_{(\vec{\lambda}, \vec{\mu}) \in \mathbb{R}_+^{2K}} \left\{ \frac{1}{2} \|\mathbf{A}^\mathsf{T}(\vec{\lambda} - \vec{\mu})\|_2^2 + \langle \mathbf{A}\vec{F}^H, \vec{\lambda} - \vec{\mu} \rangle - \langle b_{\min}, \vec{\lambda} \rangle + \langle b_{\max}, \vec{\mu} \rangle \right\}$$
$$= \min_{(\vec{\lambda}, \vec{\mu}) \in \mathbb{R}_+^{2K}} \left\{ \frac{1}{2} \|\mathbf{A}^\mathsf{T}\vec{\lambda} - \mathbf{A}^\mathsf{T}\vec{\mu}\|_2^2 - \langle \vec{\lambda}, \vec{b}_{\min} - \mathbf{A}\vec{F}^H \rangle - \langle \vec{\mu}, -\vec{b}_{\max} + \mathbf{A}\vec{F}^H \rangle \right\}$$
$$= \min_{(\vec{\lambda}, \vec{\mu}) \in \mathbb{R}_+^{2K}} \left\{ J_d(\vec{\lambda}, \vec{\mu}) \right\}.$$

Hence, we see the equivalence between our two optimization problems and note that the equation $\vec{F} = \vec{F}^H + \mathbf{A}^\mathsf{T}(\vec{\lambda} - \vec{\mu})$ allows us to find an optimal primal solution given an optimal solution to the dual. □

Although the primal problem is strictly convex and possesses a unique optimal solution, the dual formulation does not. Rather, the dual problem is convex, but not strictly convex, so multiple minima may exist. Second, our formula for reconstructing the primal solution from the dual depends on an optimal dual solution. If the solution to the dual is not optimal, the reconstruction formula may generate infeasible solutions. With these points in mind, we require two additional definitions before we may proceed to our optimization algorithm.

Definition 2 We define the diagonal operator, $\text{Diag}: \mathbb{R}^m \to \mathbb{R}^{m \times m}$, as

$$[\text{Diag}(\mathbf{x})]_{ij} = \begin{cases} \mathbf{x}_i, & i = j, \\ 0, & i \neq j. \end{cases}$$

Definition 3 For some symmetric, positive semidefinite $\mathbf{H} \in \mathbb{R}^{m \times m}$ and some $\vec{b} \in \mathbb{R}^m$, we define the operator $v_{\mathbf{H},\vec{b}} : \mathbb{R}^m \to \mathbb{R}^m$ as

$$v_{\mathbf{H},\vec{b}}(\mathbf{x}) = \begin{cases} \mathbf{x}_i, & [\mathbf{H}\mathbf{x} + \vec{b}]_i \geq 0, \\ 1, & [\mathbf{H}\mathbf{x} + \vec{b}]_i < 0. \end{cases}$$

When both $\mathbf{H}$ and $\vec{b}$ are clear from the context, we abbreviate this function as v.

In order to solve the dual optimization problem, we use a simplified version of the locally convergent Coleman-Li algorithm (Coleman and Li [7]). The key to this algorithm follows from the following lemma.

Lemma 2 *Let $\mathbf{H} \in \mathbb{R}^{m \times m}$ be symmetric, positive semidefinite and let $\vec{b} \in \mathbb{R}^m$. Then, for some $\mathbf{x}^* \geq 0$, we have that*

$$\mathbf{x}^* \in \arg\min_{x \in \mathbb{R}^m_+} \left\{ \frac{1}{2} \langle \mathbf{H}\mathbf{x}, \mathbf{x} \rangle + \langle \vec{b}, \mathbf{x} \rangle \right\} \iff \text{Diag}(v(\mathbf{x}^*))(\mathbf{H}\mathbf{x}^* + \vec{b}) = 0.$$

Proof We begin with the observation that since $\mathbf{H}$ is symmetric, positive semidefinite, the problem

$$\min_{x \in \mathbb{R}^m_+} \left\{ \frac{1}{2} \langle \mathbf{H}\mathbf{x}, \mathbf{x} \rangle + \langle \vec{b}, x \rangle \right\}$$

represents a convex optimization problem with a coercive objective and a closed, convex set of constraints. Therefore, a minimum exists and the first order optimality conditions become sufficient for optimality.

In the forward direction, we assume that we have an optimal pair $(\mathbf{x}^*, \vec{\lambda}^*)$ that satisfy the first order optimality conditions,

$$\mathbf{H}\mathbf{x}^* + \vec{b} - \vec{\lambda}^* = 0,$$

$$\mathbf{x}^* \geq 0, \quad \vec{\lambda}^* \geq 0,$$

$$\text{Diag}(\mathbf{x}^*)\vec{\lambda}^* = 0.$$

According to these equations, $\vec{\lambda}^* = \mathbf{H}\mathbf{x}^* + \vec{b}$ and $\vec{\lambda}^* \geq 0$. This implies that $\mathbf{H}\mathbf{x}^* + \vec{b} \geq 0$. Therefore, according to the definition of v, $[\text{Diag}(v(\mathbf{x}^*))]_{ii} = \mathbf{x}_i^*$ for all i. This tells us that

$$\left[\text{Diag}(v(\mathbf{x}^*))(\mathbf{H}\mathbf{x}^* + \vec{b})\right]_i = \mathbf{x}_i^*[\mathbf{H}\mathbf{x}^* + \vec{b}]_i = \mathbf{x}_i^*\vec{\lambda}_i^* = 0$$

where the final equality follows from our fourth optimality condition, complementary slackness.

In the reverse direction, we assume that $\text{Diag}(v(\mathbf{x}^*))(\mathbf{H}\mathbf{x}^* + \vec{b}) = 0$ for some $\mathbf{x}^* \in \mathbb{R}_+^m$. Since the problem

$$\min_{x \in \mathbb{R}_+^m} \left\{ \frac{1}{2}\langle \mathbf{H}\mathbf{x}, \mathbf{x}\rangle + \langle \vec{b}, \mathbf{x}\rangle \right\}$$

represents a convex optimization problem, it is sufficient to show that the first order optimality conditions hold for $\mathbf{x}^*$ and some $\vec{\lambda}^*$. Of course, we immediately see that we satisfy primal feasibility since $\mathbf{x}^* \geq 0$ by assumption.

Due to the definition of v, our initial assumption implies that $\mathbf{H}\mathbf{x}^* + \vec{b} \geq 0$. If this were not the case, then there would exist an i such that $[\mathbf{H}\mathbf{x}^* + \vec{b}]_i < 0$. In this case, we see that $[v(\mathbf{x}^*)]_i = 1$ and that $[\text{Diag}(v(\mathbf{x}^*))(\mathbf{H}\mathbf{x}^* + \vec{b})]_i = [\mathbf{H}\mathbf{x}^* + \vec{b}]_i < 0$, which contradicts our initial assumption. Therefore, $\mathbf{H}\mathbf{x}^* + \vec{b} \geq 0$. As a result, let us set $\vec{\lambda}^* = \mathbf{H}\mathbf{x}^* + \vec{b}$. This allows us to satisfy our first optimality condition, $\mathbf{H}\mathbf{x}^* + \vec{b} - \vec{\lambda}^* = 0$ as well as our third, $\vec{\lambda}^* \geq 0$.

In order to show that we satisfy complementary slackness, we combine our initial assumption as well as our knowledge that $\mathbf{H}\mathbf{x}^* + \vec{b} \geq 0$ to see that

$$0 = \text{Diag}(v(\mathbf{x}^*))(\mathbf{H}\mathbf{x}^* + \vec{b})$$
$$= \text{Diag}(\mathbf{x}^*)(\mathbf{H}\mathbf{x}^* + \vec{b})$$
$$= \text{Diag}(\mathbf{x}^*)\vec{\lambda}^*.$$

Therefore, we satisfy our final optimality condition and, hence, $\mathbf{x}^*$ denotes an optimal solution to the optimization problem. $\square$

The above lemma allows us to recast a bound-constrained, convex quadratic optimization problem into a piecewise differentiable system of equations. In order to solve this system of equations, we apply Newton's method. Before we do so, we require one additional definition and a lemma.

Definition 4 For some symmetric, positive semidefinite $\mathbf{H} \in \mathbb{R}^{m \times m}$ and some $\vec{b} \in \mathbb{R}^m$, we define the operator $K_{\mathbf{H},\vec{b}} : \mathbb{R}^m \to \mathbb{R}^{m \times m}$ as

$$[K_{\mathbf{H},\vec{b}}(\mathbf{x})]_{ij} = \begin{cases} 1, & [\mathbf{H}\mathbf{x} + \vec{b}]_i \geq 0, \\ 0, & [\mathbf{H}\mathbf{x} + \vec{b}]_i < 0. \end{cases}$$

When both $\mathbf{H}$ and $\vec{b}$ are clear from the context, we abbreviate this operator as K.

Lemma 3 *Let $\mathbf{H} \in \mathbb{R}^{m \times m}$ be symmetric, positive definite, $\vec{b} \in \mathbb{R}^m$, and define the function $J : \mathbb{R}^m \to \mathbb{R}$ as*

$$J(\mathbf{x}) = \mathrm{Diag}\bigl(v(\mathbf{x})\bigr)(\mathbf{H}\mathbf{x} + \vec{b}).$$

Then, we have that

$$J'(\mathbf{x}) = K(\mathbf{x})\mathrm{Diag}(\mathbf{H}\mathbf{x} + \vec{b}) + \mathrm{Diag}\bigl(v(x)\bigr)\mathbf{H}.$$

Proof Let us begin by assessing the derivative of v. We notice that

$$\bigl[v(\mathbf{x} + t\vec{\eta})\bigr]_i = \begin{cases} \mathbf{x}_i + t\vec{\eta}_i, & [\mathbf{H}\mathbf{x} + b]_i \geq 0, \\ 1, & [\mathbf{H}\mathbf{x} + b]_i < 0. \end{cases}$$

Therefore, from a piecewise application of Taylor's theorem, we see that

$$\bigl[v'(\mathbf{x})\vec{\eta}\bigr]_i = \begin{cases} \vec{\eta}_i, & [\mathbf{H}\mathbf{x} + b]_i \geq 0, \\ 0, & [\mathbf{H}\mathbf{x} + b]_i < 0. \end{cases}$$

Next, we apply a similar technique to J. Let us define $g : \mathbb{R}^m \to \mathbb{R}$ so that $g(\mathbf{x}) = \mathbf{H}\mathbf{x} + \vec{b}$. Then, we see that

$$\begin{aligned} J(\mathbf{x} + t\vec{\eta}) &= \mathrm{Diag}\bigl(v(\mathbf{x} + t\vec{\eta})\bigr)\bigl(\mathbf{H}(\mathbf{x} + t\vec{\eta}) + \vec{b}\bigr) \\ &= \mathrm{Diag}\bigl(v(\mathbf{x}) + tv'(\mathbf{x})\vec{\eta} + o(|t|)\bigr)(\mathbf{H}\mathbf{x} + \vec{b} + t\vec{\eta}) \\ &= \mathrm{Diag}\bigl(v(\mathbf{x})\bigr)g(\bar{x}) + t\bigl(\mathrm{Diag}\bigl(v(\mathbf{x})\bigr)\mathbf{H}\vec{\eta} + \mathrm{Diag}\bigl(v'(\mathbf{x})\vec{\eta}\bigr)g(\bar{x})\bigr) + o(|t|). \end{aligned}$$

Hence, from a piecewise application of Taylor's theorem, we have that

$$\begin{aligned} J'(\mathbf{x})\vec{\eta} &= \mathrm{Diag}\bigl(v(\mathbf{x})\bigr)\mathbf{H}\vec{\eta} + \mathrm{Diag}\bigl(v'(\mathbf{x})\vec{\eta}\bigr)(\mathbf{H}\mathbf{x} + \vec{b}) \\ &= \mathrm{Diag}\bigl(v(\mathbf{x})\bigr)\mathbf{H}\vec{\eta} + K(\mathbf{x})\mathrm{Diag}(\mathbf{H}\mathbf{x} + \vec{b})\vec{\eta}. \end{aligned}$$

Therefore, $J'(\mathbf{x}) = K(\mathbf{x})\mathrm{Diag}(\mathbf{H}\mathbf{x} + \vec{b}) + \mathrm{Diag}(v(\mathbf{x}))\mathbf{H}$. □

The preceding lemma allows us to formulate Newton's method where we seek a step $\vec{p} \in \mathbb{R}^m$ such that $J'(\mathbf{x})\vec{p} = -J(\mathbf{x})$. Although the operator $J'(\mathbf{x})$ is well structured, it is nonsymmetric. We symmetrize the system as follows.

Definition 5 For some symmetric, positive semidefinite $\mathbf{H} \in \mathbb{R}^{m \times m}$ and some $\vec{b} \in \mathbb{R}^m$, we define the operator $D_{\mathbf{H},\vec{b}} : \mathbb{R}_+^m \to \mathbb{R}^{m \times m}$ as

$$D_{\mathbf{H},\vec{b}}(\mathbf{x}) = \mathrm{Diag}\bigl(v_{\mathbf{H},\vec{b}}(\mathbf{x})\bigr)^{1/2}.$$

When both $\mathbf{H}$ and $\vec{b}$ are clear from the context, we abbreviate this operator as D.

Lemma 4 *Let* $\mathbf{H} \in \mathbb{R}^{m \times m}$ *be symmetric, positive semidefinite and let* $\vec{b} \in \mathbb{R}^m$. *Then, we have that*

$$\left(K(\mathbf{x})\mathrm{Diag}(\mathbf{Hx}+\vec{b}) + \mathrm{Diag}(v(\mathbf{x}))\mathbf{H}\right)\vec{p} = -\mathrm{Diag}(v(\mathbf{x}))(\mathbf{Hx}+\vec{b})$$
$$\iff \left(K(\mathbf{x})\mathrm{Diag}(\mathbf{Hx}+\vec{b}) + D(\mathbf{x})\mathbf{H}D(\mathbf{x})\right)\vec{q} = -D(\mathbf{x})(\mathbf{Hx}+\vec{b})$$

where $\vec{p} = D(x)\vec{q}$.

Proof Notice that

$$\begin{aligned}
0 &= \left(K(\mathbf{x})\mathrm{Diag}(\mathbf{Hx}+\vec{b}) + \mathrm{Diag}(v(\mathbf{x}))\mathbf{H}\right)\vec{p} + \mathrm{Diag}(v(\mathbf{x}))(\mathbf{Hx}+\vec{b}) \\
&= \left(K(\mathbf{x})\mathrm{Diag}(\mathbf{Hx}+\vec{b}) + D(\mathbf{x})^2\mathbf{H}\right)\vec{p} + D(\mathbf{x})^2(\mathbf{Hx}+\vec{b}) \\
&= D(\mathbf{x})\left((D(\mathbf{x})^{-1}K(\mathbf{x})\mathrm{Diag}(\mathbf{Hx}+\vec{b}) + D(\mathbf{x})\mathbf{H})\vec{p} + D(\mathbf{x})(\mathbf{Hx}+\vec{b})\right) \\
&= D(\mathbf{x})\left((D(\mathbf{x})^{-1}K(\mathbf{x})\mathrm{Diag}(\mathbf{Hx}+\vec{b}) + D(\mathbf{x})\mathbf{H})D(\mathbf{x})\vec{q} + D(\mathbf{x})(\mathbf{Hx}+\vec{b})\right) \\
&= D(\mathbf{x})\left((K(\mathbf{x})\mathrm{Diag}(\mathbf{Hx}+\vec{b}) + D(\mathbf{x})\mathbf{H}D(\mathbf{x}))\vec{q} + D(\mathbf{x})(\mathbf{Hx}+\vec{b})\right),
\end{aligned}$$

which occurs if and only if

$$0 = \left(K(\mathbf{x})\mathrm{Diag}(\mathbf{Hx}+\vec{b}) + D(\mathbf{x})\mathbf{H}D(\mathbf{x})\right)\vec{q} + D(\mathbf{x})(\mathbf{Hx}+\vec{b})$$

since $D(\mathbf{x})$ is nonsingular. □

Properly, we require a line search to ensure feasible iterates. However, we can be far more aggressive in practice. In order to initialize the algorithm, we use the starting iterate of $(\vec{\lambda}, \vec{\mu}) = (\vec{0}, \vec{0})$. This corresponds to a primal solution where $\vec{F} = \vec{F}^H$. Since the optimal solution to the primal problem is close to the target $\vec{F}^H$, we expect the optimal solution to the dual problem to reside in a neighborhood close to zero. As a result, Newton's method should converge quadratically to the solution with a step size equal to one. Therefore, we ignore the feasibility constraint and always use a unit step size. Sometimes, this allows the dual solution to become slightly infeasible, but the amount of infeasibility tends to be small. In practice, the corresponding primal solution is always feasible and produces good results. In order to allow infeasible solutions, we must use the original formulation of Newton's method rather than the symmetric reformulation. Namely, the operator D becomes ill-defined for infeasible points.

When we combine the above pieces, we arrive at the final algorithm (Table 2).

7 A Few Instructive Examples

In this section we present three numerical examples that illustrate the advantages of the OBR formulation in comparison to the M-OBR formulation. Because, as

Table 2 Dual algorithm for the solution of the remap problem

1. Define $H \in \mathbb{R}^{2K \times 2K}$ and $b \in \mathbb{R}^{2K}$ as

$$\mathbf{H} = \begin{bmatrix} \mathbf{AA}^\mathsf{T} & -\mathbf{AA}^\mathsf{T} \\ -\mathbf{AA}^\mathsf{T} & \mathbf{AA}^\mathsf{T} \end{bmatrix}, \quad \vec{b} = \begin{bmatrix} \mathbf{A}\vec{F}^H - \vec{b}_{\min} \\ -\mathbf{A}\vec{F}^H + \vec{b}_{\max} \end{bmatrix}.$$

2. Initialize $\mathbf{x} = \vec{0}$.
3. Until $\|\mathrm{Diag}(v(\mathbf{x}))(\mathbf{Hx} + \vec{b})\|$ becomes small or we exceed a fixed number of iterations.
   a. When feasible, solve

   $$\bigl(K(\mathbf{x})\mathrm{Diag}(\mathbf{Hx} + \vec{b}) + D(\mathbf{x})\mathbf{H}D(\mathbf{x})\bigr)\vec{q} = -D(\mathbf{x})(\mathbf{Hx} + \vec{b})$$

   and set $\vec{p} = D(x)\vec{q}$. Otherwise, solve

   $$\bigl(K(\mathbf{x})\mathrm{Diag}(\mathbf{Hx} + \vec{b}) + \mathrm{Diag}(v(\mathbf{x}))\mathbf{H}\bigr)\vec{p} = -\mathrm{Diag}(v(\mathbf{x}))(\mathbf{Hx} + \vec{b}).$$

   b. Set $\mathbf{x} = \mathbf{x} + \vec{p}$.

shown in Corollary 1, the solution of the M-OBR problem (43) is equivalent to the one given by the FCR algorithm, our study effectively compares and contrasts the fundamental properties of OBR and FCR; henceforth, we denote the M-OBR/FCR methods and algorithms by the common acronym M-OBR (FCR).

Most notably, the three examples reveal that the conditions on the mesh motion for OBR, given in Theorem 2, are much less restrictive than those for M-OBR (FCR). First, we demonstrate on a simple three-cell example in one spatial dimension that for certain mesh motions M-OBR (FCR) does not preserve the shape of a given density function, while OBR does. Second, we construct a related example for which M-OBR (FCR) does not preserve linear density functions under mesh motions admissible by OBR. Finally, we give a 9-cell example in two spatial dimensions for which a commonly used M-OBR (FCR) algorithm based on swept regions, see Sect. 5, does not preserve monotonicity, while OBR does. In the following, we refer to Sect. 3 for relevant notation.

We will also compare some of the numerical results with the iFCR algorithm. In particular, unless otherwise specified, iFCR(k) indicates the k-th iterate of the iFCR algorithm.

7.1 An Example of Mesh Movement in Which OBR Preserves Shape and M-OBR (FCR) Does Not

The goal of this section is to show that the smaller feasible set of the M-OBR (FCR) formulation (43) can limit its ability to accurately preserve the shape of a given density function. To this end we design a "torture" test example that shows how the shape of a given "peak" density function can be changed by M-OBR (FCR) into a step-function profile. Of course, because M-OBR (FCR) and FCR are equivalent, the same will hold true for the FCR solution.

Fig. 4 Specification of the "torture" test for shape preservation. The new mesh is defined by compressing the middle cell of the old mesh. The mean density values are subject to the conditions that $\rho_1 > \rho_3$ and that ρ_2 is the largest value. The results reported in this section correspond to $\Delta_1 = \Delta_2 = 0.14$, $\rho_1 = 80$, $\rho_2 = 100$, $\rho_3 = 0$, and $\rho_1^b = \rho_3^b = 0$

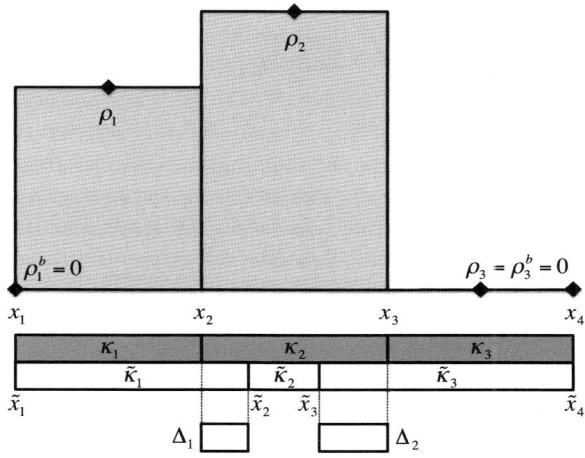

A schematic of the torture test is shown in Fig. 4. The computational domain is given by the unit interval, $\Omega = [0, 1]$. The old mesh $K_h(\Omega)$ is defined by a uniform partition of the unit interval into 3 cells using the vertices $x_1 = 0$, $x_2 = 1/3$, $x_3 = 2/3$ and $x_4 = 1$. The nodes of the new mesh $\widetilde{K}_h(\Omega)$ are set to $\widetilde{x}_1 = x_1$, $\widetilde{x}_2 = x_2 + \Delta_1$, $\widetilde{x}_3 = x_3 - \Delta_2$ and $\widetilde{x}_4 = x_4$, where $\Delta_1 > 0$ and $\Delta_2 > 0$ are such that $\Delta_1 + \Delta_2 < 1/3$; see Fig. 4. In other words, the new mesh is defined by compressing the middle cell of the old mesh. Note that $\widetilde{K}_h(\Omega)$ satisfies the locality assumption (3) and that

$$x_2 < \widetilde{x}_2 \quad \text{and} \quad \widetilde{x}_3 < x_3. \tag{51}$$

To complete the specification of the torture test we prescribe the mean density values ρ_1, ρ_2, ρ_3 on the old cells and boundary values $\rho_1^b = 0$, $\rho_3^b = 0$ at the endpoints. The mean density values are subject to the conditions

$$\rho_1 > \rho_3, \quad \rho_2 \geq \rho_1, \quad \text{and} \quad \rho_2 \geq \rho_3. \tag{52}$$

Specific numbers will be given momentarily. To explain these choices it is necessary to examine the structure of the feasible set of (37) and its modification (43), specialized to the torture test. As before, we follow the rule that the antisymmetry of fluxes and the symmetry of coefficients a_{ij} are enforced by using index pairs $\{i, j\}$ for which $i < j$. In the case of the torture test, which has three cells, there are two such pairs, given by $\{1, 2\}$ and $\{2, 3\}$. Therefore, the independent fluxes are F_{12} and F_{23},

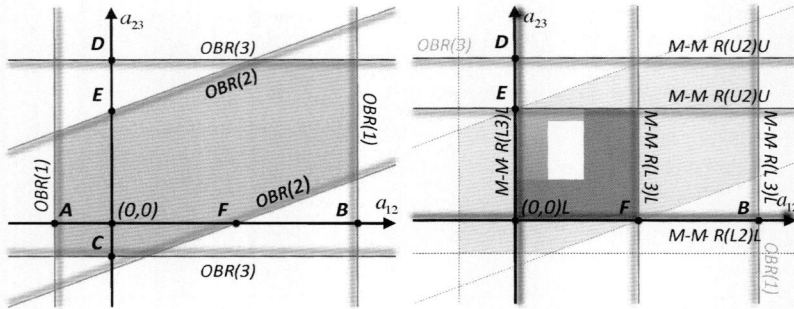

Fig. 5 Structure of the OBR (*left pane*) and M-OBR (FCR) (*right pane*) feasible sets when dF_{12} and dF_{23} are positive. The strips between the pairs of lines marked by *OBR*(1), *OBR*(2) and *OBR*(3) correspond to the three inequality constraints in (53). The lines marked by *OBR*(2) have positive slopes given by dF_{12}/dF_{23}. The lines marked by *M-OBR(U1)* and *M-OBR(U2)* represent the two upper bounds in the two inequality constraints in (54), respectively. The lower bounds correspond to the coordinate axes and are identified by *M-OBR(L1)* and *M-OBR(L2)*, respectively. The shadows point towards the interiors of the domains defined by the constraints. It is evident that the feasible set of M-OBR (FCR) is a subset of the feasible set of OBR

the unknown coefficients are a_{12} and a_{23}, and the OBR problem (37) specializes to

$$\begin{cases} \min_{a_{12}, a_{23}} \left\{ (1-a_{12})^2 (dF_{12})^2 + (1-a_{23})^2 (dF_{23})^2 \right\} \\ \text{subject to} \\ \quad (1)\ \widetilde{Q}_1^{\min} \leq a_{12} dF_{12} \leq \widetilde{Q}_1^{\max}, \\ \quad (2)\ \widetilde{Q}_2^{\min} \leq a_{23} dF_{23} - a_{12} dF_{12} \leq \widetilde{Q}_2^{\max}, \\ \quad (3)\ \widetilde{Q}_3^{\min} \leq -a_{23} dF_{23} \leq \widetilde{Q}_3^{\max}. \end{cases} \quad (53)$$

Regarding the M-OBR (FCR) formulation (37), a simple but tedious calculation shows that $dF_{12} > 0$ and $dF_{23} > 0$ whenever (i) the middle cell is compressed, i.e. (51) holds, and (ii) the first condition in (52) holds, i.e. $\rho_1 > \rho_3$. As a result, the M-OBR (FCR) problem assumes the form

$$\begin{cases} \min_{a_{12}, a_{23}} \left\{ (1-a_{12})^2 (dF_{12})^2 + (1-a_{23})^2 (dF_{23})^2 \right\} \\ \text{subject to} \\ \quad (1)\ 0 \leq a_{12} \leq \min\{D_1^+, D_2^-\}, \\ \quad (2)\ 0 \leq a_{23} \leq \min\{D_2^+, D_3^-\}. \end{cases} \quad (54)$$

The left and the right panes in Fig. 5 show cartoons of the feasible sets of (53) and (54), respectively. The horizontal and the vertical axes in these plots correspond to the unknowns a_{12} and a_{23}, respectively. The strips between the pairs of lines marked by *OBR*(1), *OBR*(2) and *OBR*(3) correspond to the three inequality constraints in (53). Note that the slope of the lines marked by *OBR*(2) is given

Table 3 Control points for the feasible sets of the OBR (53) and the M-OBR (FCR) (54) problems and their values for $\Delta_1 = \Delta_2 = 0.14$, $\rho_1 = 80$, $\rho_2 = 100$, $\rho_3 = 0$, and $\rho_1^b = \rho_3^b = 0$

Point	A	B	C	D	E	F
Definition	$\dfrac{\widetilde{Q}_1^{\min}}{dF_{12}}$	$\dfrac{\widetilde{Q}_1^{\max}}{dF_{12}}$	$\dfrac{\widetilde{Q}_3^{\max}}{-dF_{23}}$	$\dfrac{\widetilde{Q}_3^{\min}}{-dF_{23}}$	$\dfrac{\widetilde{Q}_2^{\max}}{dF_{23}}$	$\dfrac{\widetilde{Q}_2^{\min}}{-dF_{12}}$
Value	-25.04	4.10	-20.53	8.62	0.00	3.28

by dF_{12}/dF_{23} and is therefore positive. The lines marked by *M-OBR(U1)* and *M-OBR(U2)* represent the two upper bounds in the two inequality constraints in (54), respectively. The lower bounds coincide with the coordinate axes and are marked by *M-OBR(L1)* and *M-OBR(L2)*, respectively.

The relation between the two feasible sets can be understood by examining the points **A**, **B**, **C**, **D**, **E** and **F**. The first pair of points corresponds to the lower and upper bounds on a_{12} imposed by the first constraint in (53). The second pair, i.e., **C**, and **D**, corresponds to the lower and upper bounds on a_{23} imposed by the third constraint in (53). The last two points correspond to the intercepts of the lines associated with the upper and lower bounds in the second constraint in (53) with the vertical and horizontal coordinate axes, respectively. The definitions of these points and their values corresponding to the actual test data used in the study are summarized in Table 3.

To explain the construction of the torture test, note that the shape of the M-OBR (FCR) feasible set is completely determined by the positions of **E** and **F** along the vertical and the horizontal coordinate axes. This is a consequence of the worst-case analysis used to derive the constraints of (54). Consequently, by moving **E** to the origin the M-OBR (FCR) feasible set can be reduced to a line extending from the origin to point **F**. This removes the point $(1, 1)$ from the feasible set and forces the M-OBR (FCR) formulation to pick a solution that corresponds to remap by low-order fluxes. By moving **E** to the origin we also shrink the OBR feasible set. However, because the lines corresponding to the second constraint have positive slopes, they can be chosen in such a way that $(1, 1)$ remains in this feasible set.

In order to move **E** to the origin we need to set $\widetilde{Q}_2^{\max}/dF_{23} = 0$. It is not hard to see that this is true whenever (i) the middle cell is compressed, i.e. (51) holds, and (ii) the second condition in (52), i.e. $\rho_2^{\max} = \rho_2$ holds.

Figure 6 compares the OBR and M-OBR (FCR) solutions on the new mesh for $\Delta_1 = \Delta_2 = 0.14$, $\rho_1 = 80$, $\rho_2 = 100$, $\rho_3 = 0$, and boundary values $\rho_1^b = \rho_3^b = 0$. Table 4 shows the corresponding values of the lower and the upper inequality bounds as well as the values of the flux differentials in (53)–(54).

The initial density function has the shape of a "peak" and is shown in the top pane of Fig. 6. The bottom pane in Fig. 6 shows clearly that the OBR solution preserves this shape on the new mesh. However, as one can see from the middle pane in Fig. 6, the M-OBR (FCR) solution changes the shape of the peak to a step-function profile on the new mesh. We note that the iFCR(2) method delivers results identical to those of the OBR method.

Fig. 6 Initial density function (*top pane*), M-OBR (FCR) solution (*middle pane*) and OBR solution (*bottom pane*) for $\Delta_1 = \Delta_2 = 0.14$, $\rho_1 = 80$, $\rho_2 = 100$, $\rho_3 = 0$, and $\rho_1^b = \rho_3^b = 0$. The OBR solution preserves the shape of the original density function, while the M-OBR (FCR) solution does not. The iFCR(2) method delivers results identical to those of the OBR method

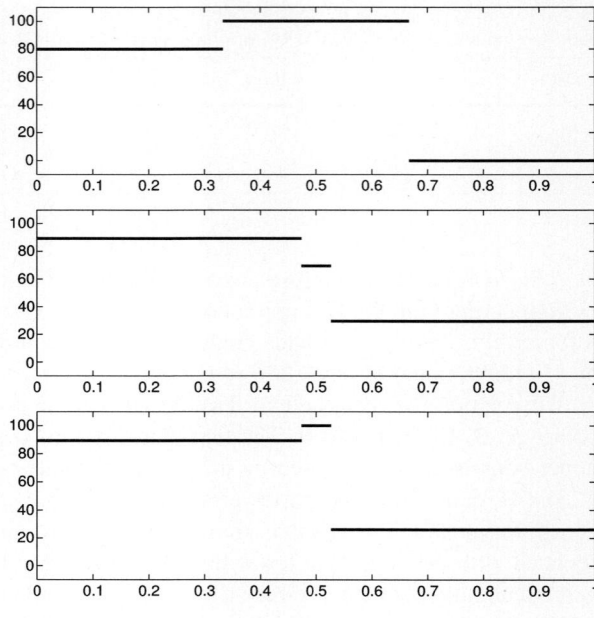

Table 4 Numerical values for the lower and the upper bounds and the flux differentials in (53)–(54) corresponding to $\Delta_1 = \Delta_2 = 0.14$, $\rho_1 = 80$, $\rho_2 = 100$, $\rho_3 = 0$, and $\rho_1^b = \rho_3^b = 0$

	$i = 1$	$i = 2$	$i = 3$
$\widetilde{Q}_i^{\min}$	-40.66	-5.33	-14.00
$\widetilde{Q}_i^{\max}$	6.66	0.00	33.33
$dF_{i,i+1}$	1.62	1.62	—

The constraint sets of (53) and (54) for this example are compared in Fig. 7. We see that $(1, 1)$ is included in the former but not in the latter. This is a consequence of the worst-case analysis used to obtain the constraint set in (54).

7.2 An Example in Which OBR Preserves Linear Densities and M-OBR (FCR) Does Not

In this section, we investigate the differences between OBR and M-OBR (FCR) concerning the preservation of linear density functions. The basic setup is closely related to the previous example. The specification of the computational mesh is identical. The density function is given by

$$\rho(x) = x, \quad 0 \le x \le 1,$$

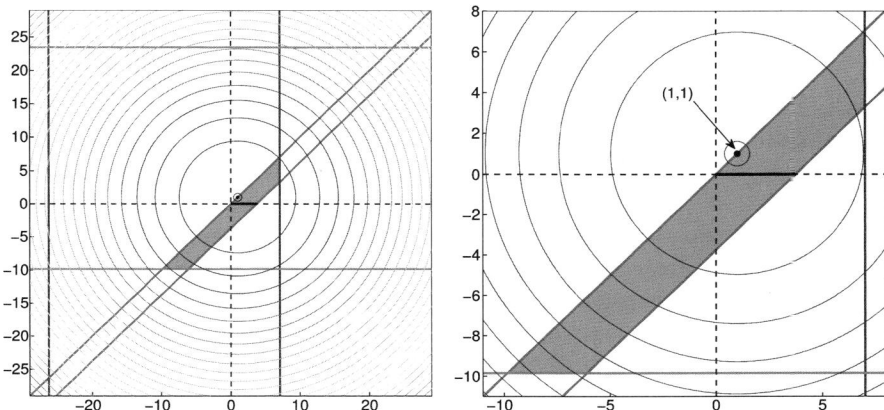

Fig. 7 Level sets of the objective functional and the feasible sets of problems (53) and (54) for $\Delta_1 = \Delta_2 = 0.14$, $\rho_1 = 80$, $\rho_2 = 100$, $\rho_3 = 0$, and $\rho_1^b = \rho_3^b = 0$. The regions between *horizontal* (*magenta*), *slanted* (*red*) and *vertical* (*blue*) lines on the left pane correspond to the first, second and third constraints in the OBR problem (53). Their intersection (*red region*) gives the OBR feasible set which contains the point (1, 1). The feasible set of M-OBR (FCR) is given by the *solid horizontal segment* (*black*) and does not contain the point (1, 1). The *right pane* shows a zoom of the OBR and M-OBR (FCR) feasible sets (Color figure online)

i.e. $\rho_1 = 1/6$, $\rho_2 = 1/2$, $\rho_3 = 5/6$, $\rho_1^b = 0$ and $\rho_2^b = 1$. We consider a series of compression increments Δ_1, Δ_2 given by

$$\Delta_1 = \Delta_2 = \frac{\ell - 1}{6\ell},$$

where $\ell = \{7, 8, 9, 10, 100, 1000\}$, resulting in ℓ-fold compressions of the middle cell.

The initial linear density function is remapped onto the compressed mesh and then back onto the original mesh, where we record the L_2 error between the thus obtained and the original density. Table 5 clearly shows that while OBR preserves linear densities for arbitrary compressions of the middle cell, M-OBR (FCR) is linearity-preserving only for $\ell \leq 8$. The iFCR algorithm, for a large number of iterations, recaptures the behavior of OBR.

The root cause for the loss of the linearity preservation in this example is the same as for the loss of shape preservation in the last section. The M-OER (FCR) problem (54) preserves linearity if and only if the unconstrained minimizer (1, 1) of the functional in (54) is included in its feasible set. When the middle cell is compressed the feasible set of M-OBR (FCR) shrinks and eventually ceases to contain the point (1, 1).

Ultimately, the loss of linearity preservation in the M-OBR (FCR) is a function of the mesh movement. To prevent the loss of this important property we recommend that M-OBR (FCR) implementations include the following test to determine the admissible mesh motions. Given a candidate new mesh, compute the quantities

Table 5 L_2 errors in the OBR, M-OBR (FCR) and iFCR remap of a linear density function in one dimension, for different compression ratios $\ell : 1$ of the middle cell. Errors close to machine precision are *highlighted*. OBR preserves linear densities for arbitrarily compressed middle cells, while M-OBR (FCR) does not. iFCR is linearity preserving given a sufficient, possibly very large, number of iterations

	$\ell = 7$	$\ell = 8$	$\ell = 9$	$\ell = 10$	$\ell = 100$	$\ell = 1000$
OBR	**1.67e−16**	**0**	**3.20e−17**	**3.58e−17**	**1.63e−16**	1.95e−14
FCR	**4.53e−17**	**3.58e−17**	2.32e−03	4.46e−03	2.09e−02	2.25e−02
iFCR(2)	**1.24e−16**	**1.57e−16**	**1.57e−16**	**1.57e−16**	3.31e−02	3.87e−02
iFCR(20)	**1.24e−16**	**1.57e−16**	**1.57e−16**	**1.57e−16**	**1.92e−16**	3.30e−02
iFCR(200)	**1.24e−16**	**1.57e−16**	**1.57e−16**	**1.57e−16**	**1.92e−16**	**1.69e−15**

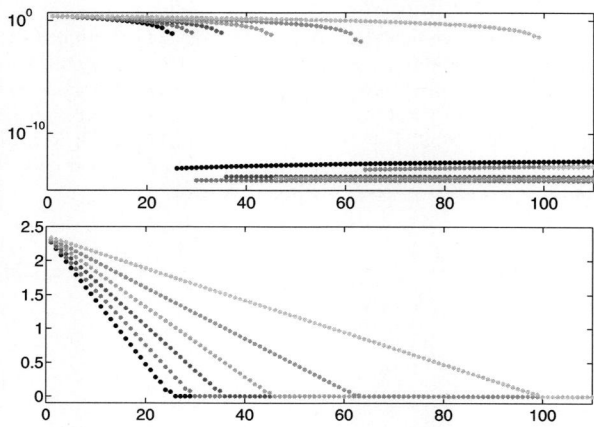

Fig. 8 Euclidean norm of the difference (*vertical axes*) between the computed iFCR and OBR fluxes for problems (53) and (54) for increasing numbers of iFCR iterations (*horizontal axes*). The compression of the middle cell is given by $\Delta_1 = \Delta_2 \in \{0.1660$ (*black*), 0.1661 (*red*), 0.1662 (*blue*), 0.1663 (*green*), 0.1664 (*gray*), 0.1665 (*cyan*)$\}$. The density profile is $\rho_1 = 80$, $\rho_2 = 100$, $\rho_3 = 0$, and $\rho_1^b = \rho_3^b = 0$. We use logarithmic (*top pane*) and linear scales (*bottom pane*). In this example, OBR recovers the high-order target flux (Color figure online)

P_i^-, D_i^- and P_i^+, D_i^+ defined in (38)–(39), for the monomial x in one dimension, monomials x and y in two dimensions and monomials x, y and z in three dimensions. Accept the mesh if and only if $D_i^- \geq 1$ and $D_i^+ \geq 1$, whenever $P_i^- < 0$ and $P_i^+ > 0$, respectively. This condition guarantees that $a_{ij} = 1$ are in the feasible set of the M-OBR (FCR) problem (46).

We also investigated whether the good performance of the iFCR algorithm for large number of iterations depends on OBR recovering the high-order flux in computations. Figures 8 and 9 show detailed computations in which it is clear that similar results between iFCR and OBR can be obtained also in the case in which OBR does not recover the target fluxes. This indicates that for a sufficiently large number

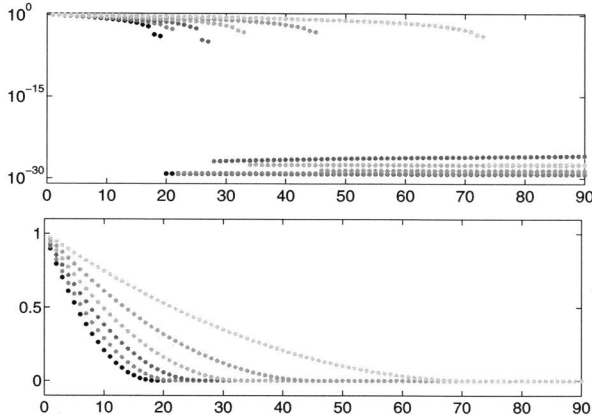

Fig. 9 Euclidean norm of the difference (*vertical axes*) between the computed iFCR and OBR fluxes for problems (53) and (54) for increasing numbers of iFCR iterations (*horizontal axes*). The compression of the middle cell is given by $\Delta_1 = \Delta_2 \in \{0.1660\ (black), 0.1661\ (red), 0.1662\ (blue), 0.1663\ (green), 0.1664\ (gray), 0.1665\ (cyan)\}$. The density profile is $\rho_1 = 80$, $\rho_2 = 82$, $\rho_3 = 0$, and $\rho_1^b = \rho_3^b = 0$. We use logarithmic (*top pane*) and linear scales (*bottom pane*). In this example, OBR does *not* recover the high-order target flux, yet the iFCR flux converges to the OBR flux, suggesting that iFCR may be a solution procedure for OBR (Color figure online)

of iterations the iFCR solution converges to the OBR solution. Our conjecture is that iFCR may be a solution procedure for OBR.

7.3 OBR Preserves Monotonicity When M-OBR (FCR) Does Not

Motivated by the one-dimensional examples we devise a simple 9-cell test that examines the fundamental properties of OBR and M-OBR (FCR) in two dimensions. The test is a tensor-product version of the one-dimensional torture test. The computational domain is given by the product of unit intervals, $\Omega = [0, 1] \times [0, 1]$. The old mesh $K_h(\Omega)$ is defined by a uniform partition of the unit intervals in x and y direction into 3 cells using the vertices $x_1 = 0$, $x_2 = 1/3$, $x_3 = 2/3$ and $x_4 = 1$ and $y_1 = 0$, $y_2 = 1/3$, $y_3 = 2/3$ and $y_4 = 1$, respectively. The nodes of the new mesh $\widetilde{K}_h(\Omega)$ are set to

$$\widetilde{x}_1 = x_1, \quad \widetilde{x}_2 = x_2 + \Delta_1^x, \quad \widetilde{x}_3 = x_3 - \Delta_2^x, \quad \widetilde{x}_4 = x_4,$$
$$\widetilde{y}_1 = y_1, \quad \widetilde{y}_2 = y_2 + \Delta_1^y, \quad \widetilde{y}_3 = y_3 - \Delta_2^y, \quad \widetilde{y}_4 = y_4,$$

where $\Delta_1^{x,y} > 0$ and $\Delta_2^{x,y} > 0$ are such that $\Delta_1^x + \Delta_2^x < 1/3$ and $\Delta_1^y + \Delta_2^y < 1/3$. In other words, as in one spatial dimension, the new mesh is defined by compressing the middle cell of the old mesh. Note that the new mesh satisfies conditions (25)–(26), i.e. is admissible by OBR.

Fig. 10 A 3×3 uniform initial grid (*left pane*) and the "compressed" grid (*right pane*) with a 4×4-fold compression of the middle cell

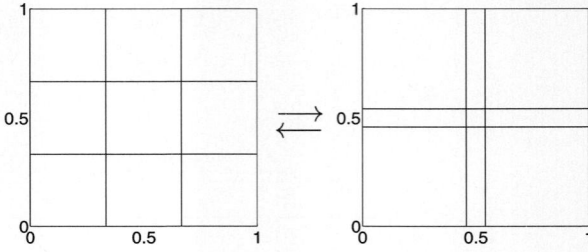

Table 6 Monotonicity of OBR, M-OBR (FCR), the donor-cell method and iFCR, implemented using swept regions, with respect to the remap of a linear density function in two dimensions, for different compression ratios $\ell \times \ell : 1$ of the middle cell. OBR is bound-preserving throughout, while M-OBR (FCR) and the donor-cell method are not. iFCR is monotone given a sufficient number of iterations. For iFCR(n) we select the smallest number of iterations n resulting in a bound-preserving remap, for compression ratios $\ell \times \ell : 1$, with $\ell \in \{15, 16, 100\}$, respectively

	$\ell=5$	$\ell=6$	$\ell=7$	$\ell=14$	$\ell=15$	$\ell=16$	$\ell=100$
OBR	✓	✓	✓	✓	✓	✓	✓
FCR	✓	✓	✓	✓	–	–	–
Donor-cell	✓	–	–	–	–	–	–
iFCR(2)	✓	✓	✓	✓	✓	–	–
iFCR(4)	✓	✓	✓	✓	✓	✓	–
iFCR(721)	✓	✓	✓	✓	✓	✓	✓

We examine both monotonicity (for OBR, M-OBR (FCR) and the donor-cell method based on swept regions) as well as the preservation of linear densities (for OBR and M-OBR (FCR)). For monotonicity studies, we employ a single remap from the original to the compressed mesh, while for the study of linearity preservation the density is additionally remapped back onto the original mesh. We point out that M-OBR (FCR) and the swept-region donor-cell method use the same computation of low-order fluxes. Monotonicity violations are detected based on the violations of inequality constraints in (20).

The density function is given by

$$\rho(x, y) = x, \quad 0 \le x, y \le 1.$$

We study a series of compression increments $\Delta_1^{x,y}$, $\Delta_2^{x,y}$ given by

$$\Delta_1^{x,y} = \Delta_2^{x,y} = \frac{\ell - 1}{6\ell},$$

where $\ell = \{5, 6, 7, 14, 15, 16, 100\}$ for the monotonicity study and for the linearity preservation study, $\ell = \{3, 4, 5, 15, 16, 100\}$, amounting to $\ell \times \ell$-fold compressions of the middle cell. An illustration for $\ell = 4$ is shown in Fig. 10.

Our first observation is that neither the donor-cell method nor M-OBR (FCR) preserve monotonicity for certain mesh motions admissible by OBR, see Table 6. This

Constrained-Optimization Based Data Transfer

Table 7 L_2 errors in the OBR, M-OBR (FCR) and iFCR remap of a linear density function in two dimensions, for different compression ratios $\ell \times \ell : 1$ of the middle cell. Errors close to machine precision are *highlighted*. OBR preserves linear densities for arbitrarily compressed middle cells, while M-OBR (FCR) does not. For iFCR(n) we select the smallest number of iterations n resulting in a linearity-preserving remap, for the compression ratios $\ell \times \ell : 1$, with $\ell \in \{4, 15, 100\}$, respectively

	$\ell = 3$	$\ell = 4$	$\ell = 5$	$\ell = 15$	$\ell = 16$	$\ell = 100$
OBR	1.36e−16	3.90e−16	2.91e−16	7.99e−16	3.33e−15	2.08e−13
FCR	1.32e−16	4.34e−03	1.06e−02	4.33e−02	4.60e−02	1.97e−01
iFCR(2)	1.36e−16	3.90e−16	6.65e−04	4.07e−02	4.36e−02	1.30e−01
iFCR(20)	1.36e−16	3.90e−16	2.91e−16	7.99e−16	3.00e−03	1.20e−01
iFCR(834)	1.36e−16	3.90e−16	2.91e−16	7.99e−16	3.33e−15	2.08e−13

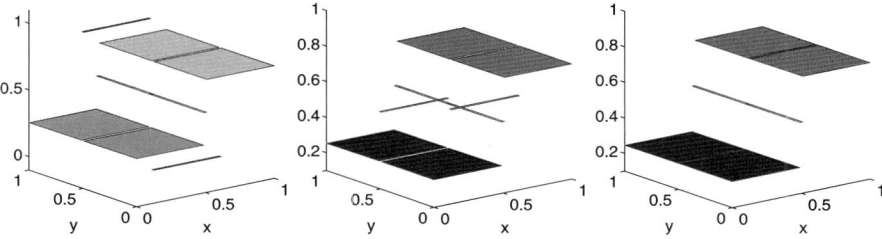

Fig. 11 Linear density $\rho(x, y) = x$ remapped from the uniform 3×3 grid to the compressed "torture" grid with $\ell = 16$. Left to right: the donor-cell method, M-OBR (FCR), OBR. It is clear that OBR gives the best density approximation

result is not surprising in view of the condition on mesh motion for the swept-region donor-cell method given in Margolin and Shashkov [17, p. 279], which is violated for meshes associated with $\ell \geq 6$. Table 6 also reveals that M-OBR (FCR) succeeds in "repairing" the loss of monotonicity inherited from the donor-cell method for $6 \leq \ell \leq 14$, but eventually loses monotonicity for $\ell \geq 15$.

Table 7 indicates that the loss of linearity preservation in M-OBR (FCR) sets in at fairly low compressions of the middle cell ($\ell \geq 4$) and is therefore not directly related to the loss of monotonicity in the swept-region low-order fluxes, which occurs at $\ell \geq 6$. This observation is in agreement with one-dimensional results, where the low-order fluxes are computed based on exact cell intersections and are therefore provably bound-preserving as long as the locality assumption (3) is satisfied, and where M-OBR (FCR) nevertheless fails to preserve linear densities for compressive mesh motions.

In contrast to M-OBR (FCR) and the donor-cell method, OBR is monotonicity and linearity preserving in all our tests. Note also that the iFCR method requires a large number of iterations to recover the OBR solution. The differences between the methods are illustrated for the compression parameter $\ell = 16$ in Figs. 11 and 12.

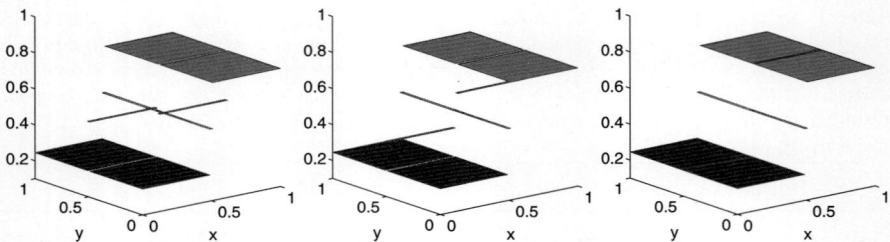

Fig. 12 Linear density $\rho(x, y) = x$ remapped from the uniform 3×3 grid to the compressed "torture" grid with $\ell = 16$. Left to right: iFCR(2), iFCR(10), iFCR(200). Given a sufficient number of iterations, iFCR recovers the OBR result from Fig. 11

8 Computational Studies

The purpose of this section is an in-depth comparison of accuracy, robustness and computational cost of OBR and M-OBR (FCR). These attributes are assessed on a series of convergence studies in two dimensions involving (i) smooth cyclic remap on grids with moderate displacements and (ii) cyclic remap on grids with large displacements.

8.1 Methodology for the Estimation of Convergence Rates of Remap Algorithms

The convergence studies in this section are designed to assess the asymptotic accuracy of the OBR and M-OBR (FCR) algorithms in the context of a continuous rezone strategy. In this case, the appropriate notion of remap error and convergence rates can be defined with the help of a *cyclic remap* test as in Margolin and Shashkov [17]. The precise methodology used in the chapter is described below.

A cyclic remap test simulates continuous rezone by performing remap over a parametrized sequence of grids $K_h^r(\Omega)$, $r = 0, \ldots, R$, such that the following three conditions are satisfied:

- Every $K_h^r(\Omega), r = 1, \ldots, R$, is topologically equivalent to the initial grid $K_h^0(\Omega)$, i.e. all grids in the sequence have the same number of cells and the same connectivity as $K_h^0(\Omega)$.
- Any two consecutive grids $K_h^{r-1}(\Omega)$, $K_h^r(\Omega)$ satisfy the locality assumption (2).
- The first and the last grids coincide, i.e., $K_h^0(\Omega) = K_h^R(\Omega)$.

The integer R is the number of remap steps. Its reciprocal $1/R$ can be thought of as a "pseudo-time" step which defines the temporal resolution of the cyclic remap test. The total resolution of the test is specified by the pair (K, R), where K is the number of cells in $K_h^0(\Omega)$.

Given a cyclic mesh sequence $\{K_h^r(\Omega)\}_{r=0}^R$, called a *cyclic grid*, with total resolution (K, R), let $\vec{\rho}^r \in \mathbb{R}^K$ denote the approximate density solution on $K_h^r(\Omega)$, and

$\|\cdot\|$ be a given norm on $\mathbb{R}^K$. The remap error on $\{K_h^r(\Omega)\}_{r=0}^R$ is defined by the norm of the density difference on the first and the last grids in the sequence, i.e

$$\mathcal{E}(\rho; \|\cdot\|, K, R) = \|\vec{\rho}^0 - \vec{\rho}^R\|. \tag{55}$$

This definition is justified by the fact that $K_h^0(\Omega) = K_h^R(\Omega)$, and so the difference between the first and last solutions provides a measure of the total error accrued by the remap algorithm.

To compute the remap error $\mathcal{E}(\rho; \|\cdot\|, K, R)$ in (55) we use three norms suggested in Margolin and Shashkov [17]. Note that in the case of cyclic remap one does not need to know the exact density distribution to compute the numerical error, which can be instead calculated by comparing the initial and final cell densities. Given an arbitrary vector $\vec{\phi} \in \mathbb{R}^K$ these norms are defined as follows:

$$\|\vec{\phi}\|_2 = \left(\sum_{i=1}^K \phi_i^2 V(\kappa_i) \right)^{1/2}, \quad \|\vec{\phi}\|_1 = \sum_{i=1}^K |\phi_i| V(\kappa_i), \quad \|\vec{\phi}\|_\infty = \max_{0 \leq i \leq K} |\phi_i|. \tag{56}$$

If $\vec{\phi}$ is a piecewise constant approximation of a given scalar function $\phi(x)$, then these norms are discrete approximations of the L_2, L_1 and L_∞ norms on Ω, respectively.

Once the appropriate notion of remap error is defined, the estimate of convergence rates proceeds in the usual fashion: we compute remap errors using a sequence of cyclic grids with increasing resolution and then estimate the slope of the curve representing the log-log plot of the remap error versus the spatial resolution of the cyclic grid. To this end we use least-squares regression fit. Specifically, for a sequence of cyclic grids with resolutions (K^q, R^q), $q = 1, \ldots, Q$ and the corresponding remap errors $\mathcal{E}^q = \mathcal{E}(\rho; \|\cdot\|, K^q, R^q)$, the rate of convergence ν^q is estimated by least-squares regression, i.e. by solving the minimization problem

$$\{\nu^q, \omega^q\} = \arg\min \sum_{i=1}^q \left(\log \mathcal{E}^q + \nu \log R^q - \omega \right)^2, \quad 1 < q \leq Q. \tag{57}$$

8.2 Smooth Cyclic Remap on Grids with Moderate Displacements

The cyclic grids and the density functions for this study are adopted from Margolin and Shashkov [17], Liska et al. [14]. Specifically, for a given number R of remap steps and $r = 0, \ldots, R$ the mesh node positions in $K_h^r(\Omega)$ are given by

$$x_{ij}^r = x(\xi_i, \eta_j, t_r), \quad y_{ij}^r = y(\xi_i, \eta_j, t_r), \quad 0 \leq i \leq N_x, \ 0 \leq j \leq N_y, \tag{58}$$

where N_x and N_y are the numbers of cells in x and y direction, respectively, $x(\xi, \eta, t)$ and $y(\xi, \eta, t)$ are coordinate maps and

$$\xi_i = \frac{i}{N_x}, \quad i = 0, \ldots, N_x; \quad \eta_j = \frac{j}{N_y}, \quad j = 0, \ldots, N_y; \quad \text{and}$$

$$t_r = \frac{r}{R}, \quad r = 0, \ldots, R,$$

are the initial (uniform) grid coordinates and the sequence of pseudo-time steps, respectively. We define two sets of coordinate maps. The first set is given by

$$x(\xi, \eta, t) = (1 - \alpha(t))\xi + \alpha(t)\xi^3; \tag{59a}$$

$$y(\xi, \eta, t) = (1 - \alpha(t))\eta + \alpha(t)\eta^2; \tag{59b}$$

$$\alpha(t) = \frac{\sin(4\pi t)}{2}. \tag{59c}$$

It generates a sequence of rectangular, tensor-product (logically Cartesian) grids. The second set is

$$x(\xi, \eta, t) = \xi + \alpha(t)\sin(2\pi \xi)\sin(2\pi \eta); \tag{60a}$$

$$y(\xi, \eta, t) = \eta + \alpha(t)\sin(2\pi \xi)\sin(2\pi \eta); \tag{60b}$$

with

$$\alpha(t) = \begin{cases} t/5 & \text{if } t \leq 5, \\ (1-t)/5 & \text{if } t \leq 5. \end{cases} \tag{60c}$$

The grids defined by (60a)–(60b) are logically Cartesian but not rectangular. One can show that for any $0 \leq t \leq 1$ the grids generated by (59a)–(59c) and (60a)–(60c) are valid (see Margolin and Shashkov [17]).

Convergence rates are estimated as follows. First, we use (58) to define a sequence of Q cyclic grids where $Q = 4$, $q = 1, \ldots, Q$, with total resolutions ($K^q \equiv N_x^q \times N_y^q$, R^q) given by $(64 \times 64, 320)$, $(128 \times 128, 640)$, $(256 \times 256, 1280)$, and $(512 \times 512, 2560)$, respectively. Thus, the total resolution is increased by a factor of $(2 \times 2, 2)$ in every subsequent set. Then, for every norm in (56) we compute the errors

$$\mathscr{E}^q = \mathscr{E}(\rho; \|\cdot\|, K^q, R^q), \quad q = 1, 2, 3, 4,$$

and solve (57) with $\{\mathscr{E}^1, \mathscr{E}^2\}$, $\{\mathscr{E}^1, \mathscr{E}^2, \mathscr{E}^3\}$, and $\{\mathscr{E}^1, \mathscr{E}^2, \mathscr{E}^3, \mathscr{E}^4\}$. This approach yields three increasingly accurate estimates of the convergence rates in each norm.

This estimation procedure is applied to three different density functions suggested in Margolin and Shashkov [17]: the "sine"

$$\rho(x, y) = 1 + \sin(2\pi x)\sin(2\pi y), \tag{61}$$

Table 8 OBR and M-OBR (FCR) errors and convergence rate estimates for the "sine" density (61) using 4 tensor-product cyclic grids defined by (59a)–(59c). The L_2 and L_∞ rates for OBR are slightly better than those for M-OBR (FCR). Additionally, we observe superconvergence for both methods in L_2 and L_1 norms

#cells	#remaps	L_2 err	L_1 err	L_∞ err	L_2 rate	L_1 rate	L_∞ rate
OBR							
64 × 64	320	6.58e−04	4.91e−04	5.78e−03	—	—	—
128 × 128	640	8.88e−05	6.16e−05	1.64e−03	2.89	3.00	1.82
256 × 256	1280	1.21e−05	7.82e−06	4.65e−04	2.88	2.99	1.82
512 × 512	2560	1.70e−06	9.89e−07	1.39e−04	2.87	2.98	1.80
FCR							
64 × 64	320	7.78e−04	4.95e−04	8.75e−03	—	—	—
128 × 128	640	1.22e−04	6.49e−05	2.81e−03	2.67	2.93	1.64
256 × 256	1280	2.00e−05	8.49e−06	8.89e−04	2.64	2.93	1.65
512 × 512	2560	3.43e−06	1.08e−06	2.84e−04	2.61	2.95	1.65

the "peak"

$$\rho(x,y) = \begin{cases} 1, & r > 0.25, \\ \max\{1.001, 4(r-0.25)+1\}, & r \leq 0.25, \end{cases} \quad (62a)$$

$$r = \sqrt{(x-0.5)^2 + (y-0.5)^2}, \quad (62b)$$

and the "shock"

$$\rho(x,y) = \begin{cases} 2, & y \geq (x-0.4)/0.3, \\ 1, & y \leq (x-0.4)/0.3. \end{cases} \quad (63)$$

Errors of the OBR and M-OBR (FCR) algorithms and the corresponding convergence rates are presented in Tables 8–10. We observe that for the peak and shock densities the OBR and M-OBR (FCR) convergence rates are virtually identical, whereas for the sine density the L_2 and L_∞ rates of OBR are better by 0.2. Intuitively this can be explained by noting that the peak and shock examples are comprised of piecewise linear functions for which the global optimization problem likely decouples into local optimization problems around the discontinuities. This diminishes the distinction between global (OBR) and local (M-OBR) optimization formulations of remap. In contrast, for the sine density, which is a globally smooth function, the feasible set of the global optimization problem remains fully coupled. In addition, we note that the L_2 and L_1 results in Table 8 are subject to superconvergence due to the choice of the grids.

Overall, these results indicate that OBR and M-OBR (FCR) have approximately the same accuracy on classical test problems. Consequently, one may wonder if the effects of the toy examples of Sect. 7 are never encountered in practice. In the

Table 9 OBR and M-OBR (FCR) errors and convergence rate estimates for the "peak" density (62a) using 4 tensor-product cyclic grids defined by (59a)–(59c). For this classical example, the convergence rates of OBR and M-OBR (FCR) are virtually identical

#cells	#remaps	L_2 err	L_1 err	L_∞ err	L_2 rate	L_1 rate	L_∞ rate
OBR							
64×64	320	6.97e−03	2.55e−03	8.00e−02	—	—	—
128×128	640	3.09e−03	8.90e−04	5.06e−02	1.17	1.52	0.66
256×256	1280	1.40e−03	3.10e−04	3.16e−02	1.16	1.52	0.67
512×512	2560	6.40e−04	1.09e−04	1.96e−02	1.15	1.52	0.68
FCR							
64×64	320	5.98e−03	2.14e−03	8.33e−02	—	—	—
128×128	640	2.54e−03	7.30e−04	5.29e−02	1.24	1.55	0.66
256×256	1280	1.11e−03	2.50e−04	3.33e−02	1.22	1.55	0.66
512×512	2560	4.98e−04	8.71e−05	2.07e−02	1.20	1.54	0.67

Table 10 OBR and M-OBR (FCR) errors and convergence rate estimates for the "shock" density (63) using 4 tensor-product cyclic grids defined by (59a)–(59c). For this classical example, the convergence rates of OBR and M-OBR (FCR) are virtually identical

#cells	#remaps	L_2 err	L_1 err	L_∞ err	L_2 rate	L_1 rate	L_∞ rate
OBR							
64×64	320	9.12e−02	2.88e−02	4.72e−01	—	—	—
128×128	640	7.12e−02	1.75e−02	4.86e−01	0.36	0.72	−0.04
256×256	1280	5.57e−02	1.06e−02	4.87e−01	0.36	0.72	−0.02
512×512	2560	4.33e−02	6.35e−03	4.98e−01	0.36	0.73	−0.02
FCR							
64×64	320	8.43e−02	2.45e−02	4.67e−01	—	—	—
128×128	640	6.57e−02	1.47e−02	4.77e−01	0.36	0.73	−0.03
256×256	1280	5.12e−02	8.87e−03	4.77e−01	0.36	0.73	−0.02
512×512	2560	3.99e−02	5.34e−03	4.88e−01	0.36	0.73	−0.02

next section we confirm, however, that important differences exist not only on toy problems; in particular, we demonstrate that OBR is more accurate and more robust than M-OBR (FCR) on grids that are of significant practical merit.

8.3 Cyclic Remap on Grids with Large Displacements

Theorem 3 asserts that the feasible set of M-OBR (FCR) is always a subset of the feasible set of the OBR formulation. This suggests that (37) may be more accurate

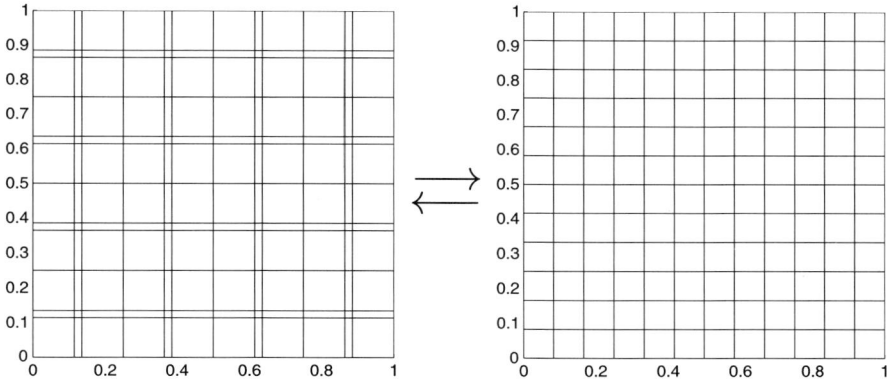

Fig. 13 Grid deformation due to local compression (*left pane*) and the 'repaired' uniform grid (*right pane*), see (64)–(65)

than (43). Examples in this section show that this is indeed the case and that the smaller feasible set of (43) can impact adversely the accuracy and, more importantly, robustness of M-OBR (FCR).

We begin with a study of accuracy. To this end, we compare convergence rates of the OBR and M-OBR (FCR) algorithms for the sine density (61) on a sequence of cyclic grids resulting from compressing every third cell equally in x and y direction, followed by a relaxation into a fully uniform grid. This mesh motion is motivated by the examples of Sect. 7 and is intended to mimic the effects of a repeated mesh repair procedure, see Fig. 13.

The 'repeated-repair' cyclic grid is given by

$$x_{ij}^r = \begin{cases} x_{ij}^0 & \text{if } r \text{ is even, for all } i, j; \text{ otherwise (when } r \text{ is odd)}: \\ x_{ij}^0 & \text{if } i \equiv 0 \pmod{3}, \text{ or if } i = N_x, \\ x_{ij}^0 + \Delta_x & \text{if } i \equiv 1 \pmod{3}, \text{ for } i < N_x, \\ x_{ij}^0 - \Delta_x & \text{if } i \equiv 2 \pmod{3}, \text{ for } i < N_x \end{cases} \quad (64)$$

and

$$y_{ij}^r = \begin{cases} y_{ij}^0 & \text{if } r \text{ is even, for all } i, j; \text{ otherwise (when } r \text{ is odd)}: \\ y_{ij}^0 & \text{if } j \equiv 0 \pmod{3}, \text{ or if } j = N_y, \\ y_{ij}^0 + \Delta_y & \text{if } j \equiv 1 \pmod{3}, \text{ for } j < N_y, \\ y_{ij}^0 - \Delta_y & \text{if } j \equiv 2 \pmod{3}, \text{ for } j < N_y. \end{cases} \quad (65)$$

The initial grid K_h^0 is a uniform grid on the unit square $[0, 1] \times [0, 1]$. We set

$$\Delta_x = \Delta_y = \frac{4}{5}(x_{10}^0 - x_{00}^0),$$

Table 11 OBR, M-OBR (FCR) and iFCR(2) errors and convergence rate estimates for the sine density (61) using 4 cyclic repeated-repair grids defined by (64)–(65). Rates expected of a second-order scheme are *highlighted*. It is evident that OBR delivers second-order accuracy, while M-OBR (FCR) exhibits a trend toward a first-order scheme. iFCR(2) gives L_1 and L_2 errors and convergence rates that are nearly identical to those given by OBR

#cells	#remaps	L_1 err	L_2 err	L_∞ err	L_1 rate	L_2 rate	L_∞ rate
OBR							
128 × 128	640	2.69e−04	3.65e−04	2.03e−03	—	—	—
256 × 256	1280	6.71e−05	9.08e−05	5.07e−04	**2.00**	**2.01**	**2.00**
512 × 512	2560	1.68e−05	2.27e−05	1.20e−04	**2.00**	**2.00**	**2.04**
1024 × 1024	5120	4.19e−06	5.66e−06	2.69e−05	**2.00**	**2.00**	**2.08**
FCR							
128 × 128	640	2.81e−04	3.47e−04	1.23e−03	—	—	—
256 × 256	1280	9.23e−05	1.19e−04	5.14e−04	1.61	1.54	1.26
512 × 512	2560	3.65e−05	5.05e−05	2.50e−04	1.47	1.39	1.15
1024 × 1024	5120	1.69e−05	2.39e−05	1.24e−04	1.35	1.28	1.10
iFCR(2)							
128 × 128	640	2.69e−04	3.64e−04	1.57e−03	—	—	—
256 × 256	1280	6.71e−05	9.07e−05	3.95e−04	**2.00**	**2.01**	**1.99**
512 × 512	2560	1.68e−05	2.27e−05	9.88e−05	**2.00**	**2.00**	**2.00**
1024 × 1024	5120	4.19e−06	5.66e−06	2.47e−05	**2.00**	**2.00**	**2.00**

resulting in a constant compression ratio of $4 \times 4 : 1$ for every third grid cell in x and y direction, whenever r is odd. For even r the grid is relaxed to its original position. See Fig. 13.

Estimates of the convergence rates of OBR and M-OBR (FCR) are presented in Table 11. The first observation is that the accuracy of the OBR algorithm on the repeated-repair cyclic grid is immune to the underlying mesh motion. In particular, the convergence rates of OBR in all three norms equals the best possible theoretical rates for a linearity-preserving scheme.

In contrast, it is clear that the convergence rates of M-OBR (FCR) suffer on the repeated-repair cyclic grid. The estimates in all three norms show a consistent trend toward a first-order scheme. We note that this is not due to a potential loss of monotonicity in low-order (donor-cell) fluxes; the compression parameters have been chosen such that the monotonicity of low-order fluxes is preserved. In other words, the loss of accuracy is purely due to a smaller feasible set employed by M-OBR (FCR). On the other hand, we observe that iFCR recovers the result of OBR at the expense of only 2 flux iterations per remap.

Our second study examines the robustness of OBR and M-OBR (FCR). To this end, we investigate the behavior of the methods on 64×64 meshes when the pseudo-time step $1/R$ is decreased significantly beyond the previously used test value of $1/320$. Table 12 displays the L_1 error in remapping the linear density $\rho(x, y) = x$

Table 12 L_1 errors in the OBR, M-OBR (FCR) and iFCR remap of a linear density function on the 64×64 tensor-product grid, for different values of the pseudo-time step $1/R$. Here iFCR^1 = iFCR(2), iFCR^2 = iFCR(20), iFCR^3 = iFCR(200) and iFCR^4 = iFCR(1000). Errors smaller than 1e−8 are *highlighted*. OBR fails to preserve linear densities at $R = 154$, while M-OBR (FCR) fails at $R = 212$, resulting in a pseudo-time step advantage for OBR of $212/154 \approx 1.4$. Beyond this point, OBR exhibits a graceful loss of accuracy; M-OBR (FCR) becomes numerically unstable. iFCR is more robust than FCR, however it does not duplicate the robustness of OBR

	$R = 213$	$R = 212$	$R = 211$	$R = 155$	$R = 154$	$R = 153$	$R = 100$	$R = 50$
OBR	**1.32e−13**	**1.42e−13**	**1.60e−13**	**4.60e−09**	4.06e−06	1.53e−05	1.97e−03	6.48e−03
FCR	**1.32e−13**	5.32e−08	1.10e−06	2.26e−03	2.35e−03	2.44e−03	5.73e+04	8.50e+11
iFCR^1	**1.32e−13**	**1.42e−13**	**1.60e−13**	1.36e−03	1.64e−03	1.39e−03	1.72e+02	1.29e+09
iFCR^2	**1.32e−13**	**1.42e−13**	**1.60e−13**	2.29e−03	1.14e−03	1.17e−03	4.61e+01	3.92e+08
iFCR^3	**1.32e−13**	**1.42e−13**	**1.60e−13**	**4.60e−09**	4.01e−05	1.32e−04	2.90e+01	1.32e+10
iFCR^4	**1.32e−13**	**1.42e−13**	**1.60e−13**	**4.60e−09**	4.01e−05	1.32e−04	2.64e+01	2.29e+08

on the tensor product cyclic grid (59a)–(59c) for varying pseudo-time steps. In the test, we choose to declare loss of linearity preservation when the L_1 error exceeds 1e−8. We note that it is expected that both OBR and M-OBR (FCR) will eventually fail to preserve linear densities due to the restrictions on admissible mesh motions, introduced earlier. OBR fails to preserve linear densities at $R = 154$, while M-OBR (FCR) fails at $R = 212$. Therefore, for this particular grid, the admissible pseudo-time step for OBR is approximately 1.4 times larger than that for M-OBR (FCR). Additionally, we observe that while OBR exhibits a graceful loss of accuracy once the OBR mesh motion conditions (25)–(26) are violated, M-OBR (FCR) becomes numerically unstable. This is most likely due to the loss of monotonicity in M-OBR (FCR) discussed and demonstrated in Sects. 5 and 7.3, respectively. We also note that iFCR is more robust than FCR, however it does not duplicate the robustness of OBR.

Similarly, Table 13 displays the L_1 error in remapping the linear density $\rho(x, y) = x$ on the smooth nonorthogonal cyclic grid (60a)–(60b). In this case OBR fails to preserve linear densities at $R = 15$, while M-OBR (FCR) fails at $R = 24$. Therefore, for this particular grid, the admissible pseudo-time step for OBR is approximately 1.6 times larger than that for M-OBR (FCR). Additionally, we observe that iFCR is more robust and more accurate than FCR, however the L_1 remap error does not converge to that of OBR as the number of iterations increases.

8.4 Computational Cost

From Theorem 4 we know that (43) decouples into a set of independent single-variable inequality-constrained optimization problems whose solution is given by (45). In other words, the computational cost of M-OBR (FCR) is quite low. On the other hand, the OBR formulation is a globally coupled inequality-constrained

Table 13 L_1 errors in the OBR and M-OBR (FCR) remap of a linear density function on the 64×64 smooth nonorthogonal grid, for different values of the pseudo-time step $1/R$. Here iFCR^1 = iFCR(2), iFCR^2 = iFCR(20), iFCR^3 = iFCR(200) and iFCR^4 = iFCR(1000). Errors smaller than 1e−8 are *highlighted*. OBR fails to preserve linear densities at $R = 15$, while M-OBR (FCR) fails at $R = 24$, resulting in a pseudo-time step advantage for OBR of $24/15 \approx$ **1.6**. iFCR is more robust than FCR, however the L_1 error does not converge to that of OBR as the number of iterations increases

	$R = 25$	$R = 24$	$R = 23$	$R = 16$	$R = 15$	$R = 14$	$R = 10$	$R = 5$
OBR	**2.32e−14**	**4.49e−14**	**2.15e−13**	**4.52e−10**	4.14e−05	5.13e−04	1.16e−03	2.45e−03
FCR	**2.32e−14**	3.63e−07	1.67e−06	8.60e−04	1.16e−03	1.69e−03	5.74e−03	1.09e−02
iFCR^1	**2.32e−14**	**4.49e−14**	**2.15e−13**	1.52e−03	3.13e−03	4.95e−03	8.08e−03	1.38e−02
iFCR^2	**2.32e−14**	**4.49e−14**	**2.15e−13**	**4.52e−10**	7.48e−05	6.89e−04	3.03e−02	7.89e−02
iFCR^3	**2.32e−14**	**4.49e−14**	**2.15e−13**	**4.52e−10**	7.48e−05	6.89e−04	1.93e−02	3.44e−02
iFCR^4	**2.32e−14**	**4.49e−14**	**2.15e−13**	**4.52e−10**	7.48e−05	6.89e−04	1.93e−02	3.39e−02

optimization problem. It is therefore of considerable practical interest to assess the performance penalty incurred by the need to solve a global optimization problem.

The algorithms used to solve M-OBR (FCR) and OBR formulations are described in Sect. 5. Table 14 presents preliminary timing results using Matlab™ implementations of M-OBR (FCR), iFCR, and OBR. For accurate estimates of the computational cost we choose the examples of Sect. 8.2. We make two observations.

First, while a direct comparison of our implementation of M-OBR (FCR) and the closely related SFCR method implemented in Fortran, see Liska et al. [14, p. 1490], is not possible, we note that the computational cost of our Matlab™ implementation is in the range of the computational cost of the Fortran implementation. We achieve this by employing only vectorized Matlab™ operations, which are delegated to fast computational kernels. The linear (ten-fold) scaling of the mesh-to-mesh computational cost reported in Liska et al. [14, p. 1490] is evident in our case when meshes are sufficiently large, i.e. when the computational overhead associated with the Matlab™ environment becomes negligible.

Second, noting that additional studies with more efficient implementations of M-OBR (FCR) and, especially, OBR are needed, we can already see that the computational cost of OBR is proportional, up to a very modest constant, to the cost of M-OBR (FCR). On average, OBR is only 2.1 times slower than M-OBR (FCR). Considering the gains in accuracy and robustness as well as the less restrictive conditions on admissible mesh motions, OBR is a strong alternative to M-OBR (FCR). Finally, for iFCR we observe that approximately 20 flux iterations per remap can be employed at the cost of OBR.

Table 14 Comparison of the computational costs (times are in seconds) of OBR, M-OBR (FCR), iFCR(2) and iFCR(20) as measured by Matlab™ wall-clock times on a single Intel Xeon X5680 3.33 GHz processor, for density functions defined in (61), (62a) and (63) and the cyclic grid (59a)–(59c). Ratios of OBR times and FCR/iFCR(2)/iFCR(20) times are also reported. The cost of OBR is proportional, up to a modest constant, to the cost of M-OBR (FCR). The average cost ratio is only **2.1**. The OBR to iFCR(2) cost ratio is **1.9**. The OBR to iFCR(20) cost ratio is **1.0**

#cells	#remaps	OBR	FCR	Ratio	iFCR(2)	Ratio	iFCR(20)	Ratio
Sine								
64×64	320	**7.3**	4.2	**1.7**	4.4	**1.7**	7.4	**1.0**
128×128	640	**49.5**	25.4	**1.9**	27.6	**1.8**	50.5	**1.0**
256×256	1280	**390.6**	176.5	**2.2**	198.9	**2.0**	387.8	**1.0**
512×512	2560	**3662.8**	1812.5	**2.0**	2156.6	**1.7**	4955.4	**0.7**
Peak								
64×64	320	**8.4**	4.9	**1.7**	5.1	**1.6**	8.7	**1.0**
128×128	640	**57.8**	28.5	**2.0**	31.0	**1.9**	55.7	**1.0**
256×256	1280	**418.6**	183.8	**2.3**	203.2	**2.1**	448.7	**0.9**
512×512	2560	**4528.6**	1832.9	**2.5**	2264.0	**2.0**	5156.8	**0.9**
Shock								
64×64	320	**9.8**	4.9	**2.0**	4.9	**2.0**	8.2	**1.2**
128×128	640	**88.9**	28.1	**3.2**	31.1	**2.9**	54.1	**1.6**
256×256	1280	**438.6**	184.7	**2.4**	220.4	**2.0**	409.0	**1.1**
512×512	2560	**3214.6**	1794.1	**1.8**	2237.4	**1.4**	4806.3	**0.7**

9 Conclusions

In this chapter we formulate and study a new class of optimization-based, conservative, bound and linearity preserving remap algorithms (OBR). The use of an optimization setting allows us to separate accuracy considerations from the enforcement of physical bounds by making the former the objective of optimization, while the latter is used to define the constraints in the optimization problem. In so doing we obtain a scheme that is provably linearity preserving and bound-preserving on arbitrary unstructured grids, including grids with non-convex polygonal or polyhedral cells.

Rigorous characterization of the relationship between the OBR and the FCR algorithm of Liska et al. [14] is another key contribution of this chapter. Specifically, we prove that the FCR is equivalent to an inequality-constrained optimization problem, termed M-OBR, which is derived from OBR by replacing its constraints by a set of simpler sufficient conditions for the local bounds. These conditions are *decoupled* box constraints derived using a worst-case local analysis to simplify the original *coupled* inequality constraints. Using the relationship between the constraints in OBR and M-OBR (FCR) we prove that the feasible set of M-OBR (FCR) is always contained in the feasible set of OBR. It follows that asymptotically OBR is at least as accurate as M-OBR (FCR). Furthermore, numerical comparison between OBR and the *iterated* FCR (iFCR) strongly suggest that the latter provides an iterative solution algorithm for the global optimization problem in the OBR formulation.

Succinctly, our theoretical and computational results establish the following hierarchy among the OBR, FCR and iFCR methods:

- OBR defines a *"master"* optimization formulation for the remap problem, characterized by a global set of linear constraints derived from physical considerations;
- FCR simplifies the master optimization problem by *decoupling* the linear constraints, which reduces the size of the feasible set;
- iFCR is an *iterative procedure* that under some conditions may recover the OBR solution.

Because FCR is motivated by flux-corrected transport (FCT), this hierarchy opens up an interesting possibility that FCT and *iterative* FCT, see Kuzmin et al. [12], may also be connected to a "master" global optimization formulation for the selection of accurate and monotone fluxes in transport algorithms.

The computational examples in this chapter provide further illustration of the hierarchy among these methods. For smooth cyclic grids with moderate displacements there are no significant differences in the accuracy and the convergence rates of M-OBR (FCR) and OBR. However, on cyclic grids with large displacements the smaller feasible set of M-OBR (FCR) can adversely impact its accuracy and robustness. In particular, we demonstrate that on such grids M-OBR (FCR) defaults to a first-order accurate scheme, while OBR achieves the theoretically best possible accuracy (second order) for a linearity-preserving scheme. Furthermore, in a series of large-displacement examples we show that the OBR formulation admits a larger pseudo-time step (1.4 to 1.6 times) and that M-OBR (FCR) can suffer numerical breakdown due to the loss of monotonicity in low-order fluxes based on swept-region computations. In contrast, OBR does not require the computation of low-order fluxes; at the same time, the employed computation of high-order fluxes using swept regions is safe because monotonicity is enforced separately, through inequality constraints. Finally, a "torture" test reveals that under certain conditions the smaller feasible set of M-OBR (FCR) can lead to the loss of qualitative information about the shape of the remapped density function.

Preliminary studies show that for a set of standard remap test problems the cost of OBR is proportional, up to a very modest constant, to the cost of M-OBR (FCR). On average, not counting potential gains from the time-step advantage of OBR, it is only about twice as expensive as FCR. This suggests that OBR can be competitive in practical applications where a (i) provably linearity-preserving (and otherwise *optimally* accurate) and (ii) bound-preserving method is desired.

The extension of the OBR approach to systems, and further theoretical and computational studies, including formal analysis of iFCR as an iterative solution algorithm for OBR will be the subject of a forthcoming paper.

Acknowledgements All authors acknowledge funding by the DOE Office of Science Advanced Scientific Computing Research (ASCR) Program. PB, DR and GS also acknowledge funding by the NNSA Climate Modeling and Carbon Measurement Project. DR and MS also acknowledge funding by the Advanced Simulation & Computing (ASC) Program.

Our colleagues Dmitri Kuzmin, Richard Liska, Kara Peterson, John Shadid, Pavel Váchal and Joseph Young provided many comments and valuable insights that helped improve this work.

Sandia National Laboratories is a multi-program laboratory operated by Sandia Corporation, a wholly owned subsidiary of Lockheed Martin Corporation, for the U.S. Department of Energy's National Nuclear Security Administration under contract DE-AC04-94AL85000. The work of MS was carried out under the auspices of the National Nuclear Security Administration of the U.S. Department of Energy at Los Alamos National Laboratory under Contract No. DE-AC52-06NA25396.

References

1. Bell, J., Berger, M., Saltzman, J., Welcome, M.: Three-dimensional adaptive mesh refinement for hyperbolic conservation laws. SIAM J. Sci. Comput. **15**(1), 127–138 (1994)
2. Berger, M., Murman, S.M., Aftosmis, M.J.: Analysis of slope limiters on irregular grids. In: Proceedings of the 43rd AIAA Aerospace Sciences Meeting, Reno, NV (2005). AIAA2005-0490
3. Berger, M.J., Colella, P.: Local adaptive mesh refinement for shock hydrodynamics. J. Comput. Phys. **82**, 64–84 (1989)
4. Bochev, P., Day, D.: Analysis and computation of a least-squares method for consistent mesh tying. J. Comp. Appl. Math **218**, 21–33 (2008)
5. Bochev, P., Ridzal, D., Scovazzi, G., Shashkov, M.: Formulation, analysis and numerical study of an optimization-based conservative interpolation (remap) of scalar fields for arbitrary Lagrangian-Eulerian methods. J. Comput. Phys. **230**(12), 5199–5225 (2011)
6. Carey, G., Bicken, G., Carey, V., Berger, C., Sanchez, J.: Locally constrained projections on grids. Int. J. Numer. Methods Eng. **50**, 549–577 (2001)
7. Coleman, T.F., Li, Y.: A reflective newton method for minimizing a quadratic function subject to bounds on some of the variables. SIAM J. Optim. **6**(4), 1040–1058 (1996)
8. Dukowicz, J.K., Kodis, J.W.: Accurate conservative remapping (rezoning) for arbitrary Lagrangian-Eulerian computations. SIAM J. Sci. Stat. Comput. **8**, 305–321 (1987)
9. Hirt, C., Amsden, A., Cook, J.: An arbitrary Lagrangian-Eulerian computing method for all flow speeds. J. Comput. Phys. **14**, 227–253 (1974)
10. Jones, P.W.: First- and second-order conservative remapping schemes for grids in spherical coordinates. Mon. Weather Rev. **127**(9), 2204–2210 (1999)
11. Kucharik, M., Shashkov, M., Wendroff, B.: An efficient linearity-and-bound-preserving remapping method. J. Comput. Phys. **188**(2), 462–471 (2003)
12. Kuzmin, D., Löhner, R., Turek, S. (eds.): Flux-Corrected Transport: Principles, Algorithms and Applications. Springer, Berlin (2005)
13. Laursen, T., Heinstein, M.: A three dimensional surface-to-surface projection algorithm for non-coincident domains. Commun. Numer. Methods Eng. **19**, 421–432 (2003)
14. Liska, R., Shashkov, M., Váchal, P., Wendroff, B.: Optimization-based synchronized flux-corrected conservative interpolation (remapping) of mass and momentum for arbitrary Lagrangian-Eulerian methods. J. Comput. Phys. **229**(5), 1467–1497 (2010)
15. Loubere, R., Shashkov, M.J.: A subcell remapping method on staggered polygonal grids for arbitrary-Lagrangian-Eulerian methods. J. Comput. Phys. **209**(1), 105–138 (2005)
16. Loubere, R., Staley, M., Wendroff, B.: The repair paradigm: New algorithms and applications to compressible flow. J. Comput. Phys. **211**(2), 385–404 (2006)
17. Margolin, L.G., Shashkov, M.: Second-order sign-preserving conservative interpolation (remapping) on general grids. J. Comput. Phys. **184**(1), 266–298 (2003)
18. Margolin, L.G., Shashkov, M.: Remapping, recovery and repair on a staggered grid. Comput. Methods Appl. Mech. Eng. **193**(39–41), 4139–4155 (2004)
19. Miller, D.S., Burton, D.E., Oliviera, J.S.: Efficient second order remapping on arbitrary two dimensional meshes. Technical Report UCRL-ID-123530, Lawrence Livermore National Laboratory (1996)

3 Flux-Corrected-Transport Algorithm

This section describes the FCT method as used in this chapter including the extensions and modifications we have made to the classic FCT algorithm. This implementation includes a DeVore type pre-limiter in lieu of Zalesak's flux pre-constraint, removal of artificial diffusion prior to the FCT flux limiter, a Jameson style artificial viscosity, a sonic fix for entropy violating rarefaction waves, and the extension of the FCT algorithm to overlapping grids. For clarity, the improvements for treating sonic points and very strong rarefactions are left to the end of the section.

3.1 Overlapping Grids and AMR

We consider the governing equations (1) and proceed with a description of the FCT method in a two dimensional overlapping grid framework. To this end, we assume the flow domain is given by Ω and is discretized using an overlapping grid $\mathscr{G}$. The overlapping grid consists of a set of component grids $\{\mathscr{G}_i\}$, $i = 1, \ldots, \mathscr{N}_g$, that cover Ω and overlap where they meet. Each component grid covers a sub-domain Ω_i. Grid points are tagged as discretization points where the governing equations are applied, ghost points used for the application of boundary conditions, interpolation points where solution values are communicated between grids, or unused points where no computation is performed which are cut out through the mesh generation procedure. The FCT stencil is 7-points wide requiring three layers of data at interpolation and physical boundaries. At interpolation boundaries, the 7-point stencil would normally require three layers of interpolation points. Although we can generate such grids, in practice we usually construct a grid with a single layer of interpolation points and obtain values at the two additional layers through extrapolation. At physical boundaries, values on the boundary and three layers of ghost points are obtained through application of the physical boundary conditions, derived compatibility conditions, and extrapolation following the approach described in [30, 31]. Note that the dependence of the solution on this final extrapolated layer is extremely weak as it can only affect whether the chosen update at the boundary is first or second order accurate (i.e. it is used only in the determination of the α in (8) below. For more details concerning general overlapping grid methods, including application of boundary conditions, see [16, 29–31]. Adaptive mesh refinement (AMR) is used in regions of the flow where the solution changes rapidly, such as near shocks and contact surfaces. We employ a block-structured AMR approach following that described originally in [10] and using modifications for overlapping grids as presented in [5, 30, 31].

3.2 FCT Discretization on a Mapped Grid

Each component grid, including base-level grids and any refined grids, is defined by a mapping from the unit square in computational space (r_1, r_2) to physical space

(x_1, x_2). In computational space, (1) becomes

$$\frac{\partial}{\partial t}\mathbf{u} + \frac{1}{J}\frac{\partial}{\partial r_1}\mathbf{F}_1(\mathbf{u}) + \frac{1}{J}\frac{\partial}{\partial r_2}\mathbf{F}_2(\mathbf{u}) = 0, \tag{3}$$

where

$$\mathbf{F}_1(\mathbf{u}) = J\left(\frac{\partial r_1}{\partial x_1}\mathbf{f}_1 + \frac{\partial r_1}{\partial x_2}\mathbf{f}_2\right), \quad \mathbf{F}_2(\mathbf{u}) = J\left(\frac{\partial r_2}{\partial x_1}\mathbf{f}_1 + \frac{\partial r_2}{\partial x_2}\mathbf{f}_2\right),$$

and

$$J = \left|\frac{\partial(x_1, x_2)}{\partial(r_1, r_2)}\right|.$$

The metrics of the mapping, $\partial x_1/\partial r_2$, $\partial x_2/\partial r_2$, etc., and the Jacobian are considered to be known for each component grid at the time of computation and can be generated analytically or approximated.

Discretization of (3) is performed using a uniform grid $(r_{1,i}, r_{2,j})$ with grid spacing $(\Delta r_1, \Delta r_2)$. The FCT method is generally considered a two-step process proceeding first with a low order update and finishing with the high-resolution FCT correction. We begin with the formulation of the low order solution update

$$\mathbf{u}_{i,j}^{\text{td},n} = \mathbf{u}_{i,j}^n - \frac{\Delta t}{J_{i,j}\Delta r_1}D_{+r_1}\mathbf{F}_{1\,i-1/2,j}^{\text{low},n} - \frac{\Delta t}{J_{i,j}\Delta r_2}D_{+r_2}\mathbf{F}_{2\,i,j-1/2}^{\text{low},n} \tag{4}$$

where D_{+r_1} and D_{+r_2} are the undivided forward difference approximations in the r_1 and r_2 directions of index space respectively. The "td" notation is consistent with [11–13, 59] and denotes "transported and diffused". For this work the HLL low order flux [26, 54] is used and for curvilinear geometries is given by

$$\mathbf{F}_{1\,i+1/2,j}^{\text{low},n} = \begin{cases} \mathbf{F}_{1\,i,j}^n & \text{if } s_- \geq 0, \\ \mathbf{F}_{1\,i+1,j}^n & \text{if } s_+ \leq 0, \\ \frac{s_+}{s_+ - s_-}\mathbf{F}_{1\,i,j}^n - \frac{s_-}{s_+ - s_-}\mathbf{F}_{1\,i+1,j}^n + \frac{s_- s_+}{s_+ - s_-}D_{+r_1}\mathbf{u}_{i,j}^n & \text{else} \end{cases} \tag{5}$$

where

$$s_- = \min\left(v_{i,j}^n - c_{i,j}^n, v_{i+1,j}^n - c_{i+1,j}^n\right)\left\|\left(\frac{\partial r_1}{\partial x_1}, \frac{\partial r_1}{\partial x_2}\right)\right\|,$$

$$s_+ = \max\left(v_{i,j}^n + c_{i,j}^n, v_{i+1,j}^n + c_{i+1,j}^n\right)\left\|\left(\frac{\partial r_1}{\partial x_1}, \frac{\partial r_1}{\partial x_2}\right)\right\|,$$

$c_{i,j}^n$ is the sound speed in a given cell, and $v_{i,j}^n$ is the component of the velocity normal to the cell face. The fluxes across other cell boundaries take similar forms.

It should be noted that in [38, 60], Zalesak suggests the use of the Rusanov flux for the low order method. This is a symmetrized version of the HLL flux resulting in further diffusion than the original HLL flux. However, the Rusanov flux as presented in [38, 60] is slightly flawed in that the selected wave speed is not sufficient to encompass the full Riemann solution for all cases. A more general Rusanov flux is

$$\mathbf{F}_{1\,i+1/2,j}^{\text{low},n} = \frac{1}{2}\big[(\mathbf{F}_{1\,i+1,j}^n + \mathbf{F}_{1\,i,j}^n) - \max(|\lambda_{i+1,j}^n|, |\lambda_{i,j}^n|) D_{+r_1} \mathbf{u}_{i,j}^n\big] \qquad (6)$$

where $\lambda_{i,j}^n$ is the largest eigenvalue (in magnitude) of the Jacobian matrix $\frac{\partial}{\partial \mathbf{u}}\mathbf{F}_1$ at a cell (i, j) and time t_n. The difference between (6) and the equation presented in [38, 60] is the use of $\max(|\lambda_{i+1,j}^n|, |\lambda_{i,j}^n|)$ rather than $1/2(|\lambda_{i+1,j}^n| + |\lambda_{i,j}^n|)$. In this work, the HLL flux is used but we have found that the Rusanov flux (6) works nearly as well and is less expensive. As presented, both of these approximate fluxes require knowledge of the eigenvalues of the Jacobian matrix. If this information were not known, a Lax-Friedrichs type flux could in principle be used instead.

The second step of the FCT algorithm requires an "anti-diffusive" flux which is defined as the difference between a high-order flux and the low-order one. In the r_1 direction of index space for example, this is

$$\mathbf{F}_{1\,i\pm1/2,j}^{\text{AD},n} = \mathbf{F}_{1\,i\pm1/2,j}^{\text{high},n} - \mathbf{F}_{1\,i\pm1/2,j}^{\text{low},n}. \qquad (7)$$

The high order flux is typically chosen to be some high-order centered flux and for this work the centered second-order flux

$$\mathbf{F}_{1\,i+1/2,j}^{\text{high},n} = \frac{1}{2}(\mathbf{F}_{1\,i,j}^n + \mathbf{F}_{1\,i+1,j}^n)$$

is chosen. The final sub-step update is now defined as

$$\begin{aligned}\mathbf{u}_{i,j}^{\text{new}} &= \mathbf{u}_{i,j}^{\text{td},n} - \frac{\Delta t}{J_{i,j}\Delta r_1} D_{+r_1}\big(\boldsymbol{\alpha}_{i-1/2,j}^n \odot \mathbf{F}_{1\,i-1/2,j}^{\text{AD},n}\big) \\ &\quad - \frac{\Delta t}{J_{i,j}\Delta r_2} D_{+r_2}\big(\boldsymbol{\alpha}_{i,j-1/2}^n \odot \mathbf{F}_{2\,i,j-1/2}^{\text{AD},n}\big)\end{aligned} \qquad (8)$$

where $\odot$ indicates component-wise multiplication. The vector of $\boldsymbol{\alpha}$'s are chosen using the FCT algorithm as described below and represent the proportion of anti-diffusive flux at each cell face that is used in the final update. Our choice of notation facilitates the use of the FCT algorithm in a method of lines type approach. By defining

$$\frac{\partial}{\partial t}\mathbf{u}_{i,j}^n = \frac{\mathbf{u}_{i,j}^{\text{new}} - \mathbf{u}_{i,j}^n}{\Delta t} \qquad (9)$$

we obtain an updated solution $\mathbf{u}_{i,j}^{n+1}$ using any ordinary differential equation (ODE) integrator we choose. Choices for ODE integrators might include Runge-Kutta methods, Adams methods, or others. For this work, we use an explicit Adams predictor-corrector method of second order to match the spatial algorithm. Detail concerning the implementation of these time integrators can be found for example in [2, 29].

Consider the determination of $\alpha_{i+1/2,j}^n$. FCT seeks to enforce solution monotonicity through the choice of α, but the property of monotonicity is valid only for characteristic variables [57]. For the non-linear Euler equations, conversion to characteristic variables requires both a linearization and an eigen-decomposition of the linearized problem. As such, we linearize about the arithmetic average $\bar{\mathbf{u}} = \frac{1}{2}(\mathbf{u}_{i,j}^{td,n} + \mathbf{u}_{i+1,j}^{td,n})$. More sophisticated choices, such as the Roe average [54], could be made but in our experience these make little difference in the eventual computed solutions. From this state, the linearized eigen-decomposition $\mathbf{T}^{-1}\Lambda\mathbf{T} = \mathbf{A} = \frac{\partial}{\partial \mathbf{u}}\mathbf{F}_1(\bar{\mathbf{u}})$ is found where we have dropped the sub- and superscripts to simplify the exposition. Whenever multiplication by $\mathbf{T}$ is performed to achieve characteristic quantities it should be understood that this implies linearization about a particular face, in this case $(i + 1/2, j)$. For two dimensions, a large number of characteristic transformations must be performed (in three dimensions the number is even larger) and this constitutes one of the most expensive parts of the FCT method.

In [19], DeVore indicates that the scheme of Zalesak does not preserve monotonicity in two dimensions and suggests limiting the fluxes using the original Boris/Book limiter [11, 13] in each direction prior to their input to the multi-dimensional limiter. This is straight forward to and we demonstrate it for $\mathbf{F}_{1\,i+1/2,j}^{AD,n}$

$$\hat{\mathbf{F}}_{1\,i+1/2,j}^{AD,n} = \mathbf{s} \odot \max\left[0, \min\left(|\mathbf{TF}_{1\,i+1/2,j}^{AD,n}|, \mathbf{s} \odot \frac{J_{i+1/2,j}\Delta r_1}{\Delta t} D_{+r_1} \mathbf{Tu}_{i+1/2,j}^{:d,n}, \right.\right.$$

$$\left.\left. \mathbf{s} \odot \frac{J_{i+1/2,j}\Delta r_1}{\Delta t} D_{+r_1} \mathbf{Tu}_{i-1/2,j}^{td,n}\right)\right],$$

where $\mathbf{s} = \text{sign}(\mathbf{TF}_{1\,i+1/2,j}^{AD,n})$ and the "hat" notation indicates that the anti-diffusive flux has been pre-limited. The other $\hat{\mathbf{F}}$ fluxes are obtained through similar formulas.

To complete the FCT algorithm, define the local maximum and minimum characteristic values as

$$\mathbf{w}_k^{\max} = \max\left(\mathbf{Tu}_{i+k-1,j}^{td,n}, \mathbf{Tu}_{i+k,j}^{td,n}, \mathbf{Tu}_{i+k+1,j}^{td,n}, \mathbf{Tu}_{i+k,j-1}^{td,n}, \mathbf{Tu}_{i+k,j+1}^{td,n}\right),$$
$$\mathbf{w}_k^{\min} = \min\left(\mathbf{Tu}_{i+k-1,j}^{td,n}, \mathbf{Tu}_{i+k,j}^{td,n}, \mathbf{Tu}_{i+k+1,j}^{td,n}, \mathbf{Tu}_{i+k,j-1}^{td,n}, \mathbf{Tu}_{i+k,j+1}^{td,n}\right),$$
(10)

where $k = 0, 1$ and the extrema are taken component-wise. The actual influx into the cells on either side of the cell face which would result from the AD fluxes is computed for example as

$$\mathbf{I}_k = \frac{1}{\Delta r_1}\left[\max\left(\frac{\hat{\mathbf{F}}_{1\,i+k-1/2,j}^{AD,n}}{J_{i+k,j}}, 0\right) - \min\left(\frac{\hat{\mathbf{F}}_{1\,i+k+1/2,j}^{AD,n}}{J_{i+k,j}}, 0\right)\right]$$
$$+ \frac{1}{\Delta r_2}\left[\max\left(\frac{\hat{\mathbf{F}}_{2\,i+k,j-1/2}^{AD,n}}{J_{i+k,j}}, 0\right) - \min\left(\frac{\hat{\mathbf{F}}_{2\,i+k,j+1/2}^{AD,n}}{J_{i+k,j}}, 0\right)\right],$$
(11)

and the maximum permissible influx such that the characteristic bounds from (10) are not violated, indicated by the tilde, is for example

$$\tilde{\mathbf{I}}_k = \frac{1}{\Delta t}\left[\mathbf{w}_k^{\max} - \mathbf{T}\mathbf{u}_{i+k,j}^{\text{td},n}\right]. \tag{12}$$

Notice in (11) that the influx into the cells from both direction of index space are considered simultaneously. This follows from [59] and reflects the fully multi-dimensional nature of this limiter as opposed to a limiter which is split along dimensional lines. Component-wise ratios of permissible to actual fluxes are then defined for the two cells as

$$\mathbf{R}_k^+ = \min\left(\frac{\tilde{\mathbf{I}}_k}{\mathbf{I}_k}, 1\right). \tag{13}$$

The quantities $\mathbf{R}_k^-$, which represent the ratio of actual AD flux leaving the cell to the maximum flux permitted to leave the cell without violation of the bounds in (10), are defined using similar reasoning. Setting

$$\mathbf{O}_k = \frac{1}{\Delta r_1}\left[\max\left(\frac{\hat{\mathbf{F}}1_{i+k+1/2,j}^{\text{AD},n}}{J_{i+k,j}}, 0\right) - \min\left(\frac{\hat{\mathbf{F}}1_{i+k-1/2,j}^{\text{AD},n}}{J_{i+k,j}}, 0\right)\right]$$
$$+ \frac{1}{\Delta r_2}\left[\max\left(\frac{\hat{\mathbf{F}}2_{i+k,j+1/2}^{\text{AD},n}}{J_{i+k,j}}, 0\right) - \min\left(\frac{\hat{\mathbf{F}}2_{i+k,j-1/2}^{\text{AD},n}}{J_{i+k,j}}, 0\right)\right], \tag{14}$$

and

$$\tilde{\mathbf{O}}_k = \frac{1}{\Delta t}\left[\mathbf{T}\mathbf{u}_{i+k,j}^{\text{td},n} - \mathbf{w}_k^{\min}\right], \tag{15}$$

we define

$$\mathbf{R}_k^- = \min\left(\frac{\tilde{\mathbf{O}}_k}{\mathbf{O}_k}, 1\right). \tag{16}$$

By choosing the most restrictive of these $\mathbf{R}$ values, the bounds from (10) are not violated. Thus we define

$$\boldsymbol{\beta} = \begin{cases} \min(\mathbf{R}_0^+, \mathbf{R}_1^-) & \text{when } J_{i,j}\hat{\mathbf{F}}1_{i+1/2,j}^{\text{AD},n} < 0, \\ \min(\mathbf{R}_1^+, \mathbf{R}_0^-) & \text{when } J_{i,j}\hat{\mathbf{F}}1_{i+1/2,j}^{\text{AD},n} \geq 0. \end{cases} \tag{17}$$

The final values for $\boldsymbol{\alpha}_{i+1/2,j}^n$ are found through component-wise inversion of the formula

$$\boldsymbol{\alpha}_{i+1/2,j}^n \odot \mathbf{F}1_{i+1/2,j}^{\text{AD},n} = \mathbf{T}^{-1}\left(\boldsymbol{\beta} \odot \hat{\mathbf{F}}1_{i+1/2,j}^{\text{AD},n}\right). \tag{18}$$

It is important to note that monotonicity of the linearized characteristic variables does not imply monotonicity of the conserved variables. Thus the final updated solution could result in a negative density, imaginary sound speed, or negative pressure. Such events do occur in the simulations we present and must be treated in a rational and reasonable way. Zalesak suggests in [38, 59] that a fail-safe limiter be employed and we take a similar approach here. At each time, if the values at a given cell (i, j) violate physically realistic bounds after advancement to $\mathbf{u}_{i,j}^{\text{new}}$ in (8), then no portion of the anti-diffusive flux is allowed at the boundaries of that cell. For such cells,

$$\alpha_{i+1/2,j}^n = \alpha_{i-1/2,j}^n = \alpha_{i,j+1/2}^n = \alpha_{i,j-1/2}^n = 0 \tag{19}$$

is enforced and the method becomes fully first order in a local region. In our experience, this fail-safe mechanism is critical for the success of the FCT algorithm. It should also be noted that after setting $\alpha_{i\pm 1/2, j\pm 1/2}^n = 0$ in one cell, the problem (negative density etc.) may then appear in a neighboring cell. In principle the result could be a cascade across all cells. These cascades are rare and do not occur for any of the simulations presented in this work.

This completes the description of the FCT algorithm itself but there is another aspect which must be addressed. In [59] it is recognized that some amount of higher order dissipation must be included to remove high frequency noise generated by the FCT procedure. In that work the high-order dissipation was added to the AD flux prior to flux correction. In our studies we found this to be unsatisfactory because the effect of the high-order dissipation is reduced by the FCT limiters. The result is unacceptable levels of numerical noise in the computed solutions. Therefore we add dissipation independently after the FCT step. To this end we implement a second-order dissipation near shocks [18, 30] to treat undamped transverse instabilities as well as a fourth-order Jameson style dissipation away from shocks [29, 33, 34]. We switch the fourth order dissipation on or off based on density variations to ensure that it is not active near shocks or contacts. One final note is that the computed solution will not violate the prescribed bounds only for CFL numbers less than 1/2 and so all FCT simulation results presented in this chapter set the CFL number to be 0.4.

3.3 Sonic Fix

As is the case for some other methods, such as Godunov's method with an approximate Roe Riemann solver [54], the FCT method can exhibit poor behavior in rarefaction waves at points where the flow speed is equal to the sound speed (sonic points). The problem is illustrated by the solution to a modified version of Sod's shock tube problem [51, 54] with left and right states given by $(\rho, u_1, u_2, p)_L = (1.0, 0.75, 0.0, 1.0)$ and $(\rho, u_1, u_2, p)_R = (0.125, 0.75, 0.0, 0.1)$, and with $\gamma = 1.4$. We compute approximations to the solution of this Riemann problem using the grid $\mathcal{L}([-1, 1], 100)$ where

$$\mathcal{L}\big([x_a, x_b], N\big) = \big\{x_i | x_i = x_a + i\Delta x,\ \Delta x = (x_b - x_a)/N,\ i = 0, 1, \ldots, N\big\}, \tag{20}$$

with the initial discontinuity located at $x = -0.4$. Figure 1 shows the results produced by the FCT method with and without our sonic fix. The problematic behavior

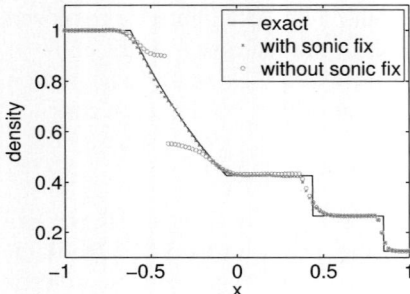
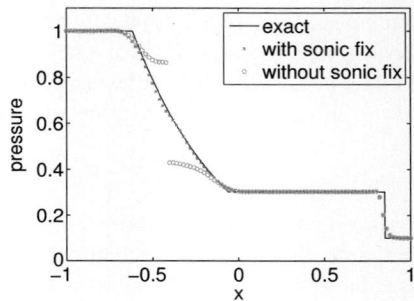

Fig. 1 FCT solution for a shock tube problem containing a sonic rarefaction with and without a sonic fix at $t = 0.5$. The *black line* represents the exact solution, the *red circles* the numerical approximation without the sonic fix and the *blue crosses* the numerical approximation with the sonic fix. The problematic behavior at the sonic point is quite clear in both the density (*left*) and pressure (*right*) (Color figure online)

of the method at the sonic point is clearly visible in the form of a rarefaction shock which represents an entropy violating weak solution.

The existence of rarefaction shocks in numerical approximations is typically the result of insufficient numerical diffusion. For FCT this is caused by the use of high-order centered fluxes. This is in contrast to Roe's method where the linearization causes the problem even at first order. The FCT method considered in this chapter uses the HLL flux (known to be devoid of rarefaction shocks [54]) for the low order update. To eliminate rarefaction shocks for FCT approximations, we rely on this fact and simply set the value for α in (8) to zero for cases where sonic rarefactions are present. This choice has implications on solution accuracy, but because sonic points exist in isolation, the impact is negligible as will be demonstrated in Sect. 4.

The anti-diffusive fluxes in (8) have associated left and right states, call these $\mathbf{u}_L$ and $\mathbf{u}_R$ respectively. For instance consider $\mathbf{F}_{1\,i+1/2,j}^{\text{AD},n}$ with $\mathbf{u}_L = \mathbf{u}_{i,j}^{\text{td},n}$ and $\mathbf{u}_R = \mathbf{u}_{i+1,j}^{\text{td},n}$. These states can be viewed as left and right states of a one dimensional Riemann problem in the direction normal to the cell face. Define the normal velocities as $v_{n,L} = (n_1, n_2) \cdot (u_{1L}, u_{2L})^T$ and $v_{n,R} = (n_1, n_2) \cdot (u_{1R}, u_{2R})^T$ where (n_1, n_2) is the unit normal to the cell face. Following the nomenclature in [54], we define the star state as the center solution to this Riemann problem (i.e. the solution between the $\mathscr{C}^+$ and $\mathscr{C}^-$ characteristics). As in [54], p^* and v_n^* can be approximated by

$$p^* = \left[\max\left(0, \left(c_L + c_R - \frac{\gamma - 1}{2}(v_{n,R} - v_{n,L})\right)\left(\frac{c_L}{p_L^z} + \frac{c_R}{p_R^z}\right)^{-1}\right) \right]^{1/z} \quad (21)$$

and

$$v_n^* = v_{n,L} + \frac{2}{\gamma - 1}(c_L - c_L^*) \quad (22)$$

where $c_L^* = c_L(p^*/p_L)^z$, $c_R^* = c_R(p^*/p_R)^z$, $z = (\gamma - 1)/(2\gamma)$, c_L is the left sound speed, and c_R is the right sound speed. These particular star states arise from the approximation of the Riemann solution by the so-called two rarefaction Riemann solver and are approximations to the true star state. Note that other choices for the star states are also acceptable. Our sonic fix defines a new value for α by

$$\alpha_{i+1/2,j}^n \leftarrow \begin{cases} 0 & \text{if } v_{n,L} - c_L \leq 0 \text{ and } v_n^* - c_L^* \geq 0, \\ 0 & \text{if } v_n^* + c_R^* \leq 0 \text{ and } v_{n,R} + c_R \geq 0, \\ \alpha_{i+1/2,j}^n & \text{else.} \end{cases}$$

The effect of these choices is to return the solver to first order accuracy near sonic points in rarefaction waves. Figure 1 shows the solution to the modified Sod's problem employing this sonic fix where it is seen that the poor behavior has been effectively eliminated apart from a small kink at the sonic point. It should be noted that the particular sonic fix demonstrated here relies on an approximate solution to the Riemann problem. For cases where this solution is not known, this fix is not applicable and sonic rarefactions must be identified in another way. For example, one might consider applying the fix wherever the flow transitions from super- to sub-sonic flow across a cell boundary.

3.4 Strong Rarefactions

In addition to the poor behavior for sonic rarefaction waves, the traditional FCT algorithm runs into difficulties for strong rarefaction waves where the velocities at which the gas is being pulled apart differ by more than the local sound speed. This is a very difficult problem for many methods because a near vacuum state is reached and failure can occur as a result of negative densities or pressures [55]. Consider the solution to a Riemann problem with left and right states $(\rho, u_1, u_2, p)_L = (1.0, -2.0, 0.0, 0.4)$ and $(\rho, u_1, u_2, p)_L = (1.0, 2.0, 0.0, 0.4)$ respectively.

Figure 2 shows the density and velocity as computed by the FCT algorithm for this case both with and without our fix. The FCT solution without any fix demonstrates oscillations in velocity close to the near vacuum state (near the origin). In order to remove this behavior a simple fix is employed which sets

$$\alpha_{i+1/2,j}^n = 0 \quad \text{if } p^* < \min(p_L, p_R) \text{ and } |v_{n_L} - v_{n_R}| \geq \max(c_l, c_r).$$

This causes the first order scheme to be used when strong rarefaction waves are present. The results shown in Fig. 2 demonstrate that the velocity from the fixed scheme is monotonic near the origin. These results are comparable to the results of Tóth in [55] but further improvements should be investigated.

3.5 A Note Concerning Monotonicity

The original FCT scheme of Boris and Book applied to 1-D linear advection problems is provably monotone. However, the extension by Zalesak to higher-

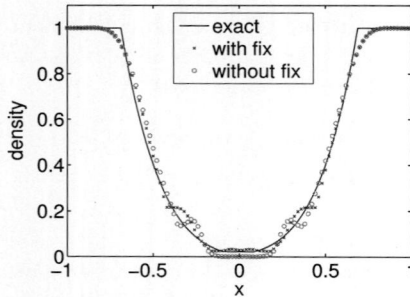 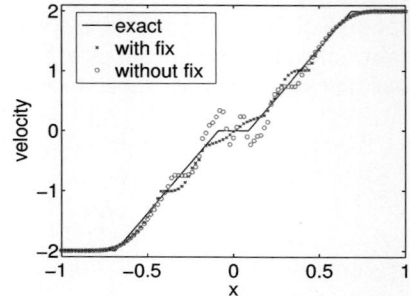

Fig. 2 Density (*left*) and velocity (*right*) for a strong rarefaction problem at $t = 0.25$. The *black line* represents the exact solution, the *red circles* the FCT approximation without a fix and the *blue crosses* the FCT solution with the fix. The oscillations in velocity for the original FCT scheme are particularly troubling but also note the undershoot of the density near the origin (Color figure online)

dimensions do not result in a monotone scheme, a fact that has apparently not been discussed in the literature. We now present a simple example to illustrate this fact. Consider linear advection with unit advection velocity,

$$\frac{\partial}{\partial t}\rho + \frac{\partial}{\partial x_1}\rho = 0.$$

We use the low-order flux given by $\mathbf{f}_{i+1/2}^{\text{low},n} = \rho_i^n$, and the second-order centered flux given by $\mathbf{f}_{i+1/2}^{\text{high},n} = \frac{1}{2}(\rho_i^n + \rho_{i+1}^n)$. At time level t^n let the approximate solution be given by

$$\rho_{-3}^n = 4.5, \qquad \rho_{-2}^n = 4, \qquad \rho_{-1}^n = 3.5, \qquad \rho_0^n = 3,$$
$$\rho_1^n = 3, \qquad \rho_2^n = 2, \qquad \rho_3^n = 1, \qquad \rho_4^n = 0.$$

Set the grid spacing as $\Delta x_1 = 1$ and the temporal spacing as $\Delta t = 0.25$. The FCT algorithm, as outlined by Zalesak [59, 60], produces the following values for α

$$\alpha_{-1/2}^n = 1, \qquad \alpha_{1/2}^n = 1, \qquad \alpha_{3/2}^n = 1.$$

By using the forward Euler time integrator (i.e. $\rho_i^{n+1} = \rho_i^{\text{new}}$), the FCT solution after a single step results in the values

$$\rho_0^{n+1} = 3.0625, \qquad \rho_1^{n+1} = 3.125.$$

The solution at time t^n was monotonically decreasing left to right while the solution for these two cells at time t^{n+1} is monotonically increasing left to right and so the violation of monotonicity is demonstrated. Many authors suggest the use of a pre-limiter, but for this case the pre-limiter suggested by Zalesak [59] and Kuzmin [38] has no effect as can be easily verified. The pre-limiter of DeVore [19], which we have adopted here, does remedy this particular problem, but a proof of monotonicity for arbitrary high order fluxes is not known.

4 Numerical Results

We now present simulation results using the FCT algorithm described in Sect. 3. The discussion centers on studying the robustness and accuracy of the overall numerical approach as well as comparing the results to those from the high-resolution Godunov method in [5, 30, 31] which uses an approximate Roe Riemann solver [48] and the MinMod limiter [54]. Of course, any comparisons presented here are only valid for these particular implementations of the FCT and Godunov methods. There are many variations to both algorithms which would change the specifics of the results. However, the present study provides a reasonable baseline comparison of the relative merits of the two schemes. Furthermore, the hope is that given the results from previous comparisons, for example in [23], one can place, in a general sense, high-resolution Godunov methods, WENO methods, and FCT in relation to each other. In fact the tests we present were largely driven by the choice of tests presented in [23] exactly for the reason that comparisons could be made.

Because the purpose of any comparisons made in this section is to provide a sense of the relative merits of the methods as they might be used in practice, the set of parameters used by each method is set to what we consider to be reasonable numbers. For the Godunov method we use CFL = 0.9 and for FCT we use CFL = 0.4. The small choice for FCT is required, as noted in Sect. 3, to ensure the desired bounds are not violated. For problems where AMR is used, the refinement criterion is the same for both schemes and is based on a weighted sum of first and second un-divided differences of the solution (see [30] for details).

We begin the discussion by establishing the expected second-order rate of convergence for the FCT method for smooth flows using the method of analytic solutions. Similar verification tests have also performed for the Godunov method as in [5], but direct comparisons are not made here because of the manufactured nature of the tests. Next we consider the solution to a series of problems including 1-D isolated contacts, isolated shocks, Sod's shock tube problem, a two-shock Riemann problem, and the Shu-Osher test case. The methods are then compared for the 2-D problems of shock impingement on a cylinder and the irregular Mach reflection of a strong shock on an inclined ramp. Finally, simulation results from the FCT method are presented for a number of more complex problems to include shock impingement on multiple fixed and movable cylinders as well as for a prototype Z-pinch implosion problem. These problems are more difficult to characterize and so no comparison to the Godunov method is provided.

4.1 Method of Analytic Solutions

We now investigate convergence of the FCT method to known smooth solutions. Smooth analytic solutions to the Euler equations are difficult to find although some do exist. One example is that of the Prandtl-Meyer fan for flow around a smoothly expanding channel. As a more general approach to constructing exact solutions, we use the method of analytic solutions, sometimes known as the method of manufactured solutions, whereby one picks an arbitrary smooth solution and includes a forcing term in (1) such that the solution to the forced set of equations is the chosen smooth solution. This approach is very general and we have found it to be an invaluable tool in verifying the implementation of a given numerical approach.

To construct such a solution we take $\mathbf{u}_s(x_1, x_2, t)$ as a known smooth function. Equation (1) is then modified to

$$\frac{\partial}{\partial t}\mathbf{u} + \frac{\partial}{\partial x_1}\mathbf{f}_1(\mathbf{u}) + \frac{\partial}{\partial x_2}\mathbf{f}_2(\mathbf{u}) = \mathbf{h}(\mathbf{u}_s), \tag{23}$$

where $\mathbf{u}, \mathbf{f}_1, \mathbf{f}_2$ are defined as before and

$$\mathbf{h}(\mathbf{u}_s) = \frac{\partial}{\partial t}\mathbf{u}_s + \frac{\partial}{\partial x_1}\mathbf{f}_1(\mathbf{u}_s) + \frac{\partial}{\partial x_2}\mathbf{f}_2(\mathbf{u}_s).$$

The boundary conditions are also modified in a similar way. Clearly one solution to the modified problem (23) is $\mathbf{u} = \mathbf{u}_s$. The particular choice of $\mathbf{u}_s$ is quite arbitrary and for the purposes of this work we use trigonometric functions in both space and time. The choice made here is

$$\left.\begin{aligned}\rho_s &= \frac{1}{8}\cos\left(\frac{\pi(x_1-5)}{10}\right)\cos\left(\frac{\pi x_2}{10}\right)\cos\left(\frac{\pi t}{10}\right) + 1 \\ u_{1,s} &= \cos\left(\frac{\pi x_1}{10}\right)\cos\left(\frac{\pi x_2}{10}\right)\cos\left(\frac{\pi t}{10}\right) \\ u_{2,s} &= \frac{1}{2}\cos\left(\frac{\pi(x_1-5)}{10}\right)\cos\left(\frac{\pi(x_2-5)}{10}\right)\cos\left(\frac{\pi t}{10}\right) \\ p_s &= \rho_s\left[\frac{1}{4}\cos\left(\frac{\pi x_1}{10}\right)\cos\left(\frac{\pi(x_2-5)}{10}\right)\cos\left(\frac{\pi t}{10}\right) + 1\right]\end{aligned}\right\} \tag{24}$$

where the conserved quantities are constructed from these given primitives. We consider the solution to (23) on two different domains, the first being a simple square with $|x_k| \leq 2, k = 1, 2$. For this domain (and for later examples) Cartesian grids are defined by

$$\mathcal{R}\big([x_{1,a}, x_{1,b}] \times [x_{2,a}, x_{2,b}], N_1, N_2\big)$$
$$= \big\{(x_{1,a} + i_1 \Delta x_1, x_{2,a} + i_2 \Delta x_2) | \Delta x_k = (x_{k,b} - x_{k,a})/N_k,$$
$$i_k = 0, 1, \ldots, N_k, k = 1, 2\big\}.$$

For this example we use $\mathcal{R}([-2, 2] \times [-2, 2], 40m, 40m)$, where m is an integer indicating grid size. The initial condition is taken to be $\mathbf{u}_s(x_1, x_2, 0)$ and the boundary conditions on the perimeter of the square are given by the exact solution for all time. The modified equations (23) are integrated numerically for $0 \leq t \leq 1$ and the

An Evaluation of a Structured Overlapping Grid Implementation of FCT   413

Table 1 Convergence results for the square domain using the FCT method (indicated by F in table headers. Maximum errors in density, velocity components and pressure at $t=1$ for grid resolutions determined by m, and the estimated convergence rates $\kappa = \log_2(e_\rho(m)/e_\rho(2m))$ as well as a least squares fits of the convergence rates over the entire refinement process $\tilde{\kappa}$ are shown

m	$e_\rho(m)$ F	κ	$e_{u_1}(m)$ F	κ	$e_{u_2}(m)$ F	κ	$e_p(m)$ F	κ
1	1.1e−3	–	4.7e−3	–	6.8e−4	–	8.6e−4	–
2	7.9e−4	.48	1.3e−3	1.9	3.9e−4	.80	2.7e−4	1.7
3	1.5e−4	2.4	3.1e−4	2.1	7.3e−5	2.4	6.4e−5	2.1
8	4.7e−5	1.7	9.4e−5	1.7	1.9e−5	1.9	1.8e−5	1.8
$\tilde{\kappa}$	1.5		1.9		1.7		1.9	

solution error is computed at the final time. Of course if the α's in (8) are not taken to be 1 the method will not be second order accurate everywhere in the domain. Still it is of substantial interest to determine the actual accuracy of the method for smooth flows such as this. Table 1 shows the maximum error in the primitive variables at $t=1$ for various grid resolutions determined by m. The convergence rate is computed from one resolution to the next as $\kappa = \log_2(e_\rho(m)/e_\rho(2m))$ as well as a least squares fit of the rates over the entire refinement process which we label $\tilde{\kappa}$. The max-norm convergence rates are generally reasonably close to second-order. Actual second order convergence is not expected because the method defaults to first order near characteristic extrema as is typical of most limited schemes.

The second domain considered is a circular disk of radius 0.8 which is discretized using an overlapping grid consisting of a background Cartesian grid given by $\mathscr{R}([-0.6, 0.6] \times [-0.6, 0.6], 30m, 30m)$, and a boundary fitted annular grid defined by $\mathscr{A}((0.0, 0.0), [0.4, 0.8], 10m, 80m)$ with

$$\mathscr{A}\big((x_{1,c}, x_{2,c}), [r_a, r_b], N_r, N_\theta\big)$$
$$= \big\{(x_{1,c}, x_{2,c}) + r_{i_r}\big(\cos(\theta_{i_\theta}), \sin(\theta_{i_\theta})\big) | r_{i_r} = r_a + i_r(r_b - r_a)/N_r,$$
$$\theta_{i_\theta} = 2\pi i_\theta / N_\theta, \; i_k = 0, 1, \ldots, N_k, \; k = r, \theta \big\}.$$

Through the use of such an overlapping mesh we provide a further check of both the implementation of the scheme on curvilinear grids as well as for the interpolation scheme at grid overlaps. Table 2 shows the maximum error in the primitive variables at $t=1$ for various resolutions of the overlapping grid. We note the near second-order convergence for each of the variables.

4.2 Isolated Contact and Shock Discontinuities

4.2.1 Contact Wave

The contact wave is a traveling discontinuous jump where characteristics run parallel to the front. As such, error can accumulate with the result that a nominally p^{th}

Table 2 Convergence results for the circular domain. Maximum errors in density, velocity components, and pressure at $t = 1$ for grid resolutions determined by m, and the estimated convergence rates $\kappa = \log_2(e_\rho(m)/e_\rho(2m))$ as well as a least squares fits of the convergence rates over the entire refinement process $\tilde{\kappa}$ are shown

m	$e_\rho(m)$ F	κ	$e_{u_1}(m)$ F	κ	$e_{u_2}(m)$ F	κ	$e_p(m)$ F	κ
1	8.6e−5	–	2.3e−4	–	7.5e−5	–	1.1e−4	–
2	2.8e−5	1.6	6.8e−5	1.8	2.3e−5	1.7	3.1e−5	1.8
3	7.2e−6	2.0	2.0e−5	1.8	6.2e−6	1.9	7.7e−6	2.0
8	2.0e−6	1.8	6.1e−6	1.7	1.7e−6	1.9	2.0e−6	1.9
$\tilde{\kappa}$	1.8		1.7		1.8		1.9	

Table 3 Convergence results for the contact wave problem using second order Godunov and FCT approximations, indicated by "F" and "G" in the headings respectively. L_1 errors in density at $t_f = 0.5$ are computed for grid resolutions determined by m. Estimated convergence rates $\kappa = \log_2(e_\rho(m)/e_\rho(2m))$ as well as a least squares fit of the convergence rates over the entire refinement process $\tilde{\kappa}$ are shown. Note that errors for velocity and pressure are identically zero

m	$e_\rho(m)$ F	κ	$e_\rho(m)$ G	κ
1	1.06e−2	–	1.39e−2	–
2	6.64e−3	.67	8.78e−3	.66
4	4.18e−3	.67	5.55e−3	.66
8	2.63e−3	.67	3.51e−3	.66
$\tilde{\kappa}$	.67		.66	

order shock capturing scheme will generally converge at the rate of $\kappa = P/(P+1)$ in the L_1 sense [6, 25, 27]. There are some so-called compressively limited schemes which can achieve $\kappa = 1$ convergence although such schemes often have other undesirable characteristics such as the artificial steepening of smooth solutions [40, 54]. The construction of the FCT method does not immediately indicate what the convergence rate should be.

The initial conditions for the contact wave consists of the left state $(\rho, u_1, u_2, p)_L = (0.1, 1.0, 0.0, 1.0)$ and the right state $(\rho, u_1, u_2, p)_R = (1.0, 1.0, 0.0, 1.0)$ with the jump at $x_0 = 0.25$. We can construct a weak solution corresponding to a vanishing viscosity solution, and we will call such solutions "exact" with the understanding that there may be many weak solutions. The exact solution to this problem consists of a propagating discontinuity moving to the right with speed 1.0. The density jumps through this discontinuity but the pressure and velocity remain constant. Simulations are performed on the grid defined by $\mathscr{L}([0.0, 1.0], 200m)$ where m is a measure of grid resolution (see (20)). A value of $\gamma = 1.4$, corresponding to a diatomic ideal gas, is chosen.

A convergence study is performed at various numerical resolutions indicated by m with the comparisons taking place at $t_f = 0.5$ using the discrete L_1 norm. Results from this study are given in Table 3. Here it is seen that both the FCT and Godunov

Table 4 Convergence results for the shock wave problem using second order Godunov (G) and FCT (F) approximations. L_1 errors in density, velocity and pressure are shown at $t_f = 0.25$ for grid resolutions determined by m. Estimated convergence rates $\kappa = \log_2(e_\rho(m)/e_\rho(2m))$ as well as a least squares fits of the convergence rates over the entire refinement process $\tilde{\kappa}$ are also shown

m	$e_\rho(m)$ F	κ	$e_\rho(m)$ G	κ	$e_{u_1}(m)$ F	κ	$e_{u_1}(m)$ G	κ	$e_p(m)$ F	κ	$e_p(m)$ G	κ
1	8.38e−3	–	7.08e−3	–	5.59e−3	–	4.83e−3	–	1.44e−2	–	1.26e−2	–
2	3.94e−3	1.0	3.65e−3	.96	2.91e−3	.94	2.76e−3	.81	6.57e−3	1.1	6.35e−3	.99
4	2.08e−3	.92	1.82e−3	1.0	1.39e−3	1.1	1.22e−3	1.2	3.63e−3	.86	3.27e−3	.96
8	9.63e−4	1.1	9.15e−4	.99	7.13e−4	.96	6.72e−4	.86	1.66e−3	1.1	1.60e−3	1.0
$\tilde{\kappa}$	1.03		.99		1.00		.97		1.02		.99	

methods attain the expected convergence rate of $\approx 2/3$ as measured by both κ and $\tilde{\kappa}$. We can also see that the FCT method captures the contact with slightly less error than the Godunov method although the results for the Godunov method are sensitive to the choice of Riemann solver and limiter [6].

4.2.2 Shock Wave

Consider a Mach 2 shock with $\gamma = 1.4$. The pre- and post-shock states are given by $(\rho, u_1, u_2, p)_L = (2.67, 1.48, 0.0, 4.5)$ and $(\rho, u_1, u_2, p)_R = (1.0, 0.0, 0.0, 1.0)$. For this nonlinear phenomenon, the characteristic curves enter into the discontinuity which acts as a natural steepening mechanism. Computations are carried out on the unit interval $x \in [0, 1]$ using mesh $\mathscr{L}([0.0, 1.0], 200m)$ with m being a measure of grid resolution. The initial jump is placed at $x_0 = 0.25$ and integration is carried out to $t_f = 0.25$ where L_1 errors are computed. The results are presented in Table 4.

Both schemes have similar L_1 errors and demonstrate the expected first order convergence with $\kappa \approx 1$ and $\tilde{\kappa} \approx 1$ for density, velocity and pressure. This implies that the number of cells for which there is $O(1)$ point-wise error is fixed which implies that the shock does not continually smear as a function of time. Contrast this to the case of the contact in Sect. 4.2.1 where the captured discontinuity contains an increasing number of grid cells even as its overall width decreased.

4.3 Sod's Shock Tube Problem (Modified)

For this example problem we investigate the behavior of the FCT and Godunov methods for a modified version of Sod's shock tube problem. This problem is designed to highlight the poor behavior of some numerical methods near sonic points in rarefaction waves and was previously discussed in Sect. 3.3 where the sonic fix for the FCT method was described. A description of sonic fixes for Godunov schemes can be found, for example, in [54]. The computational domain is again chosen to

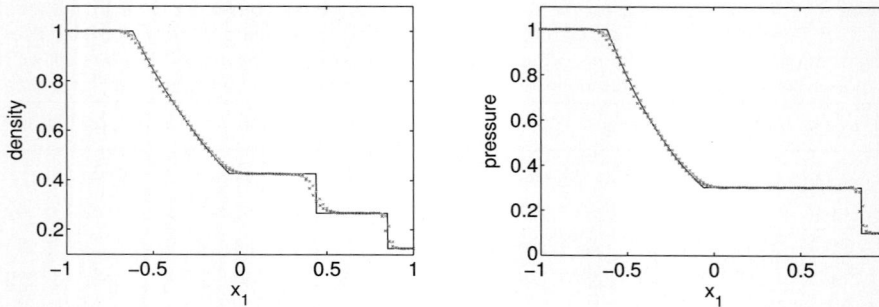

Fig. 3 Exact solution (*black line*) and numerical approximations with $m = 1$ for Godunov's method using Roe's approximate Riemann solver (*red marks*) and the FCT method (*blue marks*) for the modified Sod shock tube problem at $t_f = 0.5$. Shown here are the density (*left*) and the pressure (*right*) (Color figure online)

Table 5 Convergence results for the modified Sod shock tube problem. Discrete L_1 error and associated convergence rates for the Godunov (G) and FCT (F) schemes at selected resolutions associated with the choice of m. Apparently the mesh is of insufficient resolution for the methods to exhibit global convergence rates of 2/3 for the L_1 norm of density which is dictated by the captured contact

m	$e_\rho(m)$ F	κ	$e_\rho(m)$ G	κ	$e_{u_1}(m)$ F	κ	$e_{u_1}(m)$ G	κ	$e_p(m)$ F	κ	$e_p(m)$ G	κ
2	8.86e−3	−	9.44e−3	−	1.44e−2	−	1.44e−2	−	6.54e−3	−	6.32e−3	−
4	5.00e−3	.83	5.31e−3	.83	6.99e−3	1.0	7.51e−3	.94	3.21e−3	1.0	3.22e−3	.97
8	3.05e−3	.71	3.03e−3	.81	3.32e−3	1.1	4.08e−3	.88	1.54e−3	1.1	1.67e−3	.94
16	1.83e−3	.74	1.80e−3	.75	1.59e−3	1.1	2.42e−3	.75	7.24e−4	1.1	9.08e−4	.88
$\tilde{\kappa}$	.76		.80		1.06		.86		1.06		.93	

be $x \in [-1, 1]$, the initial jump is placed at $x_0 = -0.4$, and the governing equations (1) are integrated to $t_f = 0.5$. The computational grid for this study is given by $\mathscr{L}([-1.0, 1.0], 100m)$.

The exact density and pressure, as well as approximate results for $m = 1$ for both the Godunov and FCT methods, are shown in Fig. 3 which demonstrates the similarity of the two approximate solutions. This trend continues for all resolutions but is more easily seen for this coarse simulation where $m = 1$. Figure 3 also shows that both methods seem to be handling the sonic rarefaction. Quantitative convergence results are shown in Table 5 using the discrete L_1 norm. These results indicate that although both schemes are clearly converging to the exact solution, neither scheme is yet in the asymptotic range of convergence where the L_1 error of density will be dominated by the 2/3 convergence rate near the contact. Even so, both schemes provide similar convergence behavior with the FCT yielding slightly higher convergence rates for the pressure and velocity.

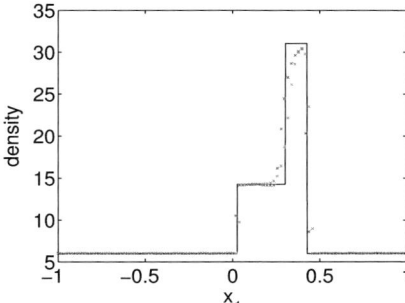
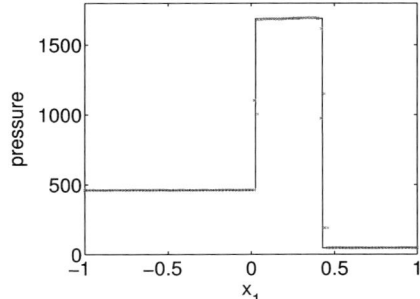

Fig. 4 Exact solution (*black line*) and approximations with $m = 1$ for Godunov's method using Roe approximate Riemann solver (*red marks*) and the FCT method (*blue marks*) for the two-shock Riemann problem at $t_f = 0.035$. Shown here are the density (*left*) and the pressure (*right*) (Color figure online)

4.4 A Two-Shock Riemann Problem

The last Riemann problem investigated in this work is commonly known as the two-shock problem. The exact solution to this problem for $\gamma = 1.4$ has a $M \approx 5.62$ shock in the rightmost characteristic field, a $M \approx 1.81$ shock in the leftmost characteristic field, and a contact wave separating the two. Left and right states are taken from [54] and given as $(\rho, u_1, u_2, p)_L = (5.99242, 19.5975, 0.0, 460.894)$ and $(\rho, u_1, u_2, p)_R = (5.99242, -6.19633, 0.0, 46.0950)$ The exact solution is determined as in [57], and results in a nearly stationary shock for the leftmost characteristic field. The actual speed of the left shock is $S \approx 0.78$, the velocity through the contact wave is $u_1 \approx 8.69$, and the rightmost shock moves with speed $S \approx 12.25$. The capturing of the nearly stationary shock proves to be one of the primary difficulties for this problem (see [1, 37] for details on slowly moving shocks). Shock capturing schemes also have difficulty representing the contact in this problem and there is a need to accurately resolve that jump before a reasonable global approximation is achieved.

The solution for this problem is approximated for $x \in [-1, 1]$ using the mesh $\mathscr{L}([-1.0, 1.0], 100m)$ and integration is carried out to a final time of $t_f = 0.035$. Figure 4 shows profiles of density and pressure for the exact solution at that time as well as the numerical approximations for $m = 1$. Qualitatively it is seen that the two schemes produce similar results, however, close inspection revels the Godunov approximation to be slightly less oscillatory particularly in the pressure while the FCT approximation shows a sharper capture of the contact wave. Table 6 shows quantitative convergence results for the two schemes using the discrete L_1 norm for the computation of the errors. This table shows that the Godunov approximations demonstrate somewhat higher convergence rates for all quantities, but that for the resolutions discussed here the FCT approximations always give smaller actual errors. In fact for the pressure and velocity, the errors in the FCT approximations are more than three times smaller than the Godunov approximations at coarse resolutions and still more than twice as small for the finest mesh considered.

Table 6 Discrete L_1 error and associated convergence rates for the two shock problem using the Godunov (G) and FCT (F) schemes at selected resolutions associated with the choice of m. Neither scheme is yet in the asymptotic range of convergence where the L_1 errors in density will be dominated by the 2/3 convergence rate at the contact

m	$e_\rho(m)$ F	κ	$e_\rho(m)$ G	κ	$e_{u_1}(m)$ F	κ	$e_{u_1}(m)$ G	κ	$e_p(m)$ F	κ	$e_p(m)$ G	κ
2	4.48e−1	–	7.53e−1	–	1.10e−1	–	3.40e−1	–	8.74e0	–	3.29e1	–
4	2.58e−1	.80	4.08e−1	.88	7.15e−2	.62	1.41e−1	1.3	6.49e0	.43	1.40e1	1.23
8	1.51e−1	.77	2.26e−1	.85	2.48e−2	1.5	7.78e−2	.86	3.34e0	.96	7.89e0	.83
16	9.00e−2	.75	1.43e−1	.66	1.66e−2	.58	4.76e−2	.71	1.82e0	.88	4.94e0	.66
$\tilde{\kappa}$	.77		.80		.91		.94		.75		.90	

4.5 Shu-Osher Problem

The final one-dimensional test case considered in this chapter is a problem originally considered by Shu and Osher [35] and subsequently by others [23, 47]. This problem consists of a $M = 3$ shock in air, $\gamma = 1.4$, traveling into unshocked air with sinusoidally perturbed density. As originally presented, the problem has a number of parameters and the specific values used here are taken from [23]. The initial setup is

$$\rho = 3.857143, \quad u_1 = 2.629369, \quad u_2 = 0, \quad p = 10.33333 \quad \text{for } x_1 < -4,$$
$$\rho = 1 - \varepsilon \sin(\lambda \pi x), \quad u_1 = 0, \quad u_2 = 0, \quad p = 1 \quad \text{for } x_1 \geq -4$$
(25)

where the parameter values are $\varepsilon = 0.2$ and $\lambda = 5$. The approximate solution is computed for $x \in [-5, 5]$ using $\mathscr{L}([-5.0, 5.0], 200m)$ and integrated to a final time $t_f = 1.8$.

When interpreting results, it is useful to understand the Riemann structure of the solution when $\varepsilon = 0$. For this case we can determine an exact solution and the waves present there give a good indication where structures in the more complicated solution will arise. When $\varepsilon = 0$, the solution consists of a $M = 3$ shock traveling with speed $S \approx 3.55$. The perturbed problem, $\varepsilon \neq 0$ and small, will have disturbances traveling along the other two characteristic fields with speeds $S \approx 2.63$ and $S \approx 0.69$. At $t = 1.8$, the lead shock will have traveled to $x_1 \approx 2.39$, the contact wave to $x_1 \approx 0.73$ and the left acoustic wave to $x_1 \approx -2.76$. For small ε it is expected that the exact solution will change character near these locations.

A reference solution, computed with $m = 128$ up to $t = 1.8$, can be seen, for example, in Fig. 5. For $x < -2.76$ the solution is the unperturbed post-shock state. For $x \in (-2.76, 0.73)$ the solution exhibits mild oscillations in all quantities. These oscillations are the result of the passage of the left acoustic wave. For $x \in (0.73, 2.39)$ the solution exhibits high frequency oscillations. Notice that for the computational resolution m, the high frequency oscillations in the density for $x \in (0.73, 2.39)$ contain approximately $2m$ grid points per wavelength. The solution with $m = 128$ uses a sufficiently fine grid to resolve these oscillations as evidenced by the fact that further refinement does not change the character of the solution, and because it results

Table 7 Convergence results for the Shu-Osher test problem using both the Godunov (G) and FCT (F) methods. Convergence rates and errors are computed with (26) and (27) using finely resolved simulations at $m = 64$ and $m = 128$

m	$e_\rho(m)$ F	κ	$e_\rho(m)$ G	κ	$e_{u_1}(m)$ F	κ	$e_{u_1}(m)$ G	κ	$e_p(m)$ F	κ	$e_p(m)$ G	κ
1	1.16e0	.75	1.20e0	.44	3.44e−1	1.1	3.02e−1	.94	2.34e0	1.1	.98e0	.92
2	9.18e−1	.86	1.01e0	.52	1.57e−1	1.1	1.55e−1	.94	1.08e0	1.1	1.08e0	.93
4	7.86e−1	1.1	8.64e−1	.67	6.35e−2	1.1	7.85e−2	.92	4.75e−1	1.1	5.67e−1	.94
8	5.98e−1	1.4	7.28e−1	.93	3.10e−2	1.1	4.22e−2	.94	2.23e−1	1.1	2.94e−1	.93
16	2.39e−1	1.4	5.00e−1	1.3	1.52e−2	1.2	2.35e−2	1.0	1.06e−1	1.2	1.70e−1	1.1
32	8.90e−2	1.4	2.19e−1	1.5	6.87e−3	1.4	1.20e−2	1.3	4.57e−2	1.3	8.38e−2	1.3

in approximately 256 cells per wavelength for $x \in (0.73, 2.39)$. For $x > 2.39$ the solution returns to the initial upstream state. The locations where the solution changes behavior are, as expected, those mentioned above in the discussion of the Riemann structure for $\varepsilon = 0$.

There is no known closed form solution to this problem and convergence results must be estimated through comparison to more finely resolved solutions. Here we use a method similar to that presented in [32]. At a given point, x_i, we assume the solution at a given resolution differs from the exact solution by

$$\mathbf{u}_e(x_i) - \mathbf{u}_m(x_i) \approx \mathbf{c}(x_i) h_m^\kappa \tag{26}$$

where $\mathbf{u}_e$ is the exact solution, $\mathbf{u}_m$ the numerical approximation, $\mathbf{c}(x_i)$ depends only on x_i, κ is the convergence rate and h_m is the grid spacing. Note that we have uniform grid spacing. From (26) one can compute

$$\left\| \mathbf{u}_{m_1}(x) - \mathbf{u}_{m_2}(x) \right\|_h \approx \left\| \mathbf{c}(x) \right\|_h \left| h_{m_1}^\kappa - h_{m_2}^\kappa \right| \tag{27}$$

using a discrete norm. Numerical approximations at three resolutions and (27) can be combined to produce two equations which define the convergence rate κ and the constant $\|\mathbf{c}(x)\|_h$. The solution error can then be approximated as $e_\mathbf{u}(m) = \|\mathbf{u}_e - \mathbf{u}_m\|_h \approx \|\mathbf{c}(x)\|_h h^\kappa$. When estimating the error and convergence rate for a given approximation with resolution given by m, we use the three approximations $\mathbf{u}_m$, $\mathbf{u}_{64}$ and $\mathbf{u}_{128}$. Table 7 shows the convergence results using the discrete L_1 norm for both the FCT and Godunov schemes. From this table it is clear that the coarser resolutions do not approximate the solution well at all, particularly for the density, and low rates of convergence are attained. Figures 5 and 6 demonstrate this graphically where the numerical approximations for $m = 1$ are plotted on top of the reference solution. Figure 5 shows the global character of the solution and Fig. 6 shows a zoom of the density in the most oscillatory region. For low resolutions, the high frequency oscillations are not well represented and both methods exhibit poor convergence properties, particularly for the density as seen in Fig. 6. This is reflected by the convergence rates which are less than 1. At some critical resolution however, both methods see a rise in convergence rates, tending to some value larger than 1.

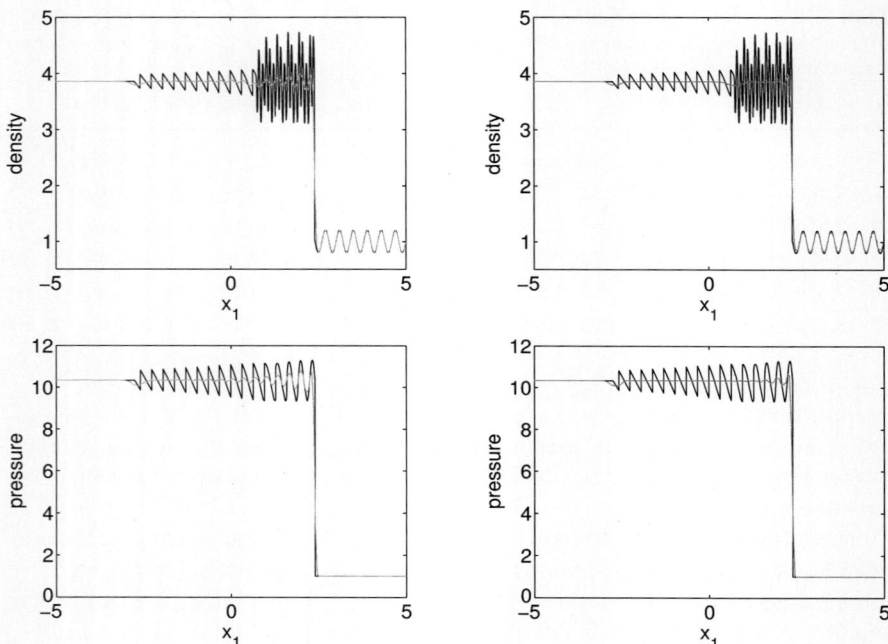

Fig. 5 Comparison of the numerical approximations with $m=1$ at $t=1.8$ for the Shu-Osher test problem. For all images the *black line* represents the reference solution with $m=128$ while the *red line* (*left*) shows the Godunov approximation and the *blue line* (*right*) shows the FCT approximation. From top to bottom are density and pressure (Color figure online)

Once this transition occurs, the high frequency oscillations begin to be well represented as shown in Figs. 7 and 8. This transition to higher convergence rates happens at lower resolution for FCT, indicating that it has more resolving power than the Godunov method. For the highest resolutions demonstrated here, both approximations are reasonably representing all structures in the flow and their convergence rates become roughly equal. However, because the FCT method experienced the transition to higher convergence rates earlier in the refinement process, the errors at the highest resolutions are smaller than for the Godunov approximations by nearly a factor of 2.

4.6 Shock Impingement on Stationary Cylinder

The first two-dimensional test problem which we consider is the impingement of a $M=2$ shock on a rigid immovable cylinder. The basic problem consists of a rigid cylinder of radius 0.5 placed in the larger domain $[-2,2] \times [-2,2]$. A Mach 2 shock initially located at $x_1 = -1.5$ runs from left to right. The computational mesh is defined as the overlapping grid constructed from an annulus

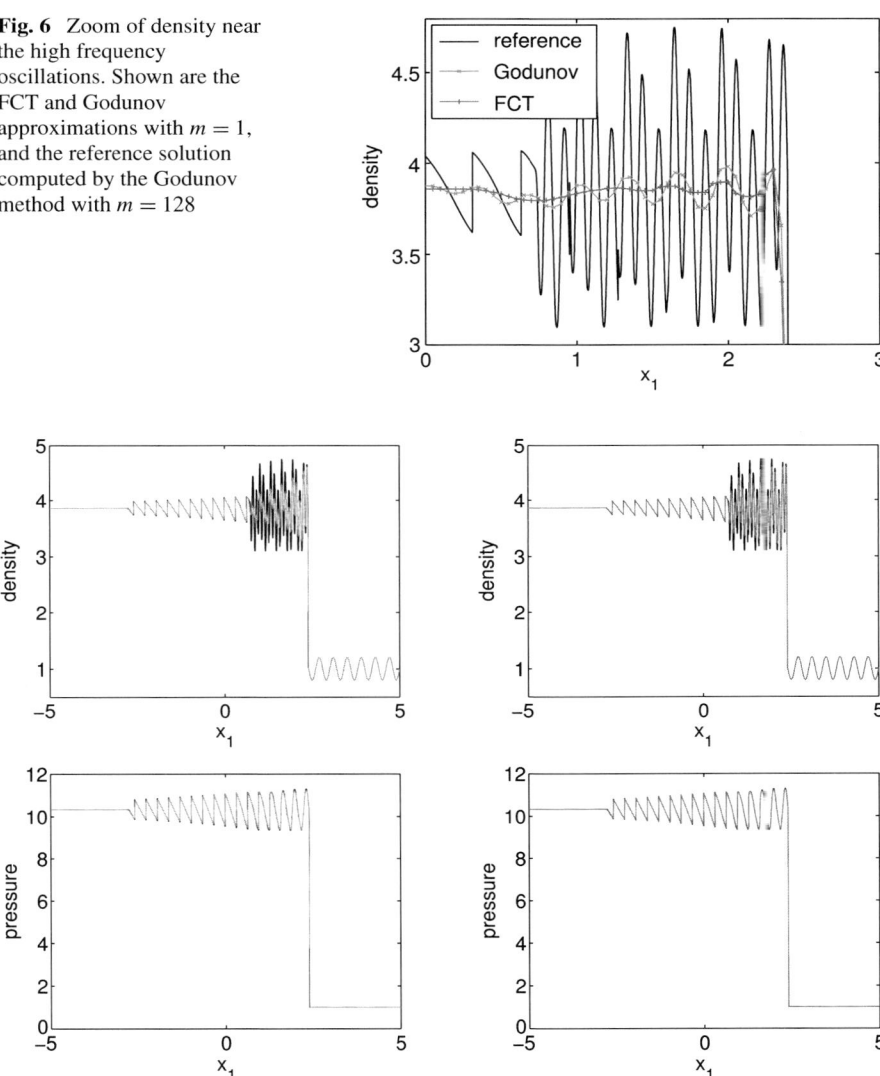

Fig. 6 Zoom of density near the high frequency oscillations. Shown are the FCT and Godunov approximations with $m = 1$, and the reference solution computed by the Godunov method with $m = 128$

Fig. 7 Comparison of the numerical approximations at $t = 1.8$ for the Shu-Osher test problem and $m = 16$. For all images the *black line* represents the reference solution with $m = 128$ while the *red line* (*left*) shows the Godunov approximation and the *blue line* (*right*) shows the FCT approximation. From top to bottom are density and pressure (Color figure online)

$\mathscr{A}((0,0), [0.5, 1.0], 10m, 80m)$ and a rectangle $\mathscr{R}([-2, 2] \times [-2, 2], 80m, 80m)$, where $\mathscr{A}$ and $\mathscr{R}$ are defined as before in Sect. 4.1. The boundary around the cylinder is defined as a slip wall (see [31]), the left boundary as an inflow, and the remaining boundaries are given outflow conditions. Phenomena of interest are limited to those associated with the shock/cylinder interaction. Provided that the simulation

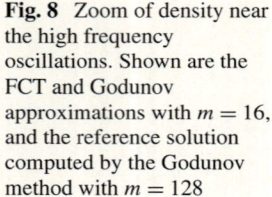

Fig. 8 Zoom of density near the high frequency oscillations. Shown are the FCT and Godunov approximations with $m = 16$, and the reference solution computed by the Godunov method with $m = 128$

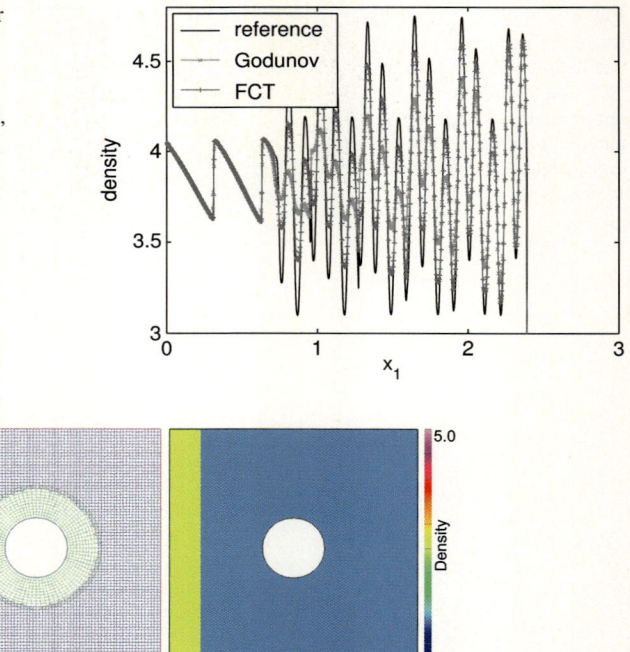

Fig. 9 Overlapping grid structure (*left*) and color contour of the initial density (*right*). The overlapping grid structure is used to capture geometry and additional adaptive grids will be dynamically added to locally increase resolution. Note that we only require one layer of interpolation points at grid overlap as discussed in Sect. 3.1. The initial density shows a $M = 2.0$ shock in air (ideal gas with $\gamma = 1.4$) moving from left to right

is not run too far in time, waves generated at the cylinder do not reach the exterior boundaries and so the exterior boundary condition choice has little influence. Figure 9 shows the computational mesh as well as color contours of density for the initial conditions. Numerical values for the initial conditions in primitive quantities, corresponding to a Mach 2 shock in air ($\gamma = 1.4$), were shown previously in Sect. 4.2.2.

The comparisons carried out in this chapter use the resolution $m = 1$ displayed in Fig. 9 for the coarse grid simulation. Adaptive mesh refinement (AMR) is then used for successive resolutions. For this test of shock interaction with a single cylinder, additional levels of AMR use a factor four refinement in each coordinate direction and so the four resolutions investigated have approximate grid spacings $h \approx 0.05$, 0.0125, 0.003125, and 0.00078125. Notice that because the initial condition uses a perfect jump, there exists numerical artifacts along the c^- characteristic and contact path. No effort is made to remove these and their contribution may be seen throughout the simulations.

Figure 10 shows the computed density using both methods for $t = 0.6$, $t = 1.0$, and $t = 1.4$ as the incident shock reflects from the cylinder boundary. Overall the

An Evaluation of a Structured Overlapping Grid Implementation of FCT 423

Fig. 10 Color contours of density for the finest resolution using FCT (*top*) and Godunov's method (*bottom*) at $t = 0.6$ (*left*), $t = 1.0$ (*middle*), and $t = 1.4$ (*right*)

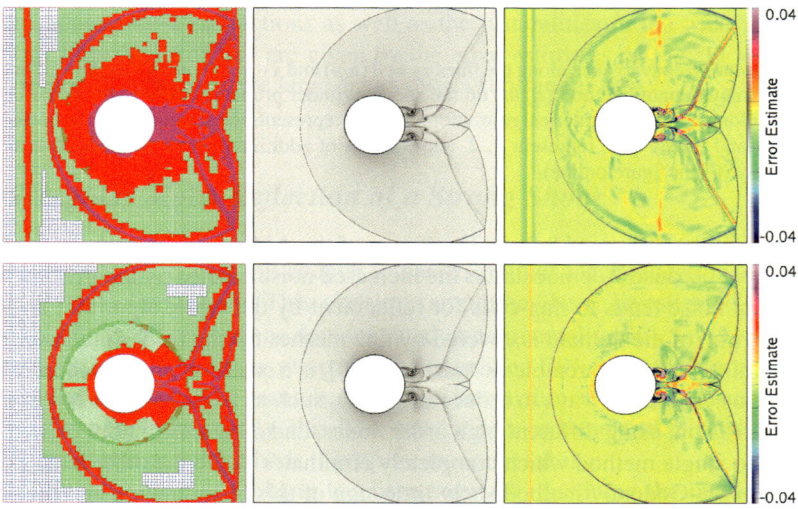

Fig. 11 AMR grid structure (*left*), numerical Schlieren images (*center*) and estimated L_1-error in density (*right*) for the FCT method (*top*) and Godunov's method (*bottom*) for the finest resolution simulation at $t = 1.4$

results show remarkably good agreement although slight differences can be seen at $t = 1.4$ in the low density wake region of the cylinder. To give a better indication of what is happening, Fig. 11 shows the AMR grid structure, numerical Schlieren images [7], and the estimated error in density at $t = 1.4$. The computation of the error

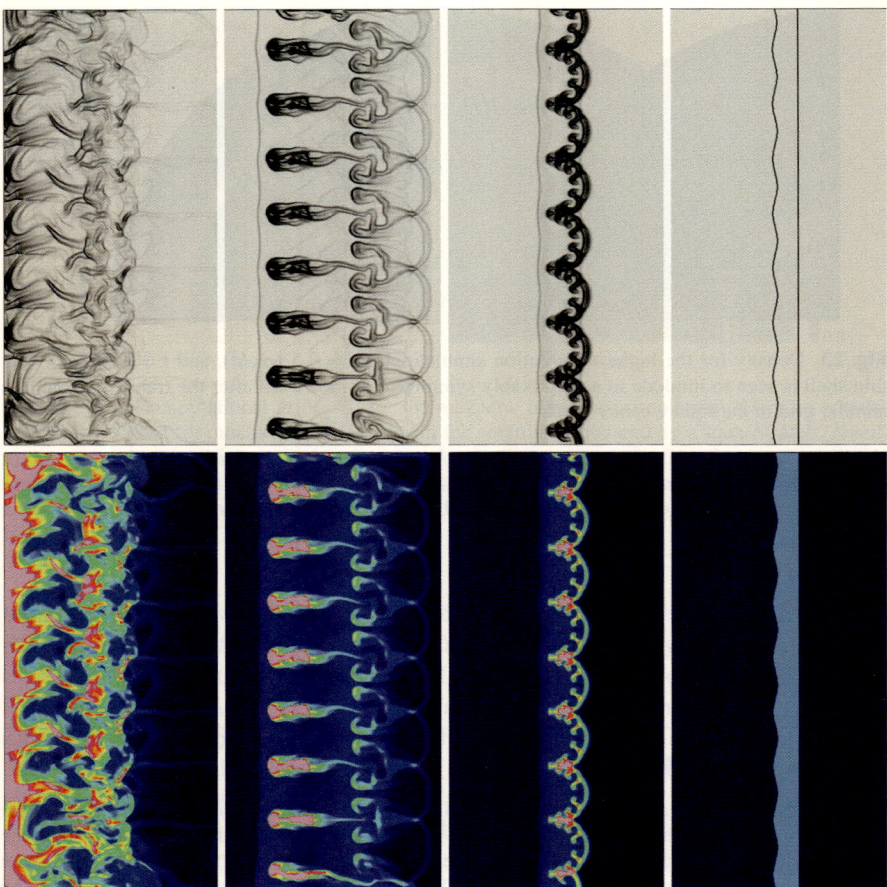

Fig. 25 A prototype Rayleigh-Taylor instability for a perturbed liner geometry for the magnetic implosion in (r, z). An initial pressure distribution to promote growth of R-T spikes ahead of the liner implosion has been selected. The instability evolves from the initial conditions on the *right* to the stagnation on the axis on the *left*. The *upper images* are a numerical Schlieren and the *lower images* show the density at times (from left to right) $t = 1.2, t = 1.0, t = 0.8$, and $t = 0.0$

density inside the linear of $\rho_{\text{pre-fill}} = 0.05$. The initial pressure is constant with a value of $p_0 = 0.01$. The base computational grid has a mesh spacing of $h \approx 0.0025$ and uses two additional levels of factor 4 AMR to give an effective grid resolution of $h \approx 0.0015625$. In this case a linear current drive $I = \sqrt{12}t$ is employed.

Figure 25 shows a numerical Schlieren of the time evolution of the initial conditions where a sinusoidal perturbation to the initial interface has been introduced. This perturbation is intended to promote the growth of a particular unstable Raleigh-Taylor mode and thus create significant structure as the liner nears stagnation. Here the perturbation has amplitude 0.005 and period 0.125 and has been introduced along the inner boundary of the conducting shell. The initial pressure distribution,

$p_0 = 0.01$, is selected to promote growth of R-T spikes ahead of the liner implosion at a sufficient rate so as to view their effect before stagnation. The instability evolves from the initial conditions on the right to the stagnation on axis on the left. The FCT AMR solution provides a very well resolved simulation of multiple unstable modes resulting in a complex pattern of R-T growth with complex interaction of shock waves at stagnation. It should be noted that the evolution of the R-T instability is qualitatively different from an actual Z-pinch system in which spikes lag behind the remaining liner material due the larger current flow in this contiguous material sheet (for example see [17, 22]). This is due to our simplified model assumption that defines the "current sheet" by the scalar λ that cannot adequately model the preferential physical current flow through the contiguous liner material over the penetrating spikes. However the magnetic force term does produce spikes and sheets of material developed by the R-T effects and R-M instabilities as the strong-shock interacts with the trailing liner material sheet. These later stages have some qualitative similarities to actual Z-pinch implosions which gives indication to why such a simple testing procedure can be very beneficial in benchmarking the flow portion of simulation tools intended for shock-hydrodynamics applications.

4.9.4 An Idealized Z-pinch Implosion with Simplified Radiation Emission and a Self Convergence Study

This final test prototype problem is intended to increase the complexity of the Z-pinch prototype problem to include a phenomenological radiation emission model and to allow the evaluation of the FCT method by considering the estimated order-of-accuracy of the method in modeling the integrated radiation output from the implosion. In actual Z-pinch modeling efforts the simulation of the temporal characteristics of the radiation output is an important and challenging goal for these types of simulations. Even with the use of such a simplified model as described below, the results produced by the Euler system solver with the $\mathbf{J} \times \mathbf{B}$ source term model produce power pulses with qualitative similarities to experimental and full computational MHD results found in the literature (see e.g. Fig. 9 in [24]).

One of the simplest radiation emission models one might consider is

$$Q_{rad} = \sigma T^4$$

where T is the temperature obtained from the equation of state (2) and the caloric equation of state for an ideal gas as given by $\rho e = \rho C_v T$, where C_v is the specific heat of the gas at constant volume. From [4]

$$\sigma = \bar{\sigma} \rho \frac{1}{T^2}$$

where $\bar{\sigma}$ is a constant chosen here to be $\bar{\sigma} = 100.0$. The radiation emission is set to be active only in the liner material by multiplication with λ so the final radiation source term becomes

$$Q_{rad} = \bar{\sigma} \lambda \rho T^2.$$

Printed by Printforce, the Netherlands